# Bioformulations: for Sustainable Agriculture

Naveen Kumar Arora •
Samina Mehnaz • Raffaella Balestrini
Editors

# Bioformulations: for Sustainable Agriculture

Springer

*Editors*
Naveen Kumar Arora
Department of Environmental Microbiology
School for Environmental Sciences,
Babasaheb Bhimrao Ambedkar University
Lucknow, Uttar Pradesh, India

Samina Mehnaz
Department of Biological Sciences
Forman Christian College (A Chartered University)
Lahore, Pakistan

Raffaella Balestrini
Istituto per la Protezione Sostenibile delle Piante (IPSP)
Consiglio Nazionale delle Ricerche (CNR)
Torino, Torino, Italy

ISBN 978-81-322-2777-9 ISBN 978-81-322-2779-3 (eBook)
DOI 10.1007/978-81-322-2779-3

Library of Congress Control Number: 2016939419



Printed on acid-free paper

This Springer imprint is published by Springer Nature
The registered company is Springer (India) Pvt. Ltd.

# Preface

Plant growth-promoting (PGP) microorganisms help the plants in a number of ways. In the last few years, several efforts have been made by the researchers to highlight the applicative potential of these beneficial soil microorganisms providing nutrients and protection to the plants. The role of such microbes in sustainable agriculture has been well documented. These microorganisms, such as plant growth-promoting rhizobacteria (PGPR) or plant growth-promoting fungi (PGPF), can be the best alternatives for chemicals (fertilizers and pesticides). Although the development and use of bioformulations with these beneficial microorganisms have seen an upward trend in the last few years, in broader perspective, the growth has not been up to mark, particularly so in the developing countries. There is a great concern for green and safe food, but a bigger question is on food security for all. We know that to fit in the shoes of chemicals, a lot more has to be done for the development of bioformulations which are reliable and as effective as in lab conditions.

To be successful, a formulation should provide the right number of beneficial microbes in good state to the soil or rhizosphere or the plant. In present-day scenario where the soils are becoming non-fertile or barren by the day, it is important that useful soil microorganisms are augmented in the fields. It is particularly important that the deficient soils are enriched with useful microorganisms such as *Rhizobium*, *Azotobacter*, PGP fluorescent pseudomonads, arbuscular mycorrhizal (AM) fungi, etc. This can be only achieved if the end users, i.e., the farmers, have confidence in the bioformulations. For this achievement, the quality of the current bioformulations has to be enhanced, involving the use of the latest technologies and understanding the ecological requirements, the mechanisms involved in the interaction with roots, and the habitat where they are going to be introduced. It is important to find out the loopholes and address them so as to enhance the credibility of the bioformulations in future. However, already we are seeing a drift or change. The bioformulations being developed now are multifaceted, involving consortia of useful microbes and also including secondary metabolites which aid the establishment in rhizosphere and help root colonization. Successful inclusion of microbial metabolites such as flavonoids, phytohormones, and lipochitooligosaccharides (LOCs) in bioformulations is showing the way. Although in recent years several improvements have been done, research in this field should be enhanced.

For further development, a fruitful partnership between industry-academia-government/regulatory authorities and the end users is required. All these issues have been elaborately discussed in the volume so as to give directions for future research and development of a consistent and globally successful product.

The book consists of 16 chapters contributed by experts from around the globe. The experts are involved in the development of bioinoculants by utilizing diverse PGP microorganisms. The book includes and discusses the history of growth of bioformulations. The process of development of bioformulations, the constraints faced, and the requirements for the future have been discussed in detail. The tome also describes the present-day market scenario of the bioformulations, including both biofertilizers and biopesticides, and how improvements can be done in the future. The diverse roles of bioformulations in reclamation of stressed and marginal soils are also discussed. Recovery of stressed and polluted soils with the help of bioformulations is a very important aspect for the future research and applicative projects. This will not only increase the arable land but will make our planet greener and also help in enhancing productivity for food security. Overall it can be said with conformity that the future of bioformulations is bright; the only thing is to achieve the goals and targets as swiftly as possible, so as to enhance the market share of these green alternatives.

The volume is a unique compilation on microbial products available and to be formulated in future for sustainability of agriculture and soils. The tome will be extremely useful for the researchers involved in the development of bioformulations/bioinoculants, agricultural sciences, microbiologists, biotechnology-related industries, end users, and regulatory authorities around the globe. The book will also be helpful for graduate and postgraduate students and faculty pursuing their career in the field of bioformulation development and use.

The editors are thankful to all the authors who have contributed in this book and have provided excellent knowledge on diverse aspects of the topic. The contributors have really made the book exceptional and novel in itself. The editors would also like to thank Dr Mamta Kapila, Senior Editor, Springer (India), for her support. It is because of her sustainable approach that such a product will see the light of the day, which will go some way to make this planet greener.

NKA is obliged to Prof RC Sobti, Vice Chancellor, BBA University, Lucknow, UP, India, for the encouragement and support. NKA would also like to thank his team of research scholars, namely, Jitendra Mishra, Rachna Singh, Sakshi Tewari, Maya, Shweta, Jai Prakash, and Sushma Verma, for helping in the compilation of the manuscript. Last but never the least, NKA is indebted to his wife Preeti and kids Pranay and Nav for bringing calmness and fulfillment.

Lucknow, Uttar Pradesh, India — Naveen Kumar Arora
Lahore, Pakistan — Samina Mehnaz
Torino, Italy — Raffaella Balestrini

# Contents

# About the Editor

**Dr. Naveen Kumar Arora, Ph.D.,** Microbiology, Associate Professor in Department of Environmental Microbiology, Babasaheb Bhimrao Ambedkar University (A Central University), Lucknow, Uttar Pradesh, India, is a renowned researcher in the field of Environmental Microbiology and Biotechnology. His specific area of research is rhizosphere biology and PGPRs. He has 38 research papers published in premium international journals and several articles published in magazines and dailies. He is editor of the books *Plant Microbe Symbiosis: Fundamentals and Advances and Plant Microbes Symbiosis: Applied Facets*, published by Springer. He is member of several national and international societies and reviewer of several international journals. He has delivered lectures in conferences and seminars around the globe. He has a long-standing interest in teaching at the PG level and is involved in taking courses in bacteriology, microbial physiology, environmental microbiology, agriculture microbiology, and industrial microbiology. He has been advisor to sixty seven postgraduate and four doctoral students. Recently he was awarded for excellence in research by the Honorable Governor of Uttar Pradesh and recieved "Young Achievers Award" from Asian PGPR Society. Although an academician and researcher by profession, he has a huge obsession for the wildlife and its conservation and has authored a book, *Splendid Wilds*. He has a website www.naveenarora.co.in dedicated for the cause of wildlife and environment conservation.

**Dr. Samina Mehnaz, Ph.D.,** Microbiology, is a Professor and Chairperson of the Department of Biological Sciences, Forman Christian College (A Chartered University), Lahore, Punjab, Pakistan. She is a researcher in the field of Microbiology. Her specific area of interest is rhizosphere microbiology, specifically PGPRs and their secondary metabolites. She has 34 research papers published in international journals and 4 book chapters (as a single author) published by Springer. Currently, she is an editor of *Archives of Phytopathology and Plant Protection* (Taylor & Francis). She is a reviewer of several international journals and is the continuous winner of the research productivity award since 2010, by Pakistan Council for Science and Technology. She is an Alexander von Humboldt (AvH) Fellow (Germany) and Natural Sciences and Engineering Research Council (NSERC) Fellow (Canada). She has been advisor to nine post graduate and three doctoral students and one post-doctorate fellow.

**Raffaella Balestrini, Ph.D.,** Biology and Biotechnology of Fungi, is Senior Researcher at the Institute for Sustainable Plant Protection, CNR. Her researches mainly focus on different aspects of plant-symbiotic fungi interactions (endo- and ectomycorrhizae). The researches she currently developed, using molecular and cellular approaches including confocal microscope, in situ hybridization, and laser microdissection (LMD), are genomics and functional genomics in mycorrhizal fungi; cellular and molecular interactions between plants and mycorrhizal fungi, with particular attention to plant and fungal cell wall modifications; plant-mycorrhizal fungi interactions to improve plant tolerance to abiotic stress; and transcriptome profiles of specific cell types involved in mycorrhizal symbiosis. She is also involved in the black truffle *Tuber melanosporum* genome project as the person in charge of the annotation of cell wall-related genes thanks to the experience in gene expression and cell biology studies in mycorrhizal associations, in post-genomic activities. She has worked with research groups from national and international institutions in joined research projects, as mirrored by the several papers in collaboration. She is referee for several international journals and project proposals. She has 68 publications in international journals and 6 book chapters to her credit.

# Part I

# Introduction

# Bioformulations for Plant Growth Promotion and Combating Phytopathogens: A Sustainable Approach

1

Jitendra Mishra and Naveen Kumar Arora

## Contents




### Abstract

The role of microbes in sustainable agriculture has provided new insights to agro-economy, and one of the direct benefits is the lesser reliance on chemical fertilizers and pesticides as the continuous application of these chemicals not only showed detrimental effects on agro-ecosystems but also resulted in health risks to humans and animals. In last few years, the development of microbial bioinoculants for enhancing plant growth and disease eradication has emerged as an alternative, but a broader aspect of their application as formulatory product has remained in infancy especially in developing countries. At the economic and social level also, this green strategy is facing hurdles and lags far behind their competitors, the synthetic fertilizers and pesticides. Most of the times it has been found that bioformulations available for a particular crop do not give good results equivalent to those in the laboratory conditions. Such and related constraints are major challenges of this greener approach. Various workers all over the world are continuously engaged in developing formulation products which could be easier to use, show enhanced activity toward phytopathogens, and may cover more target crops. Whole process of bioformulation development, from screening of microbe to product development and its implementation, need to be reviewed.



J. Mishra • N.K. Arora (✉)
Department of Environmental Microbiology, School for Environmental Sciences, Babasaheb Bhimrao Ambedkar University, Lucknow 226025, Uttar Pradesh, India
e-mail: nkarora_net@rediffmail.com




N.K. Arora et al. (eds.), *Bioformulations: for Sustainable Agriculture*,
DOI 10.1007/978-81-322-2779-3_1

In this article several aspects of formulation development have been critically discussed, and the main motive is to describe types of bioformulations being used, their efficacy in the field, and reasons which limit their wider application in field.

## 1.1 Introduction

Sustainable agriculture is proving as one of the toughest jobs in these days. There is no uniformity in agriculture practices all over the world, but one thing which is more or less common is the use of chemical pesticides and fertilizers. Around the globe about 890 synthetic chemicals are approved as pesticides, whereas marketed products are estimated to be approximately 20,700 and among them organophosphorus insecticides are the biggest group (Stenersen 2004). The use of pesticides is not evenly distributed among various crops, and data indicates that 93 % of all row crops such as corn, cotton, and soybeans are treated with some type of pesticide, whereas percentage of forage crop is less than 10 (Pimentel 1993). Albeit synthetic chemicals enjoy a great reputation in enhancing crop productivity and checking several plant diseases but are also proving as environmental havoc (Fenske and Day 2005; Colt et al. 2007), in recent years their role in damaging agro-ecosystems is very well known. The scenario in developing countries is worse and studies show that although these regions use only 20 % of the world's agrochemicals, they suffer 99 % of deaths from pesticide poisoning (Kesavachandran et al. 2009). Among them farmers are the main victims and occupational exposure is very high with lack of technical education (Konradsen et al. 2003; Coronado et al. 2004; Khan et al. 2009a). According to the World Health Organization (WHO), approximately 20,000 workers die from pesticide exposure every year (Pimentel et al. 1992). Studies conducted show that most of the agricultural pesticides now being used can have detrimental effects on human health (Reigart and Roberts 1999) that include rashes, headaches, nausea and vomiting, disorientation, shock, respiratory failure, coma, and, in severe cases, death (Moses et al.1993), whereas long-term exposure can cause cancer and neurologic and reproductive problems (Sanborn et al. 2007). Although human health is of much concern, at the same time excessive use of chemical pesticides is also deteriorating the environment and mostly aquatic systems, wildlife, and sensitive ecosystems are very much disturbed (Stoate et al. 2001; Berny 2007). Pesticides pose a drastic change in the richness of soil microbial diversity (Dorigo et al. 2009), and large-scale killing of beneficial microbes is more common in agricultural lands or the crops which are treated by their repeated use (Papp et al. 1992). Moreover it has also been found that microbial diversity is mainly affected by the type of pesticide used (Johnsen et al. 2001), and data suggest that diverse pesticides may have varied effects on residential microbial population (Spyrou et al. 2009).

Another major dependency of our agriculture is on synthetic fertilizers, used and widely accepted to maintain soil fertility and crop yields. In the last few years, a great number of long-term experiments were initiated to examine the effects of fertilization on soil fertility (Mitchell et al. 1991; Whitbread et al. 2003) and these studies concluded that in one way the use of fertilizers was necessary, and its continuous application increased the concentrations of soil organic matter (SOC), total nitrogen (TN), and other nutrients in plow layers compared with the initial value at the beginning of the experiment (Huang et al. 2010), but there were also results showing that the continuous use of fertilizers is responsible for the decline of soil quality and productivity (Kumar and Yadav 2001; Yang 2006). Recently in a study, Liang et al. (2013) also showed that excessive application of nitrogen and phosphorus fertilizers induces soil acidification and phosphorus enrichment during vegetable production. Microbial composition, population, and functions are also affected by fertilizers applied, and some workers showed that microbial activity of soil microbes increased

(Mandal et al. 2007; Ge et al. 2008), whereas other experiments confirmed little or no effect on soil microbial diversity and activities when organic or inorganic fertilizers are applied (Nakhro and Dkhar 2010). Allison and Martiny (2008) also reviewed that microbial composition is, in the majority of cases, sensitive to elevated $CO_2$, mineral fertilization, temperature changes, and C amendments. The sensitivity of soil microbial communities is also a matter of interest as it likely differs between unmanaged and agricultural ecosystems (Daniel and Kate 2014). That is why there is an emergent need of developing a greener and safer alternative for increasing crop productivity and disease control at the global level, and the use of microbe-based bioformulations is an open choice for achieving the goal of sustainability (Mishra et al. 2015).

Microbe-based formulations also known as bioformulations are more robust than synthetic chemicals as the formulation product of a single microbe may involve direct interactions with pathogens, and numerous mechanisms take part in disease suppression and plant growth promotion (Rodrigo et al. 2011). The smartness of these bioagents can be surmised by the fact that a particular strain in the vicinity of a plant is capable to control disease without producing lasting effects on the rest of the microbial community or other organisms in the ecosystem (Howarth 1991). A specific group of such useful microbes are the plant growth-promoting rhizobacteria (PGPR), characterized by their innate capability to colonize the root surface (Kloepper and Schroth 1994). PGPR are endowed with several traits that assist plant growth promotion (PGP) (Lugtenberg and Kamilova 2009). PGP activities which are very much of interest are phytohormone production (Strzelczyk et al. 1994; Suzuki et al. 2005; Tank and Sarafa 2009), secondary metabolite production such as hydrogen cyanide (Lorck 2004) and siderophores (Neilands 1995; Tian et al. 2009), ammonia production and nitrogenase activity (Glick 2012), phosphate solubilization (Whitelaw 2000; Igual et al. 2001), and several others. Amid PGPR *Rhizobium* spp., *Bradyrhizobium* spp., *Mesorhizobium* spp., *Pseudomonas* spp., *Bacillus* spp., *Azotobacter* spp., and *Azospirillum* spp. are of great interest, and now their use in making bioinoculants to promote plant growth is being promoted at the global level (Saharan and Nehra 2011). A high-impact research is being done on screening novel species of soil beneficial microbes and their use in developing microbial inoculants for combating phytopathogens and enhancing crop productivity (Hafeez et al. 2006; Glick 2012; Ahemad and Kibret 2014). Although the gradual success of such beneficial microbes in agro-ecosystems is indicating that our goal of sustainable agriculture is not a formidable task, more challengeable is to design such a state-of-the-art bioformulation technique that not only gives environmentally friendly easy-to-use product but also assures good results in field so as to displace the harmful chemicals. Besides this global market and registered companies are also trying to reduce manufacture cost which will definitely help farmers to opt these microbe-based products as suitable alternatives for their crop production. This review addresses the current status of bioformulations, their availability in the market and shortcomings, and future of these greener alternatives.

## 1.2 Brief History

The use of microbial inoculum in PGP and eradication of phytopathogens has been reported in historical events that took place in the field of soil microbiology (Davison 1988; Ehrlich 1990; Insam 2001). Although to describe all such events is a hectic task and also not the theme of this review (for this reviews by Coyne (1996) and Insam (2001) can be seen), some of the discoveries which strengthen the field of biofertilizers and biocontrol are being discussed here. In this series if we envisage the development of microbial biofertilizers, French chemist Boussingault first recognized that leguminous plants could use nitrogen ($N_2$) from the air and gave process detail in his report in 1838 (Wilson 1940). Again this finding was tested by Hellriegel and Wilfarth in Germany by performing an experiment and demonstrating that pea plants (*Pisum sativum*) in association with bacteria in their root nodules are capable of using nitrogen (Bottomley 1912). Very soon in the United States of America (USA), the first patent "Nitragin" was registered for plant inoculation with *Rhizobium* sp. for providing nitrogen source to the crop (Nobbe and Hiltner 1896). The occurrence of cyanobacteria in rice fields was first identified by Fritsch (1907), but their role in soil nitrogen supply was demonstrated by De (1939). Although the use of microbial inoculum in $N_2$ fixation was successful, the role played by other beneficial microbes was to be discovered. Gerretsen was studying Mn deficiency problem in oat and during his study he also pointed out problem of phosphate uptake by plants and first demonstrated the capability of rhizospheric bacteria in solubilizing insoluble phosphates (Mulder 1967), and later other workers also confirmed microbe-mediated solubilization of phosphates (Sperber 1957; Goldstein 1986; Subbarao 1988). During the 1960s beneficial effect of mycorrhiza in nutrient uptake was also recognized and very soon they were utilized as biofertilizers (Roger and Mosse 2004). Application of potassium (K)-solubilizing microorganisms (KSM) for increasing K availability in soils was also reported by some workers (Zahra et al.1984; Vandevivere et al. 1994; Sheng and Lin 2006). Although these events are of prime importance, a holistic approach of soil-inhabiting beneficial bacteria was first revealed by Kloepper and Schroth by identifying the potential of PGPR (Kloepper and Schroth 1978). These rhizospheric bacteria were capable of affecting plant growth both directly and indirectly, and further research showed that various crops treated by plant growth-promoting microbes not only enhanced crop productivity but also minimized the use of synthetic chemicals (Adesemoye et al. 2009). Later for large-scale use, their mass production in the form of appropriate bioformulations was required, which included the development of formulations which should be stable, have a long shelf life, and give the desired results when applied in the field with an appropriate delivery system at the lowest effective dose.

Historical trends that opened opportunities in developing biopesticides were first started when Agostino Bassi in Italy discovered that a fungus, *Beauveria bassiana*, caused an infectious disease in the silkworm (Bassi 1835). After this splendid discovery, some workers including V. Audouin in France, J. LeConte in the United States, and E. Metchnikoff in Russia also started to think that pathogens might prove to be effective agents for controlling crop pests (LeConte 1874; Steinhaus 1975). This early work was just the beginning, and some grandeur discoveries were waiting as Elie Metchnikoff was trying to control grain beetle, *Anisoplia austriaca* (Russian cereal crops were in great economic loss due to this pest) (Steinhaus 1956, 1975; McCoy et al. 1988). Indulging himself in beetle behavior and life cycle, he found a fungus that he referred to as green muscardine and named *Entomophthora anisopliae* (now known as *Metarhizium anisopliae*) that was able to kill beetles. Another milestone was the discovery of *Bacillus thuringiensis* (Bt). In 1901 a Japanese biologist Shigetane Ishiwata discovered this bacterium and gave a new facet to microbial biocontrol potential. Since then Bt has been used routinely in a variety of control programs, and after its commercial success in 1938, it was sold as a biopesticide named "Sporeine" for the first time in France (Aronson et al. 1986). In the 1950s it entered in the United States. Shortly *Bacillus popilliae* was used as another variety of bacterial biopesticide (Lord 2005). The first genetically engineered Bt crop, Bt field corn, was registered with the United States Environmental Protection Agency in 1995 (USEPA 1999). This was only the beginning as in 1965 the first fungal product "Boverin" was developed in the former USSR. This product was based on *B. bassiana*, discovered by Agostino Bassi, and used to control the Colorado potato beetle and the codling moth (de Faria and Wraight 2007). Interest in using pseudomonads for biocontrol was also initiated in the 1970s and in early 1980s (Burr et al. 1998),

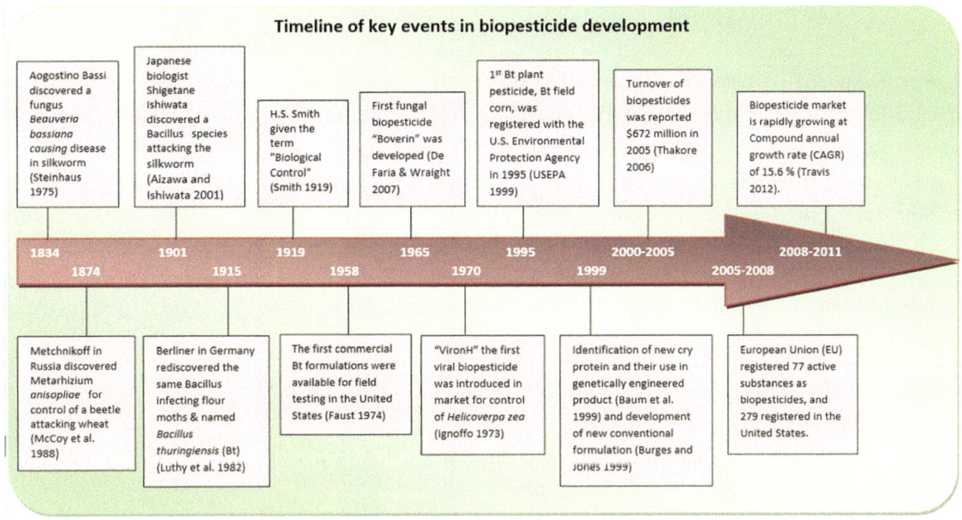


**Fig. 1.1** Time line events of biopesticide development

and later reports confirmed the role of antagonistic fluorescent pseudomonads in suppression of take-all disease of wheat and barley (Smiley 1979; Weller 1988). The role of viruses in biocontrol was first reported in 1940, but it was in the 1970s when "VironH" was for the first time introduced in the market for the control of *Helicoverpa zea* (Ignoffo 1973). Major events in biopesticide history are given in Fig. 1.1.

## 1.3 Concept of Bioformulation

Typically a formulation is a mixture of an active ingredient in a formulated product with inert (inactive) substances (http://npic.orst.edu/factsheets/formulations.html). However regarding bioformulation we see that there is no uniform definition available and various authors define it in their own way. Burges and Jones (1998) stated bioformulation comprises aids to preserve organisms, to deliver them to their targets, and once there to improve their activities, whereas Arora et al. (2010) define the term bioformulation to preparations of microorganism(s) that may be partial or complete substitute for chemical fertilization/pesticides. But any operative definition must include an active ingredient, a carrier material, and an additive. The active ingredient is mostly a viable organism; it may be live microbe or spore and its survival during storage is very essential for successful formulation development (Auld et al. 2003; Hynes and Boyetchko 2006). Suitable carrier material is inert that supports active ingredient (cells) and assures that the cells are easily established in or around the plant and provide better chances of enhancing plant growth or killing target pest. Carrier materials also increase the shelf life of the product (Burges and Jones 1998). Some inert carrier materials are fine clay, peat, vermiculite, alginate, and polyacrylamide beads, diatomaceous earth, talc, vermiculite, cellulose (carboxymethyl cellulose), and polymers specially xanthan gum (Digat 1989). Additives such as gums, silica gel, methyl cellulose, and starch protect from harsh environment conditions and improve physical, chemical, and nutritional properties of formulations (Schisler et al. 2004; Hynes and Boyetchko 2006).

## 1.4 Types of Formulation Available

Broadly two types of bioformulations are available, liquids and solids (Burges and Jones 1998), although in these days there are so many other types of bioformulation available and being used all over the world.

### 1.4.1 Solid Formulations

Solid formulations include granules (GR), microgranules (MG), wettable powders (WP), wettable/water-dispersible granules (WG, WDG), and dusts (Larena et al. 2003; Abadias et al. 2005; Guijarro et al. 2007a). They are produced by adding binder, dispersant, wetting agents, etc. (Tadros 2005; Brar et al. 2006; Knowles 2008).

#### 1.4.1.1 Granules (GR)

Granules are dry particles and contain active ingredient, binder, and carrier. Concentration of active ingredients in granules is 5–20 % (Brar et al. 2006). On the basis of particle size, they are classified as coarse particles (size range 100–1000 μm) and microgranules (size range 100–600 μm). The granules should be noncaking, non-dusty, and free flowing and should disintegrate in the soil to release the active ingredient. They are usually safer having no risk of inhalation and mostly used in soil treatment. Granular formulations are more concerned with storage and increased shelf life (Callaghan and Gerard 2005). Most commonly used granules are wheat meal granules (Navon 2000), corn meal baits, granules formed with gelatinized cornstarch or flour (Tamez et al. 1996), gluten (Behle et al. 1997), cottonseed flour and sugars (Ridgway et al. 1996), gelatin or acacia gum (Maldonado et al. 2002), sodium alginate (Guijarro et al. 2007b), diatomaceous earth (Batta 2008) and semolina (durum) wheat flour (Andersch et al. 1998). MET52 ®, a granular bioformulation of *M. anisopliae* var. anisopliae strain F52, is widely used in biocontrol of black vine weevil (*Otiorhynchus* spp.) larvae in soft fruit and ornamental crops (Ansari and Butt 2012). Sterile rice is used as organic carrier, whereas alginate prill is being utilized in "SoilGard" preparation. This granular formulation contains *Trichoderma virens* as active ingredients and marketed by Certis LLC for eradication of soilborne diseases caused by *Pythium*, *Rhizoctonia*, and *Fusarium*. Selection of different carriers may affect activity of active ingredients in field conditions. In a study Mejri et al. (2013) measured bioherbicidal activity of deleterious rhizobacterium *Pseudomonas trivialis* X33d by taking two granular formulations and found that semolina–kaolin (pesta) showed higher brome suppression activity in wheat field in comparison to kaolin–talc-based granular formulation, whereas BioShield™, formulated as a granule containing *Serratia entomophila*, is sold in New Zealand for control of grass grub larvae in established pasture (Young et al. 2010).

Although granular formulations are very effective, their application is also limited due to inactivation of active ingredient in ultraviolet (UV) light. In a study by Bailey et al. (1996), Bt product used to control apple moth caused by *Epiphyas postvittana* lost more than half of its activity within a day on exposure to sunlight, whereas BioShield, a *Serratia entomophila* containing granular formulation, is very sensitive to UV light and osmotic and desiccation stress and requires subsurface application (Johnson et al. 2001). Some UV protectants such as Tinopal, Phorwite, Intrawhite, and Leucophor; uric, folic, 2-hydroxy-4-methoxy-benzophenone, p-aminobenzoic, 2-phenylbenzimidazole-5-sulfonic acids; and dyes such as Congo red, methyl blue, safranin, brilliant yellow, and buffalo black may overcome UV inactivation of organism when added in formulation medium or coated on formulation product (Warrior et al. 2002; Cohen and Joseph 2009). Stilbene-derived optical brighteners are also more effective in baculoviruses containing formulation as these absorb UV radiation and emit visible blue wavelengths and enhance the infectivity (Goulson et al. 2003). Recently Fernandes et al. (2015) reviewed tolerance of selected entomopathogenic fungal strains to UV radiation.

#### 1.4.1.2 Wettable Powders (WPs)

Wettable powders (WPs) are one of the oldest types of formulations. They consist of 50–80 % technical powder, 15–45 % filler, 1–10 % dispersant, and 3–5 % surfactant by weight to achieve a desired potency formulation (measured in international units) (Brar et al. 2006). These dry formulations are of much interest as they are readily miscible with water and can be easily added to a liquid carrier, normally water, just before its application. WPs have a longer shelf life and by controlling moisture content, their shelf life may exceed 18 months. Longer shelf life is also related to their firm marketplace. Agricultural materials and industrial waste by-products such as wheat bran–sand mixture, sawdust–sand–molasses mixture, corn cob–sand–molasses mixture, bagasse–sand–-molasses mixture, organic cakes, cow dung–sand mixture, compost/farm manure, inert charcoal, diatomaceous earth, and fly ash (Table 1.1) can also be used to prepare powder formulations (Khan et al. 2007). Recently Cheng et al. (2015) prepared a WP containing 60 % *B. cereus* freeze-dried powder, 28.9 % diatomite as carrier, 4 % sodium lignin sulfonate as disperser, 6 % alkyl naphthalene sulfonate as wetting agent, 1 % $K_2HPO_4$ as stabilizer, and 0.1 % β-cyclodextrin as ultraviolet protectant, and in his preliminary study, they found this formulation was effective in biocontrol of postharvest disease in comparison to chemical used. Woo et al. (2014) reviewed current application of *Trichoderma*-containing products in agriculture, and it was found that 55.3 % of *Trichoderma* formulations are commercialized as WPs.

#### 1.4.1.3 Wettable/Water-Dispersible Granules (WG, WDG)

Wettable/water-dispersible granules (WG, WDG) are also known as dry flowables. They have been designed to make WPs more user and environmental friendly, non-dusty, free-flowing granules quickly dissolving in water. They contain wetting agents and dispersing agents similar to those used in WPs, but the dispersing agent is usually at a higher concentration. Like WPs, WDG also show excellent shelf life. WDG formulations have wider role in nematode control and capture 90 % of the total market available for nematode-based products. Antagonistic fungus, *Ampelomyces quisqualis*, is used to control powdery mildew caused by several pathogenic species in grapes, tomato, apples, strawberries, and cucurbits, formulated as WDG (Falk et al. 1995). Chumthong et al. (2008) produced water-soluble granules containing *Bacillus megaterium* for biological control of rice sheath blight and showed that these granule formulations exhibited good physical characteristics, such as high-water solubility and optimal viscosity, suitable for spray application.

#### 1.4.1.4 Dusts

Dusts are also one of the oldest formulation types and contain very finely ground mixture of the active ingredient (usually 10 %) with particle size ranging from 50 to 100 μm. Although they have been used since a long time and in some instances more effective in killing (Ifoulis and Savopoulou-Soultani 2004), there have always been handling and application problems associated with dusts (Harris and Dent 2000). Dust containing beauverial protein extract (weighing about 5 kDa) is also being used in biocontrol. Biofox C has been formulated as dust containing nonpathogenic *F. oxysporum* and used in basil, cyclamen, tomato and carnation (Kaur et al. 2010)

### 1.4.2 Liquid Formulations

Liquid formulations are also known as flowable or aqueous suspensions and consist of biomass suspensions in water, oils, or combinations of both (emulsions) (Schisler et al. 2004). A typical liquid formulation contains 10–40 % microorganisms, 1–3 % suspender ingredient, 1–5 % dispersant, 3–8 % surfactant, and 35–65 % carrier liquid (oil or water) (Brar et al. 2006). Liquid formulation may be of the following types.

#### 1.4.2.1 Suspension Concentrates (SCs)

SCs are produced by adding solid active ingredient(s) with poor solubility in water and satisfactory stability to hydrolysis (Tadros 2013). SCs are diluted in water before use. Their storage

**Table 1.1** Some of the WP bioformulations

| Name of product | Category of bioformulation | Target organism | Carrier material | Application method | Name of country | References |
|---|---|---|---|---|---|---|
| Gypchek | Mycoinsecticides | Gypsy moth (*Lymantria dispar*) Multicapsid nuclear polyhedrosis virus (LdMNPV) | Virus infected gypsy moth larvae processed to produce a finely ground powder | Applied using aerial or ground application equipment | USA | Reardon and Podgwaite (1992) |
| Boverin, Naturalis, Boverosil, Trichobass-P, and L BioGuard Rich Mycotrol ES® Mycotrol O® | Mycoinsecticides | *Beauveria bassiana* | Talc or other inerts such as perlite, kaolin, bentonite, starch | Used to protect agricultural, fruit, decorative, and flowering plants from beetle and moths | Russia, India, South Africa, Former USSR, Spain | de Faria and Wraight (2007) |
| Metaquino | Mycoinsecticides | *Metarhizium anisopliae* var. *anisopliae* | Autoclaved rice or wheat bran | Aerial applications used to control spittlebugs in sugarcane and pastures | Brazil | Marques et al. (1981) |
| Bio-Blast | Mycoinsecticides | *Metarhizium anisopliae* | Autoclaved rice | Termites | USA | Wraight and Carruthers (1999) |
| Biolisa Kamikiri, Melocont-Pilzgerste, | Mycoinsecticides | *Beauveria brongniartii* (Sacc.) | Barley kernels or clay granules | Used to control coleopteran, homopteran, lepidopteran, and dipteran pests in flowers, vegetables, oil palms, and other crops | Japan, Austria, Italy | Hajeck et al. (2006) |
| Bio-fungus | Fungicide | *Trichoderma* spp. | Barley kernels or clay granules | Fungi causing wilt, take-all, root rot, and wood decay | Denmark | Monte (2001) |
| Trichoject, Trichoseal, Trichoprotection ® | Fungicide | *Trichoderma harzianum* and *Trichoderma viride* | Autoclaved corn or rice seed | *Chondrostereum purpureum* and other soil and foliar pathogens | New Zealand | Burges and Jones (1998) |
| Trichodex | Fungicide | *T. harzianum* | Autoclaved corn or rice seed | As above | Israel | Monte (2001) |
| Binab® | Fungicide | *T. harzianum* | Autoclaved corn or rice seed | Control *Botrytis*, *Fusarium*, *Gaeumannomyces*, *Pythium*, *Rhizoctonia*, *Sclerotinia*, *Sclerotium*, *Verticillium*, and wood-rot fungi | Denmark, USA, Israel | |
| Binab T | | *T. harzianum and T. polysporum* | Autoclaved corn or rice seed | As above | Sweden | |
| Aspire® | Fungicide | *Candida oleophila* | Skimmed milk | Postharvest decay of citrus fruits, apples, and pears caused by *Penicillium* sp. and *Botrytis* sp. | USA | Droby et al. (1998) |

| | | | | | | |
|---|---|---|---|---|---|---|
| Yield Plus | Fungicide | *Cryptococcus albidus* | Skimmed milk | As above | South Arica | http://www.anchor.co.za/ |
| Sporidex® | Fungicide | *Pseudozyma flocculosa* | Skimmed milk | Powdery mildew cn greenhouse roses and cucumbers | Canada | Punja and Utkhede (2003) |
| DiTera™ | Nematicide | Killed *Myrothecium verrucaria* | Wheat bran, talc powder, silica | Plant parasitic nematode specially root-knot nematodes (*Meloidogyna Incognita*) | USA | Warrior et al. (1999) |
| BlightBan® A506 | Bactericide | *Pseudomonas fluorescens A506* | Talc and peat | Postharvest decay in almond, apple, cherry, wettable powder | Canada | Stockwell and Stack (2007) |
| Bio-Save 10 LP, Bio-Save 100, and Bio-Save 1000 | Fungicide | *Pseudomonas syringae* strain. ESC-100 | Talc and peat | Postharvest decay in citrus and pome fruit | USA | |
| Collego® | Bioherbicide | *Colletotrichum gloeosporioides* f. sp. aeschynomene | Undisclosed | Controls northern joint-vetch (*Aeschynomene virginica* | USA | |

and solubility can be improved by addition of surfactants and various additives. Farmers generally prefer suspension concentrates to wettable powders because they are non-dusty and easy to measure and pour into the spray tank.

#### 1.4.2.2 Oil-Miscible Flowable Concentrate (OF)

OF is stable suspension of active ingredient(s) in a fluid intended for dilution in an organic liquid before use (Singh and Merchant 2012).

#### 1.4.2.3 Ultralow Volume (ULV) Suspension (SU)

They are suspension ready for use through ULV equipment. ULV are aerial or ground spray equipment and generate extremely fine spray (Singh and Merchant 2012).

#### 1.4.2.4 Oil Dispersion (OD)

OD is a stable suspension of active ingredient (s) in water-immiscible solvent or oil (Michereff et al. 2009). ODs have validated a growing importance over the past decade. Recently Mbarga et al. (2014) developed a soybean oil based formulation and found that *Trichoderma asperellum* containing OD had great potential for the control of cacao black pod disease with increased half-life of the conidia in comparison to aqueous suspension. Some protective measures are required with regard to handling fungi containing OD formulations. As in prolonged storage, active ingredient (conidia) may be settled out of suspension or densely compacted in the bottom of the container (Butt et al. 2001).

Some of the *Trichoderma* containing liquid formulations used in biocontrol are Trichojet, Enpro-Derma, and Trichorich-L (Woo et al. 2014). Oil-based formulations have been proven better in foliar spray and considered effective in enhancing the activity of entomopathogens (Feng et al. 2004). Oil evaporates much less, so it remains in contact for greater time and can be applied as an emulsion (oil in water) (Luz and Batagin 2005) or in some cases as an invert emulsion (water in oil) (Batta 2007).

### 1.4.3 Encapsulation

Encapsulation involves coating or entrapping microbial cells within a polymeric material to produce beads which are permeable to nutrients, gases, and metabolites for maintaining cell viability within the beads (John et al. 2011). Based on the size of the polymeric bead produced, two types of techniques, i.e., macroencapsulation (size ranging from few millimeters to centimeters) and microencapsulation (size ranging from 1 to 1000 μm, generally less than 200 μm), are used (Nordstierna et al. 2010). Macroencapsulation techniques are advantageous than microencapsulation (for further details on microencapsulation review by Rathore et al. 2013 can be seen). Encapsulation provides good protection to active ingredient from harsh environmental factors. Currently, gelatin, starch, cellulose, and several other polymers are used for encapsulation of active ingredients (Amiet-Charpentier et al. 1998; Park and Chang 2000; Cheze-Lange et al. 2002). Protection may enhance to some extent by coating capsule with dyes (Cohen et al. 1990). For further detail on encapsulation, chapter by Schoebitz et al. can be seen from this very book.

Although both liquid and solid formulations have been extensively used in agrosystems, dry formulations are generally preferred over wet formulations because they provide extended shelf life and are easier to store and transport (Burges and Jones 1998). The development of a bioformulation is proving a hectic job and earlier work done in this field is not sufficient. The increasing demand for developing new formulations to replace chemical pesticides and fertilizers has created interest amongst entrepreneurs in this field, and they are funding various projects for the development of cheaper and effective technology. Some technological advances in development of Bt-based products have provided substantial aid in its commercial production. For example, Micellar-enhanced ultrafiltration (MEUF) is a technique being used to separate dissolved organic compounds like thuringiensin from aqueous streams (Tzeng et al. 1999). Similarly in situ product removal (ISPR) involves biochemical product removal during fermentation process and successfully applied in removal of Bt toxin proteins (Agrawal and Burns 1996), whereas cross-flow microfiltration (CFM) has been utilized for extraction of all kinds of proteins and harvest of recombinant yeasts (Persson et al. 2004)

## 1.5 Current Scenario/Market Trends

Microbe-based formulations for plant growth enhancement and for eradication of phytopathogens are being used all over the world (Leggett et al. 2011; Naderifar and Daneshian 2012; Gašić and Tanović 2013), but the data related to their use around the globe is fragmented. One probable reason is the discrepancy of the terminology used. Most of the developing countries are using the word "biofertilizers," whereas in the rest of the world, "bioinoculant" for enhancing crop yields is being used, but in both cases either materials derived from living organisms or organism itself are used (Vessey 2003; Chen et al. 2006) for increased absorption of nutrients in plants that assist in soil fertility and crop productivity. Many growers around the world are now routinely applying biofertilizers and biopesticides for different crops. European biofertilizer market is the most developed and widespread among all other regions and is estimated to grow from around $2566.4 million in 2012 to $4582.2 million by 2017, at a calculated annual growth rate (CAGR) of 12.3 %, from 2012 to 2017 (PRWEB 2014). In North America, biofertilizer market was highest in 2012 and is projected to grow at a rate of 14.4 % during the period of 2013–2018 (Micro Market Monitor 2015). Europe, together with North America, accounted for over 50 % of the global revenue (Grand View Research 2015). China the leading worldwide producer of rice, wheat, and many vegetables including onions and cabbage is also promoting biofertilizer use (Grand View Research 2015). There are 151 biofertilizer production units operated by government and nongovernment agencies in India (Mahajan and Gupta 2009). Among all, nitrogen-fixing biofertilizers were the most widely used, accounting for over 78 % of the global demand in 2012 (Agro news 2014). Table 1.2 provides details of some of the biofertilizers used around the globe.

In the field of biocontrol, the most successful biopesticides are Bt based and represent about 95 % of total microorganisms used (Bravo et al. 2011). Globally 322 products of Bt are generating annual revenue of $210 million (CAB International Centre 2010). But the use of other biopesticides (fungal and non-Bt) is also increasing. Recently market research survey has been carried out by various agencies to collect information about biopesticides, but reliability of such reports is of much concern and raised many questions. One of the main causes is that the criterion involved in market research may vary because some firms and agro-industries include subcategories such as microbes, biochemicals, plant growth regulators, plant-induced protectants (PIPs), insect growth regulators, essential oils, and pheromones, in the term biopesticide, whereas others use only the products of microbial origin. Gelernter (2007) endeavored to represent figures of sale in Europe and North America and estimated it to be $200 million in 2003. According to Thakore (2006), the turnover of biopesticides was $672 million in 2005, but there was no description of category included. A report by Harwood et al. (2007) showed global biopesticide market to be about $280 million in 2007. In this report true microbial agents were included. Business consultancies such as BCC Research (2010) and CPL (2006) are actively involved in direct marketing survey, so probably collecting much reliable data, and concluded that the global biopesticide market is growing at a rate of 10 % per year. Global industry analysis (2015) estimated that the annual biopesticide market could exceed $2.5 billion by 2015. Other market research reports by BCC on biopesticide represent the total sale of biopesticide to $1.2 billion in 2008 and $1.6 billion in 2009. This is planned to upturn to $3.3 billion in 2014 and $10 billion in 2017 (Marrone 2007). The

**Table 1.2** List of some of the biofertilizers used around the globe

| Name of country | Name of company | Name of product | Active ingredient | Use on the crops | References |
|---|---|---|---|---|---|
| Brazil | Embrafós, Institute of Biological Phosphate (IFB), Biophosphate of Brazil, Liderfós | BioAtivo | (Nitrogen fixation and phosphate solubilization) | Beans, maize, rice, sugarcane, soybean, eucalyptus, citrus, tomatoes, cotton, forage crops, and carrots | http://ifb.agr.br/biotecnologia/beneficios-do-bio-ativo/ |
| Central Russia | PitBioTech, JSC "Industrial Innovations" | Bamil and Omug | *Bacillus*, *Micrococcus*, and *Clavibacter* | Barley, wheat, beetroot, onion, garden radish, carrot, potato, marrow leguminous and nonleguminous | Kutyova et al. (2002) and Zhigletsova et al. (2010) |
| | | Ekud | *Bacillus* and *Staphylococcus hominis* | | |
| | | Pudret | *Bacillus* and *Staphylococcus* | | |
| | | Azotovit | *A. chroococcum* | | |
| | | Bactophosphin | *Bacillus mucilaginosus* | | |
| Southern and Eastern Russia | Natural resources LLC EM Technology | Rizotorphin | *Rhizobium* spp. | Field pea, soybean,chickpea, broad bean, narrow-leafed lupin, tomato, pepper, brinjal, sorrel, asparagus, estragon, etc. | Karimov and Zaripov (2007), Balashov et al. (2008), and Fedotova et al. (2009) |
| | | Rizoagrin | *Agrobacterium radiobacter* | | |
| | | Azotobacterin | *Azotobacter chroococcum* | | |
| | | Ekophit | *A. chroococcum* | | |
| | | Agrofil® | *Agrobacterium* sp. | | |
| | | Flavobacterin® | *Flavobacteria* spp. | | |
| | | Mizorin® | *Arthrobacter mysorens* | | |
| | | Baikal EM-1 | Lactic acid, nitrogen fixation bacterias, photosynthetic bacteria, and yeast | | |
| Japan | The Tokachi Federation of Agricultural Cooperatives (TFAC, Tokachi Nokyoren) | Mamezo | *Rhizobium* sp. | Soybeans, azuki beans, and *Phaseolus* beans | García-Fraile et al. (2015) |
| | | R-Processing seeds | Leguminous seeds inoculated with *Rhizobia* sp. | | |
| | | Hyper-coating seeds | Rhizobia within the capsule of calcium carbonate) | | |
| | EMRO (EM Research Organization) | EM•1® and EM Bokashi | *Lactobacillus plantarum*, *L. casei*, *L. fermentum*, *L. delbrueckii*, *Streptomyces cerevisiae*, and *R. palustris* | Soybeans, azuki beans, and *Phaseolus* beans | http://emrojapan.com/page/89-howtomakebokashi/ |
| | CBF China Bio-Fertilizer AG | Xin Sheng Li | (*Bacillus mucilaginosus* and *B. subtilis*) a potassium-solubilizing bacteria and a phosphorus-solubilizing bacteria | | |

| | | | | | |
|---|---|---|---|---|---|
| | Central Glass Co., Idemitsu Kosan Co., and Osaka Gas Co., | VAM | | Maize, alfalfa, onion | Ogawa (1989), Tawaraya et al. (1998), and Karasawa et al. (2001) |
| USA | TeraGanix, Inc | EM·1® | Lactic acid bacteria, yeast, and photosynthetic bacteria | Most plants (flowers, veggies, trees, etc.) | http://www.teraganix.com/EM-1-Microbial-Inoculant-Microbial-Inoculants-p/1000.htm |
| | Novozymes LLc | Nitragin Gold | *Rhizobium meliloti* | Alfalfa pea, lentil, soybean, corn, sorghum, sugar beet, winter wheat, canola sunflower,mustard | http://www.bioag.novozymes.com/en/products/unitedstates/biofertility/Pages/default.aspx |
| | | TagTeam | Combines the best in rhizobia strains with *Penicillium bilaii* | | |
| | | JumpStart | *P. bilaiae* | | |
| | | JumpStart® LCO | *P. bilaii* and LCO Promoter Technology | | |
| Argentina | Rizobacter S.A. | Rizoliq LLI | *Rizobacter* sp. | Legumes and cereal crops | http://www.rizobacter.com/argentina/productos/ |
| | | Signum | *Bradyrhizobium* sp. | | |
| | | Rizo Liq TOP | *B. japonicum* | | |
| Australia | Mapleton Agri Biotec Pty Ltd | TwinN | *Azotobacter* (soilborne species), *Azospirillum* (endophytic species), and free-living nitrogen-fixing bacteria, together in a freeze-dried form) | Legumes and cereal crops | http://www.twinn.com.au/ |
| | Vanadis Bioscience Pty. Ltd. | Catapult™ | VAM plus 2 species of *Bacillus* | | http://www.vanadisbioscience.com/ |
| Hungary | Phylazonit Kft | Phylazonit-M | *B. megaterium* and *Azotobacter chroococcum* | Rice, maize | FNCA (2006) |
| India | T. Stanes & Company Limited | Symbion-N | *Azospirillum*, *Rhizobium*, *Acetobacter*, and *Azotobacter* | Cereals, legumes, and vegetable crops | http://www.tstanes.com/products.html |
| | | Symbion-P | *B. megaterium* var. *phosphaticum* | | |
| | | Symbion-K | *Frateuria aurantia* | | |
| | | Symbion-S | *Thiobacillus thiooxidans* | | |
| | | Symbion-VAM | *Glomus fasciculatum* | | |
| | Camson Bio Technologies Limited | CALOBIUM | *Rhizobium* sp. | Legume crops, cereal crops, vegetable crops | http://www.camsonbiotechnologies.com/products/bio_fertilizers_and_stimulants.htm |
| | | CALMONAS | *Pseudomonas* sp. | | |
| | | CALSPIRAL | *Azospirillum* sp. | | |
| | | CALZOTO | *Azotobacter* sp. | | |
| | | CALTASH | K solubilizer | | |
| | | CALPHOROUS | PSB | | |

(continued)

**Table 1.2** (continued)

| Name of country | Name of company | Name of product | Active ingredient | Use on the crops | References |
|---|---|---|---|---|---|
| | Ajay Biotech | BIOPHOS | PSB | Various crops | http://www.ajaybio.in/prodpro.htm |
| | Gujarat State Fertilizers and Chemicals | Sardar Biofertilizers | *Azotobacter*, *Azospirillum*, and PSB | All types of crop | http://www.gsfclimited.com/bio_fertilizers.asp?mnuid=3&fid=32 |
| Cuba | National Institute of Agricultural Sciences | EcoMic® | *Glomus fasciculatum* | Rice, cotton, soybean, corn, coffee, sorghum | Penton et. al. (2011), and Ortega (2007) |
| | National Program Project Agricultural Biotechnology of Cuba | FOSFORINA® | *P. fluorescens* | Tomato | |
| | Institute of Ecology and Systematic of Cuba) | Micofert® | *Glomus aggregatum*, *G. manihotis*, *G. spurcum* | Sugarcane, tomato, onion, garlic, potato, banana, mandarin, malanga, yam, grains, forest | |
| Mexico | | Micofert | *G. intraradices*, *G. etunicatum*, *Gigaspora* sp. | Coffee, papaya, corn, onion, avocado | Cobos (2005) |
| | | BuRIZE1 | *Glomus intraradices* | Alfalfa, apple, tomato, orange, avocado, | BioScientific (2008) |
| | | Myconate1 | Isoflavanone and phormononetin compounds | Corn, soybean | Lara (2008) |
| Colombia | | Mycobiol1 | *Glomus* sp., *Entrophospora colombiana*, *Acaulospora mellea* | Banana, cassava, lettuce, tomato, cape gooseberry (*Physalis peruviana*) | www.corpoica.org.co |
| | | FOSFOSOL1 | *Penicillium janthinellum* | Rice | Moreno-Sarmiento et al. (2007) |
| | | Dimargon1 | *Azotobacter chroococcum* | Rice, cotton | |
| Malaysia | | UPMB 10 | *Bacillus sphaericus* | Oil palm | FNCA (2006) |
| | | MYCOgold | Species of 4 genera of AMF | Banana | Malaysian Agri Hi Tech (2008) |
| Pakistan | Government | BioPower | Multi-strain of $N_2$ fixing | Cotton, maize, rice, sugarcane, and wheat | Kennedy et al. (2004) |
| | Microbial Biotechnologies | Ferti-Bio | Multi-strain of $N_2$ fixing, P-solubilizing and growth hormone-producing bacteria | Rice, wheat, corn, cotton, sugarcane, and vegetables | |

region-based research report states that the United States represents the largest region of biopesticides worldwide, whereas Europe represents the fastest-growing regional market for biopesticides showing annual average growth rate (AAGR) of 15.0 % (Industrial Equipment News 2011). Asia-Pacific is also coming up in the market, where biopesticide sales were projected to reach US $362 million in 2012. Latin America has the smallest increase of all the regions. The market was nearly $70 million in 2005 and remained only $88 million in 2010, an AAGR of 5.0 % (Industrial Equipment News 2011).

## 1.6 Available Bioformulations

Although bioformulations have a widespread use in agriculture, they are exclusively recognized for their effective role in the field of biocontrol and in biofertilization (Trabelsi and Mhamdi 2013), and here we have focused on both the aspects.

### 1.6.1 Formulations for Nutrient Uptake

In the last few years, the use of microbial inoculants is realized as an effective way of providing nutrients to plants since it would substantially reduce the use of chemical fertilizers, and hence there are an increasing number of biofertilizers that are commercially produced for various crops (Berg 2009; Trabelsi and Mhamdi 2013). Bioformulations containing microbes for better availability of nutrient are being described here.

#### 1.6.1.1 Nitrogen (N)

N is considered as an essential plant macronutrient required in large quantities (1–3 % on a dry-weight basis) (Kraiser et al. 2011), but only a small proportion of the nitrogen fertilizers supplied to agricultural systems is utilized (Vitousek et al. 2009). According to an estimate, around 50 % is used by the crops, 25 % is emitted to the atmosphere, 20 % runs off into aquatic systems, and only 5 % is stored in the soil (Galloway 2005; Garnett et al. 2009). This is why without considering environmental impact, the amount of synthetic nitrogen applied to crops has risen dramatically, from 12 to 104 Tg/year in the last few years (Mulvaney et al. 2009). Biological nitrogen fixation (BNF) is one way of converting elemental nitrogen into plant-usable form (Gothwal et al. 2009) and has considerable advantages from ecological and economical point of view (Sainju et al. 2003), but the capability of N fixation is limited and exclusively restricted to most of the phyla of bacteria and in methanogenic archaea (Young 1992). Symbiotic fixation of nitrogen within nodules of vascular plants is done by two major groups of bacteria not phylogenetically related, rhizobia (leguminous association) and *Frankia* (nonlegume association) (Franche et al. 2009). Legumes constitute the third largest family of flowering plants (Polhill and Raven 1981) accounting for approximately 27 % of the world's crop production (Graham and Vance 2003) including important crop legumes: soybean (*Glycine max*), peanut (*Arachis hypogaea*), mung bean (*Vigna radiata*), chickpea (*Cicer arietinum*), lentil (*Lens culinaris*), common bean (*Phaseolus vulgaris*), pea (*Pisum sativum*), and alfalfa (*Medicago sativa*). About 200 species of non-legume plants distributed among 24 genera in eight angiosperm families are nodulated by *Frankia* (Huss-Danell 1997). BNF produces roughly 200 million tonnes of nitrogen annually (Peoples et al. 2009) which saves farmers millions of dollars in fertilizer costs yearly. In a study Townsend and Howarth (2010) argue that humans are now fixing nitrogen at twice the rate of natural processes. There have been many workers which confirmed the role of rhizobia in sustainable crop production and envisaged that agronomic practices using rhizobial inoculum may ensure adequate nitrogen to legumes instead of N fertilizers (Gupta 2004; Arora et al. 2010). Both leguminous seed and soil can be treated by application of legume inoculants containing live rhizobia (Martinez-Romero 2003; Deaker et al. 2004). Legume inoculants may consist of one strain or may comprise two or more strains effective on that particular host and commercially produced in powder or granular and liquid formulations (Lupwayi et al. 2006). Conventionally peat is mostly used as a carrier material in legume inoculation production (Albareda et al. 2008). But cell number is dependent on environmental conditions and rhizobial species used. In a study Albareda et al. (2008) found that growth and survival of inoculated strains were as high in peat as when compost cork and perlite were used as carrier materials. Other locally available carrier materials such as coal, bagasse, coir dust, etc., are also used (Albareda et al. 2008).

Although direct applications of rhizobia as bioinoculants have been very successful, more recent application of Nod factors or lipophilic chitin oligosaccharides (LCOs) has also showed substantial impacts on crop yield in soils containing very small populations of rhizobia (Kidaj et al. 2012). In the United States, LCO Promoter Technology® (a technique of using the beneficial effect of Nod factor molecules) has been widely popularized by Novozymes for promoting growth in economic crops, such as soybean, peanuts, and alfalfa. A liquid formulation of non-associative free-living nitrogen fixers such as *Azotobacter* and *Azospirillum* including cyanobacteria has been also marketed in various countries and showed a significant increase in crop production (Vendan and Thangaraju 2006). For sustainable rice production, southern China, Vietnam, India, and Bangladesh are also using free-floating water fern *Azolla* in which cyanobacteria *Anabaena azollae* symbiotically fixes nitrogen (Pabby et al. 2003). Recent use of endophytic nitrogen-fixing bacteria as biofertilizers is also reported, and members of *Azoarcus*, *Achromobacter*, *Burkholderia*, *Gluconacetobacter*, *Herbaspirillum*, *Klebsiella*, and *Serratia* have been identified as potent endophytic nitrogen-fixing strains (Rothballer et al. 2009; Franche et al. 2009).

#### 1.6.1.2 Phosphate

Plants obtain phosphorus as phosphate anions from the soil solution. Phosphate is being considered as probably one of the least available plant nutrients found in the rhizosphere due to its inorganic fixation and formation of organic complexes (Eswaran et al. 1997). The phosphorus content in average soils is about 0.05 % (w/w) of which only 0.1 % is available to plants (Achal et al. 2007). It has been found that the application of phosphatic fertilizers do not validate the need of the plant as a consequence of organic and inorganic fixation. Nearly 80 % of applied phosphorus may be unavailable to plants (Holford 1997). About 5.7 billion hectares of land worldwide has been reported to be phosphate deficient (Batjes 1997; Vassilev and Vassileva 2006). Biological processes in the soil, such as microbial activity, tend to control the mineralization and immobilization of organic conversion of the insoluble forms of phosphorus to an accessible form by plants (orthophosphate), which is an important trait of phosphate-solubilizing bacteria (PSB) and arbuscular mycorrhizal fungi (AMF) (Fankem et al. 2006, Khan et al. 2007). In the last few years, the development of microbial inoculum containing phosphate-solubilizing microbes (PSM) gained attention of agriculturists (Fasim et al. 2002). Application of PSM, either individually or in combined form, remained successful for increasing yield of soybean (Fernandez et al. 2007), maize (Hameeda et al. 2008), wheat (Minaxi et al. 2013), mung bean (Jha et al. 2011), and chickpea (Singh and Prakash 2012). Although phosphate-solubilizing ability of PSM was discovered very early (Mulder 1967), their commercialization in the form of bioformulations was not very successful (Rodriguez and Fraga 1999). It has been observed that quality management is essential for the development of reliable and contaminant-free bioproducts, whereas in field conditions performance is hampered by environmental variables including salinity, pH, moisture, temperature, and climatic conditions of the soil (Khan et al. 2009b). Among PSB-based biofertilizers, products having *Pseudomonas* spp., *Bacillus* spp., *Aspergillus* spp., and *Penicillium* spp. are predominantly used (Sharma et al. 2013). Phosphobacterin is one of the oldest, *Bacillus megatherium* containing biofertilizer

(Smith et al. 1961), and approximately 10 million hectares were treated with the product in Russia in 1958 (Brown 1974). In India P Sol B® has wide application in agriculture and contains *Pseudomonas striata* (NCIM 2847). FOSFOSOL® is a phosphatic biofertilizer containing *Penicillium janthinellum* and being used at very large scale in Colombia (Moreno-Sarmiento et al. 2007).

#### 1.6.1.3 Potassium

For balanced plant growth, K uptake is as important as N and phosphorus. A lot of vital functions of plant are directly or indirectly potassium dependent. This macronutrient takes part in enzyme activation of several physiological reactions including protein synthesis, photosynthesis, starch synthesis and also helps in resistance to diseases and insects (Rehm and Schmitt 2002). Although 2.5 % of the lithosphere is of K, its actual soil concentrations vary widely, ranging from 0.04 to 3 % (Sparks and Huang 1985). Soil potassium is available to plants in four different pools: (i) soil solution, (ii) exchangeable K, (iii) fixed K, and (iv) lattice K (Syers 1998). Amongst all four, soil solution and exchangeable K are directly available for plant uptake, but for a rapidly growing crop with enough K, their acquisition by this alone is not adequate (Philip et al. 2013) and require external application in the form of potassic fertilizers. At global level after the United States, China, and Brazil, India ranks fourth in total consumption of potassium fertilizers (Investing News Network 2015). It has been found that "non-exchangeable" K in soils is solubilized by the release of organic acids by some bacteria that increase the $K^+$ concentration in the soil solution (Meena et al. 2014). Their ability of solubilizing K-bearing minerals such as micas, illite, and orthoclase is of much interest in developing bioinoculants with the ability to provide soluble K to plants (Sheng and Lin 2006). In some countries, especially in China and South Korea, K biofertilizers have been tested (Basak and Biswas 2008). Most of the work in developing K biofertilizers involves utilization of those PSB which can also solubilize K-bearing minerals (Ahmed and El-Araby 2012). Recently *Frateuria aurantia* has been recognized as very effective K-mobilizing bacterium and used in commercial production of Symbion-K, Biosol-K, and K Sol B® biofertilizers.

#### 1.6.1.4 Iron

Almost all life forms require iron in the form of various proteins and pigments (Escolar et al. 1999). In soil its concentration ranges from 7000 to 500,000 mg $kg^{-1}$ mainly in the insoluble Fe (III) (ferric) form which hydrolyzes readily to give Fe $(OH)_2{}^+$, Fe $(OH)_3$, and Fe $(OH)_4{}^+$. Although in soil ferric iron dominates, plant uptake of iron occurs in ferrous (II) form and availability of both form of Fe depends on pH and oxygen level in the soil (Fageria et al. 1990). It has been found that the microorganisms inhabiting around growing roots drop redox potential in the rhizosphere because of the microbial oxygen demand, and this serves to increase concentrations of Fe (II) ions for plant uptake (Nikoloic and Romheld 1999). This microbe-mediated iron uptake is facilitated by a low molecular weight iron chelators termed siderophores (Garibaldi and Neilands 1956). There are so many siderophores identified from rhizospheric region (Saha et al. 2015), but those produced by pseudomonads are known for their high affinity to the ferric ion (Meyer 2000), for example, pyoverdines produced by fluorescent pseudomonads show affinity for $Fe^{3+}$ with a stability constant of about $10^{32}$ (Meyer and Abdallah 1978). Such a high concentration of siderophores produced by plant growth-promoting microbes confers them competitive advantage in comparison to other microorganisms (Saha et al. 2015). Microbe-mediated iron uptake has been identified as potential tool for providing efficient nutrient uptake to plant, and many studies showed application of microbial inoculants having capability to chelate iron even at low concentrations to enhance plant productivity (Fageria 2009). One more attribute of good siderophore-producing microorganisms is that they may suppress soil-borne fungal pathogens by scavenging available iron and rendering it unavailable to other organisms (Beneduzi et al. 2012). Although there are various reports indicating the use of microbe-mediated iron uptake to plants (Saha et al. 2015), commercially available bioformulations exclusively used in iron uptake are rare. In India Fe Sol B ®, a product of Agri Life Bio Solutions, is recently recognized as iron-mobilizing biofertilizer used for various crops.

### 1.6.2 Formulations for Biocontrol: Biopesticides

Worldwide, approximately 1400 biopesticide products are being sold (Marrone 2007), and the number of registered products is increasing day by day. Three types of microbes are always considered in developing a useful biopesticide formulation which are bacteria, fungi, and viruses. Although biopesticide formulations are dependent on the type of organism being used, the ultimate goal is that it must ensure that the agent is delivered in a form that is viable, virulent, and with sufficient inoculum potential to be effective in the field (Ash 2010). Many workers have discussed details of production methods of formulations (Burges and Jones 1998; Couch 2000; Ehlers and Shapiro-Ilan 2005).

#### 1.6.2.1 Bacteria

Amongst all known microbe-based biopesticides, Bt and its subspecies enjoy a great reputation (Raddadi et al. 2008; Bravo et al. 2011). Great literature is available that confirms success of Bt in biopesticide market (Melnick et al. 2009; Sansinenea 2012). Rosas-Garcia (2009) classified Bt-based products: first-generation products contain mixture of spores and crystals from a native strain and govern mainstream of commercial products. By the advent of molecular biology techniques, it was possible to construct genetically engineered Bt strains (Cerda and Maurizio 2004) carrying numerous insecticidal crystal proteins (ICPs); these are second-generation products. Such formulation products have proved beneficiary because of their selective action on target pests. Recombinant *P. fluorescens* cells transformed with genes coding for Cry delta-endotoxins (Young et al. 2008) are classified as third-generation products. In such products engineered *P. fluorescens* expressing Bt Cry delta-endotoxin are cultivated and then chemically killed to fix the toxin within the cells, subsequently microencapsulates are prepared which not only stabilize the toxin but also reduce degradation when applied to plant leaves, so there is an improvement in the storage life of the product. Commercially Bt can be produced by (semi)solid state fermentation, whereas industrial production of Bt is performed by liquid state fermentation, but semisolid and solid state fermentation on small scale is also common in developing countries (Devi et al. 2005). Details of the production have been reviewed by Lisansky et al. (1993), Couch (2000), and El-Bandary (2006). Improvements have been suggested in experimental and commercial formulations of *Bacillus* spp. and other biocontrol agents (Schisler et al. 2004). In these days commercially available Bt products are proteins (ICP), viable spores, and enzyme systems (proteases, unknown virulent factors along with inerts/adjuvants (Brar et al. 2006).

*Pseudomonas* spp. strains are also extensively used in biocontrol of different phytopathogens (Ortet et al. 2011; Mishra and Arora 2012; Loper et al. 2012; Tewari and Arora 2014). For example, different strains of *Pseudomonas fluorescence* known to produce a variety of antibiotics or antifungal metabolites, are directly involved in suppression of diseases (Weller 2007). Bio-Save, BlightBan, Cedomon, Biocoat, and Victus are *Pseudomonas* spp.-based bioformulations and used in biocontrol of various plant diseases. Besides these *Agrobacterium radiobacter*, *Burkholderia cepacia*, and *Streptomyces griseoviridis* have been found in controlling soilborne seedling diseases and fruit and vegetable pathogens (Leonard and Julius 2000). Mostly entomopathogenic bacteria can be easily produced in vitro systems except *B. popilliae* and its close relatives which can only be produced in its natural host. Couch (2000) discussed detailed industrial fermentation and formulation of entomopathogenic bacteria.

Yan et al. (2007) reported the use of wastewater sludge to support growth of diverse *B. thuringiensis* serovars, yielding lower cell counts but higher entomotoxicity per spore compared to synthetic media. Similarly various other alternative raw materials have been also tested to minimize the cost of production (Ravensberg 2011).

### 1.6.2.2 Fungal

Since the discovery of *M. anisopliae* and *B. bassiana*, there have been plentiful attempts to develop commercial formulations based on fungi, but success is not as in the case of bacterial formulations (McCoy 1990). *B. bassiana* formulations alone contribute to 34 % of the commercial mycoinsecticides available in the market worldwide (de Faria and Wraight 2007). At least 750 species of fungi are known to be entomopathogenic, but very few have been established as control agents of insect pests (Copping 2009). Various reasons are responsible for these discrepancies, but mainly the shelf life of fungal formulations is short and mass production costs are high. One more reason is that preservation of active ingredients beyond a certain limit of time is low due to the fragile nature of the conidia and hyphae. Many of the registered formulations contain *Trichoderma harzianum*, *Trichoderma asperellum*, *Trichoderma gamsii*, *Coniothyrium minitans*, *Aspergillus flavus*, and *Chondrostereum purpureum* (Auld 2002). *Trichoderma* spp. are among the most studied fungal biocontrol agents and their commercialization is increasing day by day (Vinale et al. 2008), and new techniques are being developed for mass production of fungi. Genetic manipulation has also been used for fungi to increase their action on target pests. Techniques including $Ca^{2+}$ polyethylene glycol (PEG)-mediated protoplast transformation, electroporation, and particle bombardment are best suited for fungal transformations (Bianca and Berg 2013). Mass production of entomopathogenic fungi by solid state fermentation (SSF), also called solid substrate fermentation (SSF), and by liquid state fermentation (LSF), also called submerged culture fermentation (SCF), is described by Jaronski (2014). A wide variety of organic materials have been used as substrates for mass production of entomopathogenic fungi. Among them broken rice, cassava chips, cotton and coconut cake, finger and kodo millet, wheat bran, and rice husk are more common (Jaronski 2014). Recently several inorganic substrates such as calcined diatomaceous earth (diatomite) (Jaronski and Jackson 2012) and open-pored clay granules have been also used (Jaronski 2014). As much of the work is done on Bt in bacterial formulations, development in case of fungal formulation *B. bassiana* is also done by solid-substrate production of aerial conidia (Feng et al. 1994). Mycotech Corp. of Montana has gained attention in *B. bassiana* production due to higher production of conidia ($10^{13}$ conidia $kg^{-1}$) (Bradley et al. 1992). Production of *B. bassiana* and *Metarhizium* conidia by growing on a carrier in plastic autoclavable bags or in trays is common in developing countries (Mendonça 1992). Solid state fermentation by complex bioreactor is also used by some world famous companies: Koppert, Laverlam, and Prophyta (Bradley et al. 1992). Many fungi that are difficult to produce efficiently on solid substrates can be readily cultured in liquid media. In submerged cultures, fungi may produce blastospores or submerged conidia and/or mycelial parts or pellets. This depends on the species, the strain, and the production medium and parameters (McCoy et al. 1988). Industrial wastewaters and dewatered sludge have been also reported as rich nutrient sources for production and formulation of fungi (Verma et al. 2007). Jaronski and Jackson (2012) described a medium that works very well with a wide range of *Beauveria and Metarhizium* isolates.

### 1.6.2.3 Viral

Seven families of viruses, *Baculoviridae*, *Reoviridae*, *Iridoviridae*, *Poxviridae*, *Parvoviridae*, *Picornaviridae*, and *Rhabdoviridae*, cause diseases in insects. These viruses infect insects and form occlusion bodies which confirm their role in biocontrol (Kalawate 2014), but *Baculoviridae* is commonly used (Harrison and Hoover 2012). The family *Baculoviridae* have four genera: *Alphabaculovirus* (containing nucleopolyhedroviruses (NPVs) that infect lepidopteran insects), *Betabaculovirus* (containing the granuloviruses (GVs) found in lepidopteran insects), *Gammabaculovirus* (NPVs infecting hymenopteran insects), and *Deltabaculovirus* (NPVs infecting insects in the order Diptera) (Reid et al. 2014). According to an estimate, more than 20 species and 30 different products of baculoviruses have been registered as commercially available insecticides (Rao et al. 2015). China is the world's largest viral insecticide producer (more than 32 registered products) (Sun 2015). Europe and the United States also have good market for viral insecticides. Here they are sold by the trade name Madex 3 (Andermatt Biocontrol), Granupom (AgrEvo), Carpovirusine (NPP-Calliope), Carposin (Agrichem), Virin-Gyap (NPO Vector), and CYD-X (Thermo Trilogy). Several recombinant baculoviruses have also been developed experimentally for insect control (Inceoglu et al. 2006), and some success was attained as the expression of insect-specific toxin in *Autographa californica* NPV (AcNPV) (Zlotkin et al. 1971) results in hyperactivity of the nervous system and musculature, cessation of feeding, and paralysis of the pest. Other recombinant systems involve insertion of juvenile hormone esterase (JHE) in AcNPV, gene of diuretic hormone of the tomato hornworm (*Manduca sexta* Joh), silkworm (*Bombyx mori* L.) virus, and straw itch mite (*Pyemotes tritici*) toxin TxP-1 into AcNPV (Copping and Menn 2000). The use of recombinant baculoviruses is also reviewed by Szewczyk et al. (2006). Viral biopesticides can be produced in vitro and in vivo. In in vivo production, original host is used and further descriptions of in vivo production methods in different insects are given by several workers (Tani et al. 2003; van Beek and Davis 2007). The main difficulty with in vivo production is contamination with microorganisms, mainly bacteria, and degeneration of the virus (Tani et al. 2003). Problem associated with in vivo production of viruses, specially control of process, resulted in recommendation of in vitro production which was found to be much better (Szewczyk et al. 2006; Nguyen et al. 2011), although requirements of productive insect cell lines and culture media are also a matter of consideration for in vitro production of baculovirus (Baines 2002; Verónica et al. 2006). Pawar and Thombre (1992) also gave detailed account on mass production of baculoviruses, whereas Rhodes (1996) provided all the economic aspects related to in vitro production. Commercially, viral insecticides are produced in the form of concentrated wettable powders apart from liquid or granular forms.

### 1.6.3 Consortia-Based Inoculants

Although most of the available bioformulations contain single strain, mixed cultures or co-inoculation with other microorganisms is proving to be better approach for overall plant growth and development. Practices such as use of rhizobial co-inoculation with mycorrhiza showed better results with legumes. This dual association not only improves nutritional status of nodulated plants but also reported to increase drought or osmotic tolerance in lucerne (Ardakani et al. 2009), soybean (Gao et al. 2012), broad bean (Jia et al. 2004), chickpea (Tavasolee et al. 2011), and pigeon pea (Bhattacharjee and Sharma 2012). Studies also confirmed that combined application of PSB and nodule-forming bacteria in legumes stimulated plant growth (Messele and Pant 2012). Phosphate and potassium rocks are a cheaper source of P and K, and integrated application of PSB with the co-inoculation of K-solubilizing bacteria may provide a faster and more continuous supply of these nutrients for optimal plant growth and best regarded for sustainable crop production (Xiufang et al. 2006). Recently consortia formulations have been developed by various workers and patents are also filed (Paikray and Malik 2010). In a study conducted by Maiyappan et al. (2010), a consortium bioformulation containing nine strains of the genera *Bacillus* spp., *Streptomyces* spp., *Azotobacter* spp., and *Frauteria* spp. was prepared as a wettable powder and proved to be beneficial to black gram. In a similar study, consortium bioformulation of *Burkholderia* sp. MSSP with three other PGP bacteria were tested for growth enhancement of *Cajanus cajan* by using various carrier materials including sugarcane bagasse, sawdust, cocoa peat, rice husk, wheat bran, charcoal, rock phosphate, and paneer whey as liquid carrier, and their finding confirmed increase growth of pigeon pea plant when consortium was used in the formulation (Pandey and Maheshwari 2007). Tajini et al. (2012) showed that combined inoculation of AM fungi and rhizobia provides higher N and P accumulation in the shoots of common bean plants compared with single inoculum. Recently Zayadan et al. (2014) recommended that consortia of cyanobacteria, microalgae, and *Azotobacter* can be used as a biostimulator and biofertilizer for crops. BioGro™ is a consortium-based biofertilizer used in Vietnam, containing strains of *P. fluorescens*, two bacilli, and a soil yeast (Cong et al. 2009).

## 1.7 Conclusion and Future Steps

A bioformulation is not effective until it does not have an impact in field conditions, market existence and reliability and cost-effectiveness (Brar et al. 2006). Social and public interactions toward Bt-based biopesticides are given by Navon (2000). He concluded that the toxicity of protein as an oral insecticide and environmental conditions reduces the efficacy of the product. Production of bioformulation is not only dependent on the detailed knowledge of microbial as well as plant physiology, but a number of technological challenges are also involved such as fermentation process, formulation type, population of microbe, and delivery systems (Malusá et al. 2012). Barea (2015) suggested that prior to formulation development and application, it is necessary to understand the ecology of the PGPR–host–pest interaction. Fermentation depends on the medium used in the production process and concentration of constituents, oxygen transfer, incubation temperature, time of harvest, and postharvest treatments (Montazeri and Greaves 2002). Delivery system also decides usability of formulation and any bioformulation will not sustain until its delivery is not proper. For Bt water-miscible formulations and oil-miscible flowable (OF) are more preferable which are sprayed in ultralow volume (ULV). Himel et al. (1990) and Bateman (1993) described spray droplet size in detail and embodied all the mechanical aspects of droplet application. For foliar spray environmental factors, temperature, dew period, UV irradiation, and desiccation are involved (Bailey et al. 1996). In this direction technological advantage is preparation of bioformulation with very small particles or droplet size is now feasible and getting very good results when used as suspension or emulsion (Peng and Wolf 2011).

Impetus of Hynes and Boyetchko (2006) was to introduce research initiatives in the art and science of formulation development. They addressed various obstacles responsible in formulation technology development, including insufficient literature, registration process, and existing formulation technology. Current scenario toward bioformulation development and application is not satisfactory. The bioformulation policy including both for biofertilizers and biopesticides needs to be revived, and this is why Greaves and Grant (2010) have mentioned that the "biological control industry has the weakest policy network" and confirmed that the technical knowledge is not always matched by an understanding of political processes.

Albeit the research is ongoing, we have not succeeded in producing such an elite formulation which not only has broad spectrum activity but also fulfills economic challenges. Although many microbial inoculants have been developed, very few products are found to be promising. At present development of bioformulation involves collective effort of both microbiological and technological aspects, and vigorous research efforts are required for technological part (Arora 2015). Besides this there is urgent requirement in the field of bioformulation technology to reevaluate the whole process and plugging of loop holes. We have to consider every step such as selection of organism, production method, delivery system, application technology, factors affecting development, persistence in the environment, and ultimately market availability of product.

## References


Abadias M, Teixido N, Usall J, Solsona C, Vinas I (2005) Survival of the postharvest biocontrol yeast *Candida* sake CPA–1 after dehydration by spray–drying. Biocontrol Sci Tech 15:835–846

Achal V, Savant VV, Sudhakara Reddy M (2007) Phosphate solubilization by wide type strain and UV–induced mutants of *Aspergillus tubingensis*. Soil Biol Biochem 39:695–699

Adesemoye AO, Torbert HA, Kloepper JW (2009) Plant growth–promoting rhizobacteria allow reduced application rates of chemical fertilizers. Microb Ecol 58:921–929

Agrawal A, Burns MA (1996) Application of membrane–based preferential transport to whole broth processing. Biotechnol Bioeng 55:581–591

Agro news (2014) Biofertilizers market–global industry analysis, size, share, growth, trends and forecast, 2013–2019. Available online http://news.agropages.com/News/NewsDetail——11612–e.htm

Ahemad M, Kibret M (2014) Mechanisms and applications of plant growth promoting rhizobacteria: current perspective. J King Saud Univ Sci 26:1–20

Ahmed HFS, El–Araby MMI (2012) Evaluation of the influence of nitrogen fixing, phosphate solubilizing and potash mobilizing biofertilizers on growth, yield, and fatty acid constituents of oil in peanut and sunflower. Afr J Biotechnol 11:10079–10088

Aizawa K, Ishiwata S (2001) His discovery of sottokin (*Bacillus thuringiensis*) in 1901 and subsequent investigations in Japan. In: Ohba M, Nakmura O, Mizuki E, Akao T (Eds) Proceedings of a centennial symposium commemorating Ishiwata's discovery of *Bacillus thuringiensis*, Kurume, pp 1–14

Albareda M, Rodríguez–Navarro DN, Camacho M et al (2008) Alternatives to peat as a carrier for rhizobia inoculants: solid and liquid formulations. Soil Biol Biochem 40:2771–2779

Allison SD, Martiny JBH (2008) Resistance, resilience, and redundancy in microbial communities. Proc Natl Acad Sci USA 105:11512–11519

Amiet–Charpentier C, Gadille P, Digat B, Benoit JP (1998) Microencapsulation of rhizobacteria by spray–drying: formulation and survival studies. J Microencapsul 15:639–659

Andersch W, Hain R, Kilian M (1998) Granulates containing microorganisms. US Patent 5,804,208 filed 15 Nov 1995:8

Ansari MA, Butt TM (2012) Evaluation of entomopathogenic fungi and a nematode against the soil–dwelling stages of the crane fly *Tipula paludosa*. Pest Manag Sci 68:1337–1344

Ardakani MR, Pietsch G, Moghaddam A, Raza A, Friedel JK (2009) Response of root properties to tripartite symbiosis between lucerne (*Medicago sativa* L.), rhizobia and mycorrhiza under dry organic farming conditions. Am J Agric Biol Sci 4:266–277

Aronson AI, Beckman W, Dunn P (1986) *Bacillus thuringiensis* and related insect pathogens. Microbiol Rev 50:1–24

Arora NK (ed) (2015) Plant microbe symbiosis: applied facets. Springer, New Delhi, p 383

Arora NK, Khare E, Maheshwari DK (2010) Plant growth promoting rhizobacteria: constraints in bioformulation, commercialization, and future strategies. In: Maheshwari DK (ed) Plant growth and health promoting bacteria. Springer–Verlag, Berlin, pp 97–116

Ash GJ (2010) The science, art and business of successful bioherbicides. Biol Control 52:230–240

Auld B (2002) Fungi as biocontrol agents: progress, problems and potential. Plant Pathol 51:518–518

Auld BA, Hetherington SD, Smith HE (2003) Advances in bioherbicide formulation. Weed Biol Manag 3:61–67

Bailey P, Baker G, Caon G (1996) Field efficacy and persistence of *Bacillus thuringiensis* var. kurstaki against *Epiphyas postvittana* (Walker) (Lepidoptera: CAPS PRA: *Epiphyas postvittana* 7 Tortricidae) in relation to larval behaviour on grapevine leaves. Aust J Entomol 35:297–302

Baines D (2002) New approaches to insect tissue culture. In: Vlak JM, de Gooijer CD, Tramper J, Miltenburger HG (eds) Insect Cell Culture: fund applied aspects, Kluwer Academic Publishers, New York, vol 2, pp 13–22

Balashov AV, Molchanov VN, Naboychenko KV (2008) The response of winter wheat varieties to bacterial fertilizer "Rizoagrin". Proc Lower Volga Agro-Univ Complex 12:7–11

Barea JM (2015) Future challenges and perspectives for applying microbial biotechnology in sustainable agriculture based on a better understanding of plant–microbiome interactions. J Soil Sci Plant Nutr 15:261–282

Basak BB, Biswas DR (2008) Influence of potassium solubilizing microorganism (*Bacillus mucilaginous*) and waste mica on potassium uptake dynamics by Sudan grass (*Sorghum vulgare* Pers) grown under two Alfisols. Plant Soil 317:235–255

Bassi A (1835) And the sign of the plaster or disease that dormice afflicts bugs daseta Part–I, Tip Terica Orcesi Lod, pp 1–67

Bateman R (1993) Simple, standardized methods for recording droplet measurements and estimation of deposits from controlled droplet applications. Crop Prot 12:201–206

Batjes NH (1997) A world data set of derived soil properties by FAO–UNESCO soil unit for global modelling. Soil Use Manag 13:9–16

Batta YA (2007) Biocontrol of almond bark beetle (*Scolytus amygdali Geurin–Meneville*, Coleoptera: Scolytidae) using *Beauveria bassiana* (Bals.) Vuill. (Deuteromycotina: Hyphomycetes). J Appl Microbiol 103:1406–1414

Batta YA (2008) Control of main stored–grain insects with new formulations of entomopathogenic fungi in diatomaceous earth dusts. Int J Food Eng 4:9

BCC Research (2010) Biopesticides: the global market report CHM029C. Wellesley

Behle RW, McGuire MR, Gillespie RL, Shasha BS (1997) Effects of alkaline gluten on the insecticidal activity of *Bacillus thuringiensis*. J Econ Entomol 90:354–360

Beneduzi A, Ambrosini A, Pasagila LMP (2012) Plant growth promoting rhizobacteria (PGPR): their potential as antagonistic and biocontrol agents. Genet Mol Biol 35:1044–1051

Berg G (2009) Plant–microbe interactions promoting plant growth and health: perspectives for controlled use of microorganisms in agriculture. Appl Microbiol Biotechnol 84:11–18

Berny P (2007) Pesticides and the intoxication of wild animals. J Vet Pharmacol Ther 30:93–100

Bhattacharjee SG, Sharma D (2012) Effect of dual inoculation of *Arbuscular Mycorrhiza* and *Rhizobium* on the chlorophyll, nitrogen and phosphorus contents of pigeon pea (*Cajanus cajan* L.). Adv Microbiol 2:561–564

Bianca G, Berg MVD (2013) Transformation of filamentous fungi in microtiter plate. In: Gupta VK, Tuohy MG (eds) Laboratory protocols in fungal biology: current methods in fungal biology. Springer Science + Business Media LLC, New York, pp 343–348

BioScientific Inc (2008) BuRIZE—VA–mycorrhizal soil and root inoculant, for use on production agricultural crops. http://www.biosci.com/brochure/BRZBro.pdf

Bottomley WB (1912) Root nodules of Myrica gale. Ann Bot 26:111–117

Bradley C, Black W, Kearns R, Wood P (1992) Role of production technology in mycoinsecticide development. In: Leatham G (ed) Frontiers of Industrial Mycology. Springer, Boston, pp 160–173

Brar SK, Verma M, Tyagi RD, Valéro JR (2006) Recent advances in downstream processing and formulations of *Bacillus thuringiensis* based biopesticides. Process Biochem 41:323–342

Bravo A, Likitvivatanavong S, Gill SS, Sobero NM (2011) *Bacillus thuringiensis*: a story of a successful bioinsecticide. Insect Biochem Mol Biol 41:423–431

Brown ME (1974) Seed and root bacterization. Ann Rev Phytopathol 12:181–197

Burges HD, Jones KA (eds) (1998) Formulation of microbial biopesticides: beneficial microorganisms, nematodes and seed treatments. Kluwer Academic Publishers, Dordrecht, p 411

Burr TJ, Schroth MN, Suslow T (1998) Increased potato yields by treatment of seed pieces with specific strains of *Pseudomonas fluorescens* and *P. putida*. Phytopathology 68:1377–1383

Butt TM, Jackson C, Magan N (2001) Introduction–fungal biocontrol agents: progress, problems and potential. In: Butt TM, Jackson C, Magan N (eds) Fungi as biocontrol agents: progress, problems and potential. CABI International, Oxon

CAB International Centre (2010) The 2010 world wide biopesticides market summary. CAB International Centre, Wallingford

Callaghan MO, Gerard FM (2005) Establishment of *Serratia entomophila* in soil from a granular formulation. N Z Plant Prot 58:122–125

Cerda H, Maurizio GP (2004) Genetic engineering with *Bacillus thuringiensis* and conventional approaches for insect resistance in crops. Crit Rev Plant Sci 23:317–323

Chen YP, Rekha PD, Arun AB, Shen FT, Lal WA, Young CC (2006) Phosphate solubilizing bacteria from subtropical soil and their tricalcium phosphate solubilizing abilities. Appl Soil Ecol 34:33–41

Cheng H, Li L, Hua J, Yuan H, Cheng S (2015) A preliminary preparation of endophytic bacteria CE3 wettable powder for biological control of postharvest diseases. Not Bot Hortic Agrobot 43:159–164

Cheze-Lange H, Beunard D, Dhulster P, Guillochon D, Caze A-M, Morcellet M, Saude N, Junter GA (2002) Production of microbial alginate in a membrane bioreactor. Enzym Microb Technol 30:656–661

Chumthong A, Kanjanamaneesathian M, Pengnoo A, Wiwattanapatapee R (2008) Water–soluble granules containing *Bacillus megaterium* for biological control of rice sheath blight: formulation, bacterial viability and efficacy testing. World J Microbiol Biotechol 24:2499–2507

Cobos C (2005) Crean con hongos nuevo fertilizante. http://fox.presidencia.gob.mx/buenasnoticias/. Accessed 20 Nov 2008

Cohen E, Joseph T (2009) Photostabilization of *Beauveria bassiana* conidia using anionic dyes. Appl Clay Sci 42:569–574

Cohen S, Bano MC, Visscher KB, Chow M, Allcock HR, Langer R (1990) Ionically crosslinkable polyphosphazene: a novel polymer for microencapsulation. J Am Chem Soc 112:7832–7833

Colt JS, Cyr MJ, Zahm SH, Tobias GS, Hartge P (2007) Inferring past pesticide exposures: a matrix of individual active ingredients in home and garden pesticides used in past decades. Environ Health Perspect 115:248–254

Cong PT, Dung TD, Hien TM (2009) Inoculant plant growth–promoting microorganisms enhance utilisation of urea–N and grain yield of paddy rice in southern Vietnam. Eur J Soil Biol 45:52–61

Copping LG (ed) (2009) The manual of biocontrol agents: a world compendium. British Crop Protection Council, Alton Hampshire, p 851

Copping LG, Menn JJ (2000) Biopesticides: a review of their action, applications and efficacy. Pest Manag Sci 56:651–676

Coronado GD, Thompson B, Strong L, Griffith WC, Islas I (2004) Agricultural task and exposure to organophosphate pesticides among farm workers. Environ Health 112:142–147

Couch TL (2000) Industrial fermentation and formulation of entomopathogenic bacteria. In: Charles JF, Delécluse A, Nielsen–LeRoux C (eds) Entomopathogenic bacteria: from laboratory to field application. Kluwer, Dordrecht, pp 297–316

Coyne MS (1996) A cartoon history of soil microbiology. J Nat Res Life Sci Educ 25:30–36

CPL (2006) Biopesticides 2007. CPL Business Consultants, Wallingford

Daniel G, Kate MS (2014) Long–term effects of mineral fertilizers on soil microorganisms. Soil Biol Biochem 75:54–63

Davison J (1988) Plant beneficial bacteria. Biotechnology 6:282–286

De PK (1939) The role of blue-green algae in nitrogen fixation in rice fields. Proc R Soc Lond 127:121–139

de Faria MR, Wraight SP (2007) Mycoinsecticides and mycoararicides: a comprehensive list with worldwide coverage and international classification of formulation types. Biol Control 43:237–256

Deaker R, Roughley RJ, Kennedy IR (2004) Legume seed inoculation technology–a review. Soil Biol Biochem 36:1275–1288

Devi PSV, Ravinder T, Jaidev C (2005) Cost–effective production of *Bacillus thuringiensis* by solid–state fermentation. J Invertebr Pathol 88:163–168

Digat B (1989) Strategies for seed bacterization. Acta Hortic 253:121–130

Dorigo U, Lefranc M, Leboulanger C, Montuelle B, Humbert JF (2009) Spatial heterogeneity of periphytic microbial communities in a small pesticide–polluted river. FEMS Microbiol Ecol 67:491–501

Droby S, Cohen L, Daus A et al (1998) Commercial testing of Aspire: a yeast preparation for the biological control of postharvest decay of citrus. Biol Control Theory Appl Pest Manag 12:97–101

Ehlers RU, Shapiro–Ilan DI (2005) Mass production. In: Grewal PS, Ehlers RU, Shapiro–Ilan DI (eds) Nematode biocontrol agents. CAB International, Wallingford, pp 65–78

Ehrlich HL (1990) Geomicrobiology. Dekker, New York, p 646

Escolar L, Pérez–Martín J, de Lorenzo V (1999) Opening the iron box: transcriptional metalloregulation by the Fur protein. J Bacteriol 181:6223–6229

Eswaran H, Reich P, Beigroth F (1997) Global distribution of soils with acidity. In: Moniz AC et al (eds) Plant soil plant interactions at low pH. Brazilian Soil Science Society Vicosa, Campines Brazil, pp 159–164

Fageria NK (2009) The use of nutrients in crop plants. CRC Press, Boca Raton

Fageria NK, Baligar VC, Edwards DG (1990) Soil–plant nutrient relationships at low pH stress. In: Baligar VC, Duncan RR (eds) Crops as enhancers of nutrient use. Academic, San Diego, pp 475–507

Falk SP, Gadoury DM, Cortesi P, Pearson RC, Seem RC (1995) Partial control of grape powdery mildew by the mycoparasite *Ampelomyces quisqualis*. Plant Dis 79:483–490

Fankem H, Nwaga A, Dueubel A, Dieng L, Merbach W, Etoa FX (2006) Occurrence and functioning of phosphate solubilizing microorganisms from oil palm tree (*Elaeis guineensis*) rhizosphere in Cameroon. Afr J Biotechnol 5:2450–2460

Fasim F, Ahmed N, Parsons R, Gadd GM (2002) Solubilization of zinc salts by a bacterium isolated from the air environment of a tannery. FEMS Microbiol Lett 213:1–6

Faust PG (1974) *Bacillus thuringiensis*: its history and mode of action. Dev Ind Microbiol 15:195–198

Fedotova LS, Kravchenko AV, Gavrilov AN (2009) The importance of bacterial fertilizers in biological production of potato. Achi Sci Technol 3:28–30

Feng MG, Poprawski TJ, Khachatourians GG (1994) Production, formulation and application of the entomopathogenic fungus *Beauveria bassiana* for insect control: current status. Biocontrol Sci Technol 4:3–34

Feng MG, Pu XY, Ying SH, Wang YG (2004) Field trials of an oil–based emulsifiable formulation of *Beauveria bassiana* conidia and low application rates of imidacloprid for control of false–eye leafhopper *Empoasca vitis* on tea in southern China. Crop Prot 23:489–496

Fenske RA, Day EW (2005) Assessment of exposure for pesticide handlers in agricultural, residential and institutional environments. In: Franklin CA, Worgan JP (eds) Occupational and residential exposure assessment for pesticides. Wiley, Chichester

Fernandes ÉK, Rangel DN, Braga GL, Roberts D (2015) Tolerance of entomopathogenic fungi to ultraviolet radiation: a review on screening of strains and their formulation. Curr Genet 61:427–440

Fernández LA, Zalba P, Gómez MA, Sagardoy MA (2007) Phosphate–solubilization activity of bacterial strains in soil and their effect on soybean growth under greenhouse conditions. Biol Fertil Soils 43:805–809

FNCA (2006) Biofertilizer manual. Japan Atomic Industrial Forum, Tokyo

Franche C, Lindström K, Elmerich C (2009) Nitrogen–fixing bacteria associated with leguminous and non–leguminous plants. Plant Soil 321:35–59

Fritsch FE (1907) The subaerial and freshwater algal flora of the tropics. A phytogeographical and ecological study. Ann Bot 21:235–275

Galloway JN (2005) The global nitrogen cycle: past, present and future. Sci China Ser C 48:669–677

Gao X, Lu X, Wu M, Zhang H, Pan R, Tian J et al (2012) Co–inoculation with rhizobia and AMF inhibited soybean red crown rot: from field study to plant defense–related gene expression analysis. PLoS One 7:e33977

García–Fraile P, Menéndez E, Rivas R (2015) Role of bacterial biofertilizers in agriculture and forestry. Bioengineering 2:183–205

Garibaldi JA, Neilands JB (1956) Formation of iron–binding compounds by micro–organisms. Nature 177:526–527

Garnett T, Conn V, Kaiser BN (2009) Root based approaches to improving nitrogen use efficiency in plants. Plant Cell Environ 32:1272–1283

Gasic S, Tanovic B (2013) Biopesticide formulations, possibility of application and future trends. Pestic Phytomed 28:97–102

Ge Y, Zhang JB, Zhang LM, Yang M, He JZ (2008) Long–term fertilization regimes affect bacterial community structure and diversity of an agricultural soil in northern China. J Soils Sediments 8:43–50

Gelernter WD (2007) Microbial control in Asia: a bellwether for the future? J Invertebr Pathol 95:161–167

Glick BR (2012) Plant growth–promoting bacteria: mechanisms and applications. Hindawi Publishing Corporation, Article ID 963401, Scientifica, pp 15. http://dx.doi.org/10.6064/2012/963401

Global Industry Analysts (2015) Global biopesticides market to reach US$2.8 billion by 2015, according to a new report by global industry analysts, inc. Available online http://prweb.com/printer/8041130.htm

Goldstein AH (1986) Bacterial phosphate solubilization: historical perspective and future prospects. Am J Altern Agric 2:57–65

Gothwal R, Nigam V, Mohan M, Sasmal D, Ghosh P (2009) Screening of nitrogen fixers from rhizospheric bacterial isolates associated with important desert plants. Appl Ecol Environ Res 6:101–109

Goulson D, Derwent LC, Penagos DI, Williams T (2003) Effects of optical brighteners included in biopesticide formulations on the growth of crops. Agric Ecosyst Environ 95:235–240

Graham PH, Vance CP (2003) Legumes: importance and constraints to greater use. Plant Physiol 131:872–877

Grand View Research (2015) Biofertilizers market analysis by product (nitrogen fixing, phosphate solubilizing), by application (seed treatment, soil treatment) and segment forecasts To 2022. Available online. http://www.grandviewresearch.com/industry–analysis/biofertilizers–industry

Greaves J, Grant W (2010) Underperforming policy networks: the biopesticides network in the United Kingdom. Br Polit 5:14–40

Guijarro B, Melgarejo P, De Cal A (2007a) Effect of stabilizers on the shelf–life of *Penicillium frequentans* conidia and their efficacy as a biological agent against peach brown rot. J Food Microbiol 113:117–124

Guijarro B, Melgarejo P, Torres R, Lamarca N, Usall J, De Cal A (2007b) Effects of different biological formulations of *Penicillium frequentans* on brown rot of peaches. Biol Control 42:86–96

Gupta AK (2004) The complete technology book on biofertilizers and organic farming. National Institute of Industrial Research Press, Delhi

Hafeez FY, Yasmin S, Ariani D, Mehboob–UR–Rahman ZY, Malik KA (2006) Plant growth promoting bacteria as biofertilizer. Agro Sustain Dev 26:143–150

Hajek AE, Huang B, Dubois T, Smith MT, Li Z (2006) Field studies of control of *Anoplophora glabripennis* (Coleoptera: Cerambycidae) using fiber bands containing the entomopathogenic fungi *Metarhizium anisopliae* and *Beauveria brongniartii*. Biocontrol Sci Technol 16:329–343

Hameeda B, Harini G, Rupela OP, Wani SP, Reddy G (2008) Growth promotion of maize by phosphate–solubilizing bacteria isolated from composts and macrofauna. Microbiol Res 163:234–242

Harris J, Dent D (2000) Priorities in biopesticide research and development in developing countries. CABI Publishing, Wallingford

Harrison R, Hoover K (2012) *Baculoviruses* and other occluded insect viruses. In: Kaya FEVK (ed) Insect pathology. Academic, San Diego, pp 73–131

Harwood RWJ, Lee MSK, Lisansky SG, Quinlan R (2007) Current worldwide markets for biopesticides and success factors for the business. In: Proceedings of the XVIth international plant protection congress/ BCPC international congress. Crop Science & Technology, Glasgow, pp 598–599

Himel CM, Loats H, Bailey GW (1990) Pesticide sources to the soil and principles of spray physics. In: Cheng HH (ed) Pesticides in the soil environment: processes, impacts, and modelling, vol 2, Soil science of America book series. Soil Science Society of America, Wisconsin, pp 7–50

Holford ICR (1997) Soil phosphorus: its measurement, and its uptake by plants. Aust J Soil Res 35:227–239

Howarth FG (1991) Environmental impacts of classical biological control. Annu Rev Entomol 36:485–509

Huang JK, Mi JW, Lin H, Wang ZJ, Chen RJ, Hu RF, Rozelle S, Pray C (2010) A decade of Bt cotton in Chinese fields: assessing the direct effects and indirect externalities of Bt cotton adoption in China. Sci China Life Sci 53:981–991

Huss–Danell K (1997) Actinorhizal symbioses and their $N_2$ fixation. New Phytol 136:375–405

Hynes RK, Boyetchko SM (2006) Research initiatives in the art and science of biopesticide formulations. Soil Biol Biochem 38:845–849

Ifoulis AA, Savopoulou-Soultani M (2004) Biological control of *Lobesia botrana* (Lepidoptera: Tortricidae) larvae by using different formulations of Bacillus thuringiensis in 11 vine cultivars under field conditions. J Econ Entomol 97:340–343

Ignoffo CM (1973) Development of a viral insecticide: concept to commercialization. Exp Parasitol 33:380–406

Igual J, Valverde MA, Cervantes E, Velázquez E (2001) Phosphate–solubilizing bacteria as inoculants for agriculture use of updated molecular techniques in their study. Agronomy 21:561–568

Inceoglu AB, Kamita SG, Hammock BD (2006) Genetically modified *baculoviruses*: a historical overview and future outlook. Adv Virus Res 68:323–360

Industrial Equipment News (2011) Biopesticides market to reach $1 billion in 2010. Available online http://www.ien.com/article/biopesticides–market–to/8648

Insam H (2001) Developments in soil microbiology since the mid–1960s. Geoderma 100:389–402

Investing News Network (2015) What is Potash? From types of potash to potash stocks. Available online http://investingnews.com/daily/resource–investing/agriculture–investing/potash–investing/2013–top–potash–p

Jaronski ST (2014) Mass production of entomopathogenic fungi: state of the art. In: Morales–Ramos JA, Rojas MG, Shapiro–Ilan DI (eds) Mass production of beneficial organisms: invertebrates and entomopathogens. Elsevier, London, pp 357–414

Jaronski ST, Jackson MA (2012) Mass production of entomopathogenic Hypocreales. In: Lacey LA (ed) Manual of techniques in invertebrate pathology. Academic, New York, pp 255–284

Jha A, Sharma D, Saxena J (2011) Effect of single and dual phosphate solubilizing bacterial strain inoculations on overall growth of mung bean plants. Arch Agron Soil Sci 58:967–981

Jia Y, Gray VM, Straker CJ (2004) The influence of *Rhizobium* and *arbuscular mycorrhizal* fungi on nitrogen and phosphorus accumulation by *Vicia faba*. Ann Bot 94:251–258

John RP, Tyagi RD, Brar SK, Surampalli RY, Prévost D (2011) Bio–encapsulation of microbial cells for targeted agricultural delivery. Crit Rev Biotechnol 31:211–226

Johnsen K, Jacobsen CS, Torsvik V (2001) Pesticide effects on bacterial diversity in agricultural soils: a review. Biol Fertil Soils 33:443–453

Johnson VW, Pearson JF, Jackson TA (2001) Formulation of *Serratia entomophila* for biological control of grass grub New Zealand. Plant Protect 54:125–127

Kalawate A (2014) Microbial viral insecticides. In: Sahayaraj K (ed) Basic and applied aspects of biopesticides. Springer, New Delhi, pp 47–68

Karasawa T, Kasahara Y, Takebe M (2001) Variable response of growth and arbuscular mycorrhizal colonization of maize plants to preceding crops in various types of soils. Biol Fertil Soils 33:286–293

Karimov KZ, Zaripov NV (2007) Bacterial nitrogen fertilizers in cropping spring wheat. Her Kazan State Agrar Univ 6:64–65

Kaur R, Kaur J, Singh RS (2010) Nonpathogenic *Fusarium* as a biological control agent. J Plant Pathol 9:79–91

Kennedy IR, Choudhury ATMA, Kecskes ML (2004) Non–symbiotic bacterial diazotrophs in crop–farming systems: can their potential for plant growth promotion be better exploited? Soil Biol Biochem 36:1229–1244

Kesavachandran C, Pathak MK, Fareed M, Bihari V, Mathur N, Srivastava AK (2009) Health risks of employees working in pesticide retail shops: an exploratory study. Indian J Occup Environ Med 13:121–126

Khan MS, Zaidi A, Wani PA (2007) Role of phosphate-–solubilizing microorganisms in sustainable agriculture–a review. Agron Sustain Dev 27:29–43

Khan DA, Hashmi I, Mahjabeen W, Naqvi TA (2009a) Monitoring health implications of pesticide exposure in factory workers in Pakistan. Environ Monit Assess 168:231–240

Khan MS, Zaidi A, Wani PA (2009b) Role of phosphate solubilizing microorganisms in sustainable agriculture. In: Lictfouse E et al (eds) Sustainable agriculture. Springer, Dordrecht, pp 551–570

Kidaj D, Wielbo J, Skorupska A (2012) Nod factors stimulate seed germination and promote growth and nodulation of pea and vetch under competitive conditions. Microbiol Res 167:144–150

Kloepper JW, Schroth MN (1978) Plant growth–promoting rhizobacteria in radish. In: Proceedings of the 4th international conference on plant pathogenic bacteria. Ed. Station de pathologic Vegetal et Phytobacteriologic. Agners France, vol 2, pp 879–882

Kloepper JW, Schroth MN (1994) Plant growth promoting rhizobacteria (other systems). In: Okon Y (ed) Azospirillum? Plant Associations, Boca Raton/Ann Arbor/London/Tokyo, pp 135–166

Knowles A (2008) Recent developments of safer formulations of agrochemicals. Environmentalist 28:35–44

Konradsen F, Vander Hoek W, Cole DC, Hutchinson G, Daisley H, Singh S, Eddleston M (2003) Reducing acute poisoning in developing countries–options for restricting the availability of pesticides. Toxicology 192:249–261

Kraiser T, Gras DE, Gutiérrez AG, González B, Gutiérrez RA (2011) A holistic view of nitrogen acquisition in plants. J Exp Bot 62:1455–1466

Kumar A, Yadav DS (2001) Long–term effects of fertilizers on the soil fertility and productivity of a rice–wheat System. J Agron Crop Sci 186:47–54

Kutyova TY, Durinina EP, Muravyova NE, Sheyko AV (2002) Microbal fertilizers Bamil, Omug, Ekud, Pudret their properties, influence on soil and crops. Soil Science Series, vol 4. Herald of Moscow State University Publication, Moscow, pp 40–46

Lara AM (2008) Nueva herramienta en el tratamiento de semillas de granos y cereales. http://www.phcmexico.com.mx/phcsemillas1.html

Larena I, Melgarejo P, De Cal A (2003) Drying of conidia of *Penicillium oxalicum*, a biological control agent against *Fusarium* wilt of tomato. J Phytopathol 151:600–606

LeConte JL (1874) Hints for the promotion of economic entomology. Am Assoc Adv Sci 22:11–22

Leggett M, Leland J, Kellar K, Epp B (2011) Formulation of microbial biocontrol agents–an industrial perspective. Can J Plant Pathol 33:101–107

Leonard GC, Julius JM (2000) Review biopesticides: a review of their action, applications and efficacy. Pest Manag Sci 56:651–676

Liang LZ, Zhao X, Yi XY, Chen ZC, Dong XY, Chen RF, Shen RF (2013) Excessive application of nitrogen and phosphorus fertilizers induces soil acidification and phosphorus enrichment during vegetable production in Yangtze River Delta, China. Soil Use Manag 29:161–168

Lisansky SG, Quinlan RJ, Tassoni G (1993) The Bacillus thuringiensis production handbook. CPL Scientific, Newbury. El–Bandary MA

Loper JE, Hassan KA, Mavrodi DV, Davis EW II, Lim CK, Shaffer BT et al (2012) Comparative genomics of plant–associated *Pseudomonas* spp. Insights into diversity and inheritance of traits involved in multitrophic interactions. PLoS Genet 8:1002784

Lord JC (2005) From Metchnikoff to Monsanto and beyond: the path of microbial control. J Invertebr Pathol 89:19–29

Lorck H (2004) Production of hydrocyanic acid by bacteria. Plant Physiol 1:142–146

Lugtenberg B, Kamilova F (2009) Plant–growth—promoting rhizobacteria. Annu Rev Microbiol 63:541–556

Lupwayi NZ, Clayton GW, Rice WA (2006) Rhizobial inoculants for legume crops. J Crop Improv 15:289–321

Luthy P, Cordier JL, Fischer HM (1982) *Bacillus thuringiensis* as a bacterial insecticide: basic considerations and application. In: Kurstak E (ed) Microbial and viral pesticides. Marcel Dekker, New York, pp 35–74

Luz C, Batagin I (2005) Potential of oil–based formulations of *Beauveria bassiana* to control *Triatoma infestans*. Mycopathologia 160:51–62

Mahajan A, Gupta RD (2009) Bio–fertilizers: their kinds and requirement in India. In: Mahajan A, Gupta RD (eds) Integrated nutrient management (INM) in a sustainable rice–wheat cropping system. Springer, Dordrecht, pp 75–100

Maiyappan S, Amalraj ELD, Santhosh A, Peter AJ (2010) Isolation, evaluation and formulation of selected microbial consortia for sustainable agriculture. J Biofertil Biopestic 2:109

Malaysian Agri Hi Tech (2008) Mycogold. http://www.alibaba.com/catalog/11813495/Mycorrhiza_

Maldonado Blanco MG, Galan Wong LJ, Rodriguez Padilla C, Quiroz Martinez H (2002) Evaluation of polymer–based granular formulations of *Bacillus thuringiensis* israelensis against larval *Aedes aegypti* in the laboratory. J Am Mosq Control Assoc 18:352–358

Malusá E, Sas–Paszt L, Ciesielska J (2012) Technologies for beneficial microorganisms inocula used as biofertilizers. Sci World J 2012:491206

Mandal A, Patra AK, Singh D, Swarup A, Masto RE (2007) Effect of long–term application of manure and fertilizer on biological and biochemical activities in soil during crop development stages. Biores Technol 98:3585–3592

Marques EJ, Vilas Boas VAM, Pereira CEF (1981) Bol Teenico PLANALSUCAR 3:5–23

Marrone PG (2007a) Barriers to adoption of biological control agents and biological pesticides. CAB Rev: Perspect Agri Vet Sci Nutr Natur Resour, CAB International Wallingford, pp 2–51

Marrone PG (2007b) Barriers to adoption of biological control agents and biological pesticides. CAB Rev: Perspect Agric Vet Sci Nutr Nat Resour 2:15

Martinez–Romero E (2003) Diversity of Rhizobium—Phaseolus vulgaris symbiosis: overview and perspectives. Plant Soil 252:11–23

Mbarga JB, Begoude BAD, Ambang Z, Meboma M, Kuate J, Schiffers B, Hoopen GMT (2014) A new oil–based formulation of *Trichoderma asperellum* for the biological control of cacao black pod disease caused by *Phytophthora megakarya*. Biol Control 77:15–22

McCoy CW (1990) Entomogenous fungi as microbial pesticides. In: Baker RR, Dunn PE (eds) New direction in biological control: alternatives for suppressing agricultural pests and diseases. Alan R Liss Inc, New York, pp 139–159

McCoy CW, Samson RA, Boucias DG (1988) In: Ignoffo CM, Mandava NB (eds) Handbook of natural pesticides: microbial pesticides, part A. Entomogenous protozoa and fungi. CRC Press, Boca Raton, pp 151–236

Meena VS, Maurya BR, Verma JP (2014) Does a rhizospheric microorganism enhance $K^+$ availability in agricultural soils? Microbiol Res 169:337–347

Mejri D, Gamalero E, Souissi T (2013) Formulation development of the deleterious rhizobacterium *Pseudomonas trivialis* X33d for biocontrol of brome (*Bromus diandrus*) in durum wheat. J Appl Microbiol 114:219–228

Melnick R, Stoneman B, Marrone PG, Messerschmidt O, Donaldson M (2009) Biopesticide use and attitudes survey results. Biopesticide Industry Alliance. http://www.ipmcenters.org/ipmsymposium09/37–2_Stoneman.pdf. Accessed 25 July 2011

Mendonca AF (1992) Mass production, application and formulation of *Metarhizium anisopliae* for control of sugarcane grasshopper, *Mahanarva posticata* in Brazil. In: Prior C, Lomer CJ (eds) Biol cont Locusts Grasshopper. CAB International, Wallingford, pp 239–244

Messele B, Pant LM (2012) Effects of inoculation of *Sinorhizobium ciceri* and phosphate solubilizing bacteria on nodulation, yield and nitrogen and phosphorus uptake of chickpea (*Cicer arietinum* L.) in Shoa Robit Area. J Biofertil Biopestic 3:1000129

Meyer JM (2000) Pyoverdines: pigments, siderophores and potential taxonomic markers of fluorescent *Pseudomonas* species. Arch Microbiol 174:135–142

Meyer JM, Abdallah MA (1978) The fluorescent pigment of *Pseudomonas fluorescens*: biosynthesis, purification and physicochemical properties. J Gen Microbiol 107:319–328

Michereff FM, de Faria MR, Wraight SP, Silva KFAS (2009) Micoinseticidas emicoacaricidas no Brasil: Como estamos após quatro décadas? Arq Inst Biol 76:769–779

Micro Market Monitor (2015) North America biofertilizer market by application (cereals & grains, fruits & vegetables, pulses & oilseeds), by type (nitrogen fixing biofertilizers, phosphate solubilizing biofertilizers, potash mobilizing biofertilizers), by source, by geography –analysis and forecast to 2019. Available online http://www.micromarketmonitor.com/market/north–america–bio–fertilizer–5250154124.html

Minaxi SJ, Chandra S, Nain L (2013) Synergistic effect of phosphate solubilizing rhizobacteria and arbuscular mycorrhiza on growth and yield of wheat plants. J Soil Sci Plant Nutr 13:511–525

Mishra S, Arora NK (2012) Management of black rot in cabbage by rhizospheric *Pseudomonas* species and analysis of 2, 4–diacetylphloroglucinol by qRT–PCR. Biol Control 61:32–29

Mishra J, Tewari S, Singh S, Arora NK (2015) Biopesticides: where we stand? In: Arora NK (ed) Plant microbes symbiosis: applied facets. Springer, New Delhi, pp 37–76

Mitchell CC, Westerman RL, Brown JR, Peck TR (1991) Overview of long–term agronomic research. Agron J 83:24–25

Montazeri M, Greaves MP (2002) Effects of culture age, washing and storage conditions on desiccation tolerance of *Colletotrichum truncatum* Conidia. Biocontrol Sci Technol 12:95–105

Monte E (2001) Understanding *Trichoderma*: between biotechnology and microbial ecology. Int Microbiol 4:1–4

Moreno–Sarmiento N, Moreno–Rodriguez LF, Uribe D (2007) Biofertilizantes para la agricultura en Colombia. In: Izaguirre–Mayoral ML, Labandera C, Sanjuan J (eds) Biofertilizantes en Iberoamerica: Visionte´ cnica, cientificay empresarial. Denad Internacional, Montevideo, pp 38–45

Moses M, Johnson ES, Anger WK, Burse VW, Horstman SW, Jackson RJ et al (1993) Environmental equity and pesticide exposure. Toxicol Health 9:913–959

Mulder EG (1967) Dr. F C Gerretsen 1889–1966. Plant Soil 1:1–4

Mulvaney RL, Khan SA, Ellsworth TR (2009) Synthetic nitrogen fertilizers deplete soil nitrogen: a global dilemma for sustainable cereal production. J Environ Qual 38:2295–2314

Naderifar M, Daneshian J (2012) Effect of different nitrogen and biofertilizers effect on growth and yield of *Brassica napus* L. Int J Agric Crop Sci 4:478–482

Nakhro N, Dkhar MS (2010) Impact of organic and inorganic fertilizers on microbial populations and biomass carbon in paddy field. Soil J Agron 9:102–110

Navon A (2000) *Bacillus thuringiensis* application in agriculture. In: Charles JF, Delécluse A, Roux CL (eds) Entomopathogenic bacteria: from laboratory to field application. Springer, Dordrecht, pp 355–369

Neilands JB (1995) Siderophores: structure and function of microbial iron transport compounds. J Biol Chem 270:26723–26726

Nguyen Q, Qi YM, Wu Y, Chan LC, Nielsen LK, Reid S (2011) *In vitro* production of *Helicoverpa baculovirus* biopesticides—automated selection of insect cell clones for manufacturing and systems biology studies. J Virol Methods 175:197–205

Nikoloic M, Römheld V (1999) Mechanism of Fe uptake by the leaf symplast: is Fe inactivation in leaf a cause of Fe deficiency chlorosis? Plant Soil 215:229–237

Nobbe F, Hiltner L (1896) Inoculation of the soil for cultivating leguminous plants. US Patent 570 813

Nordstierna L, Abdalla AA, Nordin M, Nyden M (2010) Comparison of release behavior from microcapsules and microspheres. Prog Org Coat 69:49–51

Ogawa M (1989) Inoculation methods of VAM fungi: charcoal ball method and rice hulls method. In: Hattori T et al (eds) Recent advances in microbial ecology. Japan Scientific Societies Press, Tokyo, pp 247–252

Ortega E (2007) Una visio´n del uso de fertilizantes en Cuba. In: Izaguirre–Mayoral ML, Labandera C, Sanjuan J (eds) Biofertilizantes en Iberoamerica: Visio´n te´cnica, cientı´fica y empresarial, vol 1. Denad Internacional, Montevideo, pp 46–51

Ortet P et al (2011) Complete genome sequence of a beneficial plant root–associated bacterium, *Pseudomonas brassicacearum*. J Bacteriol 193:3146

Pabby A, Prasanna R, Nayak S, Singh PK (2003) Physiological characterization of the cultured and freshly isolated endosymbionts from different species of *Azolla*. Plant Physiol Biochem 41:73–79

Paikray S, Malik V (2010) Microbial formulation for widespread used in agricultural practices: google patents

Pandey P, Maheshwari DK (2007) Bioformulation of *Burkholderia* sp. MSSP with a multispecies consortium for growth promotion of *Cajanus cajan*. Can J Microbiol 53:213–222.

Papp ILG et al (1992) Manual do Produtor do Algodão, Bolsa de Mercadorias & Futuros. São Paulo, p 158

Park JK, Chang HN (2000) Microencapsulation of microbial cells. Biotechnol Adv 18:303–319

Pawar VM, Thombre UT (1992) Prospects of *baculovirus* in integrated pest management of pulses. In: Ananthakrishnan TN (ed) Emerging trends in biocontrol of phytophagous insects. Oxford and IBH Publishing Company Private Limited, New Delhi, pp 253–258

Peng G, Wolf TM (2011) Synergy between synthetic and microbial herbicides for weed control. Pest Technol 5:18–27

Penton G et al (2011) Using EcoMic® Pectimorf® bioactive product and the establishment of two forage species. Pastures Forages 34:281

Peoples MB et al (2009) The contributions of nitrogen–fixing crop legumes to the productivity of agricultural systems. Symbiosis 48:1–17

Persson A, Jonsson AS, Zacchi G (2004) Use of microfiltration as first step in recovery of protein from a fermentation broth. Appl Biochem Biotechnol 112:151–162

Philip JW, Timothy SG, Lionel XD, Alison JK, Tracy AV, Wiesel L, Wishart J (2013) Root traits for infertile soils. Front Plant Sci 4:193

Pimentel D, Acquay H, Biltonen M, Rice P, Silva M, Nelson J (1992) Environmental and economic costs of pesticide use. Bioscience 42:750–760

Pimentel D (1993) Habitat factors in new pest invasions. In: Kim KC, Mc Pheron BA (eds) Evolution of insect pests–patterns of variation. Wiley, New York, pp 165–181

Polhill RM, Raven PH (1981) Advances in legume systematics, parts 1 and 2. Royal Botanic Gardens, Kew

PRWEB (2014) Europe bio fertilizer market is expected to reach $4,582.2 million in 2017 new report by MicroMarket Monitor. Available online http://www.micromarketmonitor.com/market/europe–bio–fertilizer–4637178345.html

Punja ZK, Utkhede RS (2003) Using fungi and yeasts to manage vegetable crop diseases. Trends Biotechnol 21:400–407

Raddadi N, Cherif A, Boudabous A, Daffonchio D (2008) Screening of plant growth promoting traits of *Bacillus thuringiensis*. Ann Microbiol 58:47–52

Rao GVR, Kumar CS, Sireesha K, Kumar PL (2015) Role of *nucleopolyhedroviruses* (NPVs) in the management of lepidopteran pests in Asia. In: Sree KS, Varma A (eds) Biocontrol of lepidopteran pests, vol 43, Soil biology. Springer, Cham, pp 11–52

Rathore S, Desai PM, Liew CV, Chan LW, Heng PWS (2013) Microencapsulation of microbial cells. J Food Eng 116:369–381

Ravensberg WJ (2011) Critical factors in the successful commercialization of microbial pest control products. In: Ravensberg WJ (ed) A roadmap to the successful development and commercialization of microbial pest control products for control of arthropods, vol 10, Progress in biological control. Springer Science Business Media, Dordrecht, pp 295–356

Reardon R, Podgwaite J (1992) The gypsy moth nucleopolyhedrosis virus product. USDA Forest Service. NA–TP–02–92. Northeastern Area State and Private Forestry, Radnor, p 9

Rehm G, Schmitt M (2002) Potassium for crop production. Regents of the University of Minnesota, Saint Paul

Reid S, Chan L, Van Oers M (2014) Production of entomopathogenic viruses. In: Morales–Ramos JA, Rojas MG, Shapiro–Ilan DI (eds) Mass production of beneficial organisms. Elsevier, Amsterdam, pp 437–482

Reigart JR, Roberts JR (1999) Recognition and management of pesticide poisonings. Available online http://www2.epa.gov/pesticide–worker–safety

Rhodes DJ (1996) Economics of *baculovirus* – insect cell production systems. Cytotechnology 20:291–297

Ridgway RL, Illume VL, Farrar RR, Calvin DD, Fleischer SJ, Inscoe MN (1996) Granular matrix formulation of *Bacillus thuringiensis* for control of the European corn borer (Lepidoptera: Pyralidae). J Econ Entomol 89:1088–1094

Rodrigo M et al (2011) Deciphering the rhizosphere microbiome for disease–suppressive bacteria. Science 332:1097–1100

Rodriguez H, Fraga R (1999) Phosphate solubilizing bacteria and their role in plant growth promotion. Biotechnol Adv 17:319–339

Roger TK, Mosse B (2004) A history of research on arbuscular mycorrhiza. Mycopathology 14:145–163

Rosas–Garcia NM (2009) Biopesticide Production from *Bacillus thuringiensis*: an environmentally friendly alternative. Recent Pat Biotechnol 3:28–36

Rothballer M, Schmid M, Hartmann A (2009) Diazotrophic bacterial endophytes in *Gramineae* and other plants. In: Pawlowski K (ed) Prokaryotic symbionts in plants. Springer, Berlin, pp 273–302

Saha M, Sarkar S, Sarkar B, Sharma BK, Bhattacharjee S, Tribedi P (2015) Microbial siderophores and their potential applications: a review. Environ Sci Pollut Res. doi:10.1007/s11356-015-4294-0

Saharan BS, Nehra V (2011) Plant growth promoting rhizobacteria: a critical review. Life Sci Med Res 21:1–30

Sainju UM, Whitehead WF, Singh BP (2003) Agricultural management practices to sustain crop yields and improve soil and environmental qualities. Sci World J 3:768–789

Sanborn M, Kerr KJ, Sanin LH, Cole DC, Bassil KL, Vakil C (2007) Non–cancer health effects of pesticides: systematic review and implications for family doctors. Can Fam Physician 53:1712–1720

Sansinenea E (2012) Discovery and description of *Bacillus thuringiensis*. In: Sansinenea E (ed) *Bacillus thuringiensis* biotechnology. Springer Science, Dordrecht, pp 3–18

Schisler DA, Slininger PJ, Behle RW, Jackson MA (2004) Formulation of *Bacillus spp.* for biological control of plant diseases. Phytopathology 94:1267–1271

Sharma SB, Sayyed RZ, Trivedi MH, Gobi TA (2013) Phosphate solubilizing microbes: sustainable approach for managing phosphorus deficiency in agricultural soils. Springer Plus 2:587

Sheng XF, Lin YH (2006) Solubilization of potassium–bearing minerals by a wild–type strain of *Bacillus edaphicus* and its mutants and increased potassium uptake by wheat. Can J Microbiol 52:66–72

Singh KN, Merchant K (2012) The agrochemical industry. In: Kent JA (ed) Handbook of industrial chemistry and biotechnology. Springer Science+Business Media, New York, pp 643–699

Singh M, Prakash TN (2012) Characterization of phosphate solubilising bacteria in sandy loam soil under chickpea cropping system. Indian J Microbiol 52:167–173

Smiley RW (1979) Wheat–rhizoplane pseudomonads as antagonists of *Gaeumannomyces graminis*. Soil Biol Biochem 11:371–376

Smith HS (1919) On some phases of insect control by the biological method. J Econ Entomol 12:288–292

Smith JH, Allison FE, Soulides DA (1961) Evaluation of phosphobacterin as a soil inoculant. Soil Sci Soc Am Proc 25:109–111

Sparks DL, Huang M (1985) Physical chemistry of soil potassium. In: Munson RD (ed) Potassium in agriculture. ASA. CSSA.SSSA, Madison, pp 201–276

Sperber JI (1957) Solution of mineral phosphates by soil bacteria. Nature 180:994–995

Spyrou IM, Karpouzas DG, Menkissoglu–Spiroudi U (2009) Do botanical pesticides alter the structure of the soil microbial community? Microb Ecol 58:715–727

Steinhaus EA (1956) Microbial control–the emergence of an idea. A brief history of insect pathology through the nineteenth century. Hilgardia 26:107–160

Steinhaus EA (1975) Disease in a minor chord. Ohio State University Press, Columbus

Stenersen J (2004) Chemical pesticides: mode of action and toxicology. CRC Press, Boca Raton

Stoate C, Boatman ND, Borralho RJ, Rio Carvalho C, de Snoo GR, Eden P (2001) Ecological impacts of arable intensification in Europe. J Environ Manag 63:337–365

Stockwell VO, Stack JP (2007) Using *Pseudomonas* spp. for integrated biological control. Phytopathology 97:244–249

Strzelczyk E, Kamper M, Li C (1994) Cytokinin–like–-substances and ethylene production by *Azospirillum* in media with different carbon sources. Microbiol Res 149:55–60

Subbarao NS (1988) Phosphate solubilizing micro–organism. In: Biofertilizer in agriculture and forestry. Regional Biofert Dev Centre, Hissar, pp 133–142

Sun X (2015) History and current status of development and use of viral insecticides in China. Viruses 7:306–319

Suzuki A, Akune M, Kosigo M, Imagama K, Osuki K, Uchiumi T, Higashi S, Han S, Yoshida S, Asami T, Abe M (2005) Control of nodule number by phytohormone abscisic acid in the roots of two leguminous species. Plant Cell Physiol 45:914–922

Syers JK (1998) Soil and plant potassium in agriculture, Proceedings no. 411. The International Fertiliser Society, York, p 32

Szewczyk BL, Hoyos-Carvajal M, Hoyos-Carvajal M, Paluszek I, Skrzecz Lobo de Souza M (2006) *Baculoviruses*–re–emerging biopesticides. Biotechnol Adv 24:143–160

Tadros F (2005) Applied surfactants, principles and applications. Wiley, Weinheim, pp 187–256

Tadros T (2013) Suspension concentrates. In: Tadros T (ed) Encyclopedia of colloid and interface science. Springer, Berlin/Heidelberg, pp 1334–1334

Tajini F, Trabelsi M, Drevon JJ (2012) Combined inoculation with *Glomus intraradices* and *Rhizobium tropici* CIAT899 increases phosphorus use efficiency for symbiotic nitrogen fixation in common bean (*Phaseolus vulgaris* L.). Saudi J Biol Sci 19:157–163

Tamez-Guerra P, McGuire MR, Medrano-Roldan H, Galan-Wong LJ, Shasha BS, Vega FE (1996) Sprayable granule formulations for *Bacillus thuringiensis*. J Econ Entomol 89:1424–1430

Tani H, Limn CK, Yap CC, Onishi M, Nozaki M, Nishimune et al (2003) *In vitro* and *in vivo* gene delivery by recombinant baculoviruses. J Virol 77:9799–9808

Tank N, Sarafa M (2009) Salinity–resistant plant growth promoting rhizobacteria ameliorates sodium chloride stress on tomato plants. J Plant Interact 5:51–58

Tavasolee AN, Aliasgharzad G, Salehijouzani MM, Asgharzadeh A (2011) Interactive effects of arbuscular mycorrhizal fungi and rhizobial strains on *Chickpea* growth and nutrient content in plant. Afr J Microbiol 10:7585–7591

Tawaraya K, Hashimoto K, Wagatsuma T (1998) Effect of root exudate fractions from P–deficient and P–sufficient onion plants on root colonisation by the arbuscular mycorrhizal fungus *Gigaspora margarita*. Mycorrhiza 8:67–70

Tewari S, Arora NK (2014) Multifunctional exopolysaccharides from *Pseudomonas aeruginosa* pf23 involved in plant growth stimulation, biocontrol and stress amelioration in sunflower under saline conditions. Curr Microbiol 69:484–494

Thakore Y (2006) The biopesticide market for global agricultural use. Ind Biotechnol 2:194–208

Tian F, Ding Y, Zhu H, Yao L, Du B (2009) Genetic diversity of siderophore–producing bacteria of tobacco rhizosphere. Braz J Microbiol 40:276–284

Townsend A, Howarth RW (2010) Human acceleration of the global nitrogen cycle. Sci Am 302:32–39

Trabelsi D, Mhamdi R (2013) Microbial inoculants and their impact on soil microbial communities: a review. Bio Med Res Int 86:32–40

Travis G, Caradus J, Gelernter W, Jackson T, Keyhani N, Köhl J, Marrone P, Morin L, Stewart A (2012) Have biopesticides come of age? Trends Biotechnol 5:250–258

Tzeng YM, Tsun HY, Chang YN (1999) Recovery of thuringiensin with cetylpyridinium chloride using micellar–enhanced ultrafiltration process. Biotechnol Prog 15:580–586

USEPA (1999) Meeting summary: EPA–USDA Bt crop insect resistance management workshop. Available online http://www.epa.gov/oppbppd1/biopesticides/summary826.htm

Van Beek N, Davis DC (2007) *Baculovirus* insecticide production in insect larvae. Methods Mol Biol 388:367–378

Vandevivere P, Welch SA, Ullman WJ, Kirchman DL (1994) Enhanced dissolution of silicate minerals by bacteria at near–neutral pH. Microb Ecol 27:241–251

Vassilev SV, Vassileva CG (2006) A new approach for the classification of coal fly ashes based on their origin, composition, properties, and behavior. Fuel 86:1490–1512

Vendan R, Thangaraju M (2006) Development and standardization of liquid formulation for *Azospirillum* bioinoculant. Indian J Microbiol 46:379–387

Verma M, Brar S, Tyagi R, Surampalli R, Valero J (2007) Industrial wastewaters and dewatered sludge: rich nutrient source for production and formulation of biocontrol agent, *Trichoderma viride*. World J Microbiol Biotechnol 23:1695–1703

Verónica VG, Volker J, Juan DC (2006) Growth, metabolism and baculovirus production in suspension cultures of an Anticarsia gemmatalis cell line. Cytotechnology 52:113–124

Vessey JK (2003) Plant growth promoting rhizobacteria as biofertilizers. Plant Soil 255:571–586

Vinale F, Sivasithamparam K, Ghisalberti EL, Marra R, Woo SL, Lorito M (2008) *Trichoderma*–plant–pathogen interactions. Soil Biol Biochem 40:1–10

Vitousek PM et al (2009) Nutrient imbalances in agricultural development. Science 324(1519):1520

Warrior P, Rehberger LA, Beach M, Grau PA, Kirfman GW, Conley JM (1999) Commercial development and introduction of DiTeraTM, a new nematicide. Pest Sci 55:376–379

Warrior P, Konduru K, Vasudevan P (2002) Formulation of biological control agents for pest and disease management. In: Gnanamanickam SS (ed) Biological control of crop diseases. Marcel Dekker Inc, New York

Weller DM (1988) Biological control of soil borne plant pathogens in the rhizosphere with bacteria. Annu Rev Phytopathol 26:379–407

Weller DM (2007) *Pseudomonas* biocontrol agents of soil borne pathogens: looking back over 30 years. Phytopathology 97:250–256

Whitbread A, Blair G, Konboon Y, Lefroy R, Naklang K (2003) Managing crop residues, fertilizers and leaf litters to improve soil C, nutrient balances and the grain yield of rice and wheat cropping systems in Thailand and Australia. Agric Eco Environ 100:251–263

Whitelaw MA (2000) Growth promotion of plants inoculated with phosphate solubilizing fungi. Adv Agron 69:99–151

Wilson PW (1940) The biochemistry of symbiotic nitrogen fixation. The University of Wisconsin Press, Madison, p 302

Woo SL, Ruocco M, Vinale F, Nigro M, Marra R, Lombardi N, Pascale A, Lanzuise S, Manganiello G, Lorito M (2014) *Trichoderma*-based products and their widespread use in agriculture. Open Mycol J 8:71–126

Wraight SP, Carruthers RI (1999) Production, delivery, and use of mycoinsecticides for control of insects pests on field crops. In: Hall FR, Menn JJ (eds) Biopesticides: use and delivery. Humana Press, Totowa, pp 233–269

Xiufang HC, Jishuang J, Guo (2006) Two phosphate and potassium solubilizing bacteria isolated from Tianmu Mountain, Zhejiang China. World J Microbiol Biotechnol 22:983–990

Yan S, Mohammedi S, Tyagi R, Surampalli R, Valéro J (2007) Growth of four Serovar of *Bacillus thuringiensis* (Var. Kurstaki, Israelensis, Tenebrionis, and Aizawai) in Wastewater Sludge. Pract Per Haz Toxic Rad Waste Manag 11:123–129

Yang S (2006) Effect of long–term fertilization on soil productivity and nitrate accumulation in Gansu oasis. Agric Sci China 5:57–67

Young JPW (1992) Phylogenetic classification of nitrogen–fixing organisms. In: Stacey G, Burris RH, Evans HJ (eds) Biological nitrogen fixation. Chapman and Hall, New York, pp 43–86

Young J, Choi J, Gringorten L, Bélanger L, Morel L, Bourque D, Masson L, Groleau D, Míguez CB (2008) Production of an insecticidal crystal protein from *Bacillus thuringiensis* by the Methylotroph *Methylobacterium extorquens*. Appl Environ Microbiol 74:5178–5182

Young SD, Townsend RJ, Swaminathan J, Callaghan MO (2010) *Serratia entomophila*–coated seed to improve ryegrass establishment in the presence of grass grubs New Zealand. Plant Prot 63:229–234

Zahra MK, Monib M, Abdel-Al SI, Heggo A (1984) Significance of soil inoculation with silicate bacteria. Zentralbl Mikrobiol 139:349–357

Zayadan BK, Matorin DN, Baimakhanova GB, Bolathan K, Oraz GD, Sadanov AK (2014) Promising microbial consortia for producing biofertilizers for rice fields. Microbiology 83:391–397

Zhigletsova SK, Dunajtsev IA, Besaeva SG (2010) Possibility of application of microorganisms for solving problems of ecological and food safety. Agrochemical 6:83–96

Zlotkin E, Rochat H, Kopeyan C, Miranda F, Lissitzky S (1971) Purification and properties of the insect toxin from the venom of the *scorpion Androctonus australis*. Biochemistry 53:1073–1078

# Formulation Technology of Biocontrol Agents: Present Status and Future Prospects

2

Chetan Keswani, Kartikay Bisen, Vivek Singh, Birinchi Kumar Sarma, and Harikesh Bahadur Singh

## Contents




C. Keswani • K. Bisen • V. Singh • B.K. Sarma • H.B. Singh (✉)
Department of Mycology and Plant Pathology, Institute of Agricultural Sciences, Banaras Hindu University, Varanasi 221005, India
e-mail: hbs1@rediffmail.com



**Abstract**

Recent shift in trends of agricultural practices from application of synthetic fertilizers and pesticides to organic farming has brought into focus the use of microorganisms those carryout analogous functions. Formulations of rhizomicroorganisms available in global markets range from talc-based and liquid and secondary metabolite-based formulations. The ideal conditions required for development of high efficiency formulations of biopesticides include selection of potent strains, shelf life, storage, application technology, quality control, biosafety, and registration. In this chapter, we will discuss the constraints associated with development and commercialization of bioinoculants. Moreover, special emphasis will be on the next generation of antimicrobial secondary metabolite formulations which will not only have a much longer shelf life but also a higher efficiency against soilborne phytopathogens particularly against bacteria; also, a consortium of antimicrobial metabolites against individual pathogens could be formulated and used regardless of geographic location where the incidence of that particular disease is high. This approach would be unsurpassed by current technology, as the formulation would specifically target a particular pathogen while remaining soil microbiota would remain unaffected.




N.K. Arora et al. (eds.), *Bioformulations: for Sustainable Agriculture*,
DOI 10.1007/978-81-322-2779-3_2

## 2.1 Introduction

Conventional farming around the world is primarily based on chemical fertilizers and pesticides for plant nutrition and disease management, a practice which pounded huge negative impacts on human and environment health. Globally, rising awareness of the hazardous effects of synthetic pesticides has increased the demand for safer alternatives. Various microorganisms are currently being explored and utilized as biological control agents (BCAs) or biopesticides. Popular BCAs include *Trichoderma* spp., *Pseudomonas fluorescence*, *Bacillus* spp., *Ampelomyces quisqualis*, *Agrobacterium radiobacter*, nonpathogenic *Fusarium*, *Coniothyrium*, and atoxigenic *Aspergillus niger* (Singh 2006; Keswani et al. 2014, 2015; Mishra et al. 2015). Approximately, there are 1400 biopesticide products being sold worldwide (NAAS 2013). The total market share of biopesticides was around 0.2 % of the total pesticides' market during the year 2000, and it amplified to 4.5 % by 2010. The market value is expected to reach around US$ 1 billion (Singh et al. 2012). In India, currently 34 microorganisms have been included in the schedule of Gazette of India for registration as biopesticide with Central Insecticide Board, Faridabad, under sections 9 (3B) and 9(3) of the Insecticides Act, 1968 (Table 2.1) (Keswani et al. 2013a).

Various classes of biocontrol agents have shown significant antagonism to a range of phytopathogens in vitro, but generally they have irregular performance in field conditions. Various factors are responsible for inconsistent performance which includes poor shelf life, susceptibility of microbial strain to various abiotic stresses, and low organic carbon content in the soil. Formulation technologies are used for stabilizing the microorganisms during production, storage, and

**Table 2.1** Microbes listed in the Gazette of India for production of biopesticides and registration under sections 9 (3B) and 9(3) of the Insecticides Act, 1968

| Bacterial | Fungal | Viral |
|---|---|---|
| *Burkholderia cepacia* | *Verticillium chlamydosporium* | Granulosis viruses |
| *Agrobacterium radiobacter* strain 84 | *Streptomyces griseoviridis* | Nuclear polyhedrosis viruses (NPV) |
| *Agrobacterium tumefaciens* | *Streptomyces lydicus* | |
| *Erwinia amylovora* (hairpin protein) | *Candida oleophila* | |
| *Alcaligenes* spp. | *Fusarium oxysporum* (non pathogenic) | |
| *Photorhabdus luminescences akhurustii strain K-1* | *Penicillium islanidicum* (for groundnut) | |
| *Photorhabdus luminescences* | *Pythium oligandrum* | |
| *Serratia marcescens GPS 5* | *VAM (fungus)* | |
| *Bacillus subtilis* | *Trichoderma* spp. | |
| *Pseudomonas fluorescens* | *Aspergillus niger – strain AN27* | |
| | *Gliocladium* spp. | |
| | *Beauveria bassiana* | |
| | *Verticillium lecanii* | |
| | *Metarhizium anisopliae* | |
| | *Nomuraea rileyi* | |
| | *Hirsutella* sp. | |
| | *Ampelomyces quisqualis* | |
| | *Phlebia gigantean* | |
| | *Coniothyrium minitans* | |
| | *Chaetomium globosum* | |
| | *Myrothecium verrucaria* | |
| | *Paecilomyces lilacinus* | |
| | *Piriformospora indica* | |

distribution, aiding in the application and handling, protecting the microorganism from damaging environmental factors, and enhancing the activity of the organism (Jones and Burges 1997). In a microbial formulation, the major focus is to preserve microorganism for enhancing their antagonism against target pathogens. Potency of microbial formulation is primarily dependant on the strain of microorganism used, though there may be crucial physical and nutritional requirements of the microorganism to remain active for longer time. Beneficial microorganisms are considered as eco-friendly, and it is mandatory that any additives in the formulation should be eco-friendly.

Commercial success of these formulations is based on the capability of a microorganism to survive and proliferate in the field condition, shelf life, and efficiency to control pest and disease, market price, ease of handling, and application (Lisansky 1985). Decision of selecting formulation depends fundamentally on the target organism to be managed, as well as on the ecology and biology of the biocontrol agents and host plant (Jacobsen and Backman 1993). Moreover, the best feature of this approach is that it can easily be integrated with different pest management modules. BCAs have been formulated in various ways such as wettable powders, liquid, and granules for application such as sprays, seed treatments, drenches, and dips and incorporation into soil and pot mix.

## 2.2 Types of Formulations

Although pesticides are formulated in various ways including dry formulations such as dusts (DP), granules (GR), and microgranules (MG); seed dressing formulations such as powders for seed dressing (DS); dry formulations for dilution in water including dispersible granules (WG) and wettable powders (WP); liquid formulations for dilution in water such as emulsions, suspension concentrates (SC), oil dispersions (OD), and capsule suspensions (CS); and ultralow volume formulations (Knowles 2005, 2006) however, globally biopesticides available in the market are in wettable powder, liquid, and granular formulations (Singh et al. 2012, 2014) (Table 2.2).

*Dusts (DP)* are formulated by adding an active ingredient on fine solid mineral powder such as clay and talc with particle size ranging from 50 to 100 μm. Dusts are applied directly to the target, either manually or mechanically. Inert ingredients used for dust formulation are anticaking agents, ultraviolet protectants, and adhesive materials to enhance adsorption. Dusts usually contain <10 % of microorganisms by weight.

*Granule (GR)* particles are heavier and larger compared to dust. Microgranules (100–600 μm) and coarse particles size (100 1000 μm) arc made from mineral materials such as silica, kaolin, starch, attapulgite, polymers, ground plant residues, and dry fertilizers (Tadros 2005). Concentration of microorganisms in granules ranges from 5 to 20 %. Three types of granule formulations are currently available: (1) the microorganism is sprayed on a rotating granular carrier without a sticker, (2) the microorganism is attached to the outer surface of a granular carrier by a sticker, and (3) the microorganism is incorporated into a carrier paste or powder as a matrix.

### 2.2.1 Wettable Powders and Liquids

BCAs such as *Bacillus subtilis*, *Pseudomonas putida*, and *Trichoderma* spp. are used to control various diseases and are generally applied as dip or drenches and sprays to fruit after harvest (Tronsmo and Dennis 1983; Colyer and Mount 1984; Pusey and Wilson 1984; Wilson and Pusey 1985). Spraying of *Penicillium* sp. to pineapple fruit resulted in reduced postharvest diseases (Lim and Rohrbach 1980).

### 2.2.2 Granular Formulations

Lignite silage was applied to produce granules containing *Trichoderma harzianum* and *Gliocladium roseum* to control *Rhizoctonia solani* in soil causing damping-off of peanut (Jones et al. 1984). Lignite was grinded to produce granules of 425–2000 μm in diameter and

**Table 2.2** Types of pesticide formulations (Modified from Patanjali and Raza 2013; Jones and Burges 1997)

| Formulation | Abbreviation | Features |
|---|---|---|
| *Formulations diluted in water* | | |
| Emulsifiable concentrate | EC | Emulsion formed when added to spray tank |
| Water-in-oil emulsion | EO | Preformed emulsion |
| Oil-in-water emulsion | EW | Preformed emulsion |
| Suspension concentrate | SC | Suspended insoluble AI |
| Capsule suspension | CS | AI contained in capsules |
| Soluble concentrate | SL | Used for water-soluble AI |
| Water-soluble powder | SP | Powder soluble, but may contain inert ingredients |
| Water-soluble granule | SG | Used for water-soluble AI |
| Tablet | TB | Used for portable water-soluble AI |
| Briquette | BR | Controlled-release formulation |
| Wettable powder | WP | Typically consist of AI, clay carrier, and surfactants |
| Water-dispersible granule | WG | AI dispersed, but not dissolved, in water |
| *Formulations diluted with organic solvents* | | |
| Oil-miscible liquid | OL | AI dissolved in organic solvent |
| Oil-miscible flowable | OF | Suspension in organic liquid |
| Oil-dispersible powder | OP | Powder to be applied in oil |
| *Formulations applied undiluted* | | |
| Dustable powder | DP | AI carried on free-flowing powder |
| Encapsulated granule | CG | Controlled-release granule |
| Microgranule | MG | Diameter below O.6 mm |
| Electro-chargeable liquid | ED | Used with electrostatic spray equipment |
| Spreading oil | SO | Applied to water surface |
| Ultralow volume liquid | UL | Applied through UL V sprayers |
| Ultralow volume suspension | SU | As above |
| Granule | GR | Applied to soil and water |
| *Seed treatments* | | |
| Powder for dry seed treatment | DS | Liquid suspension |
| Flowable concentrate | FS | |
| Solution for seed treatment | LS | |
| Coated seed | PS | |
| Water-dispersible powder | SS | |
| *Miscellaneous* | | |
| Bait concentrate | CB | Bait diluted before application |
| Bait | RB | |

then amended with the product of sorghum fermentation. Isolates of *T. harzianum* and *G. roseum* were allowed growing on these granules for 7 days. These granules were allowed to air-dry followed by incubation before application in *R. solani*-infested soil. Trapping of antagonistic microorganism in calcium alginate granules, also known as prill, has been used widely used (Connick 1988). Though most of the commercially available alginates are derived from kelp, other organisms are also reported to produce alginates. Alginates produced by *Azotobacter vinelandii* can be used in place of kelp alginate for the production of microbial formulation to control plant pathogens (De Lucca et al. 1990). Further research may lead to less costly alginates. Sodium alginates showing great variation in viscosity and purity are available commercially. More granular sodium alginates like Kelgin HV, Kelgin, and sodium alginate IG-350 are easier to handle than the more powdery alginic acids.

Daigle and Cotty (1995) reported that 5 % gluten from wheat grains improved the performance of a toxigenic *Aspergillus flavus* but that concentrations were very complex to process. The final formulated product contained 5 % gluten, 1 % sodium alginate, and 5 % corn cob grits. Likewise, in case of *Streptomyces* spp. formulation, clay was handy to keep in suspension, formulated in alginate with carrier polyamide.

## 2.3 Biocontrol Products Containing Fungi and Their Formulations

Plethora of antagonistic fungi and their products have been registered as commercial biopesticides globally (Table 2.3). Different types of formulations of these antagonists are currently available in markets and are successfully used against various phytopathogens. *Trichoderma* spp., *Gliocladium* spp., *Coniothyrium*, and nonpathogenic strains of fungal genus are available as different formulations.

Various biopesticide products (Binab-T, Bio-Fungus, Supresivit, RootShield, T-22HB, T-22G, Trichodex, Trichoseal, Trichopel, Trichodowels, BioMax, BioVam, Trichoject, *Trichoderma* 2000) contain *Trichoderma* spp. to control a variety of pathogens, including *Fusarium*, *Botrytis*, *Gaeumannomyces*, *Rhizoctonia*, *Pythium*, *Sclerotinia*, *Verticillium*, *Sclerotium*, and wood-rot fungi. *Trichoderma* product formulation varies considerably. Combination of *Trichoderma viride* and *T. harzianum* is formulated differently, for example, as a pellet for soil (Trichopel), as dowels for injecting in woods (Trichodowels), as a wettable powder for resuspension to apply on wounds with brush (Trichoseal), and as a wettable powder in syringe (Trichoject). Bio-Fungus is available as a granule, as crumbles for soil incorporation, and as wettable powder impregnated in sticks. *Gliocladium virens* formulation is available in alginate prill under the trade name SoilGard and effectively used against *R. solani* and *Pythium* spp. Nonpathogenic strain of *Fusarium oxysporum* has been used against pathogenic *F. oxysporum* and *Fusarium moniliforme* on carnation, tomato, basil, and cyclamen. Commercially available products of nonpathogenic *F. oxysporum*, namely, Biofox C, are formulated as alginate prill or dust, and Fusaclean is available as microgranule. Aspire containing yeast *Candida oleophila*, is formulated as wettable powder for postharvest application in citrus and pome fruit to control *Penicillium* spp. and *Botrytis*. *Ampelomyces quisqualis*, parasite of powdery mildew fungi, is commercially available as water-dispersible granule (AQ10) and applied to the leaves of apples, grapes, strawberries, tomatoes, cucurbits, and ornamentals. *Pythium oligandrum* has been formulated as granule or powder for use in seed treatment for management of pathogenic *Pythium* spp. (Jones and Burges 1997).

Lewis and Lumsden (2001) prepared a solid matrix formulation of *Gliocladium* and *Trichoderma* using wheat bran and vermiculite effective against *R. solani*. Formulated product was applied at a rate of 1.0 % (w/w). The product was tested, and a significant reduction in damping-off of pepper seedlings was observed. Efficacy of *Trichoderma* spp. against *Sclerotinia sclerotiorum* causing sunflower head rot was evaluated in the field. *Trichoderma* formulation contained viable hyphal fragments, *Trichoderma* conidia, milled corn kernels, and industrial talc. Sunflower heads were infected with *S. sclerotiorum* after 2 days of the first delivery of *Trichoderma* formulation by honeybees. Head rot incidence was reduced in sunflower when 100 g formulation was carried by honeybees in a 10 h per day period (Escande et al. 2002). An invert emulsion based on soybean and coconut oils offered the lowest viscosity (27 ± 0.81 cps) and most stable emulsion layer (93 % v/v) for formulating *T. harzianum* conidia (Batta 2004). In this formulation, 36 months of shelf life of conidia was recorded with 50 % declined viability at 20 ± 1 °C after 5 months. *Botrytis* sporulation on the fruit lesion surface was also inhibited after 10 days of inoculation. Biocontrol potential of *Trichoderma* isolated from rhizosphere of *Musa* sp. was evaluated against *F. oxysporum* in vitro. Among different isolates, *T. harzianum* Th-10 was found most significant

**Table 2.3** Some commercially available fungal biocontrol products

| Organism | Product name | Disease against used | Formulation | Application |
|---|---|---|---|---|
| *Ampelomyces quisqualis* | AQ10 Biofungicide | Powdery mildew | Water-dispersible granule | Spray |
| *Candida oleophila* | Aspire | *Penicillium* spp. *Botrytis* spp. | Wettable powder | Postharvest application to fruit as drench, drip, or spray |
| *Trichoderma harzianum*, *T. polysporll* | Binab-T | Pathogenic fungi causing wilt, take-all, root rot, internal decay of wood products, and decay in tree wounds | Wettable powder | Postharvest application to fruit as drench, drip, or spray |
| *Fusarium oxysporum* | BiofoxC | *F. oxysporum*, *F. moniliforme* | Dust or alginate granule | Seed treatment soil or incorporation |
| *Trichoderma* spp. | Bio-Fungus | *Sclerotinia*, *Rhizoctonia solani*, *Pythium* spp., *Fusarium*, *Verticillium*, *Phytophthora* | Granules, wettable powders, sticks, and crumbles | Applied after fumigation, incorporated in soil sprayed or injected |
| *Coniothyrium minitans* | Contans | *Sclerotinia sclerotiorum* and *S. minor* | | Spray |
| *Myrothecium verrucaria* (killed cells) | DiTera | Root-knot, citrus cyst, stubby root, sting, lesion, and burrowing nematodes | Wettable powder, emulsifiable liquid or granule | |
| *Fusarium oxysporum* (nonpathogenic) | Fusaclean | *F. oxysporum* | Spores, microgranule | In drip to rock wool; incorporate in potting |
| *Phlebia gigantean* | Rotstop | *Heterobasidion annosum* | Spores in inert powder | Spray, chain saw oil |
| *T. harzianum* strainT-22 | Root Shield | *Pythium* spp., *R. solani*, *Fusarium* spp. | Granules | Mix with soil or potting medium |
| *Gliocladium virens* | Soil Gard (formerly GlioGard) | *Rhizoctonia solani*, *Pythium* spp. | Granule ($1 \times 10^6$ cfu $g^{-1}$) | Granules are incorporated in soil or soilless growing media prior to seedlings |
| *T. harzianum* | Supresivit | Various fungi | | |
| *T. harzianum* and *T. viride* | T-22G 19, T-22 HB | *Pythium* spp., *R. solani*, *Fusarium* spp., *Sclerotinia homeocarpa* | Granules or dry powder (both at $1 \times 10^7$cfu $g^{-1}$) | Granules added in furrow with granular applicator, by broadcast application to turf, mixed with green house soil |
| *T. harzianum* | Trichodex | *Botrytis cinerea*, *Colletotrichum* spp., *Plasmopara viticola*, | Wettable powder | Spray |
| *T. harzianum* and *T. viride* | Trichopel, Trichoject, Trichodowel Trichoseal | *Armillaria*, *Botryosphaeria*, *Fusarium*, *Chondrostereum* | | |
| *Trichoderma* sp. | *Trichoderma* 2000 | *R. solani*, *S. rolfsii*, *Pythium* sp. | | Incorporated into soil or potting medium |
| *T. harzianum* NBRI-1055 | Sardar Eco Green Biofungicide | *Pythium* sp., *R. solani*, *S. rolfsii* | Wettable powder | Spray |
| *T. harzianum* NBRI-1055 | Tricha | *Pythium* sp., *R. solani*, *S. rolfsii* | Wettable powder | Spray |

in pathogen inhibition. Five organic substrates including rice chaffy grain, rice bran, banana pseudostem, farmyard manure, and dried banana leaf were tested for the mass production, and dried banana leaf was found the best carrier material for *T. harzianum* growth. Dried banana leaves were colonized within a few days by strain Th-10 and produced propagules of high density ($4.6 \times 10^{32}$ cfu $g^{-1}$ of leaf). Furthermore, addition of jaggery (10 % w/v) to the dried leaves enhances the growth of *T. Harzianum*, and more than 6 months of survival was recorded (Thangavelu et al. 2004).

*Trichoderma atroviride* isolate C52 was observed on onion roots when inoculated in soil with various formulations (McLean et al. 2005). Pellet formulation product maintained the concentration of *Trichoderma* up to $10^5$ cfu $g^{-1}$soil in comparison to $10^1$ cfu and $10^4$ $g^{-1}$ soil concentrations that were maintained by seed coating and solid substrate formulations, respectively. *Trichoderma* isolate C52 was inoculated into *Sclerotium cepivorum*-infested soil as both solid substrate and pellet formulations, and no difference was observed in disease control, but more healthy plants were observed in the pellet treatment. Increased root and shoot lengths, plant height, and dry weight were recorded after treatment of plants with *T. viride* formulated in talc. Application of *T. viride* formulation also resulted in a significant reduction of sheath blight caused by *R. solani* (Mathivanan et al. 2005). New carrier formulation was developed by using *T. harzianum* M1, and resistance to carbendazim showed inhibitory effect against *Pythium aphanidermatum*. Different formulations including lignite, talc, wettable powder, lignite + fly ash-based powder formulation, bentonite paste, gelatin-glycerin gel, and polyethylene and glycol paste were developed for seed treatment. Shelf life of the microbial formulations was assessed at 24 °C for 9 months. Up to 74 % reduction in disease incidence was recorded when *Trichoderma* formulation was applied as seed treatment. Additionally, enhanced plant biomass under field and greenhouse conditions was also recorded (Jayaraj et al. 2006). Pelletized formulations of kaolin clay and wheat bran in an alginate gel containing conidia, fermentor biomass, or chlamydospores of *G. virens* and *Trichoderma* spp. were prepared (Lewis and Papavizas 2007). Higher population densities of *Trichoderma* and *Gliocladium* were observed when soil is incorporated with alginate pellets containing chlamydospores rather than condia and bran rather than kaolin as the carriers.

A new *Trichoderma asperellum* formulation was developed using soybean oil dispersion. Complete inhibition of *Phytophthora megakarya* causing cacao black pod disease was recorded when formulation was applied to the pods. Ninety percent prevention of infection in treated pods was recorded after 1 week, and 50 % reduction after 3 weeks was recorded when formulation was sprayed on cacao clones susceptible to *P. megakarya*. The formulations showed a significant effect in disease management (Mbarga et al. 2014). Talc-based formulation of a novel *T. viride*, BHU-2953, successfully controlled damping-off of chili caused by *P. aphanidermatum* and tomato wilt caused by *F. oxysporum*. Significant reduction in diseases was observed when *T. viride* formulated as 2 % wettable powder was applied to the seeds and furrow (Singh et al. 2014). Formulations of *Trichoderma* with mixture of 1 % w/v Sure-Jell, 1 % w/v PDB, and 0.3 ml $L^{-1}$ of the surfactant BreakThru 100SL (BT) and an invert oil emulsion of 50 % v/v corn oil, 2.5 % w/v lecithin, and 0.5 % w/v PDB (COP) were evaluated against frosty pod rot pathogen *Moniliophthora roreri*. *T. harzianum* DIS 219f and *Trichoderma ovalisporum* DIS 70a were applied (180 ml $tree^{-1}$, $2.46 \times 10^7$ conidia $ml^{-1}$) in the field. COP/DIS-70a formulation resulted in a maximum increase in yield as compared to other treatments (Crozier et al. 2015).

## 2.4 Biocontrol Products Containing Bacteria and Their Formulation

Several bacterial species having biocontrol potential have been formulated and are commercially available (Table 2.4). Nonpathogenic

**Table 2.4** Some commercially available bacterial biocontrol products

| Biocontrol agent | Trade name | Target pathogen | Formulation | Application |
|---|---|---|---|---|
| *Pseudomonas syringae* ESC 10 | Bio-Save10 | *B. cinerea*, *Penicillium* spp., *Mucor piriformis*, *Geotrichum candidum* | Wettable powder | Postharvest application to fruit as drench, dip, or spray |
| *P. syringae* ESC 11 | Bio-Save11 | *B. cinerea*, *Penicillium* spp., *M. piriformis*, *G. candidum* | Wettable powder | Postharvest application to fruit as drench, dip, or spray |
| *P. fluorescens* A506 | Blight Ban A 506 | Frost, *Erwinia amylovora* | Wettable powder | Drench dip or spray |
| *P. fluorescens* NCIB | Conquer | *P. tolaasii* | Aqueous biomass suspension | Spray |
| *B. subtilis* | Epic II | *R. solani*, *Fusarium*, *Alternaria* spp., and *Aspergillus* spp. | Dry powder ($5.5 \times 10^{10}$ spores $g^{-1}$) | Added to a slurry, mix with a chemical fungicide |
| *Agrobacterium radiobacter* | Galltrol-A13 | Crown gall disease *A. tumefaciens* | Petri dishes with pure culture grown on agar ($1.2 \times 10^{11}$ cfu $plate^{-1}$) | Root dips, drench |
| *Pseudomonas cepacia* | Intercept 14 | *R. solani*, *Fusarium* spp., *Pythium* sp. | Wettable powder | Drench dip or spray |
| *A. radiobacter* | Nogall, Diegall 16 | *A. tumefaciens* | Washed plates, culture suspension | Root dips |
| *P. solanacearum* (nonpathogenic) | PSSOL 12 | *P. solanacearum* | | |
| *P. fluorescens* NCIB 12089 | Victus | *R. solani*, *Fusarium* spp. | Aqueous suspension | Spray |
| *B. subtilis* | System3 | *S. rolfsii*, *S. sclerotiorum* | Dust | Seed treatment in planter box |
| *B. cepacia*-type Wisconsin M36 | Blue Circle | *Fusarium*, *Pythium*, spiral, lesion, lance, and sting nematode | Peat carrier or liquid | Seed treatment or drip irrigation |

*A. radiobacter* is commercially available under various trade names such as Nogall, Galltrol-A, Diegall, and Norbac 84C. Bacteria are generally suspended in non-chlorinated water and applied as dips and sprays to cuttings and stems or as soil drench. *Pseudomonas syringae* formulated as wettable powders is commercially available under trade names Bio-Save 10 and Bio-Save 11 and used for postharvest application to citrus and pome fruit for management of *Botrytis*, *Mucor*, *Penicillium*, and *Geotrichum*.

*P. fluorescens* strains isolated from rhizosphere of various crops with antagonistic potential against *Fusarium* spp. were formulated as talc-based and peat-based products. In talc-based and peat-based formulations, *P. fluorescens* survived for a maximum of 240 days. When formulation was applied to the chick pea seed, the total shelf life of the bacteria was recorded at 180 days (Vidhyasekaran and Muthamilan 1995). Peat-based formulation of *P. fluorescens* strains PfALR2 was developed and assessed for root treatment, seed treatment, foliar spraying, and soil application. All four treatments in combination resulted in the significant control of sheath blight in greenhouse condition (Rabindran and Vidhyasekaran 1996). *P. fluorescens* strain PF-1, isolated from rhizosphere of maize roots, showed antagonistic potential against *R. solani* f. sp. *sasakii* causing banded leaf and sheath blight of maize. Among the different carriers, talc and peat maintained the population at $18.3 \times 10^7$ and $19.5 \times 10^7$ cfu $g^{-1}$ of the bacterium, respectively, after 40 days. Significant control of disease was recorded after the seed treatment with peat-based formulation (Sivakumar et al. 2000). Formulations of *B. subtilis* AF 1 showed both plant growth promotion and biocontrol potential when prepared in peat. The formulation was supplemented with *A. niger* mycelium, 0.5 % chitin, and organic compost from cultivation of *Agaricus bisporus*. Biocontrol potential of formulated products was evaluated against two pathogens on groundnut and pigeon pea. Chitin, *A. niger* mycelium, and *A. bisporus* compost were used as supplement for improving the growth rate of *B. subtilis* AF 1A. Peat formulation supplemented with chitin when used for seed treatment demonstrated better control of wilt in pigeon pea caused by *Fusarium udum* and *A. niger* responsible for crown rot disease in groundnut (Manjula and Podile 2001). *P. fluorescens* strains FP7 and PF1 and their consortium were formulated as talc-based product and mixed both with and without chitin were assessed individually against sheath blight of rice. Significant reduction in disease incidence was recorded after the application of formulated product through seed, soil, root, and foliar spray. In field as much as 62.1 % reduction in sheath blight incidence was observed in the consortium treatment containing chitin (Commare et al. 2002). 0.1 % calcium hydroxide significantly promoted the growth of *B. amyloliquefaciens* strain B190 used against *Botrytis elliptica* in lily. Spraying *B. amyloliquefaciens* B190 mixed with 0.05 % sodium carbonate, 0.025 % calcium hydroxide, or 0.025 % ammonium nitrate suppressed the gray mold on lily. Concentration of adjuvant was kept below 0.1 % (v/v); carboxymethyl cellulose (CMC) and Tween 80 were effective to *B. amyloliquefaciens* B190 formulation against lily gray mold (Chiou and Wu 2003). The efficacy of talc-based formulations of *T. viride* and *P. fluorescens* alone and in combination on sheath blight disease, crop growth, and yield in rice was studied in field experiments. Application of formulated product of *P. fluorescens* and *T. viride* either alone or in combination resulted in increase in root and shoot lengths and plant height when compared with control. Significant reduction in sheath blight incidence was also recorded after application of *P. fluorescens* and *T. viride* (Mathivanan et al. 2005).

*Bacillus licheniformis* strain N1 exhibiting the biocontrol activity against *Botrytis cinerea* was formulated using fermentation of the bacterial culture in Biji medium. Wettable powder formulation of antagonist based on olive oil corn and starch was selected for evaluation of the disease control. A dose of 100-fold-diluted *B. licheniformis* N1E was found to be the optimum spray formulation and significantly reduced the disease. 90.5 % reduction in disease by formulated product was recorded in comparison to the 77 % reduction by synthetic fungicide including carbendazim and diethofencarb. Results of this study indicated that the olive oil- and corn starch-based formulation of *B. licheniformis* using liquid fermentation will be effective against tomato gray mold (Lee et al. 2006). Nineteen isolates of antagonistic *Pseudomonas* and twelve isolates of yeast were screened for the biocontrol activity against *Colletotrichum musae* causing banana anthracnose. *P. fluorescens* strain FP7 showed a maximum inhibition of *C. musae* mycelial growth. Water-in-oil formulation of *P. fluorescens* FP7 was formulated by adding various oils such as rice bran (28.50 %), coconut (28.50 %), and castor (28.50 %) separately to the bacterial culture and bacterial populations were reported to survive for 210 days of storage. The application of water-in-oil formulation of bacterium significantly reduced the disease incidence (Peeran et al. 2014). *B. subtilis* strain BY-2 was unable to colonize the leaf surface and stem in oilseed rape when applied as pellet. Populations of BY-2 declined from $10^8$ CFU seed$^{-1}$ to $10^4$ CFU g root$^{-1}$ and $\leq 10^{23}$ CFU g stem$^{-1}$ after 60 days. Significant reduction in disease was observed when compared to control (Hu et al. 2014). Efficacy of aqueous suspension (Serenade ASO or QRD 145) and foliar sprays of wettable powder formulation (Serenade MAX or QRD 141) of *B. subtilis* QST 713 alone and in combination with copper hydroxide was investigated against bacterial spot disease of tomato. The aqueous suspension of *B. subtilis* QST 713 alone significantly reduced bacterial spot on tomato foliage when compared to control. The wettable powder of *B. subtilis* QST

713 alone did not reduce bacterial spot, but in combination with copper hydroxide it reduced disease severity and enhanced the total fruit yield (Abbasi and Weselowski 2015).

## 2.5 Seed Treatment

The application of microbes to seed surface requires few technical considerations. Significant amount of the inoculums must survive the application procedure and must have the capacity to grow in vicinity of seed. Since seeds are at low moisture levels for most of the time during storage, microorganisms must have the ability to survive under low water activity. Microorganisms might likewise need to be mixed with other active ingredients, for example, insecticides and fungicides. These aspects raise issues of formulation stability and strain selection. Seed treatment with beneficial microorganisms has been a prime area of investigation for many years (Bisen et al. 2015). Microorganisms with various properties have been applied to seeds to perform various functions, including plant growth promotion, nitrogen fixation, phosphate solubilization, and biological control of plant pathogens (Keswani et al. 2013b). BCAs are applied to the seeds for protection of seed and seedlings from effects of various seed-borne and soilborne plant pathogens. For the successful biological control, applied microorganism must grow and colonize the rhizosphere in order to protect the plant, thus the release of microbes from the formulation and prerequisite condition for growth are of paramount importance.

The commercial wettable powder product Mycostop (Kemira AgroOy, Finland), containing *Streptomyces griseoviridis* strain K61, with $10^8$ cfug$^{-1}$ is available for seed treatment. The powder is used at 5–8 g kg$^{-1}$ seeds for control of seed-borne and soilborne fungal phytopathogens in herbs, vegetables, and ornamentals (Tomlin 1994). The shelf life of formulation is estimated to be 6 months at 8 °C or for 12 months at −12 °C and must be stored in airtight containers. Seed inoculation with bacteria to protect the seed from soilborne pathogens and promote plant has been well investigated (Merriman et al. 1974). Kodiak, a *B. subtilis*-based biopesticide, was used as seed inoculant for peanut, cotton, and beans to control the root diseases caused by *Fusarium* and *Rhizoctonia* (Mahaffee and Backman 1993). Quantum-4000, a *B. subtilis* strain A13-based product, is commercially available as seed inoculants for peanut, and another biopesticide based on strain GB 07 named Epic is available for cotton. Bacterial BCAs actively colonize the rhizosphere and compete with other microorganisms including pathogens (Bisen et al. 2015). Nonspore-forming bacterium *Enterobacter cloacae* showed antagonistic activity against *Pythium* spp. causing seed rot in cotton seeds. Simple seed inoculation technique with (CMC) as sticker in cotton seeds was carried out successfully (Nelson 1988). Liquid culture of *E. cloacae* was applied by solid matrix priming to tomato and cucumber seeds in combination with a fungicide (Harman and Taylor 1988).

Generally, in seed treatment, microbial formulation is applied to seed as powder or liquid. For example, *P. fluorescens* and *Burkholderia cepacia* were applied to pea seeds with or without captan for control of *Aphanomyces* root rot and *Pythium* damping-off (Parke et al. 1991). Potato tubers have been treated with ascospores of *Talaromyces flavus* in a pyrophyllite carrier (Fravel et al. 1985). In order to ensure the better adhesion of bioinoculants on seed surface, stickers are added to microbial formulations for seed treatment. To control *Pythium ultimum* attack in pea and soybean seeds, PelGel has been used to treat the seed with *P. putida* (Paulitz et al. 1992). In cucumber seed treatment, PelGel has also been used with *T. harzianum* (Harman 1991; Taylor et al. 1991). Another sticker Polynox-N-I0 was used for seed treatment with shale and then applied on bean seeds with conidia of *T. harzianum*, and efficient disease control was observed. Various concentrations of plant gum, methyl cellulose, and xanthane gum with talc have been used for treatment of potato tuber with PGPR (Kloepper and Schroth 1981). At 40 °C, population of rhizobacteria in talc and 20 % xanthane gum did not decline for 2 months and resulted in better increase in plant development in field potato. However, rhizobacteria did not last longer in formulation containing gum tragacanth or gum karaya. CMC was applied with microbial agents to seeds to control *R. solani*. One percent CMC with clay carrier was applied to *Verticillium biguttatum* and other biocontrol agents to potato tubers (Jager and Velvis 1985). Various antagonists have been applied to treat the sugar beet seeds using either gum xanthan or methylcellulose in combination with a neutralized talc or peat carrier (Suslow and Schroth 1982). Similarly, to control take-all disease in wheat, surface-disinfested seeds were coated with *P. fluorescens* in combination with 1 % methylcellulose (Weller and Cook 1983). In seed treatment of Chinese aster with *T. flavus*, a polymer binder was used in quartz flour to pellet seeds (Nagtzaam and Bollen 1994), and antagonist *T. flavus* was isolated from seeds after 17 years. Strains of *E. cloacae* and *Trichoderma* were delivered to cucumber and tomato seeds through solid matrix priming (Harman and Taylor 1988). During priming seeds are brought to a certain moisture level just below the required level for germination and then mixed with moistened *Trichoderma* or *E. cloacae*, shale, sphagnum moss, or bituminous coal. The seeds and carriers are then mixed with water and incubated before planting.

## 2.6 Carriers and Adjuvant Used in Microbial Formulations

Variations in effectiveness of microbial formulations from beneficial microorganisms are credited to three main causes: (1) presence of the microorganism prior to inoculation in field, (2) poor survival of the organism in the environment, and (3) low quality of the microbial product itself. Success of biopesticides depends on the delivery of viable, active microorganisms in high numbers to the field which requires high-quality inoculants. The carrier substrate is the most critical part of the microbial formulation, and it must be capable of supporting high numbers of microbe. Carriers are inert ingredients, and they do not have biocontrol potential; however, they can affect the efficacy of the product and shelf life of microorganism (Table 2.5). Better survival of *Pseudomonas* spp. on minerals with small particle size, such as zeolite, montmorillonite, and vermiculite, than on minerals with larger particle size such as talc, pyrophyllite, and kaolinite was observed during storage at 20 °C (Dandurand et al. 1994). Backman and Rodriguez-Kabana (1975) compared diatomaceous earth and attapulgus clay for their various physical properties including water-holding capacity and strength after autoclaving. Attapulgus clay granules swelled in water and lost their integrity after autoclaving, whereas, diatomaceous earth granules did not swell in water and remained intact after autoclaving thus making it to absorb a molasses-based medium for delivery of *T. harzianum* to soil. Wheat bran-perlite mixture and poplar bark compost have been used as carrier for nonpathogenic *Fusarium* strains mixed with soil that induced resistance to *F. oxysporum* f. sp. *dianthi* (Garibaldi et al. 1987). Fine-ground tree barks have also

**Table 2.5** Various adjuvants used in microbial formulations

| Adjuvant | Type | Function |
|---|---|---|
| Oils | Mineral oils | Improve uptake, photostability |
| | Crop oil | |
| Surfactant | Wetting agents | Improve spreading, wetting, or dispersion |
| | Spreaders | |
| | Penetrants | |
| Stabilizing agents | Emulsifiers | Maintain stability during application |
| | Dispersants | |
| | Anti-flocculating agents | |
| Solvents | Cosolvents | Maintain AI in solution |
| | Coupling agents | |
| Hygroscopic agents | Humectants | Prevent premature drying of deposit |

been used as a carrier (Stack et al. 1988). Peat is generally used as carrier for *Rhizobium* spp. and is also useful for soil applications and seed coating of biocontrol agents. Huber et al. (1989) used fine-ground peat with a methylcellulose sticker for wheat seed treatment with bacteria for control of *Gaeumannomyces graminis*. Alder bark has been used as a carrier to apply *T. flavus* to potato seed pieces (Keinath et al. 1990).

One of the mechanisms involved in biocontrol is the production of hydrolytic enzymes by the antagonistic microorganism. The amount and type of nutrients in the formulation must allow ample production of hydrolytic enzymes (Stack et al. 1988). Increased molar concentrations of carbon and nitrogen sources (0.02–0.18 M maltose and 0.006–0.024 M arginine) and increased carbon/nitrogen ratios (12:1–80:1) enhanced the proliferation of *Thielaviopsis basicola* and *Trichoderma* spp. on lignite granules (Stack et al. 1987). Likewise, *T. flavus* was formulated with eight different organic carriers and used against *Verticillium dahliae* on eggplant, and a maximum inhibition of pathogen was recorded in treatments with the highest carbon/nitrogen ratios (159:1 for pyrophyllite, 97:1 for corn cobs) (Fravel et al. 1985). As new information on mechanism of biocontrol revealed, it may be possible to express desirable biocontrol traits by manipulating nutrient composition of formulations. For example, biocontrol potential of *G. virens* GL-21 depends on the form of nitrogen in the formulation. Alginate prill of *G. virens* with wheat bran as a carrier resulted in significant control of *Sclerotium rolfsii* in comparison to vermiculite plus wheat bran (Ristaino et al. 1994).

## 2.7 Basic Information Required for Microbial Product Registration

Certain information about the various parameters is prerequisite for the biopesticide registration. These parameters include microorganism strain specifications, cfu count of microorganism used, target microorganism, moisture content of the product, type of formulation, and technical bulletin/product profile. Strain specifications include information on genus and species, rhizosphere competence, biological control potential, growth promotion potential, and growth parameters like pH and temperature.

### 2.7.1 Strain Specifications

Selection of potential antagonist strain under field and lab conditions ensures the effective and consistent performance of bioagent in field. Screening of effective strain can be done in different ways: selection of potential strain in relation to phytopathogens, screening of isolates with high biotechnological application, or search for economical viable substrates which are suitable for mass production of bioagent (Singh et al. 2003, 2006).

### 2.7.2 Shelf Life and Storage

For commercialization of a microbial product in the market, it is essential that it has long shelf life and can be stored at room temperature. It has been suggested that shelf life of a minimum 18 months is acceptable for commercial microorganism-based product. Storage at room temperature is an essential condition as the farmers cannot afford the equipment to keep the product at any temperature. According to the recent guideline of the Central Insecticide Board and Registration Committee, the data required for claiming 1-year shelf life of the product is for 15 months for talc-based formulation, i.e., the microbe should remain viable for 15 months.

### 2.7.3 CFU Count

According to the Central Insecticide Board and Registration Committee guideline, colony-forming unit (cfu) count should not be less than $2 \times 10^6$ spores $ml^{-1}$ or $g^{-1}$ on selective media (SM) for antagonistic fungi, and CFU count on selective medium should be a minimum of $1 \times 10^8$ $ml^{-1}$ or $g^{-1}$ for antagonistic bacteria (Singh 2012).

### 2.7.4 Pathogenic Contamination

The pathogenic contaminants such as *Salmonella*, *Shigella*, and *Vibrio* should not be present. Other microbial contaminants must not to exceed $1 \times 10^4$ counts $ml^{-1}$or $g^{-1}$.

### 2.7.5 Moisture Content

Maximum moisture content of the product should not exceed more than 8 % for dry formulation of fungi and 12 % for bacteria (www.cibrc.nic.in/2.1.22011.doc).

## 2.8 Constraints in the Production of Microorganism-Based Biopesticides

The main reasons for slow growth of microbial-based biopesticide industry include inconsistent performance of final product in field condition, short shelf life of microorganism in formulation, possibilities of contamination with other plant and human pathogens, lack of suitable application technology, small market size, and lack of proper knowledge about the biopesticides in farmers. Research should be focused on the development of superior formulation to protect the reliability of the product, because a single failure will jeopardize the whole trade's reputation. Production of biopesticides is a long process which includes selection of suitable strain for formulation, mass production of selected strain, screening of microorganism for suitable carrier for formulation, assessment of the shelf life of microorganism in selected formulation, and product efficacy in field condition. Several reports on contamination and low population of microorganism in biopesticides being sold in the market were registered (Singh et al. 2004a; Alam 2000; Arora et al. 2010). Due to low microbial count, it is obvious that their performance in the field is inconsistent and poor. The unpredictable seasonal nature of the existing demand needs capable storage for biopesticides. The storage requires sophisticated and special facilities which are generally not available to most producers, sellers, and farmers. Shelf life is a result of combining several factors including production technology, material used as carrier and packaging, and transport. The mass production of significant numbers of viable, efficient, and stable propagules of the microorganism is a prerequisite in biopesticide development (Singh et al. 2004b). Submerged fermentation system has been traditionally followed by the producers over the solid substrate fermentation because of its cost-effectiveness and easy technology (Churchill 1982; Stowell 1991). However, new large-scale production systems are required for bacteria and fungi that do not like to produce spores in liquid media. Unfortunately fermentation systems for mass production based on solid substrates are not frequently available (Connick et al. 1990). Submerged fermentation methods are generally well adopted for the mass production of secondary metabolites, antibiotics, organic acids, and bacteria; however, it is not suitable for the production of viable filamentous fungi. Therefore, selection of the cost-effective fermentation technology for the mass production viable and efficient propagules is a matter of concern. Choice of carriers and adjuvant used in the formulation is another technical problem in the development of stable and effective biopesticides. One of the major goals in formulating biopesticides is to maintain the viability and effectiveness of the active ingredient for a possible duration, preferably 2 years. After the production of microorganism, the main challenge faced by producers in formulation development is the shelf life of microorganism during the storage period. If the product carries less numbers of active ingredients due to shorter shelf life, the overall performance of the formulation will be affected.

## 2.9 Future Prospects

The necessity for more safe products for plant disease management prompts an inclination of microbial biopesticide formulations with efficient antagonism and good stability. Biopesticides give eco-friendly alternatives to synthetic pesticides, yet they confront various difficulties in their production, formulation, and application. Since biopesticide generally contains live organism, maximum care is needed to maintain the microbial population and efficacy from beginning to the end use. Study of their formulation and production could enormously help in the commercialization of biopesticides. It appears to be that biopesticides will have a more extensive use in the future as their application techniques enhance as less expensive inert materials are recognized for different formulations. Introduction of new adjuvants showed significant increase in activity of microbes and proposed a new area of research. Selection of proper formulation may enhance the product stability, larger shelf life, and performance of microbes in field conditions. Biopesticides offer a more balanced plant protection product application, and in the future formulation products should have more balance between production cost and efficiency (El–Sayed 2005; Rao et al. 2007; Glare et al. 2012; Khater 2012). Development related to the formulation type would possibly shift from dusts to granules, from suspension concentrates and wettable powders to water-dispersible granules, and from single microorganism-based product to microbial consortium-based formulation. With the advances in nanotechnology science, different new microbial formulations such as nanosuspension, nanoemulsion, and nanocapsule suspension with superior efficiency will be released in the market (Rao et al. 2007; Ghormade et al. 2011; Glare et al. 2012). Significant advancement has been made in the production of new formulation products and application methods; however, there is still much work to be done. For further research to improve production and application techniques, scientist and researchers are likely to provide safe and effective products for plant disease management.

Eco-friendly and safer biopesticides play an important role in modern agriculture; however, their major drawbacks have led to the use of nanotechnology in agriculture (Mishra and Singh 2015). Silver nanoparticles (AgNPs) are the most frequently used metallic nanoparticles in various sector including agriculture (Jo et al. 2009; Kim et al. 2012; Mishra et al. 2014). Many workers have reported the antimicrobial activity of AgNPs against a vast range of phytopathogens. A nanosized silica-silver particle formulation was developed and showed significant antimicrobial activity against a wide range of phytopathogens including *Colletotrichum* sp., *Pythium* sp., *P. syringae*, *Xanthomonas compestris*, etc. (Park et al. 2006). Biosynthesized AgNPs of bacterium *Serratia* sp. BHU-S4 showed significant antifungal activity against *Bipolaris sorokiniana* (Mishra et al. 2014).

Recovering the real connections of AgNPs with agroecosystems including soil, soil biota, and plants and their poisonous quality level can be infer whether AgNP application could be helpful for agroecosystems. Considering the elements deciding destiny, transport, portability, and poisonous quality of AgNPs in soil, it is additionally accepted that the size of the poisonous quality of these particles could really be evaded by controlling them and henceforth requires more profound exploration. Specifically, research concentrating on this methodology utilizing biosynthesized AgNPs ought to get more thoughtfulness regarding an efficient comprehension of how incorporating biosynthesized AgNPs in agrarian applications contributes toward the farmer's benefits. One important point is that biosynthesized AgNPs have demonstrated their worth for agricultural applications by performing two noteworthy undertakings, viz., plant development upgrade and plant malady management. Hence, now it is high time when future studies must be coordinated toward upgrading the utility of biosynthesized AgNPs in regular environments keeping in mind the end goal to predict their future agricultural extension.

## 2.10 Conclusion

In recent past, disease management strategies have been inclined to much safer alternatives due to concerns over hazardous effects of chemical pesticides on human and plant health. Biological control of plant pathogens employing living microorganisms offers such safe alternative to the chemicals. In order to improve the efficiency and shelf life of biocontrol agents, various formulations based on solid and liquid carriers have been developed. Maximum care is required in biopesticide production and formulation as it contains live organisms. It is quite a confirmation that biopesticides will have a more extensive use and share greater market space in the future. Specific procedures and technologies have been recently developed that would significantly affect biopesticide formulations. Selection of abiotic stress tolerance or rhizospheric competence could allow a wider range of applications (Keswani 2015). Moreover, fungicide tolerance screening would promote integrated management with reduced chemical inputs.

**Acknowledgments** HBS is grateful to the Department of Biotechnology (BT/PR5990/AGR/5/587/2012), New Delhi, for providing financial support. KB is grateful to BHU for financial assistance. BKS and CK are thankful to ICAR-AMAAS (Project No. P27/131) for financial support.

## References

Abbasi PA, Weselowski B (2015) Efficacy of *Bacillus subtilis* QST 713 formulations, copper hydroxide, and their tank mixes on bacterial spot of tomato. Crop Prot 74:70–76

Alam G (2000) A study of biopesticides and biofertilizers in Haryana, India, Gatekeeper series no. 93. IIED, London

Arora NK, Khare E, Maheshwari DK (2010) Plant growth promoting rhizobacteria: constraints in bioformulation, commercialization, and future strategies. In: Maheshwari DK (ed) Plant growth and health promoting bacteria. Springer, Berlin, pp 97–116

Backman PA, Rodriguez–Kabana R (1975) A system for the growth and delivery of biological control agents to the soil. Phytopathology 65:819–821

Batta YA (2004) Postharvest biological control of apple gray mold by *Trichoderma harzianum* Rifai formulated in an invert emulsion. Crop Prot 23:19–26

Bisen K, Keswani C, Mishra S, Saxena A, Rakshit A, Singh HB (2015) Unrealized potential of seed biopriming for versatile agriculture. In: Rakshit A, Singh HB, Sen A (eds) Nutrient use efficiency: from basics to advances. Springer, New Delhi, pp 193–206

Chiou AL, Wu WS (2003) Formulation of *Bacillus amyloliquefaciens* B190 for control of lily grey mould (*Botrytis elliptica*). J Phytopathol 151:13–18

Churchill BW (1982) Mass production of microorganisms for biological control. In: Charudattan R, Walker HL (eds) Biological control of weeds with plant pathogens. Wiley, New York, pp 139–156

Colyer PD, Mount MS (1984) Bacterization of potatoes with *Pseudomonas putida* and its influence on post–-harvest soft rot diseases. Plant Dis 68:703–706

Commare RR, Nandakumar R, Kandan A, Suresh S, Bharathi M, Raguchander T, Samiyappan R (2002) *Pseudomonas fluorescens* based bio–formulation for the management of sheath blight disease and leaf folder insect in rice. Crop Prot 21:671–677

Connick WJ Jr (1988) Formulation of living biological control agents with alginate. In: Cross, Scher HB (eds) Pesticide formulations: innovations and developments, vol 371, ACS symposium series. American Chemical Society, Washington, DC, pp 241–250

Connick WJ Jr, Lewis JA, Quimby PC (1990) Formulation of biocontrol agents for use in plant pathology. In: Baker RR, Dunn PE (eds) New directions in biological control: alternatives for suppressing agricultural pests and diseases. Alan R. Liss, New York, pp 345–372

Crozier J, Arroyo C, Morales H, Melnick RL, Strem MD, Vinyard BT, Bailey BA (2015) The influence of formulation on *Trichoderma* biological activity and frosty pod rot management in *Theobroma cacao*. Plant Pathol 64(6):1385–1395

Daigle DJ, Cotty PJ (1995) Formulating atoxigenic *Aspergillus flavus* for field release. Biocontrol Sci Technol 5:175–184

Dandurand LM, Morra MJ, Chaverra MH, Orser CS (1994) Survival of *Pseudomonas* spp. in air dried mineral powders. Soil Biol Biochem 26:1423–1430

DeLucca AJ, Connick WJ II, Fravel DR Jr (1990) Use of bacterial alginates to prepare biocontrol formulations. J Indust Microbiol 6:129–134

El–Sayed W (2005) Biological control of weeds with pathogens: current status and future trends. J Plant Dis Prot 112:209–221

Escande AR, Laich FS, Pedraza MV (2002) Field testing of honeybee-dispersed *Trichoderma* spp. to manage sunflower head rot (*Sclerotinia sclerotiorum*). Plant Pathol 51:346–351

Fravel DR, Marois JJ, Lumsden RD, Connick WJ (1985) Encapsulation of potential biocontrol agents in alginate. Phytopathology 75:774–777

Garibaldi A, Brunatti F, Gullino ML (1987) Evaluation of several antagonists and different methods of application against fusarium wilt of carnation. OEPP/EPPO Bull 17:625–629

Ghormade V, Deshpande MV, Paknikar KM (2011) Perspectives for nano–biotechnology enabled protection and nutrition of plants. Biotechnol Adv 29:792–803

Glare T, Caradus J, Gelernter W, Jackson T, Keyhani N, Köhl J, Stewart A (2012) Have biopesticides come of age. Trend Biotechnol 30:250–258

Harman GE (1991) Seed treatments for biological control of plant disease. Crop Prot 10:166–171

Harman GE, Taylor AG (1988) Improved seedling performance by integration of biological control agents at favourable pH levels with solid matrix priming. Phytopathology 78:520–525

Hu X, Roberts DP, Xie L, Maul JE, Yu C, Li Y, Liao X (2014) Formulations of *Bacillus subtilis* BY–2 suppress *Sclerotinia sclerotiorum* on oilseed rape in the field. Biol Control 70:54–64

Huber DM, El–Nasshar H, Moore LW (1989) Interaction between a peat carrier and bacterial seed treatments evaluated for biological control of the take–all disease of wheat (*Triticum aestivum* L.). Biol Fert Soils 8:166–171

Jacobsen BJ, Backman A (1993) Biological and cultural plant disease controls: alternatives and supplements to chemicals in IPM systems. Plant Dis (USA) 77:311–315

Jager G, Velvis H (1985) Biological control of *Rhizoctonia solani* on potatoes by antagonists. 4. Inoculation of seed tubers with *Verticillium biguttatum* and other antagonists in field experiments. Neth J Plant Pathol 91:49–63

Jayaraj J, Radhakrishnan NV, Velazhahan R (2006) Development of formulations of *Trichoderma harzianum* strain M1 for control of damping–off of tomato caused by *Pythium aphanidermatum*. Arch Phytopathol Plant Protec 39:1–8

Jo YK, Kim BH, Jung G (2009) Antifungal activity of silver ions and nanoparticles on phytopathogenic fungi. Plant Dis 93:1037–1043

Jones KA, Burges HD (1997) Product stability: from experimental preparation to commercial reality. In: Evans HF (ed) Microbial insecticides: novelty or necessity? Symposium proceedings no. 68. British Crop Protection Council, Farnham, pp 163–171

Jones RW, Pettit RE, Taber RA (1984) Lignite and silage: carrier and substitute for application of fungal biocontrol agents to soil. Phytopathology 74:1167–1170

Keinath AP, Fravel DR, Papavizas GC (1990) Evaluation of formulations of *Talaromyces flavus* for biocontrol of *Verticillium* wilt of potato. Biol CultTests 7:27

Keswani C (2015) Ecofriendly management of plant diseases by biosynthesized secondary metabolites of *Trichoderma* spp. J Brief Idea. doi:10.5281/zenodo.15571

Keswani C, Singh SP, Singh HB (2013a) *Beauveria bassiana*: status, mode of action, applications and safety issues. Biotech Today 3:16–20

Keswani C, Singh SP, Singh HB (2013b) A superstar in biocontrol enterprise: *Trichoderma* spp. Biotech Today 3:27–30

Keswani C, Mishra S, Sarma BK, Singh SP, Singh HB (2014) Unraveling the efficient application of secondary metabolites of various *Trichoderma*. Appl Microbiol Biotechnol 98:533–544

Khater HF (2012) Prospects of botanical biopesticides in insect pest management. J Appl Pharm Sci 2:244–259

Kim SW, Jung JH, Lamsal K, Kim YS, Min JS, Lee YS (2012) Antifungal effects of silver nanoparticles (AgNPs) against various plant pathogenic fungi. Mycobiology 40:53–58

Kloepper JW, Schroth MN (1981) Development of a powder formulation of rhizobacteria for inoculation of potato seed pieces. Phytopathology 71:590–592

Knowles A (2005) New developments in crop protection product formulation, Agrow reports. T and F Informa UK Ltd, London, pp 153–156

Knowles A (2006) Adjuvants and additives, Agrow reports. T&F Informa UK Ltd, London, pp 126–129

Lee JP, Lee SW, Kim CS, Son JH, Song JH, Lee KY, Moon BJ (2006) Evaluation of formulations of *Bacillus licheniformis* for the biological control of tomato gray mold caused by *Botrytis cinerea*. Biol Control 37:329–337

Lewis JA, Lumsden RD (2001) Biocontrol of damping–off of greenhouse–grown crops caused by *Rhizoctonia solani* with a formulation of *Trichoderma* spp. Crop Prot 20:49–56

Lewis JA, Papavizas GC (2007) Characteristics of alginate pellets formulated with *Trichoderma* and *Gliocladium* and their effect on the proliferation of the fungi in soil. Plant Pathol 34:571–577

Lim TK, Rohrbach KG (1980) Role of *Penicillium funiculosum* strains in the development of pineapple fruit diseases. Phytopathology 70:663–665

Lisansky SG (1985) Production and commercialization of pathogens. In: Scopes N, Hussey NW (eds) Biological pest control. Blanford Press, Poole, pp 210–218

Mahaffee WF, Backman PA (1993) Effect of seed factors on spermosphere and rhizosphere colonization of cotton by *Bacillus subtilis* GB03. Phytopathology 83:1120–1125

Manjula K, Podile AR (2001) Chitin–supplemented formulations improve biocontrol and plant growth promoting efficiency of *Bacillus subtilis* AF 1. Can J Microbiol 47:618–625

Mathivanan N, Prabavathy VR, Vijayanandraj VR (2005) Application of talc formulations of *Pseudomonas fluorescens* Migula and *Trichoderma viride* Pers. ex SF Gray decrease the sheath blight disease and enhance the plant growth and yield in rice. J Phytopathol 153:697–701

Mbarga JB, Begoude BAD, Ambang Z, Meboma M, Kuate J, Schiffers B, Ten Hoopen GM (2014) A new oil–based formulation of *Trichoderma asperellum* for the biological control of cacao black pod disease caused by *Phytophthora megakarya*. Biol Control 77:15–22

McLean KL, Swaminathan J, Frampton CM, Hunt JS, Ridgway HJ, Stewart A (2005) Effect of formulation on the rhizosphere competence and biocontrol ability of *Trichoderma atroviride* C52. Plant Pathol 54:212–218

Merriman PR, Price RD, Kollmorgen JF (1974) Effect of seed inoculation with *Bacillus subtilis* and *Streptomyces griseus* on the growth of cereals and carrots. Aust J Agric Res 25:219–226

Mishra S, Singh HB (2015) Biosynthesized silver nanoparticles as a nanoweapon against phytopathogens: exploring their scope and potential in agriculture. App Microbiol Biotechnol 99:1097–1107

Mishra S, Singh BR, Singh A, Keswani C, Naqvi AH, Singh HB (2014) Biofabricated silver nanoparticles act as a strong fungicide against *Bipolaris sorokiniana* causing spot blotch disease in wheat. PLoS One 9:97881

Mishra S, Singh A, Keswani C, Saxena A, Sarma BK, Singh HB (2015) Harnessing plant–microbe interaction for enhanced protection against phytopathogens. In: Arora NK (ed) Plant microbes symbiosis: applied facets. Springer, New Delhi, pp 111–125

NAAS (2013) Biopesticides – quality assurance. Policy paper No. 62. National Academy of Agricultural Sciences, New Delhi, p 20

Nagtzaam MPM, Bollen GJ (1994) Long shelf life of *Talaromyces flavus* in coating material of pelleted seed. Eur J Plant Pathol 100:279–282

Nelson EB (1988) Biological control of *Pythium* seed rot and pre emergence damping–off of cotton with *Enterobacter cloacae* and *Erwinia herbicola* applied as seed treatments. Phytopathology 72:140–142

Park HJ, Kim SH, Kim SJ, Choi SH (2006) A new composition of nanosized silica–silver for control of various plant diseases. Plant Pathol J 22:295–302

Parke JL, Rand RE, Joy AE, King EB (1991) Biological control of *Pythium* damping–off and *Aphanomyces* root rot of peas by application of *Pseudomonas cepacia* or *P. fluorescens* to seed. Plant Dis 75:987–992

Patanjali PK, Raza SK (2013) Recent advances in pesticide formulation technology. In: Abstract of international conference on emerging issues on health effects of pesticides residues in food and environment– ultimate challenges and research opportunities, Food and Drug Toxicological Research Institute, Hydrabad, 12–13 December 2013

Paulitz TC, Anas O, Fernando DG (1992) Biological control of *Pythium* damping–off by seed treatment with *Pseudomonas putida*: relationship with ethanol production by pea and soybean seeds. Biocontrol Sci Technol 2:193–201

Peeran FM, Krishnan N, Thangamani PR, Gandhi K, Thiruvengadam R, Kuppusamy P (2014) Development and evaluation of water–in–oil formulation of *Pseudomonas fluorescens* (FP7) against *Colletotrichum musae* incident of anthracnose disease in banana. Eur J Plant Pathol 138:167–180

Pusey PL, Wilson CL (1984) Postharvest biological control of stone fruit brown rot by *Bacillus subtilis*. Plant Dis 68:753–756

Rabindran R, Vidhyasekaran P (1996) Development of a formulation of Pseudomonas *fluorescens* PFALR2 for management of rice sheath blight. Crop Prot 15:15–721

Rao GVR, Rupela OP, Rao VR, Reddy YVR (2007) Role of biopesticides in crop protection: present status and future prospects. Ind J Plant Prot 35:1–9

Ristaino JB, Lewis JA, Lumsden RD (1994) Influence of *Gliocladium virens* and delivery systems on biological control of southern blight on carrot and tomato in the field. Plant Dis 78:153–156

Singh HB (2006) *Trichoderma*: a boon for biopesticides industry. J Mycol Plant Pathol 36:373–384

Singh HB (2012) Regulatory requirements, quality control, commercialization and IPR related issues by bio–pesticides. In: Singh HP, Chowdappa P, Chakraborty, Podile AR (eds) Molecular approaches for plant fungal disease management. Westville, New Delhi, pp 228–224

Singh HB, Singh A, Nautiyal CS (2003) Commercialization of biocontrol agents: problems and prospects. In: Rao GP, Manohachari C, Bhat DJ, Rajak RC, Lakhnapal TN (eds) Frontiers of fungal diversity in India. International Book Distributing Company, Lucknow, pp 847–861

Singh HB, Singh A, Singh SP, Nautiyal CS (2004a) Commercialization of biocontrol agents: the necessity and its impact on agriculture. In: Singh SP, Singh HB (eds) Eco–agriculture with bioaugmentation: an emerging concept. Rohitashwa printers, Lucknow, pp 1–20

Singh HB, Singh S, Singh A, Nautiyal CS (2004b) Mass production, formulation and delivery system of fungal and bacterial antagonistic organisms in India. In: Singh SP, Singh HB (eds) Eco–agriculture with bioaugmentation: an emerging concept. Rohitashwa printers, Lucknow, pp 53–69

Singh HB, Ahmad A, Srivastav S, Nautiyal CS (2006) A synergistic composition useful as bioinoculant (PCT No. WO 2007/11068A2)

Singh HB, Singh BN, Singh SP, Sarma BK (2012) Exploring different avenues of *Trichoderma* as a potent bio–fungicidal and plant growth promoting candidate–an overview. Rev Plant Pathol 5:315–426

Singh HB, Singh A, Sarma BK, Upadhyay DN (2014) *Trichoderma viride* 2% WP (Strain No. BHU–2953) formulation suppresses tomato wilt caused by *Fusarium oxysporum* f. sp. *lycopersici* and chilli damping–off caused by *Pythium aphanidermatum* effectively under different agroclimatic conditions. Int J Agri Environ Biotechnol 7:313–320

Sivakumar G, Sharma RC, Rai SN (2000) Biocontrol of banded leaf and sheath blight of maize by peat based *Pseudomonas fluorescens* formulation. Indian Phytopathol 53:190–192

Stack JP, Kenerley CM, Pettit RE (1987) Influence of carbon and nitrogen sources, relative carbon, and nitrogen concentrations and soil moisture on the growth in nonsterile soil of soilborne fungal antagonists. Can J Microbiol 33:626–631

Stack JP, Kenerley CM, Pettit RE (1988) Application of biocontrol agents, in biocontrol of plant diseases. CRCPress, Boca Raton, pp 43–54

Stowell LJ (1991) Submerged fermentation of biological herbicides. In: TeBeest D (ed) Microbial control of weeds. Chapman and Hall, New York, pp 225–261

Suslow T, Schroth MN (1982) Rhizobacteria of sugar beets: effects of seed application and root colonization on yield. Phytopathology 72:199–206

Tadros F (2005) Applied surfactants, principles and applications. Wiley–VCH Verlag GmbH and Co. KGA, Weinheim, pp 187–256

Taylor AG, Min TG, Harman GE, Jin X (1991) Liquid coating formulation for the application of biological seed treatments of *Trichoderma harzianum*. Biol Control 1:16–22

Thangavelu R, Palaniswami A, Velazhahan R (2004) Mass production of *Trichoderma harzianum* for managing fusarium wilt of banana. Agri Eco Environ 103:259–263

Tomlin C (ed) (1994) The pesticide manual. British Crop Protection Council/Royal Society of Chemistry, Farnham

Tronsmo A, Dennis C (1983) The use of *Trichoderma* species to control strawberry fruit rots. Neth J Plant Pathol 83:449–455

Vidhyasekaran P, Muthamilan M (1995) Development of formulations of *Pseudomonas fluorescens* for control of chickpea wilt. Plant Dis 79:782–786

Weller DM, Cook RJ (1983) Suppression of take–all of wheat by seed treatments with fluorescent *Pseudomonads*. Phytopathology 73:463–469

Wilson CL, Pusey PL (1985) Potential for biological control of post harvest diseases. Plant Dis 69:375–378

# Beneficial Microorganisms: Current Challenge to Increase Crop Performance

3

Márcia do Vale Barreto Figueiredo, Aurenivia Bonifacio, Artenisa Cerqueira Rodrigues, Fabio Fernando de Araujo, and Newton Pereira Stamford

## Contents



**Abstract**

The major goal of agricultural microbiology is a comprehensive analysis of beneficial microorganisms. Fundamental knowledge of the ecology and evolution of interactions could enable the development of microbe-based sustainable agriculture. Plant growth-promoting bacteria (PGPB) have gained worldwide importance and acceptance for their agricultural benefits. This is due to the emerging demand to reduce dependence on synthetic chemical products within a holistic vision of developing and focalizing environmental protection. Beneficial microorganisms also help to solubilize mineral phosphates and other nutrients, enhance resistance to stress, stabilize soil aggregates, improve soil structure and organic matter content, and inhibit phytopathogens. Several efforts have been made in research to clarify definitions as well as develop commercial inoculants using these organisms, with a special emphasis on formulations that interact synergistically and are currently being devised. In addition, numerous recent studies indicate increased crop performance with the use of these commercial inoculants. In this chapter, the progress to date in the use of beneficial microbes for agricultural applications is summarized and discussed.

M.d.V.B. Figueiredo (✉)
National Research and Technological Development, Agronomical Institute of Pernambuco (IPA/SEAGRI), 1371, Gen. San Martin Avenue, Recife, 50761-000 PE, Brazil
e-mail: mbarreto@elogica.com.br

A. Bonifacio • A.C. Rodrigues
Federal University of Piaui (UFPI), Teresina, PI, Brazil
e-mail: bonifacio.a@live.com; artenisacerqueira@hotmail.com

F.F. de Araujo
Faculty of Agrarian Sciences, University of West Paulista (UNOESTE), Presidente Prudente, SP, Brazil
e-mail: fabio@unoeste.br

N.P. Stamford
National Research and Technological Development, Department of Agronomy, Federal Agricultural University of Pernambuco (UFRPE), Recife, PE, Brazil
e-mail: newtonps@depa.ufrpe.br


N.K. Arora et al. (eds.), *Bioformulations: for Sustainable Agriculture*,
DOI 10.1007/978-81-322-2779-3_3

## 3.1 Introduction

Agricultural practices are the major factors underlying many environmental problems, and thus, contemporary agriculture faces enormous challenges (Hazell and Wood 2008). All around the world, agriculture is frequently produced using intensive production methods and indiscriminate agrochemicals that increase production costs and contribute to increased environmental pollution that significantly affects human and animal health (Javaid 2010). Therefore, alternatives to chemical-based conventional agriculture that increase a farmer's profit and that are environmentally safe are required (Singh et al. 2011).

Considering that agricultural productivity is strongly related to microbial activity in the soil system (Chaparro et al. 2014), the use of beneficial microorganisms makes a positive contribution to environmentally safe agriculture (Kloepper et al. 1980; Chanway 1998; Vessey 2003; Gray and Smith 2005; Figueiredo et al. 2012). In addition, the presence of these microorganisms in the soil is a known characteristic of pathogen-suppressive soils (Fliessbach et al. 2009). In biological studies, crustacean chitosan is frequently used to increase plant resistance against pathogens (Goy et al. 2009). On the other hand, fungal chitosan has antimicrobial properties and increases the availability of nutrients (Franco et al. 2011; Berger et al. 2013).

The modes of action of the microorganisms and their various benefits to plants, as described in the literature, range from the simple occupation of biological empty spaces to ecological relationships such as antibiosis, competition, predation, and symbiosis, among others (Kilian et al. 2000; Kloepper et al. 2004; Avis et al. 2008). In addition, activities related to the production of hormones and enzymes, that are important for plants, occur in the soil or phylloplane (Araujo et al. 2005; Raaijmakers et al. 2009). The use of selected microorganisms may represent an important biotechnological approach to decrease the deleterious effects of stress in crops (Egamberdieva et al. 2013; Nadeem et al. 2014).

Studies have also shown that the growth-promoting ability of some bacteria may be highly specific to a certain plant species, cultivar, or genotype (Bashan 1998; Figueiredo et al. 2010). Many of these microorganisms synthesize extracellular polysaccharides or exopolysaccharides (EPS) with a commercially significant applicability (Nwodo et al. 2012). The formulation step is a crucial aspect of producing microbial inoculants, and it determines the success of a biological agent (Brahmaprakash and Sahu 2012).

In recent years, the strong potential of biopolymers to be used as inoculants has been studied (Rodrigues 2012; Bashan et al. 2014). Another possibility for the development of new inoculants or biofertilizers is the use of biofilm (Seneviratne et al. 2009). Furthermore, the role of these compounds in stress adaptation may be an important criterion for the selection of inoculant strains to increase plant productivity through biological nitrogen fixation (BNF) under different soil and climatic conditions (Bomfeti et al. 2011; Sharmila et al. 2014).

## 3.2 Combinations of Beneficial Microorganisms: Keys for Effective Use in Agriculture

The rhizosphere microbiome consists of three types of microbe groups (Figueiredo et al. 2012; Chaparro et al. 2014): microorganisms that are beneficial to plant growth, plant pathogenic microorganisms, and human pathogenic bacteria. Beneficial microorganisms can fix atmospheric nitrogen, decompose organic wastes and residues, detoxify pesticides, suppress plant diseases and soilborne pathogens, improve nutrient cycling, and produce bioactive compounds that stimulate plant growth (Singh et al. 2011). Thus, the use of beneficial microorganisms present in the rhizosphere is crucial for the development and health of plants and is advantageous to soil fertility (Chaparro et al. 2014). The group of microbes that is beneficial to plants consists of a wide variety of microorganisms, such as nitrogen-fixing bacteria, plant growth-promoting bacteria (PGPB), and endo- and ectomycorrhizal fungi (Mendes et al. 2013).

Nitrogen-fixing bacteria constitute a microorganism group that carries out BNF, an important process in ecosystems that supplies nitrogen to plants in a utilizable form (Peix et al. 2015). These microorganisms are also called diazotrophic bacteria and may act as free-living bacteria or form a symbiosis with legumes and establish root nodules where BNF occurs (Bonfante and Anca 2009; Chaparro et al. 2014; Peix et al. 2015). Rhizobia, a group of associative diazotrophic bacteria, are often used in combination with other microbes, either bacteria or fungi, to improve the growth and yield of many plant species (Figueiredo et al. 2008, 2010; Hungria et al. 2013; Rodrigues et al. 2013a).

Various bacteria distributed across different genera, such as *Azospirillum*, *Azotobacter*, *Bacillus*, *Burkholderia*, *Enterobacter*, *Klebsiella*, *Pseudomonas*, *Serratia*, and *Variovorax*, are collectively designated as PGPB due to their noticeable positive effects on plant metabolism (Nadeem et al. 2014). PGPB are commonly used as plant biostimulants (Bhattacharyya and Jha 2012) and increase proliferation and elongation of roots, facilitate nutrient acquisition, modulate plant hormone levels, and act as biocontrol bacteria (Ali et al. 2014; Bashan et al. 2014). Considering these positive effects, PGPB are commonly used to improve crop yields and in combination with rhizobia may constitute an alternative to increase crop plant performance under normal and stress conditions (Figueiredo et al. 2010; Bhattacharyya and Jha 2012; Rodrigues et al. 2013b; Bashan et al. 2014).

Under stress situations, plants trigger defense mechanisms that involve gene expression that is linked with hormonal balance between roots and shoots (Munns and Tester 2008). In this sense, PGPB may provide beneficial effects for plant drought tolerance due to modulation stimulated by hormones, mainly abscisic acid, ethylene, and cytokinins (Bhattacharyya and Jha 2012; Ali et al. 2014). The inoculation of wheat with the PGPB *Azospirillum lipoferum* alleviated plant drought stress and increased the growth and yield (Arzanesh et al. 2011). Drought situations may trigger oxidative stress in plants and induce a decline in nitrogen fixation rates (Arzanesh et al. 2011; Rodrigues et al. 2013b; Larrainzar et al. 2014). To cope with detrimental oxidative situations, a triple inoculation with rhizobia, *Paenibacillus graminis* and *P. durus*, was used in cowpea, which improved symbiotic performance and BNF as well as decreased the deleterious effects of the oxidative stress (Rodrigues et al. 2013a).

The use of selected PGPB may represent an important biotechnological approach to decrease the deleterious effects of salt stress in crops (Egamberdieva et al. 2013; Nadeem et al. 2014). Approximately 45 million hectares of irrigated land worldwide are affected by salinity, which represents a limiting factor to plant growth and reduces crop productivity (Munns and Tester 2008). Under salt stress, the PGPB *Bacillus amyloliquefaciens* promotes the growth of rice plants (Nautiyal et al. 2013). Moreover, positive effects on plant biomass were also observed in maize inoculated with the PGPB *Azotobacter chroococcum* when cultivated under saline stress (Rojas-Tapias et al. 2012). Additionally, the combined action of rhizobia and PGPB *Pseudomonas* spp. alleviated deleterious symptoms of salt stress in *Galega officinalis* and improved root and shoot growth (Egamberdieva et al. 2013).

Among PGPB, *Pseudomonas*, *Bacillus*, and *Azospirillum* are the most studied genera, and of these, *Azospirillum* is considered the most important due to the strong improvement in plant growth (Bashan et al. 2014). *Azospirillum* are free-living PGPB that are able to colonize root surfaces and often possess endophytic ability (Okon and Labandera-Gonzalez 1994). Moreover, these PGPB are capable of positively affecting the yield of many plants growing in different soils and regions, and co-inoculation with other microorganisms improves their beneficial effects (Mendes et al. 2013; Bashan et al. 2014). The co-inoculation of *Azospirillum* with phosphate-solubilizing bacteria allows an improvement in absorption of mineral nutrients

(Singh et al. 2011). Additionally, co-inoculation of *Azospirillum* and rhizobia induces positive responses in plant growth (Star et al. 2012; Hungria et al. 2013).

The rhizobia-PGPB symbiosis is often utilized to improve plant nutrition in terms of macro- and micronutrients. The presence of the microorganisms in the soil can efficiently affect the solubility and therefore the availability of several nutrients (Mendes et al. 2013); this is termed biofertilization (Adesemoye and Kloepper 2009). The use of a biofertilizer containing beneficial microbes, i.e., a mix of *A. chroococcum*, *Bacillus megaterium*, and *Bacillus mucilaginosus*, promoted the growth of maize and ameliorated the soil properties (Wu et al. 2005). In common bean (*Phaseolus vulgaris*), dual and triple inoculations with *Rhizobium*, *Bacillus subtilis*, and *B. megaterium* significantly increased the uptake of macro- and micronutrients (Elkoca et al. 2010). The co-inoculation of rhizobia with phosphate-solubilizing microbes, such as *Aspergillus*, *Pseudomonas*, and *Trichoderma*, increased the nodulation and availability of phosphorus to plants (Yadav et al. 2013).

*Trichoderma* are filamentous fungi that are widely used in agriculture as biocontrol agents due to their antagonistic abilities against phytopathogenic fungi, mainly *Fusarium* sp. and *Rhizoctonia* sp. (Brotman et al. 2013). Inoculation with *Trichoderma* induces systemic resistance to diseases in many crop plants that results in beneficial effects on plant growth (Hermosa et al. 2012). In Brazil, ten products containing *Trichoderma* are registered and commercially available to control various types of crop diseases. *Trichoderma* strains are frequently applied singly to many plants and induce resistance to biotic and abiotic stresses (Brotman et al. 2013; Colla et al. 2014). Nevertheless, the combination of *Trichoderma* with other microbes potentiates their beneficial effects, as shown by Colla et al. (2014) who studied many vegetable crops that were simultaneously co-inoculated with *T. atroviride* and *Glomus intraradices*, an arbuscular mycorrhizal fungi (AMF).

The term "mycorrhiza" (pl. mycorrhizae) is used to indicate nonpathogenic symbiotic associations between soil fungi and plant roots (Bonfante and Genre 2008). Mycorrhizal symbiosis plays important ecological roles that include the ability to increase nutrient uptake and protect plants from numerous stress situations (Figueiredo et al. 2012). Mycorrhizae are divided into ecto- and endomycorrhizae, and in the latter, the hyphae penetrate the root cells and establish an intracellular symbiosis (Bonfante and Anca 2009). Among the endomycorrhizae, AMF have great importance in tropical ecosystems and are the fungi that are most frequently found in the roots of crop plants (Bonfante and Genre 2008). AMF are a group of soil-dwelling fungi that are obligatory symbionts belonging to the *Glomeromycota* phylum; these fungi have a substantial influence on plant productivity (Redecker et al. 2013).

The positive effects of AMF colonization are commonly attributed to an improvement in nutrient uptake, mainly phosphorus, and to an increased pathogen resistance and herbivore tolerance in plants (Bonfante and Genre 2008; Smith and Smith 2012). Other advantageous effects of AMF-plant interactions are the increase in plant tolerance to stressful environments, phytoremediation promotion, and improvement of soil stability (Meier et al. 2012; Smith and Smith 2012; Estrada et al. 2013). Some rhizobacteria favor the formation of mycorrhizae on roots and are commonly referred to as "mycorrhiza helper bacteria" (MHB) (Frey-Klett et al. 2007). These MHB increase the spore germination and mycelial growth of AMF and are involved in the establishment of AMF-plant symbiosis. The MHB group includes nitrogen-fixing and PGPB (Table 3.1).

MHB aid AMF establishment in roots and the associations between AMF and MHB induce various positive aspects of plant development, mainly increases in the tolerance against drought and salt stress in plants (Gamalero et al. 2009; Lingua et al. 2013). For example, the co-inoculation with a mixture of AMF from the genera *Glomus*, *Gigaspora*, and *Acaulospora* and the rhizobia *Sinorhizobium terangae* resulted in a positive osmotic adjustment and an improved tolerance of *Acacia saligna* to saline soil

**Table 3.1** Examples of the mycorrhiza helper bacteria (MHB) associated with arbuscular mycorrhizal fungi (AMF) identified in different host plants (Frey-Klett et al. 2007; Soliman et al. 2012)

| Genres and groups of the MHB | | Species of the AMF | Host plant |
|---|---|---|---|
| Gram-negative *Proteobacteria* | *Agrobacterium*, *Alcaligenes*, *Azospirillum*, *Azotobacter*, *Burkholderia*, *Bradyrhizobium*, *Enterobacter*, *Klebsiella*, *Pseudomonas*, *Rhizobium*, and *Sinorhizobium* | *Endogone* sp., *Acaulospora* sp., *Gigaspora margarita*, *Glomus caledonium*, *G. clarum*, *G. deserticola*, *G. fasciculatum*, *G. fistulosum*, *G. intraradices*, and *G. mosseae* | *Medicago sativa*, *Morus alba*, *Carica papaya*, *Zea mays*, *Solanum tuberosum*, *Allium cepa*, *Anthyllis cytisoides*, *Hordeum vulgare*, *Triticum aestivum*, *Trifolium* sp., *Acacia saligna*, *Cucumis sativum*, *Pennisetum americanum*, *Ipomea batatas*, *Uniola paniculata*, and *Licopersicum esculetum* |
| Gram-positive *Firmicutes* | *Bacillus*, *Brevibacillus*, and *Paenibacillus* | *G. etunicatum*, *G. fasciculatum*, *G. mosseae*, and *G. intraradices* | *Sorghum bicolor* |
| Gram-positive actinomycetes | *Streptomyces* | *Glomus intraradices* | *Sorghum* sp. |

(Soliman et al. 2012). In cowpea co-inoculated with rhizobia, AMF *Glomus etunicatum* and the PGPB *Paenibacillus brasiliensis*, it was observed that AMF positively affected the nitrogen acquisition by rhizobia and increased the symbiotic efficiency, which enhanced plant growth (Lima et al. 2011). For all the above mentioned reasons, the use of beneficial microbes must be increased and disseminated considering that this is a sustainable and environmentally friendly agricultural practice.

## 3.3 Prospective Biocontrol Agents of Plant Diseases

Beneficial microorganisms have been widely used in agriculture to control diseases and pests of different plant species, in addition to other benefits they provide to plants. Within this group, we can highlight the bacteria of the genera *Bacillus*, *Azospirillum*, *Pseudomonas*, and *Rhizobium* and fungi of the genera *Beauveria*, *Gliocladium*, *Metarhizium*, and *Trichoderma*, as well as actinomycetes of the genus *Streptomyces*.

Within the microorganism groups in the soil, rhizobacteria can be highlighted, which develop a close relationship with plants and in most cases occupy the space known as the rhizosphere (Raaijmakers et al. 2009). In this environment, which is rich in sugars, amino acids, flavonoids, proteins, and fatty acids (Badri et al. 2009), there is a higher fungal and bacterial presence because these microorganisms are attracted by these compounds. However, there occurs a greater prevalence of rhizobacteria, which in most cases are able to form a biofilm that protects the root surface against pathogens (Bogino et al. 2013).

The maintenance of these beneficial microorganisms is supported by the host plant; the microorganisms drain carbon-rich compounds (exudates) produced by the roots that are essential for microbial nutrition in the rhizosphere (Vandenkoornhuyse et al. 2007). However, exudates may also act as antimicrobial molecules, inhibiting the growth of some species. This indicates that the host plant can select or attract microorganisms for their rhizosphere colonization (Chaparro et al. 2012), which has led to the occurrence of plant species that respond more to the presence of growth-promoting microorganisms in the soil.

Some practices, such as tillage, organic fertilization, crop rotation, and residue management, can also increase or decrease microbial activity in the rhizosphere (Raaijmakers et al. 2009). Chemical factors, such as pH and organic matter nutrient availability, are considered important factors for maintaining microbial activity (Chaparro et al. 2012).

The presence of beneficial microorganisms in the soil is a known characteristic of pathogen-suppressive soils. These soils can maintain conditions favorable for diverse microfauna, preventing the prevalence of pathogens. It has also been found that soils rich in microbial biomass and diversity have been less responsive to microbial inoculant introduction, indicating that this introduction is more suitable for poor soils in this regard (Fliessbach et al. 2009).

Recent studies indicate that plants that interact with rhizobacteria are able to increase their photosynthetic capacity (Xie et al. 2009), soil salinity tolerance (Dimkpa et al. 2009), disease suppression, and iron absorption efficiency (Zhang et al. 2009). These results confirm the potential use of inoculants containing rhizobacteria in agriculture. Plants play a critical role in influencing the microbial community composition of the rhizosphere. Some species can attract larger amounts of microorganisms by releasing signaling substances in the rhizosphere. It has also been observed that depending on the plant growth stage, the signaling substances can influence the microbial community structure present in the rhizosphere (Chaparro et al. 2014).

The multiple beneficial effects related to rhizosphere microorganism activities are well explained by the diagram presented in Fig. 3.1. In this representation, it can be seen that microorganisms considered to possess biological control activity can play indirect roles in growth promotion, such as increasing the supply of macro- and micronutrients to plants. Moreover, it has been found that nitrogen-fixing species, such as *Rhizobium* sp., can play a role as biological control agents (Avis et al. 2008).

The main plant growth promotion forms are based on three important actions, i.e., phytostimulation, biofertilization, and pathogen control. Some microbial species exhibit good development of these three activities in the soil, indicating good potential for the market. Some of these species and the previously mentioned activities described in the literature are presented in Table 3.2.

In the case of *B. subtilis*, pathogen control is performed by different modes of action,

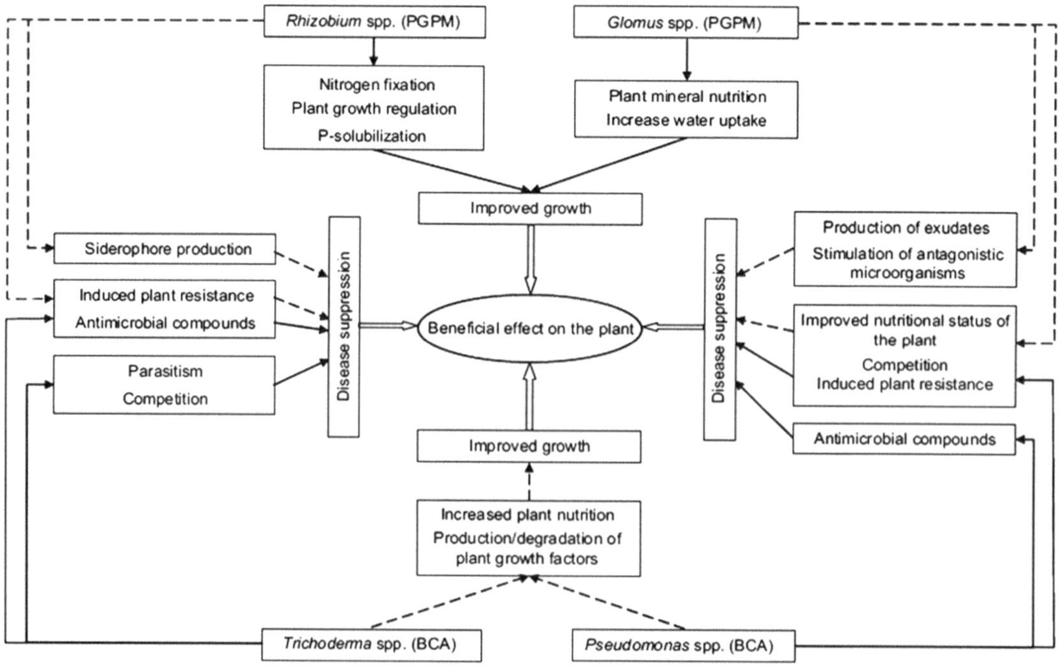


**Fig. 3.1** Potential modes of action of plant growth-promoting and biological control agents (Avis et al. 2008)

**Table 3.2** Simultaneous actions developed by different species of soil microorganisms within the beneficial activities for plants

| Species | Biofertilization | Phytostimulation | Pathogen control |
|---|---|---|---|
| *Azospirillum brasilense*[a] | Nitrogen fixation | Auxins and gibberellins | Induced systemic resistance (ISR) and siderophores |
| *B. subtilis*[b] | Increase in phosphorus and nitrogen | Auxins | Antibiotics |
| *Pseudomonas fluorescens*[c] | Increase in phosphorus | Auxins | 2,4-diacetyl-phloroglucinol |
| *Streptomyces griseus*[d] | Increase in phosphorus | Auxins | Chitinases |
| *Trichoderma harzianum*[e] | Increase in phosphorus and nitrogen | Auxins | Mycoparasitism and enzymes |

[a]Dobbelare et al. (1999)
[b]Araujo et al. (2005) and Araujo (2008)
[c]Raaijmakers et al. (1997), Fernández et al. (2012), and Pallai et al. (2012)
[d]Hamdali et al. (2008)
[e]Contreras-Cornejo et al. (2009), Molla et al. (2012), and Vos et al. (2014)

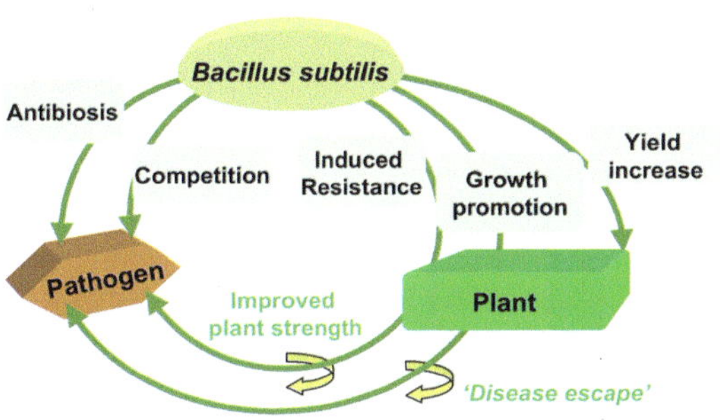


**Fig. 3.2** Modes of action of *B. subtilis* and the interaction between *Bacillus*-plant pathogen (Kilian et al. 2000)

especially antibiosis, competition for ecological niches, and resistance induction in the host (Fig. 3.2). These main modes, in addition to some secondary modes, provide effective control of pathogens and reduce the possibility of the pathogens developing resistance. These secondary modes of action reinforce the immense potential that bioinoculants have in agriculture. The applications of different mixtures of rhizobacterial species to the soil can result in systemic resistance induction, which increases the efficiency against various pathogens (Ramamoorthy et al. 2001). Taking into account both the primary and secondary benefits related to the microorganisms mentioned above (Table 3.2), the addition of multiple microorganisms to agricultural production systems appears attractive (Avis et al. 2008). Many of the individual effects of co-inoculated microorganisms have been enhanced, resulting in synergistic effects in many cases (Kohler et al. 2007).

At present, the use of bacterial consortia in bioinoculant formulation has been highlighted with good prospects for the future, which may reflect multiple benefits in crop yield and the biological balance of the soil (Pindi 2012). This line of action of combining different microbial species with multiple functions into one product was recently used for the development of so-called biofilms (Triveni et al. 2012). In this case, there is the possibility of greater microorganism protection and survival with the use of the final product. The formulation of biofilms typically involves the combination of fungi and bacteria in a product developed in the laboratory. However, the analysis of this new technology is limited in recent studies that have examined the use and development of new inoculants.

Currently, more than 78 brands of products with a microbiological origin are being used in agriculture, covering more than 5 million ha. Some of these products are in the registration process or are marketed only to the organic agricultural market. It is also noted that most of these products are related to pest control. Worldwide, several commercial products developed using rhizobacteria are on the market and are mainly indicated for plant disease control (Table 3.3).

**Table 3.3** Commercial products developed using different PGPR species

| Species | Products | Intended crop |
|---|---|---|
| *B. subtilis* | Epic, Kodiak, Rhizo-Plus, Serenade, and Subtilex | Fruit, cereals, ornamentals, trees, and forage crops |
| *P. fluorescens* | BlightBan, Conquer, and Victus | Apple, cherry, peach, potato, strawberry, and tomato |
| *T. harzianum* | Custom and RootShield | Turf and ornamentals |
| *A. brasilense* | Azo Green | Forage crops |
| *S. griseus* | Mycostop | Ornamental and vegetable crops |

The prospects for the future market are very promising due to a greater receptivity of organic products by farmers and also due to higher environmental and health restrictions on the registration and authorization of conventional chemicals used for disease control in plants. In this context, organic products are used more often in integrated control strategies.

## 3.4 Microbial Polysaccharides: Production and Application as a Vehicle for Inoculation

The term "microbial polysaccharides" includes polysaccharides produced by fungi and bacteria, which may be located inside the cell, on the cell wall or yet be released out of the cell (Donot et al. 2012). Polysaccharides are classified by the sugar composition in homopolysaccharides, which enclose a single type of monosaccharide and are usually neutral glucan, and heteropolysaccharides, which are composed of different sugar residues and usually display a regular backbone structure with a linear or branched repetitive unit (Delbarre-Ladrat et al. 2014). Heteropolysaccharides may also contain organic or inorganic substituents, such as phosphate, sulfate, acetate, and pyruvate, which confer polyanionic characteristics to these molecules (Schmid et al. 2011). Collectively, the sugar sequence in the molecule, the presence or absence of substituents, and how the chains intertwine influence the physicochemical features of polysaccharides (Rehm 2010).

Some bacteria and fungi secrete polysaccharides as an evolutionary adaptation to help them adhere to different surfaces or for adaptation to stressful environments (Sharmila et al. 2014). Polysaccharides protect against dehydration, serve as barriers to prevent viruses and antibodies from binding to specific sites on the cell wall, neutralize toxins, act as a carbon source, and also interact with plant cells in specific pathogenic or symbiotic relationships (Badel et al. 2011; Donot et al. 2012). In addition, microbial polysaccharides, especially those released to the external environment, have been reported to protect microbial cells against heavy metals or environmental stresses (Poli et al. 2011). For example, *Pseudomonas* strains survive under stress conditions due to polysaccharides released to the outside of the cell, which protect them from water stress and then regulate the diffusion of carbon sources in the microbial environment (Sandhya et al. 2009).

In addition to singular biological characteristics, polysaccharides exhibit chemical and physical properties, and these peculiar features represent advantages compared with plant polysaccharides for their use at large scales (Mahapatra and Banerjee 2013). Polysaccharides are generally nontoxic, biodegradable compounds (Rehm 2010) and are oftentimes named according to the biopolymers or gums. In contrast to traditional gum, microbial polysaccharide production is advantageous because it is more rapid and requires less space for manufacture (Badel et al. 2011). Among the polysaccharide types, the extracellular polysaccharides or EPS are the most interesting due the possibility of recovering these directly from the environment in which they are secreted, enabling higher productivity. Fungi and bacteria release EPS in extracellular media, and these EPS are useful to promote cell-cell recognition (Donot et al. 2012).

**Table 3.4** Some extracellular polysaccharides (EPS) produced by fungi or bacteria with commercial and economic importance (Freitas et al. 2011; Delbarre-Ladrat et al. 2014)

| EPS | Microorganism | Application |
|---|---|---|
| *Homopolysaccharides* | | |
| Curdlan | *Agrobacterium*, *Rhizobium*, and *Alcaligenes* | Food additive, applied in pharmaceutical industries, bioremediation |
| Dextran | *Leuconostoc* and *Streptococcus* | Food industry, biomedical as plasma volume expander, biotechnological support for separation |
| Pullulan | *Aureobasidium pullulans* | Food, adhesive, and cosmetic additives; flocculant; thickener; viscosity stabilizer in preparation of nontoxic, biodegradable, edible plastic materials; health care |
| Scleroglucan | Fungi of the genus *Sclerotium* | Oil, ink, cosmetic, and pharmaceutical industries; animal feed; food additive |
| *Heteropolysaccharides* | | |
| Alginate | *Pseudomonas aeruginosa* and *Azotobacter vinelandii* | Food hydrocolloid, wound care, drug encapsulating agent |
| Gellan | *Sphingomonas paucimobilis* and *Pseudomonas elodea* | Gelling in culture medium, food additive |
| Succinoglycan | *Alcaligenes*, *Agrobacterium*, and *Rhizobium* | Food and pharmaceutical industries, oil recovery |
| Xanthan | *Xanthomonas campestris* | Viscosifying and texturizing agent in various foods, used for oil recovery in petroleum industry, health care |

Many microorganisms synthesize EPS with commercially significant applications (Nwodo et al. 2012). Biopolymers derived from natural resources, such as microbial EPS, have a competitive advantage due their sustainable production, biodegradability, and, often, biocompatibility (Rehm 2010). Due to their ability to alter the rheology of aqueous solutions and their potential use as a biomaterial, EPS possess great economic importance (Table 3.4) and wide applicability in various industry sectors with significant commercial value (Freitas et al. 2011; Finore et al. 2014). EPS show huge production in a short time as well as ease of isolation and purification in relation to intracellular and cell wall polysaccharides (Delbarre-Ladrat et al. 2014). They can function as thickeners, gelling, emulsifiers, stabilizers, lubricants, or as suspending agents and film formers (Freitas et al. 2011; Mahapatra and Banerjee 2013).

Although they exhibit a wide range of biological functions, the production of EPS by fungi remains poorly studied (Giavasis 2014). The EPS produced by fungi are frequently highly hygroscopic β-glucans (Mahapatra and Banerjee 2013; Sharmila et al. 2014) that possess numerous applications in the food and pharmaceutical industries. Basidiomycetes, filamentous fungi, and yeasts from different ecological niches are known as being capable of synthesizing EPS in different production systems (Delbarre-Ladrat et al. 2014). The production of fungal EPS depends on the fungal strain used and the physicochemical conditions applied inside the fermentative broth (Mahapatra and Banerjee 2013). Collectively, the oxygen levels, carbon and nitrogen sources, pH, and temperature influence the production and composition of fungal EPS (Badel et al. 2011; Mahapatra and Banerjee 2013; Zhang et al. 2013; Giavasis 2014).

Scleroglucan, schizophyllan, and pullulan are types of fungal EPS commonly produced at a large scale and that possess singular rheological properties. Scleroglucan is an extracellular glucan excreted by *Sclerotium glucanicum* or *S. rolfsii* that is used frequently in petroleum recovery and has great potential for use in the agricultural, food, and pharmaceutical industries (Ansari et al. 2012). Schizophyllan is an EPS secreted by the basidiomycete *Schizophyllum commune* that is used in many fields, such as the food industry and pharmaceuticals (Zhang et al. 2013). Pullulan is a water-soluble glucan gum commonly used as a food additive and is

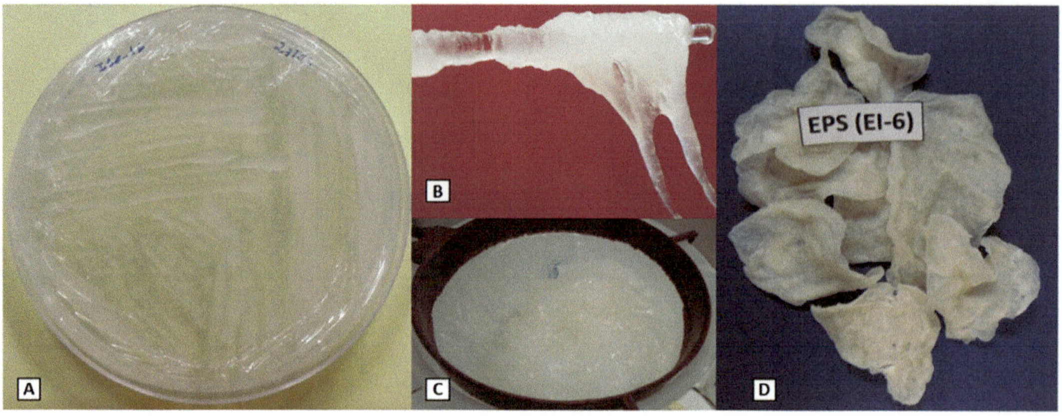


**Fig. 3.3** Exopolysaccharides (EPS) produced by the EI-6 strain of *R. tropici*: (**a**) culture medium showing the EPS production, (**b**) EPS recovered from the fermentation broth, (**c**) EPS washed with 70 % alcohol, (**d**) dry EPS after 3 days in the stove at 30 °C (Photos courtesy of Artenisa Cerqueira Rodrigues)

produced by *Aureobasidium pullulans* (Singh et al. 2008). This compound has adhesive properties and the ability to form fibers and thin biodegradable films that are transparent and waterproof to oxygen because of their physical properties and distinctive linkage patterns (Cheng et al. 2011).

Researchers have extensively studied bacterial EPS due their structural variability that offers a large range of physicochemical and biological properties (Badel et al. 2011; Poli et al. 2011). Numerous bacterial EPS were reported in recent decades as biopolymers with industrial importance and significant commercial value, particularly xanthan, which occupies a prominent place in the market by presenting very different and unusual rheological properties (Freitas et al. 2011). Many soil bacteria from different habitats are recognized to produce complex and diverse EPS, such as *Rhizobium*, *Bradyrhizobium*, *Mesorhizobium*, and *Sinorhizobium* strains (Albareda et al. 2008; Staudt et al. 2012; Castellane et al. 2014). In a biological context, bacterial EPS are necessary to mediate *Rhizobium*-legume symbioses and are indispensable to promote the formation of efficient nodules on different hosts (Bomfeti et al. 2011; Castellane et al. 2014).

The viability of rhizobia in the field is maintained by their production of EPS (Staudt et al. 2012), and EPS can protect nitrogenase enzyme involved in the nitrogen-fixing process from high oxygen concentration in the nodules (Vu et al. 2009). Rhizobia are among the well-known EPS producers and can excrete large amounts of these polysaccharides in the rhizosphere (Albareda et al. 2008; Serrato et al. 2008; Bomfeti et al. 2011; Nwodo et al. 2012; Castellane et al. 2014) or culture medium (Fig. 3.3). Rhizobial inoculants have been used for many years to obtain greater yields in legumes and are often produced with peat, a material with a fossil origin (Rivera et al. 2014). Peat mines are situated in preserved environments where extraction is forbidden, and a lack of natural peat deposits exists in some countries (Albareda et al. 2008). The use of peat as a bacterial carrier, although established, has decreased, and alternative materials are being sought.

Alternative materials for bacterial inoculation can maintain the quality and efficiency of the inoculum and reduce production costs and negative environmental impacts compared with peat inoculants (Fernandes Júnior et al. 2009, 2012; Herrmann and Lesueur 2013; Rivera et al. 2014). For example, the combination of carboxymethylcellulose (CMC) and starch forms a polymeric carrier that adequately maintains rhizobial cell viability and has the same performance as peat inoculants (Fernandes Júnior et al. 2009, 2012). Rivera et al. (2014) tested eight polymers to use

**Table 3.5** Number of nodules and amount of nitrogen accumulated in the shoots of cowpea plants inoculated with strain BR 3267 of *Bradyrhizobium* sp. using different inoculants (carriers): peat, a combination of the carboxymethylcellulose (CMC) and starch, and EPS produced by the EI-6 strain of *R. tropici*

| Carrier | Number of nodules (nodule $plant^{-1}$) | Nitrogen in shoot dry mater (mg N $SDM^{-1}$) |
|---|---|---|
| Peat[a] | 78.0 | 78.7 |
| Combination of CMC and starch[a] | 77.0 | 92.0 |
| EPS from by *R. tropici* (EI-6)[b] | 94.0 | 145.2 |

[a]Fernandes Júnior et al. (2009)
[b]Rodrigues (2012)

as vehicles for bacterial inoculation, and among them, sodium alginate and hydroxypropyl methylcellulose (HPMC) were able to provide a higher viability for strain G58 of *Rhizobium* sp. in comparison to the control (peat), without affecting the physiological process of nodulation in cowpea roots.

CMC and HPMC are competent bacterial inoculants; however, these compounds are semi-synthetic polymers (Fernandes Júnior et al. 2012; Rivera et al. 2014). Because of their biological nature, EPS synthesized by rhizobia are currently considered as excellent carriers for inoculation (Table 3.5). EPS can encapsulate living bacterial cells and gradually release these to the soil in large quantities, which confirms that EPS are degraded by soil microorganisms (Albareda et al. 2008). *Rhizobium leguminosarum*, *R. meliloti*, and *R. tropici* (SEMIA 4080 and EI-6 strains) are known to produce significant amounts of rhizobial EPS (Rodrigues 2012; Rodrigues et al. 2013a, b; Castellane et al. 2014). The composition of the EPS produced by *R. tropici* SEMIA 4080 shows relatively higher contents of glucose and galactose combined with small amounts of mannose, rhamnose, and glucuronic and galacturonic acids (Castellane et al. 2014).

The EI-6 strain of *R. tropici* was first isolated from root nodules of cowpea (*Vigna unguiculata*) cultivated in an arid environment (Figueiredo et al. 1999; Oliveira et al. 2012) and is known to produce significant amounts of EPS with good quality (Xavier 2009; Santos 2010; Oliveira 2011; Rodrigues 2012). The EPS produced by the EI-6 strain of *R. tropici* (already shown in Fig. 3.3.) are a polysaccharide of glucose and galactose without the presence of uronic acids (Rodrigues 2012). Additionally, the EPS possess acetyl and pyruvate residues and a higher amount of sodium and potassium ions (Xavier 2009; Rodrigues 2012). This feature resulted in the EPS produced by *R. tropici* EI-6 being classified as polyanionic heteropolysaccharides (Rodrigues et al. 2013a). Currently, after intensive studies on *R. tropici*, it is possible to confirm that their EPS have high potential for use in agriculture.

## 3.5 Biofertilizers: Agricultural Innovations

Soluble fertilizers are of great importance for plant growth and production, but their use by low-income farmers is limited due to the high price. Furthermore, the fertilizers contain high amounts of soluble nutrients and may easily undergo lixiviation in the deeper soil layers and promote plant damage through soil and water contamination (Van Straaten 2007). In modern and sustainable agriculture, the application of slowly soluble fertilizers is required to economically increase food production and soil fertility while minimizing environmental damage (Stamford et al. 2008). Biofertilizers produced from phosphate and potash rocks mixed with elemental sulfur and inoculated with sulfur-oxidative bacteria (*Acidithiobacillus*) may be an alternative for effective and economic fertilization and have considerable influence on the availability of elements derived from rocks through the metabolic effect of sulfuric acid (Stamford et al. 2008), as shown in the equation below:

$$S + H_2O + 1.5\ O_2 \xrightarrow{Acidithiobacillus} H_2SO_4$$

Nitrogen (N) is one of the most important nutrients for plant growth and yield, and due to its role in some chemical compounds such as proteins, nucleic acids, and many other components, it is necessary for all types of life in the world (Berger et al. 2013). However, phosphorus (P) and potassium (K) rock biofertilizers do not contain nitrogen for plants and microbial organisms in the soil, and the production of sulfuric acid may reduce soil pH, which damages and reduces plant growth. Therefore, rock biofertilizers can be mixed with organic matter (e.g., earthworm compound) that has a high pH (pH 7.9) and inoculated with selected free-living diazotrophic bacteria to promote N enrichment through the process of BNF (Lima et al. 2010).

In biological studies, crustacean chitosan is frequently used to increase plant resistance against pathogens. The chitosan displays better chelating properties compared with other polymers and can release nutrients into the environment (Boonlertnirun et al. 2008; Goy et al. 2009). The use of chitosan from fungal biomass has great advantages compared with crustacean chitosan, such as the independence of seasonal factors and the simultaneous extraction of chitin and chitosan (Franco et al. 2004). Mixed biofertilizers (NPKB) provide nutrients for plants, especially when inoculated with fungal chitosan such as *Cunninghamella elegans,* which also increases inorganic phosphate (Franco et al. 2004). Furthermore, the bioprotector (NPKP) has antimicrobial properties, promotes plant protection against pathogens, and increases the availability of nutrients (Berger et al. 2013; Franco et al. 2011).

### 3.5.1 Production of Biofertilizer (NPKB) and Bioprotector (NPKP)

The biofertilizer (NPKB) was produced from phosphate and potassium rock biofertilizers (PKB) mixed with organic matter (OM) in a 1:3 ratio and incubated for 30 days. The PK rock biofertilizers were produced at the Horticultural Experimental Station of the Federal Agricultural University of Pernambuco (UFRPE). Two furrows (10 m long, 1.0 m wide, 0.5 m deep) were used following the procedure described by Stamford et al. (2007).

The sulfur-oxidative bacteria were grown in Erlenmeyer flasks that contained 1000 mL of culture-specific medium (El Tarabily et al. 2006) and sterilized for 30 min at 120 °C. The Erlenmeyer flasks were shaken (150 rpm) for 5 days at 30 °C. The materials (phosphate and potash rocks plus elemental sulfur) were incubated for 60 days, and the humidity was maintained at a level close to field holding capacity. To avoid excessive humidity due to rain and to increase the efficiency of the sulfur-oxidative bacteria, the furrows were covered with black plastic (Fig. 3.4).

The biofertilizer (NPKB) was processed by mixing PK rock biofertilizers with organic biofertilizer (earthworm compound) enriched in N by inoculation with selected free-living diazotrophic bacteria (NFB 10001), as reported by Lima et al. (2010). The bioprotector (NPKP) represents the biofertilizer (NPKB) inoculated with *C. elegans* (UCP 542), which are fungi that contain chitosan in their cellular wall (Franco et al. 2004). The *C. elegans* fungus was purified in potato dextrose agar medium and grown for 10 days at 28 °C. A monosporic culture of *C. elegans* was obtained by growing the *Mucorales* fungus in potato dextrose medium using Erlenmeyer flasks (containing 1000 mL) maintained under shaking conditions (180 rpm, 96 h, 28 °C). The culture diluted in distilled water (20 $L^{-1}$) was applied using manual irrigation. The chemical differences between the P and K biofertilizers and the biofertilizer (NPKB) obtained following the procedure of Stamford et al. (2007) are presented in (Table 3.6).

The results of several experiments carried out in Brazilian soils cropped with different economic cultures such as cowpea, yam bean, grapes, melon, sugarcane, green pepper, and tomato showed that the biofertilizer (NPKB) and bioprotector (NPKP) increased crop performance and soil nutrients compared with the mineral soluble fertilizer. The NPKB and NPKP may be alternatives to replace conventional NPK soluble fertilizer and pesticides.

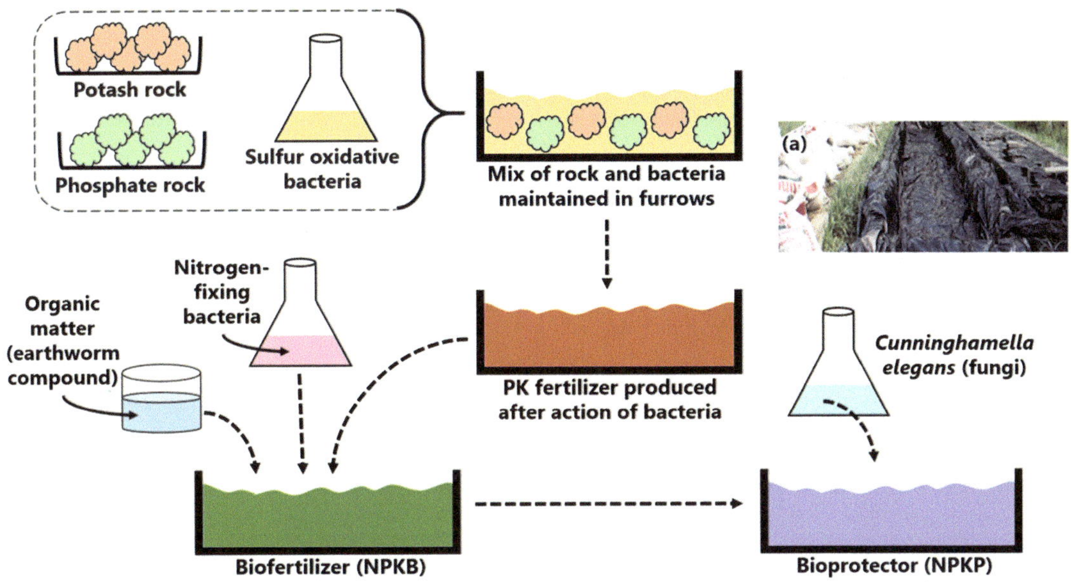


**Fig. 3.4** Simplified schematic of the production of the biofertilizer (NPKB) and bioprotector (NPKP). The furrows covered with *black* plastic under field conditions are shown in (**a**)

**Table 3.6** Chemical analyses of the P and K biofertilizers produced with sulfur-oxidative bacteria and of the biofertilizer (NPKB) after inoculation with selected free-living bacteria and mixed with organic matter as described in the text

| Biofertilizer type | pH | Phosphorus (P) | Potassium (K) | Nitrogen (N) |
|---|---|---|---|---|
| P biofertilizer | 3.8 | 60.0 g kg$^{-1}$ | – | – |
| K biofertilizer | 3.3 | – | 10.0 g kg$^{-1}$ | – |
| PK rock after bacterial inoculation (NPKB) | 6.4 | 21.0 g kg$^{-1}$ | 19.0 g kg$^{-1}$ | 20.0 g kg$^{-1}$ |

## 3.6 Conclusion

Beneficial microorganisms can be a potential tool for sustainable agriculture as well as a trend for the future. The beneficial effects include biological control of diseases, promotion of plant growth, increases in crop yield, and quality improvement. Knowledge of the complex environment of the rhizosphere, the mechanisms of action, and the practical aspects of inoculant formulation using these microorganisms is important in the search for new products to increase crop performance. New biotechnological methods for crop protection are based on the use of beneficial microorganisms applied as biofertilizers and bioprotectors. This approach represents an important tool for plant disease control and can lead to a substantial reduction of synthetic chemical products, which are important source of environmental pollution. Future studies are needed to identify management conditions that can contribute to the optimization of several mechanisms of the plant-microorganism interrelationship in real-life agricultural systems.

## References


Adesemoye AO, Kloepper JW (2009) Plant-microbes interactions in enhanced fertilizer-use efficiency. Appl Microbiol Biotechnol 85(1):1–12

Albareda M, Rodríguez-Navarro DN, Camacho M, Temprano FJ (2008) Alternatives to peat as a carrier for rhizobia inoculants: solid and liquid formulations. Soil Biol Biochem 40:2771–2779

Ali S, Charles TC, Glick BR (2014) Amelioration of high salinity stress damage by plant growth-promoting

bacterial endophytes that contain ACC deaminase. Plant Physiol Biochem 80:160–167
Ansari SA, Matricardi P, Di Meo C, Alhaique F, Coviello T (2012) Evaluation of rheological properties and swelling behavior of sonicated scleroglucan samples. Molecules 17:2283–2297
Araujo FF (2008) Seed inoculation with *Bacillus subtilis*, formulated with oyster meal and growth of corn, soybean and cotton. Ciênc Agrotecnol 32(2):456–462
Araujo FF, Henning AA, Hungria M (2005) Phytohormones and antibiotics produced by *Bacillus subtilis* and their effects on seed pathogenic fungi and on soybean root development. World J Microbiol Biotechnol 21:1639–1645
Arzanesh MH, Alikhani HA, Khavazi K, Rahimian HA, Miransari M (2011) Wheat (*Triticum aestivum* L.) growth enhancement by *Azospirillum* sp. under drought stress. World J Microbiol Biotechnol 27(2):197–205
Avis TJ, Gravel V, Antoun H, Tweddell RJ (2008) Multifaceted beneficial effects of rhizosphere microorganisms on plant health and productivity. Soil Biol Biochem 40:1733–1740
Badel S, Bernardi T, Michaud P (2011) New perspectives for *Lactobacilli* exopolysaccharides. Biotechnol Adv 29:54–66
Badri DV, Weir TL, van der Lelie D, Vivanco JM (2009) Rhizosphere chemical dialogues: plant-microbe interactions. Curr Opin Biotechnol 20:642–650
Bashan Y (1998) Inoculants of plant growth-promoting bacteria for use in agriculture. Biotechnol Adv 16:729–770
Bashan Y, De-Bashan LE, Prabhu SR, Hernandez JP (2014) Advances in plant growth-promoting bacterial inoculant technology: formulations and practical perspectives (1998–2013). Plant Soil 378:1–33
Berger LRR, Stamford NP, Santos CERS, Freitas ADS, Franco LO, Stamford TCM (2013) Plant and soil characteristics affected by biofertilizers from rocks and organic matter inoculated with diazotrophic bacteria and fungi that produce chitosan. J Soil Sci Plant Nutr 13:592–603
Bhattacharyya PN, Jha DK (2012) Plant growth-promoting rhizobacteria (PGPR): emergence in agriculture. World J Microbiol Biotechnol 28(4):1327–1350
Bogino PC, Oliva MM, Sorroche SG, Giordano W (2013) The role of bacterial biofilms and surface components in plant-bacterial associations. Int J Mol Sci 14:15838–15859
Bomfeti CA, Florentino LA, Guimarães AP, Cardoso PG, Guerreiro MC, Moreira FMS (2011) Exopolysaccharides produced by the symbiotic nitrogen-fixing bacteria of Leguminosae. Rev Bras Ciênc Solo 35:657–671
Bonfante P, Anca IA (2009) Plants, mycorrhizal fungi, and bacteria: a network of interactions. Annu Rev Microbiol 63:363–383
Bonfante P, Genre A (2008) Plants and arbuscular mycorrhizal fungi: an evolutionary-developmental perspective. Trends Plant Sci 13(9):492–498
Boonlertnirun S, Boonraung C, Suvanasara R (2008) Application of chitosan in rice production. J Min Met Mat S 18(2):47–52
Brahmaprakash GP, Sahu PK (2012) Biofertilizers for sustainability. J Indian Inst Sci 92:37–69
Brotman Y, Landau U, Cuadros-Inostroza Á, Takayuki T, Fernie AR, Chet I, Viterbo A, Willmitzer L (2013) *Trichoderma*-plant root colonization: escaping early plant defense responses and activation of the antioxidant machinery for saline stress tolerance. PLoS Pathog 9(3):1003221
Castellane TCL, Lemos VFM, Lemos EGM (2014) Evaluation of the biotechnological potential of *Rhizobium tropici* strains for exopolysaccharide production. Carbohydr Polym 111:191–197
Chanway CP (1998) Bacterial endophytes: ecological and practical implications. Sydowia 50:149–170
Chaparro JM, Sheflin AM, Manter DK, Vivanco JM (2012) Manipulating the soil microbiome to increase soil health and plant fertility. Biol Fertil Soils 48:489–499
Chaparro JM, Badri DV, Vivanco JM (2014) Rhizosphere microbiome assemblage is affected by plant development. ISME J 8(4):790–803
Cheng KC, Demirci A, Catchmark JM (2011) Pullulan: biosynthesis, production, and applications. Appl Microbiol Biotechnol 92(1):29–44
Colla G, Rouphael Y, Di Mattia E, El-Nakhel C, Cardarelli M (2014) Co-inoculation of *Glomus intraradices* and *Trichoderma atroviride* acts as a biostimulant to promote growth, yield and nutrient uptake of vegetable crops. J Sci Food Agric. doi:10.1002/jsfa.6875
Contreras-Cornejo HA, Macías-Rodríguez L, Cortés-Penagos C, López-Bucio J (2009) *Trichoderma virens*, a plant beneficial fungus, enhances biomass production and promotes lateral root growth through an auxin-dependent mechanism in Arabidopsis. Plant Physiol 149:1579–1592
Delbarre-Ladrat C, Sinquin C, Lebellenger L, Zykwinska A, Colliec-Jouault S (2014) Exopolysaccharides produced by marine bacteria and their applications as glycosaminoglycan-like molecules. Front Chem 2:85
Dimkpa C, Weinand T, Asch F (2009) Plant-rhizobacteria interactions alleviate abiotic stress conditions. Plant Cell Environ 32:1682–1694
Dobbelare S, Croonenborghs A, Thys A, van de Broek A, Vanderleyden J (1999) Phytostimulatory effect of *Azospirillum brasilense* wild type and mutant strains altered in IAA production on wheat. Plant Soil 212:155–164
Donot F, Fontana A, Baccou JC, Schorr-Galindo S (2012) Microbial exopolysaccharides: main examples of synthesis, excretion, genetics and extraction. Carbohydr Polym 87:951–962
Egamberdieva D, Berg G, Lindström K, Räsänen LA (2013) Alleviation of salt stress of symbiotic *Galega officinalis* L. (goat's rue) by co-inoculation of

*Rhizobium* with root-colonizing *Pseudomonas*. Plant Soil 369(1–2):453–465

El Tarabily KA, Soaud A, Saleh M, Matsumoto S (2006) Isolation and characterization of sulfur bacteria, including strains of *Rhizobium* from calcareous soils and their effects on nutrient uptake and growth of maize (*Zea mays* L.). Austr J Agric Res 57:101–111

Elkoca E, Turan M, Donmez MF (2010) Effects of single, dual and triple inoculations with *Bacillus subtilis*, *Bacillus megaterium* and *Rhizobium leguminosarum* bv. Phaseoli on nodulation, nutrient uptake, yield and yield parameters of common bean (*Phaseolus vulgaris* L. cv. 'elkoca-05'). J Plant Nutr 33(14):2104–2119

Estrada B, Aroca R, Barea JM, Ruiz-Lozano JM (2013) Native arbuscular mycorrhizal fungi isolated from a saline habitat improved maize antioxidant systems and plant tolerance to salinity. Plant Sci 201–202:42–51

Fernandes Júnior PI, Rohr TG, Oliveira PJ, Xavier GR, Rumjanek NG (2009) Polymers as carriers for rhizobial inoculant formulations. Pesq Agrop Brasileira 44:1184–1190

Fernandes Júnior PI, Silva Júnior EB, Silva Júnior S, Santos CES, Oliveira PJ, Rumjanek NG, Martins LMV, Xavier GR (2012) Performance of polymer compositions as carrier to cowpea rhizobial inoculant formulations: survival of rhizobia in pre-inoculated seeds and field efficiency. African J Biotechnol 11 (12):2945–2951

Fernández L, Agaras B, Zalba P, Wall L, Valverde C (2012) *Pseudomonas* spp. isolates with high phosphate-mobilizing potential and root colonization properties from agricultural bulk soils under no-till management. Biol Fertil Soils 48:763–773

Figueiredo MVB, Vilar JJ, Burity HA, França FP (1999) Alleviation of water stress effects in cowpea by *Bradyrhizobium* spp. inoculation. Plant Soil 207:67–75

Figueiredo MVB, Burity HA, Martinez CR, Chanway CP (2008) Alleviation of drought stress in the common bean (*Phaseolus vulgaris* L.) by co-inoculation with *Paenibacillus polymyxa* and *Rhizobium tropici*. Appl Soil Ecol 40:182–188

Figueiredo MVB, Seldin L, Araujo FF, Mariano RLR (2010) Plant growth promoting rhizobacteria: fundamentals and applications. In: Maheshwari DK (ed) Plant growth and health promoting bacteria. Springer-Verlag, Berlin

Figueiredo MVB, Kuklinsky-Sobral J, Lima CEP, Araújo ASF (2012) Ecological agriculture: strategy for sustainable development. In: Thangadurai D, Busso C, Arenas LGA, Jayabalan S (eds) Frontiers in biodiversity studies. IK International, New Delhi

Finore I, Di Donato P, Mastascusa V, Nicolaus B, Poli A (2014) Fermentation technologies for the optimization of marine microbial exopolysaccharide production. Mar Drugs 12:3005–3024

Fliessbach A, Winkler M, Lutz MP, Oberholzer HR, Mader P (2009) Soil amendment with *Pseudomonas fluorescens* CHA0: lasting effects on soil biological properties in soils low in microbial biomass and activity. Microb Ecol 57:611–623

Franco LO, Maia RCC, Porto ALF, Messias AS, Fukushima K, Takaki GMC (2004) Heavy metal biosorption by chitin and chitosan isolated from *Cunninghamella elegans* (IFM 46109). Braz J Microbiol 35:243–247

Franco LO, Albuquerque CDC, Stamford NP, Lima MAB, Takaki-Campos GM (2011) Evaluation of acid and alkaline activity and accumulation of inorganic phosphate in samples with *Cunninghamella elegans*. Analytica 54:70–78

Freitas F, Alves VD, Reis MAM (2011) Advances in bacterial exopolysaccharides: from production to biotechnological applications. Trends Biotechnol 29:388–398

Frey-Klett P, Garbaye JA, Tarkka M (2007) The mycorrhiza helper bacteria revisited. New Phytol 176 (1):22–36

Gamalero E, Lingua G, Berta G, Glick BR (2009) Beneficial role of plant growth promoting bacteria and arbuscular mycorrhizal fungi on plant responses to heavy metal stress. Can J Microbiol 55(5):501–514

Giavasis I (2014) Bioactive fungal polysaccharides as potential functional ingredients in food and nutraceuticals. Curr Opin Biotechnol 26:162–173

Goy RC, Britto D, Assis OBG (2009) A review of the antimicrobial activity of chitosan polymers. Sci Technol 9:241–247

Gray EJ, Smith DL (2005) Intracellular and extracellular PGPR: commonalities and distinctions in the plant-bacterium signaling processes. Soil Biol Biochem 37:395–412

Hamdali H, Hafidi M, Virolle MJ, Ouhdouch Y (2008) Rock phosphate-solubilizing Actinomycetes: screening for plant growth-promoting activities. World J Microbiol Biotechnol 24:2565–2575

Hazell P, Wood S (2008) Drivers of change in global agriculture. Phil Trans R Soc B 363(1491):495–515

Hermosa R, Viterbo A, Chet I, Monte E (2012) Plant-beneficial effects of *Trichoderma* and of its genes. Microbiology 158:17–25

Herrmann L, Lesueur D (2013) Challenges of formulation and quality of biofertilizers for successful inoculation. Appl Microbiol Biotechnol 97:8859–8873

Hungria M, Nogueira MA, Araujo RS (2013) Co-inoculation of soybeans and common beans with rhizobia and azospirilla: strategies to improve sustainability. Biol Fertil Soils 49(7):791–801

Javaid A (2010) Beneficial microorganisms for sustainable agriculture. In: Lichtfouse E (ed) Genetic engineering, biofertilisation, soil quality and organic farming sustainable agriculture reviews. Springer-Verlag, Berlin

Kilian M, Steiner U, Krebs B, Junge H, Schmiedeknecht G, Hain R (2000) Fzb24 *Bacillus subtilis* – mode of action of a microbial agent enhancing plant vitality. Planzenschutz-Nachrich Bayer 1/00 (1):72–93

Kloepper JW, Schroth MN, Miller TD (1980) Effects of rhizosphere colonization by plant growth-promoting rhizobacteria on potato plant development and yield. Phytopathology 70:1078–1082

Kloepper JW, Ryu CM, Zhang S (2004) Induced systemic resistance and promotion of plant growth by *Bacillus* spp. Phytopathology 94:1259–1226

Kohler J, Caravaca F, Carrasco L, Rolda A (2007) Interactions between a plant growth-promoting rhizobacterium, an AM fungus and a phosphate-solubilizing fungus in the rhizosphere of *Lactuca sativa*. Appl Soil Ecol 35:480–487

Larrainzar E, Molenaar JA, Wienkoop S, Gil-Quintana E, Alibert B, Limami AM, Arrese-Igor C, González EM (2014) Drought stress provokes the down-regulation of methionine and ethylene biosynthesis pathways in *Medicago truncatula* roots and nodules. Plant Cell Environ 37:2051–2063

Lima FS, Stamford NP, Sousa CS, Lira Junior MA, Malheiros S, van Straaten P (2010) Earthworm compound and rock biofertilizer enriched in nitrogen by inoculation with free-living diazotrophic bacteria. World J Microbiol Biotechnol 26:1769–1777

Lima AST, Xavier TF, Lima CEP, Oliveira JP, Mergulhão ACES, Figueiredo MVB (2011) Triple inoculation with *Bradyrhizobium*, *Glomus* and *Paenibacillus* on cowpea (*Vigna unguiculata* [L.] walp.) development. Braz J Microbiol 42(3):919–926

Lingua G, Bona E, Manassero P, Marsano F, Todeschini V, Cantamessa S, Copetta A, D'Agostino G, Gamalero E, Berta G (2013) Arbuscular Mycorrhizal fungi and plant growth-promoting pseudomonads increases anthocyanin concentration in strawberry fruits (*Fragaria* x *ananassa* var. selva) in conditions of reduced fertilization. Int J Mol Sci 14(8):16207–16225

Mahapatra S, Banerjee D (2013) Fungal exopolysaccharide: production, composition and Applications. Microbiol Insights 6:1–16

Meier S, Borie F, Bolan N, Cornejo P (2012) Phytoremediation of metal-polluted soils by arbuscular mycorrhizal fungi. Crit Rev Environ Sci Technol 42(7):741–775

Mendes R, Garbeva P, Raaijmakers JM (2013) The rhizosphere microbiome: significance of plant beneficial, plant pathogenic and human pathogenic microorganisms. FEMS Microbiol Rev 37(5):634–663

Molla AH, Haque M, Haque A, Ilias GNM (2012) *Trichoderma*-enriched biofertilizer enhances production and nutritional quality of tomato (*Lycopersicon esculentum* Mill.) and minimizes NPK fertilizer use. Agric Res 1:265–272

Munns R, Tester M (2008) Mechanisms of salinity tolerance. Annu Rev Plant Biol 59:651–681

Nadeem SM, Ahmad M, Zahir ZA, Javaid A, Ashraf M (2014) The role of mycorrhizae and plant growth promoting rhizobacteria (PGPR) in improving crop productivity under stressful environments. Biotechnol Adv 32(2):429–448

Nautiyal CS, Srivastava S, Chauhan PS, Seem K, Mishra A, Sopory SK (2013) Plant growth-promoting bacteria *Bacillus amyloliquefaciens* NBRISN13 modulates gene expression profile of leaf and rhizosphere community in rice during salt stress. Plant Physiol Biochem 66:1–9

Nwodo U, Green E, Okoh A (2012) Bacterial EPS: functionality and prospects. Int J Mol Sci 13:14002–14015

Okon Y, Labandera-Gonzalez CA (1994) Agronomic applications of *Azospirillum*: an evaluation of 20 years worldwide field inoculation. Soil Biol Biochem 26(12):1591–1601

Oliveira JP (2011) Genetic characterization, production of extracellular biopolymers and proteome of diazotrophic bacteria. Thesis, Federal University of Ceará

Oliveira JP, Figueiredo MVB, Silva MV, Mendes MMC, Vendrusculo C, Burity HA (2012) Production of extracellular biopolymers and identification of intracellular proteins and *Rhizobium tropici*. Curr Microbiol 65 (6):686–691

Pallai R, Hynes RK, Verma B, Nelson LM (2012) Phytohormone production and colonization of canola (*Brassica napus* L.) roots by *Pseudomonas fluorescens* 6–8 under gnotobiotic conditions. Can J Microbiol 58 (2):170–178

Peix A, Ramírez-Bahena MH, Velázquez E, Bedmar EJ (2015) Bacterial associations with legumes. Crit Rev Plant Sci 34(1–3):17–42

Pindi PK (2012) Liquid microbial consortium – a potential tool for sustainable soil health. J Biofertil Biopestici 3:1–9

Poli A, Donato P, Abbamondi G, Nicolaus B (2011) Synthesis, production and biotechnological applications of exopolysaccharides and polyhydroxyalkanoates by Archaea. Archaea 693253:1–13

Raaijmakers JM, Weller DM, Thomashow LS (1997) Frequency of antibiotic-producing *Pseudomonas* spp. in natural environments. Appl Environ Microbiol 63:881–887

Raaijmakers JM, Paulitz TC, Steinberg C, Alabouvette C, Moënne-Loccoz Y (2009) The rhizosphere: a playground and battlefield for soilborne pathogens and beneficial microorganisms. Plant Soil 321:341–361

Ramamoorthy V, Viswanathan R, Raguchander T, Prakasam V, Samiyappan R (2001) Induction of systemic resistance by plant growth promoting rhizobacteria in crop plants against pests and diseases. Crop Prot 20:1–11

Redecker D, Schüßler A, Stockinger H, Stürmer SL, Morton JB, Walker C (2013) An evidence-based consensus for the classification of arbuscular mycorrhizal fungi (Glomeromycota). Mycorrhiza 23:515–531

Rehm BHA (2010) Bacterial polymers: biosynthesis, modifications and applications. Nat Rev 8:578–592

Rivera D, Obando M, Barbosa H, Tapias DR, Buitrago RB (2014) Evaluation of polymers for the liquid rhizobial formulation and their influence in the *Rhizobium*-Cowpea interaction. Univ Sci 19(3):265–275

Rodrigues AC (2012) Interrelationship *Bradyrhizobium* and PGPB and Cowpea: evaluation of the enzymatic activity and symbiotic performance. Thesis, Federal Agricultural University of Pernambuco

Rodrigues AC, Silveira JAG, Bonifacio A, Figueiredo MVB (2013a) Metabolism of nitrogen and carbon: optimization of biological nitrogen fixation and cowpea development. Soil Biol Biochem 67:226–234

Rodrigues AC, Bonifacio A, Antunes JEL, Silveira JAG, Figueiredo MVB (2013b) Minimization of oxidative stress in cowpea nodules by the interrelationship between *Bradyrhizobium* sp. and plant growth-promoting bacteria. Appl Soil Ecol 64:245–251

Rojas-Tapias D, Moreno-Galván A, Pardo-Díaz S, Obando M, Rivera D, Bonilla R (2012) Effect of inoculation with plant growth-promoting bacteria (PGPB) on amelioration of saline stress in maize (*Zea mays*). Appl Soil Ecol 61:264–272

Sandhya V, Ali SZ, Grover M, Reddy G, Venkateswarlu B (2009) Alleviation of drought stress effects in sunflower seedlings by exopolysaccharides producing *Pseudomonas putida* strain P45. Biol Fertil Soil 46:17–26

Santos AA (2010) Polysaccharides production aiming to obtain biological inoculants of interest to agriculture. Dissertation, Federal Agricultural University of Pernambuco

Schmid J, Meyer V, Sieber V (2011) Scleroglucan: biosynthesis, production and application of a versatile hydrocolloid. Appl Microbiol Biotechnol 91:937–947

Seneviratne G, Thilakaratne RMS, Jayasekara APDA, Seneviratne KACN, Padmathilake KRE, Silva MSDL (2009) Developing beneficial microbial biofilms on roots of non-legumes: a novel biofertilizing technique. In: Khan MS, Zaid A, Musarrat J (eds) Microbial strategy for crop improvement. Springer-Verlag, Berlin

Serrato RV, Sassaki GL, Gorin PAJ, Cruz LM, Pedrosa FO, Choudhury B, Carlson RW, Iacomini M (2008) Structural characterization of an acidic exoheteropolysaccharide produced by the nitrogen-fixing bacterium *Burkholderia tropica*. Carbohydr Polym 73:564–572

Sharmila K, Thillaimaharani KA, Durairaj R, Kalaiselvam M (2014) Production and characterization of exopolysaccharides (EPS) from mangrove filamentous fungus, *Syncephalastrum* sp. Afr J Microbiol Res 8(21):2155–2161

Singh RS, Saini GK, Kennedy JF (2008) Pullulan: microbial sources, production and applications. Carbohydr Polym 73(4):515–531

Singh JS, Pandey VC, Singh DP (2011) Efficient soil microorganisms: a new dimension for sustainable agriculture and environmental development. Agric Ecosyst Environ 140(3):339–353

Smith SE, Smith FA (2012) Fresh perspectives on the roles of arbuscular mycorrhizal fungi in plant nutrition and growth. Mycologia 104(1):1–13

Soliman AS, Shanan NT, Massoud ON, Swelim DM (2012) Improving salinity tolerance of *Acacia saligna* (Labill.) plant by arbuscular mycorrhizal fungi and *Rhizobium* inoculation. Afr J Biotechnol 11 (5):1259–1266

Stamford NP, Santos PR, Santos CERS, Freitas ADS, Dias SHL, Lira Junior MA (2007) Agronomic effectiveness of biofertilizers with phosphate rock, sulphur and *Acidithiobacillus* in a Brazilian tableland acidic soil grown with yam bean. Biores Technol 98:1311–1318

Stamford NP, Lima RA, Lira Junior MA, Santos CERS (2008) Effectiveness of phosphate and potash rocks with *Acidithiobacillus* on sugar cane yield and their effects in soil chemical attributes. World J Microbiol Biotechnol 24:2061–2066

Star L, Matan O, Dardanelli MS, Kapulnik Y, Burdman S, Okon Y (2012) The *Vicia sativa* spp. *nigra-Rhizobium leguminosarum* bv. *viciae* symbiotic interaction is improved by *Azospirillum brasilense*. Plant Soil 356 (1–2):165–174

Staudt AK, Wolfe L, Shrout JD (2012) Variations in exopolysaccharide production by *Rhizobium tropici*. Arch Microbiol 194:197–206

Triveni S, Prasanna R, Saxena AK (2012) Optimization of conditions for in vitro development of *Trichoderma viride*-based biofilms as potential inoculants. Folia Microbiol 57:431–437

Van Straaten P (2007) Agrogeology – the use of rocks for crops. Enviroquest, Canada

Vandenkoornhuyse P, Mahe S, Ineson P, Staddon P, Ostle N, Cliquet JB, Francez AJ, Fitter AH, Young JP (2007) Active root-inhabiting microbes identified by rapid incorporation of plant derived carbon into RNA. Proc Natl Acad Sci U S A 104:16970–16975

Vessey JK (2003) Plant growth-promoting rhizobacteria as biofertilizers. Plant Soil 255:571–586

Vos CMF, Cremer KD, Cammue BPA, Coninck BD (2014) The toolbox of *Trichoderma* spp. in the biocontrol of *Botrytis cinerea* disease. Mol Plant Pathol (online). doi:10.1111/mpp.12189

Vu B, Chen M, Crawford RJ, Ivanova EP (2009) Bacterial extracellular polysaccharides involved in biofilm formation. Molecules 14:2535–2554

Wu SC, Cao ZH, Li ZG, Cheug KC, Wong MH (2005) Effects of biofertilizer containing N-fixer, P and K solubilizers and AM fungi on maize growth: a greenhouse trial. Geoderma 125:155–166

Xavier TF (2009) Production and characterization of exopolysaccharides (EPSs) synthetized by diazotrophic microorganisms. Dissertation, Federal Agricultural University of Pernambuco

Xie X, Zhang H, Paré PW (2009) Sustained growth promotion in *Arabidopsis* with long-term exposure to the beneficial soil bacterium *Bacillus subtilis* (GB03). Plant Signal Behav 4:948–953

Yadav SK, Dave A, Sarkar A, Singh HB, Sarma BK (2013) Co-inoculated biopriming with *Trichoderma*, *Pseudomonas* and *Rhizobium* improves crop growth in

*Cicer arietinum* and *Phaseolus vulgaris*. Int J Agric Environ Biotechnol 6(2):255–259

Zhang H, Sun Y, Xie X, Kim M, Dowd SE, Paré PW (2009) A soil bacterium regulates plant acquisition of iron via deficiency inducible mechanisms. Plant J 58:568–577

Zhang Y, Kong H, Fang Y, Nishinari K, Phillips GO (2013) Schizophyllan: a review on its structure, properties, bioactivities and recent developments. Bioact Carbohydr Diet Fibre 1(1):53–71

# The Production and Potential of Biofertilizers to Improve Crop Yields

4

Didier Lesueur, Rosalind Deaker, Laetitia Herrmann, Lambert Bräu, and Jan Jansa

## Contents



D. Lesueur (✉) • L. Herrmann
CIRAD, UMR Eco&Sols – Ecologie Fonctionnelle & Biogéochimique des Sols & Agroécosystèmes (SupAgro–CIRAD–INRA–IRD), Land Development Department, Office of Science for Land Development, Paholyothin Road, Chatuchak, Bangkok 10900, Thailand

School of Life and Environmental Sciences, Faculty of Science and Technology, Deakin University, 221 Burwood Hwy, Burwood, Victoria 3125, Australia
e-mail: didier.lesueur@cirad.fr; dllesueurdidier@gmail.com

R. Deaker
Faculty of Agriculture and Environment, University of Sydney, Level 4 Biomedical Building, 1 Central Ave, Eveleigh, NSW 2015, Australia

L. Bräu
School of Life and Environmental Sciences, Faculty of Science and Technology, Deakin University, 221 Burwood Hwy, Burwood, Victoria 3125, Australia

J. Jansa
Institute of Microbiology, Academy of Sciences of the Czech Republic, Vídeňská 1083, Praha 4 – Krč, Czech Republic

**Abstract**

Extensive interactions of plant roots with soil microorganisms affect plant nutrition either directly by influencing mineral nutrient availability or indirectly through enhanced uptake efficiency via plant root growth promotion. Beneficial microbial interactions with roots may be either endophytic or associative and can be symbiotic, mutualistic, or incidental in nature. The increased understanding of the role of root – or rhizosphere –associated with microbes in the nutrition and/or yield of agricultural crops in particular has resulted in promotion of their use in agricultural production as alternatives or supplements to mineral or organic fertilizers. Despite this, there is an obvious lack of market penetration of microbial inoculants. This review specifically focuses on microbial inoculants, collectively termed biofertilizers, used to improve nutrition and yields of grain, legume, oil, tuber,


N.K. Arora et al. (eds.), *Bioformulations: for Sustainable Agriculture*,
DOI 10.1007/978-81-322-2779-3_4

and other crops. A vast number of commercial biofertilizers are available worldwide; however, the quality and efficacy of many of them are not proven or tested. In the absence of efficacious biofertilizers of good and consistent quality, the dependence on the use of mineral fertilizers is not likely to decrease. Thus the availability of high-quality biofertilizers must be priority particularly in countries where crop plant production plays a key role in the economy and food security.

## 4.1 Introduction

### 4.1.1 Rhizosphere: Strong Interactions Between Plant Roots and Soil Microorganisms

The small volume of soil surrounding and under the direct influence of plant roots is called the "rhizosphere" and sustains great portion of life on Earth. Soil has the highest microbial diversity on Earth with thousands of species of prokaryotes and hundreds of species of eukaryotes in each gram. The rhizosphere probably represents the most dynamic habitat and the most important zone in terms of defining the quality and quantity of our terrestrial food resource (Hinsinger et al. 2009). Plant roots interact with the physical, chemical, and biological properties of soil and encounter an enormous variety of organisms in the rhizosphere (Roesch et al. 2007). Structural and functional characteristics of roots contribute to so-called rhizosphere processes that have a significant influence on the capacity of plants to acquire nutrients (Vacheron et al. 2013; Lamont et al. 2014). Extensive interactions of roots with soil microorganisms further affect plant nutrition either directly by influencing mineral nutrient availability and/or uptake or indirectly through plant (root) growth promotion enhancing uptake efficiency (Richardson et al. 2009). These "root–microorganism" interactions are ubiquitous across various trophic levels and are essential components of ecosystem functions. It has become increasingly evident that they are intricate and involve highly complex communities that function in very heterogeneous environments (Giri et al. 2005).

Microorganisms are a significant component in cycling/recycling of mineral nutrients and carbon in the soil and through their activities; nutrients can become more (through solubilization, mineralization) or less (through adsorption, immobilization) available to plants (Morgan et al. 2005; Leake et al. 2006). Through associations with a range of microorganisms, plants may also be protected from a variety of biotic and abiotic stresses including interplant competition (Morgan et al. 2005; Sanon et al. 2009; Smith et al. 2010). Beneficial microbial interactions with roots may be either endophytic or associative and can be symbiotic, mutualistic, or incidental in nature. Figure 4.1 shows the main interactions between plants and soil microorganisms. A few organisms, such as nodule-forming rhizobia and mycorrhizal fungi, have developed intimate symbiotic relationships with plants. Arbuscular mycorrhizal fungi (AMF) have become completely dependent on plant carbon and are unable to complete their life cycle without association with their host plant (van der Heijden et al. 2015). Other microorganisms are generally less dependent on the plant for carbon source, though in many instances, they derive a significant portion of their carbon from the plants (Drigo et al. 2010; Leake et al. 2006).

Plant-associated microorganisms trade plant carbon for mineral nutrients gathered in the soil. Examples include the mycorrhiza-assisted plant uptake of nutrients such as phosphorus (P) and zinc (Zn) that normally have low diffusion rates in the soil, plant sulfur (S) supply by both bacteria and mycorrhizal fungi, or nitrogen (N) transfer to plants after fixation from the atmospheric $N_2$ pool (Lendenmann et al. 2011; Gao et al. 2012; Smith and Smith 2012; Gahan and Schmalenberger 2014). Plant growth benefits are not only achieved through improved availability of limiting nutrients such as N and/or P but also through hormonal and enzyme-induced

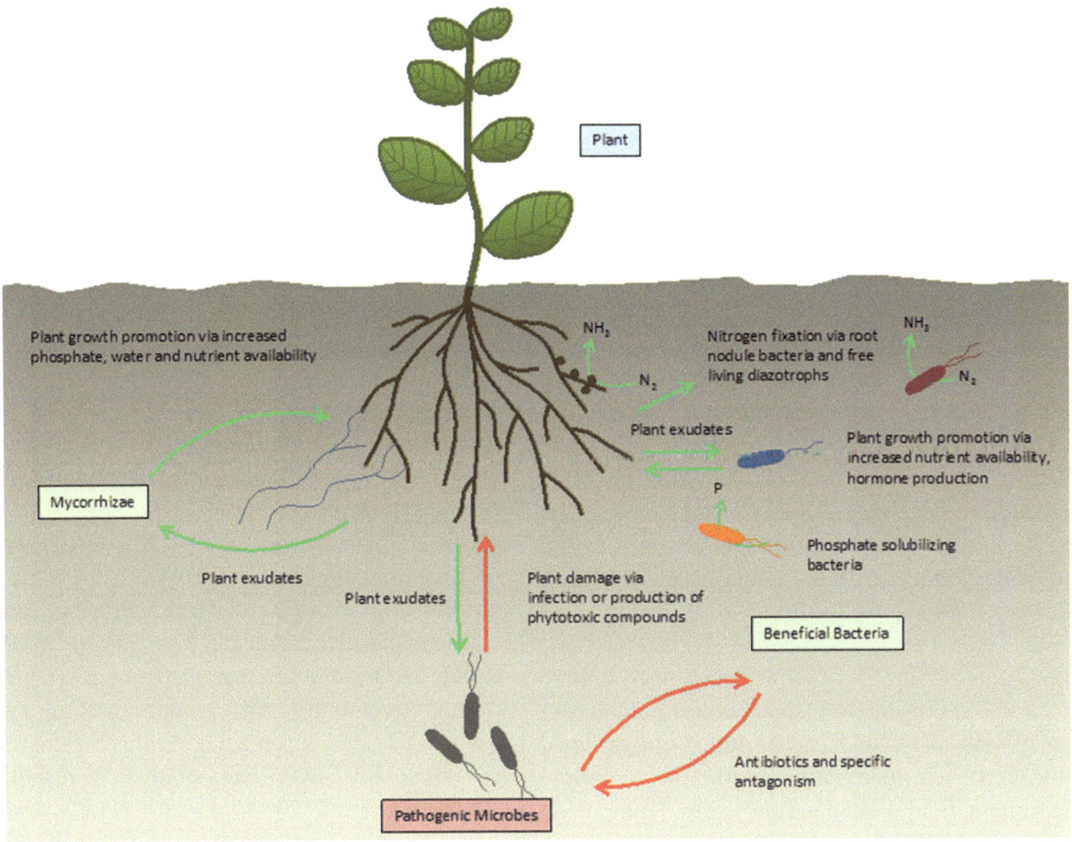


**Fig. 4.1** Main beneficial interactions between plants and soil microorganisms (Figure modified from Richardson et al. 2009)

enhancement of root growth which improves nutrient uptake efficiency (Glick 2012; Hayat et al. 2010). Halpern et al. (2015) reviewed the mechanisms involved in the enhancement of plant nutrient uptake by various biostimulants. Biostimulants are defined as substances or materials except nutrients and pesticides which, when applied to seeds or plants, have the capacity to modify physiological processes in plants in a way that provides potential benefits to growth, development, or stress response. Included in this group are the plant growth-promoting rhizobacteria (PGPR) that facilitate atmospheric $N_2$ fixation and P and iron solubilization and can also induce changes in root morphology.

Microbes have also been shown to reduce the incidence of plant diseases by acting as antagonists of root pathogens or by inducing systemic resistance in plants (Babalola 2010; Choudhary et al. 2011; Singh et al. 2011). Berendsen et al. (2012) described that upon pathogen or insect attack, plants are able to recruit protective microorganisms and enhance microbial activity to suppress pathogens in the rhizosphere. As shown by Yang et al. (2009), PGPR can also elicit an "induced systemic tolerance" to abiotic stresses such as salinity and drought and provide protection to maintain plant health. Other mechanisms are likely to be involved in the PGPR–plants interactions but are still largely unexplored. Recently, Palacios et al. (2014) highlighted the potential role of vitamins in the interactions of plants with PGPR that would also require a deeper exploration.

This chapter will focus on microbial inoculants that influence plant nutrition and grain yields, collectively termed biofertilizers, and not on microbially mediated disease suppression in plants that has been reviewed extensively elsewhere (Howell 2003; Harman et al. 2004; Compant et al. 2005; Avis et al. 2008; Berg 2009; Dutta and Podile 2010; Saharan and Nehra 2011).

### 4.1.2 The Potential of Soil Microorganisms to Improve Plant Nutrition

N and P are the most important and often the two most limiting mineral nutrients for living organisms. In terrestrial ecosystems, P is often considered as the most limiting nutrient, frequently more than N. Although most agricultural soils contain large amounts of inorganic and organic P, most of it is unavailable for plants. Even following the addition of P fertilizers to soils, plants may not access enough P as it is easily immobilized, becoming sparingly soluble and less accessible to plants. Hence, only a very small portion of soil P can be directly assimilated by plants and many soils are unable to supply adequate P to P-demanding crops (Bünemann et al. 2011; Dumas et al. 2011).

Economically mineable P deposits are finite and a better management of the P cycle is becoming increasingly important (Cordell et al. 2009). The world's main source of P is rock phosphate, a nonrenewable resource, and the mining and trading of rock phosphate contributes to global energy consumption, which is harmful to the environment and is extremely inefficient. The quality of rock phosphate, especially in terms of cadmium contamination, is continuously decreasing as cheap, high-quality reserves are becoming increasingly scarce and a major P shortage is predicted within few decades (Frossard et al. 2009). Currently, agricultural systems rely strongly on imported rock phosphate as well as various other rock phosphate derivatives from politically unstable regions such as Western Sahara. For example, while the European Union (EU) has for many decades been considered as a secure region in terms of food production, a report prepared by Schröder et al. (2010) demonstrates that the EU food system is actually highly vulnerable to future P scarcity, leading to decreased agricultural production (see Table 4.1 for world P demand).

Similar observations have been made concerning the use of N fertilizers. Global use of synthetic N fertilizers increased by 800 % between the years 1960 and 2000, and recently the anthropogenic N input was shown to be more than double the amount of N cycling through the biosphere (Canfield et al. 2010). Vast quantities of synthetic nitrogenous fertilizers are used worldwide and the Food and Agriculture Organization (FAO) estimates that nitrogenous fertilizer demand exceeds 130 million tones of N per year (Table 4.1). Our dependence on these resources is not only environmentally devastating (Rockström et al. 2009) but also economically shortsighted as synthetic N fertilizer production depends primarily on the use of fossil fuels (Canfield et al. 2010). While P is a limited natural resource, N occurs naturally in the atmosphere as molecular di-nitrogen ($N_2$) that can become a virtually unlimited resource after transformation into plant-available forms by a range of specific microorganisms (e.g., through biological nitrogen fixation (BNF); Fig. 4.2). BNF is a natural process of a prime importance in world agriculture (Werner and Newton 2005; Dakora et al. 2008; Canfield et al. 2010), and anything that can be done to increase BNF utilization will have global benefits. Herridge et al. (2008) reviewed and updated long-standing estimates

**Table 4.1** World total demand for primary nutrients, 2011–2015 (thousand tons) – FAO current world fertilizer trends and outlook to 2015

| Year | 2011 | 2012 | 2013 | 2014 | 2015 |
|---|---|---|---|---|---|
| Nitrogen (N) | 130,844 | 133,951 | 136,846 | 139,517 | 141,682 |
| Phosphate ($P_2O_5$) | 48,119 | 49,256 | 50,323 | 51,180 | 51,940 |
| Potash ($K_2O$) | 32,190 | 33,361 | 34,359 | 35,482 | 36,367 |

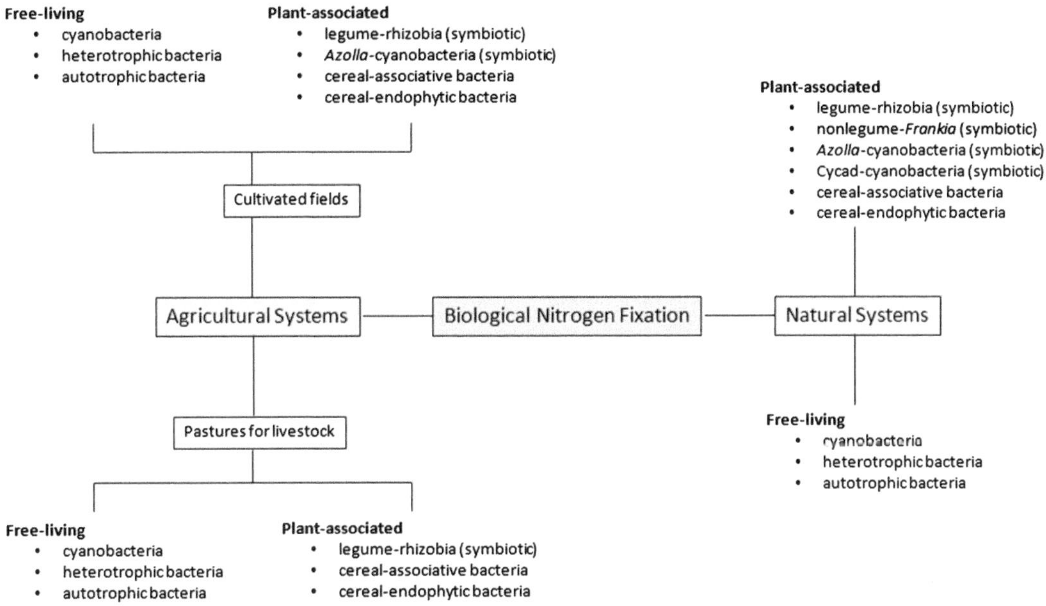


**Fig. 4.2** Biological N-fixing agents in agricultural and terrestrial natural systems (Figure modified from Herridge et al. 2008)

of BNF in different agricultural systems, and the most important $N_2$-fixing agents are the symbiotic associations between legumes and rhizobia. Annual inputs of fixed N are calculated to be 2.95 Tg for pulses and 18.5 Tg for oilseed legumes. Soybean [*Glycine max* (L.) Merrill] is the dominant crop legume worldwide, representing 50 % of global area planted with crop legumes and 68 % of global legume crop production. It was estimated that soybean alone fixes up to 16.4 Tg N annually, representing 77 % of the N fixed by the crop legumes.

The increased understanding of the role of root- or rhizosphere-associated microbes in nutrition and/or growth, biomass accumulation and fitness of plants in general, and yield of agricultural crops in particular has resulted in their increased promotion for use in agricultural production as alternatives or supplements to mineral or organic fertilizers (Adesemoye and Kloepper 2009; Lugtenberg and Kamilova 2009; Compant et al. 2010; Miransari 2011; Tikhonovich and Provorov 2011; De–Bashan et al. 2012; Saikia et al. 2012). This aims to offset not only the increase in costs (while increasing the productivity of crops) but also the environmental and political issues related to the use of chemical fertilizer inputs. Inoculation of agricultural crops with different microorganisms such as rhizobia (for legumes), AMF, or other PGPR such as P-solubilizing microbes and associative diazotrophs has been practiced for many years with variable successes (Khan et al. 2007; Rooney et al. 2009; Sharma et al. 2012; Pérez–Montaño et al. 2014). The first patent for a biofertilizer (rhizobia inoculant) was registered more than a century ago (Nobbe and Hiltner 1896). In 2000, the area of legumes treated with commercial biofertilizers was more than 40 million hectares annually (Phillips 2004), with about half of this figure being represented by soybean production (Catroux et al. 2001). The commercial production of rhizobial inoculants has thus been in place for a number of decades, partially replacing the need for mineral N fertilizers on legume crops in many countries (Rodríguez–Navarro et al. 2011). In some regions and some crops (e.g., peas in North America), the proportion of inoculated acreage can reach almost 100 % of the total acreage for the particular crop in that region, whereas for most crops, the figures are below 50 % of the planted acreage. In poor regions of Africa and elsewhere, application of rhizobia to leguminous crops still represents a negligible fraction of the overall legume production mostly due to lack of local research, information, infrastructure, and markets (Jansa et al. 2011). In contrast to rhizobial inoculants, P-solubilizing bacteria (e.g., *Bacillus* or *Pseudomonas*), root-associated diazotrophs (e.g., *Azospirillum*), and AMF have been used less frequently and on much lesser scale than the rhizobial inoculants. We estimate that no more than few thousand hectares are treated annually with non-rhizobial biofertilizers including AMF inoculants. This suggests that commercial non-rhizobial biofertilizer application does not affect global food production significantly, although specific inoculants (especially the *Azospirillum*-based products) may be important locally, e.g., in Cuba, Mexico, and Brazil (Saikia et al. 2010). The global agricultural crop production including legumes is estimated at some 1.6 billion hectares (www.fao.org/nr/solaw/solaw–home/en), but there is an obvious lack of market penetration by non-rhizobial biofertilizers in spite of decades of research. This indicates issues in the uptake and use of these products and is in contrast with the well-documented roles of microbes in specific aspects of plant nutrition (Douds et al. 2005; Cardoso and Kuyper 2006; Verbruggen et al. 2013) and with multiple research programs promoting large-scale application of such biofertilizers. Moreover, application of microbial inoculants does not always result in the desired effects as other factors such as abiotic and biotic soil conditions and climatic factors can play large and to a great extent unknown roles in the interplay between plants, microbes, and their shared environment (Wakelin et al. 2005). For example, Pereg and McMillan (2015) discussed the potential uses of beneficial microorganisms for increasing productivity in cotton cropping systems and pointed out that Australian cotton industry could greatly benefit from research into isolation of crop-specific beneficial microbes. While commercial biofertilizers are currently available for the cotton industry, they are not systematically applied because of inconsistent results. The authors also highlighted the need for taking into account the indigenous microbes because they often proved to be the most effective and most adapted to the environmental conditions in the cropping system for which they are intended. Some of these issues will be explored in this chapter, with a special focus on the biofertilizers that are used to improve nutrition and yields of grain, legume, oil, tuber, and other crops.

## 4.2 What Are the Biofertilizers?

Several definitions have been proposed over the years for the term "biofertilizer," and it generally refers to products containing one or more living microorganisms able to stimulate plant growth and development. Biofertilizers can increase the plant availability and/or uptake of mineral nutrients and are ideally formulated in an easy-to-use and economical carrier material (Arora et al. 2011; Malusa et al. 2012). According to Vessey (2003), a biofertilizer is a substance that contains living microorganisms which, when applied to seed, plant surfaces (leaves), roots or soil, colonize the rhizosphere or the interior of the plant and promote growth by several mechanisms that increase the supply or availability of primary nutrients to the host plant. Biofertilizers are also called microbial inoculants or bioformulations (Arora et al. 2011), but the term should not be used interchangeably with green manure, manure, intercrop, or organic-supplemented chemical fertilizer (Bhattacharyya and Jha 2012; Halpern et al. 2015). This term was first coined in 1978 for a group of microorganisms now known as PGPR (Kloepper et al. 1989) to define the free-living soil, rhizosphere, rhizoplane, and phyllosphere bacteria that, under some conditions, are beneficial for plants (Bashan and de Bashan 2005; Andrews et al. 2012). It has recently been proposed that the classification of PGPR should be confined to microbial strains that fulfill at least two of the three criteria such as aggressive colonization, plant growth stimulation, and biocontrol (Khalid et al. 2006; Bhattacharyya and Jha 2012). This chapter focuses on the biofertilizers including rhizobia, AMF, and other nonsymbiotic microorganisms (PGPR) specifically involved in plant nutrition, but does not address the ectomycorrhizal symbiosis, though this may be relevant for mineral nutrition and growth of some energy crops (willows, poplars) but not food crops and production of edible mushrooms including truffles.

## 4.3 Current Contribution of Biofertilizers (Successes and Failures) to Agricultural Production

The most well-known success story of the use of microbial inoculants is clearly the inoculation of legumes with rhizobia. Inoculation of legumes with $N_2$-fixing rhizobia has been used in agriculture for more than 100 years (Deaker et al. 2004; Sindhu et al. 2010). It has largely transformed legume production systems so that high production rates are not totally dependent on N fertilizer. Theoretically, high $N_2$ fixation rates of 1–2 kg N $ha^{-1}$ $day^{-1}$ should be possible in all legumes and would satisfy most of the legume's needs. Alves et al. (2003), Hungria et al. (2006), and Melchiorre et al. (2011) have shown that the grain yields obtained in Brazil, Argentina, and the USA, respectively, can reach up to 4 t $ha^{-1}$ per growing season through BNF and by the use of legume inoculants that contain both rhizobia and active compounds such as lipo-chito oligosaccharides (LCOs) which signal the symbiosis of rhizobia with legumes and the formation of nitrogen-fixing root nodules. There have been considerable efforts over many decades to enhance endophytic and associative nitrogen fixation in cereals and grasses through inoculation with free-living diazotrophs, most notably *Azotobacter*, *Azospirillum*, and more recently *Herbaspirillum*, *Gluconacetobacter*, and *Burkholderia* (Van Dommelen and Vanderleyden 2007; da Silva et al. 2012; Vargas et al. 2012). In Brazilian sugarcane plantations, nitrogen fixation rates of 40 kg N $ha^{-1}$ $year^{-1}$ have been measured although the contribution of nonsymbiotically fixed N to plants remains small and difficult to establish (Unkovich and Baldock 2008).

Promotion of plant growth by bacteria from the genus *Azospirillum* is largely attributed to the production of phytohormones, predominantly indole-3-acetic acid (IAA); but gibberellins and cytokinins have also been implicated (reviewed by Baca and Elmerich (2007) and Halpern et al. (2015)). Plants inoculated with *Azospirillum* have been observed to increase root respiration rates and water and mineral uptake (reviewed by Dobbelaere et al. 2001). A selection of reviews on field application of *Azospirillum* covering 34 years and a range of sites worldwide indicated a high proportion (70 %) of increased plant growth and yield with reported averages typically just below 10 % (Table 4.2). However, publication of negative results may be suppressed and this may strongly affect the accuracy of the reported proportion of positive responses. The review by Dobbelaere et al. (2001) concluded that beneficial effects of *Azospirillum* are observed on lighter soils with low organic matter and intermediate levels of fertilizer and water. Early effects on plant growth do not necessarily result in increased yield, which is dependent on growing conditions during later stages of plant and grain development. Inoculants used in the described studies included liquid- and peat-based formulations of various species and strains of *Azospirillum* applied to both seeds, and soil and plant growth promotion was generally observed only after consideration was given to inoculum potential and appropriate formulation. Recent analyses of other field experiments indicate similar rates of increased yield (Díaz–Zorita and Fernández–Canigia 2009; Veresoglou and Menexes 2010).

Ferreira et al. (2013) evaluated maize response to *Azospirillum brasilense* inoculation and nutrient (macronutrients and micronutrients) applications under field conditions in clay and sandy soils of the Brazilian Cerrado. Maize growth was significantly affected when inoculated

**Table 4.2** Summary of reviews on 34 years of field application of *Azospirillum* to cereal crops

| Authors (year of publication) | Review period | Sites | Outcomes/conditions |
|---|---|---|---|
| Okon and Labanderra-Gonzalez (1994) | 1974–1994 | USA, India, Thailand, Israel, Egypt, Italy, France, Brazil, Mexico, Uruguay, Argentina | Significant yield increases 5–30 % |
| | | | Positive yield responses in 60–70 % of trials |
| Dobbelaere et al. (2001) | 1994–2001 | Belgium, Uruguay, Mexico, Israel | Positive responses dependent on conditions |
| | | | Lighter soils, intermediate levels of fertilizer and water |
| Díaz-Zoríta and Fernandez-Canígia (2009) | 2002–2006 | Trials at 297 sites in Pampas region of Argentina | Yield increased 8 % at 70 % of sites |
| | | | Response was independent of fertilization and other crop management practices |
| Veresoglou and Menexes (2010) | 1981–2008 | Meta-analysis of 59 articles | Yield increased 8–9 % effect greatest at low N responses varied with plant and microbial species |

with *A. brasilense*, but the response was dependent on the soil type. The dual application of N fertilizer and *A. brasilense* increased grain yield by up to 29 % when compared to N fertilization alone. All these results on the field inoculation of crops with *Azospirillum* strains highlight the complexity of these plant–microbe interactions in the field and the number of factors playing a role on plant growth and grain yield. Marks et al. (2013) successfully tested the use of signaling molecules produced by rhizobia on both *Bradyrhizobium* spp. and *A. brasilense* field inoculation of soybean and maize, respectively. These results emphasize the potential of using secondary metabolites of rhizobia together with biofertilizers to improve the growth and yield of grain crops. According to Saikia et al. (2010), to gain the full potential of this association, the *Azospirillum* research should proceed toward more basic understanding of the underlying fundamental components of the system and less toward full-scale field experiments.

There is no doubt that bacterial biofertilizers can increase the yield of various crops significantly also through improved P acquisition (Hinsinger et al. 2015). However, field results are generally inconsistent despite some recent encouraging field inoculation studies. Krey et al. (2013) showed that maize plants inoculated with a P-solubilizing *Pseudomonas fluorescens* strain grew better and were significantly more infected by AMF than non-inoculated control plants. Kaur and Reddy (2014) identified two P-solubilizing bacterial strains playing an important role in plant growth promotion and improvement of soil fertility in different agroclimatic regions. Owen et al. (2015) reviewed the terminology, composition, and functions of P solubilizers and the many factors that impact their efficacy for increasing P availability in different soil and plant environments. These authors concluded that the effectiveness of the commercial biofertilizers is currently confounded by inadequate quality control standards and insufficient knowledge of the underlying mechanisms, which have led to contradicting reports on field performance.

AMF have been widely acknowledged as beneficial for plant P nutrition. Smith and Smith (2012) highlighted their capacity to very efficiently take up P from the soil to make it available for the associated plant. The outcome of this beneficial association will depend upon the soil characteristics and the environmental parameters such as temperature and soil moisture, as well as availability of other nutrients such as N (Johnson et al. 2015). For example, beneficial dual inoculation (AMF and rhizobia) has been successfully tested in the field to enhance woody legume growth (Lesueur et al. 2001; Lesueur and Duponnois 2005; Mortimer et al. 2013). Yet despite this, such kind of commercial biofertilizer remains rarely available on international markets, due mostly to production constraints and the lack of applied research.

Moreover, among the new biofertilizers that have recently appeared on various markets, many have not been subjected to scientific scrutiny. A 3-year Bill and Melinda Gates funded project (2009–2011) entitled "Evaluation and scaling up new chemical and biological commercial products for improving and sustaining crop yield in selected agro–ecological zones in sub Saharan Africa" (COMPRO) screened and evaluated several commercial biofertilizers from different countries of production (including rhizobia-, PGPR-, and AMF-based products) on major legume, cereal, and banana crops across diverse agroecological conditions in Kenya, Ethiopia, and Nigeria and under laboratory, greenhouse, and field conditions. Among the rhizobial inoculants, several were found very effective in increasing nodule biomass on soybean and increasing grain yield by up to 30 %, achieving a benefit–cost ratio of up to 5.0 (Thuita et al. 2012). The impact of inoculation was particularly significant in soil with low N content and when yields obtained without inoculation were low (0.5–1.0 t $ha^{-1}$). The efficiency seemed to be independent of the soybean variety and soil type. Results demonstrated economic returns of US$ 4 for every dollar invested for soybean production. However, when these effective rhizobial commercial inoculants were combined with biofertilizers containing AMF and/or P-solubilizing microorganisms, almost no additional effects were observed on crop yields (Jefwa et al. 2014). Commercial biofertilizers containing only PGPR bacteria were totally ineffective on major legumes, cereals, and banana crops not only under controlled conditions but also across diverse agroecological areas. Kavoo–Mwangi et al. (2013) demonstrated that tissue culture banana could benefit from application of AMF to improve survival and growth during the nursery phase as well as to enhance plant performance under field conditions. However, the effect of AMF was less evident in the field on both maize and soybean yields, particularly in the absence of mineral fertilizer application.

In the literature, a number of other studies have also looked at the effects of AMF inoculation, both under glasshouse and field conditions, to assess their impact on crop yields. The results were variable and often the direct impact of the AMF could not be quantified, as it is still difficult to (i) filter the contribution of the indigenous AMF out and (ii) measure the survival/function of the introduced genotypes under field conditions.

Other projects are currently ongoing in Africa aimed at improving the utilization of legumes and rhizobial inoculants to promote maize yields (SIMLESA project funded by ACIAR concerning East and Southern Africa http://aciar.gov.au/simlesa) and at promoting the utilization of Biological Nitrogen Fixation in African agricultural systems ($N_2$ Africa project funded by BMGF http://www.n2africa.org/). These projects should provide relevant results and technical protocols accessible to local African farmers for using appropriate and cheap biofertilizers to partially replace rare and expensive mineral fertilizers. Moreover, the information garnered from these projects should be transferable to other agricultural systems, particularly where a combination of high edaphic pressures and limited farmer access to fertilizer inputs results in poor crop yields.

## 4.4 Persistence of the Inoculated Microorganisms Contained Within Biofertilizers and Competition with Autochthonous Microbes

The soil is an extremely complex environment, inhabited by heterogeneous microbial communities interacting and communicating with each other via different chemical compounds secreted into the soil. Rhizobacteria constitute "common soil inhabitants" in all climate zones, from arctic to tropical areas, and are commonly found in all types of soils. After inoculation into soils, microorganisms contained in the biofertilizers must compete with the indigenous population for survival and are obviously disadvantaged as compared to the well-adapted native populations.

In the case of rhizobia, it has been shown that the presence, level, and diversity of native rhizobia in the soil are critical for inoculant performance. Successful inoculations were observed in soils with low native rhizobial populations and with a low N level. However, benefits are much less common in soils having strong established rhizobial populations, and inoculation may sometimes fail to produce the expected outcomes on nodule occupancy, BNF, and grain yield (McInnes and Haq 2007; Althabegoiti et al. 2008). Historically it was considered that only the presence of effective rhizobia in sufficient numbers in close proximity to the legume seed could ensure effective nodulation, BNF, and grain yield (Kaschuk et al. 2010; Torres et al. 2012). Depending on soil type, the required quantity of inoculated rhizobia in most cases was about $10^2$–$10^5$ cells inoculated $g^{-1}$ of soil (Andrade et al. 2002). There are numerous reports describing the effects of rhizobial inoculants in soils with a low level of soil N and few indigenous rhizobia. Skorupska et al. (2010) have made suggestions to explain why root nodules of legumes form a microenvironment not accessible to all rhizobia inhabiting rhizosphere: (i) the relatively high plant host–microsymbiont specificity as the nodules can be induced and consequently colonized only by rhizobia which recognize and exchange suitable molecular signals with the host and (ii) the variation in secreted Nod factors. Rhizobia recognized as "suitable microsymbionts" are subjected to a selection process during plant tissue invasion and colonization in two ways: they are exposed to intensive competition from other strains, and they are under some selective pressure from the host plant.

Soybean is one of the most important food crops in the world and provides an interesting example of whether or not to systematically employ rhizobial inoculation. High soybean grain yields require large amounts of N. To reach 1000 kg $ha^{-1}$ of grain produced, the plant requires approximately 80 kg of N $ha^{-1}$ although this varies between the different plant genotypes (Hungria et al. 2006). However, yields higher than 4000 kg $ha^{-1}$ have been obtained in both Brazil and Argentina using BNF as the main source of N (Hungria et al. 2006; Rodríguez–Navarro et al. 2011). Soybean nodulation requires specific *Bradyrhizobium* species, and populations of these bradyrhizobia are seldom available in soils where soybean has not been grown previously (Abaidoo et al. 2007). In an attempt to find a solution to this problem, the soybean breeding program at the International Institute of Tropical Agriculture (IITA) in Nigeria developed soybean genotypes designated TGx (tropical *glycine* cross) able to nodulate effectively with indigenous *Bradyrhizobium* spp. populations (Kueneman et al. 1984). These TGx soybean genotypes (also called promiscuous varieties) have been tested in many western, eastern, and southern African countries without N fertilizer or *Bradyrhizobium* inoculation, and results indicate that indigenous strains do not always meet the N requirements of the plants despite good levels of nodulation (Mpepereki et al. 2000; Musiyiwa et al. 2005; Wasike et al. 2009; Thuita et al. 2012). The benefits of promiscuous soybean varieties are still being debated in the absence of efficient strains of rhizobia in soils. Previous studies carried out under greenhouse conditions with promiscuous soybean varieties showed that inoculation with selected commercial biofertilizers increased nodulation, N fixation, and biomass yield (Thuita et al. 2012). This illustrates the importance of evaluating the symbiotic performance of any new proposed strain in a field environment for suitability and adaptability before being recommended for use in an inoculant.

Mycorrhizal inoculants are also exposed to competition with autochthonous genotypes of AMF. In theory, these fungi already inhabit every soil on Earth (Jansa et al. 2006). However, solid information on competition introduced with native fungi is scarce due to the fact that, until recently, it was difficult to track the inoculant genotypes of AMF in non-sterile soil or the roots of field plants. The main reasons were the multi-genomic nature of these fungi, the lack of cultivability, and the slow growth of these microbes, limiting the amounts of pure DNA for molecular genetic studies (Thonar et al. 2012). Some developments in sequencing AMF genomes and shifting focus from multi-allelic nuclear to more homogeneous mitochondrial genome (Lang and Hijri 2009) have enabled the design and use of specific molecular markers to identify individual genotypes (strains) of AMF, applicable both in the glasshouse and in the field (Boerstler et al. 2010; Kiers et al. 2011). However, not all results are yet available in a fully quantitative mode as some studies only apply specific genotype/strain markers using endpoint PCR (Sýkorová et al. 2012), whereas only few very recent papers employed specific markers in quantitative PCR (Kiers et al. 2011; Krak et al. 2012; Walker et al. 2012; Couillerot et al. 2013). Available data show that the inoculant strains sometimes survive for up to two seasons (Sýkorová et al. 2012). However their abundance generally rapidly decreases with time, and their beneficial effects established under model conditions (sterile soil) are masked due to intensive competition with native fungi (Biró et al. 2000), i.e., lack of long-term establishment of inoculants in the field. Yet another study substantiated that the identity of fungi applied on non-AMF seedlings greatly affected the community composition of AMF communities after transplantation into the field soils (Mummey et al. 2009), which may indicate that even a transient presence of the AMF inoculant may have long-lasting effects on the rhizosphere composition and functioning. Verbruggen et al. (2013) identified three factors determining inoculation success and AMF persistence in soils after inoculation: (i) species compatibilities, (ii) - field-carrying capacity, and (iii) priority effects. These authors explored how these factors can be employed for the establishment and persistence of AMF. Inoculant choice, plant choice, management practices, and timing of inoculation came up as being very important for the successful manipulation of the resulting AMF community. Lehmann et al. (2012) realized a meta-analysis study (from 1981 to 2010, 39 publications working on 320 different crop plant genotypes) about mycorrhizal responsiveness trends in annual crop plants and their wild relatives. They concluded that new cultivars were more mycorrhiza responsive (and possibly dependent) compared to ancestral genotypes. It means that no evidence was found indicating that new plant genotypes lost their ability to respond to AMF inoculation due to agricultural and breeding practices. The controversy about the low impact of AMF inoculation on crop yields thus cannot be explained by the utilization of these new plant genotypes.

## 4.5 Production of the Biofertilizers

The aim of inoculation is to supply a high number of viable, effective microorganisms to the rhizosphere soon after seed germination to optimize competition by introduced microorganisms with less effective resident strains. Development of adequate inoculants technologies to support the broad needs of current and potential biofertilizer microorganisms is challenging. Inoculant formulations must be designed to provide a suitable temporary microenvironment to support high numbers of viable cells and prevent the rapid decline of these populations during delivery under field conditions to maximize their effects on plants. A good inoculant must be nontoxic for plants, animals, and humans, permit optimal growth and survival of selected organisms during storage and inoculation (optimal water capacity, pH, composition), be made of a uniform and readily available carrier, and meet a low production cost. Different dispersal agents have been formulated depending on the microorganisms they contain, the target crop, the environmental conditions in which the products are used, the production cost, and the market availability and market demand. Four main dispersal products are distinguished currently in the market according to the kind of carrier materials used: powders, slurries, granules, and liquids. For each of them, a wide range of carriers are available. Most of the literature relating to commercial biofertilizers refers to bacterial inoculants, mainly due to the success story of rhizobial inoculants. However, there is a rapidly increasing interest in other types of inoculants, particularly incorporating AMF and other microorganisms showing root growth promotion properties. The number of companies producing inoculum of mycorrhizal fungi has increased around the world over the last decade. Currently, in Europe, approximately 20 companies actively produce and distribute mycorrhizal biofertilizers, meaning that there are real market opportunities for their application (Vosátka et al. 2008). Companies have taken different market approaches, ranging from products with a single AMF strain for specific targets to mixed products for more general markets. AMF inoculants are generally provided as granular substrates made of peat, compost, vermiculite, perlite, sand, and/or expanded clay and contain spores, colonized roots, hyphal segments, or mixture of the three (Dalpé and Monreal 2004). Baar (2008) also described liquid formulations but these often contained only spores. The most frequently used AMF species for commercial inoculum are *Rhizophagus* sp. (formerly *Glomus intraradices*) and other *Glomus* species. Interest in the *Gigaspora*, *Scutellospora*, and *Acaulospora* genera is gradually increasing for commercial inoculum production (Dalpé and Monreal 2004; Baar 2008). Due to their obligate symbiotic status, AMF need to be associated with plants for growth and proliferation (Dalpé and Monreal 2004), and this remains a major obstacle for their commercialization since their high-scale production requires more specific technical skills and substantial infrastructures (IJdo et al. 2011). AMF production is generally performed using pot cultures, but other soilless techniques such as NFT (nutrient film technique), aeroponic, and monoxenic cultures are sometimes used (Dalpé and Monreal 2004; IJdo et al. 2011; Malusa et al. 2012). An in vitro method using split-plate cultures and Ri T-DNA transformed roots of carrots was also developed and resulted in very high concentrations of spores and hyphae in a very limited space (Dalpé and Monreal 2004).

In addition of the technical manual published by Deaker et al. (2011) entitled "Practical methods for the quality control of inoculant biofertilizers," two recent review papers have described in details the challenges of formulation and quality of biofertilizers for successful inoculation (Herrmann and Lesueur 2013) and the advances in biofertilizer technology (Bashan et al. 2014). Bashan et al. (2014) made a critical point on the practical aspects of bacterial inoculants for both agriculture and environmental restoration. They suggested that new developments should use renewable, non-contaminated natural resources opening up new avenues for research. Herrmann and Lesueur (2013) emphasized the importance of a suitable carrier to protect the microorganisms from the often–harsh conditions during storage and to ensure their survival and establishment after introduction into soils as well as the need for adequate quality controls, both during production and storage.

Despite 100 years of experience, 90 % of all rhizobial inoculants commercially produced worldwide are recognized to have little or no demonstrated impacts whatsoever on the productivity of the legumes for which they are used (Brockwell and Bottomley 1995; Catroux et al. 2001). Similar numbers are true for other kinds of inoculants (Gianinazzi and Vosátka 2004), although the knowledge is not assembled in as clear way as for the rhizobial products. As stated by Herridge et al. (2002), assessment of inoculant efficacy should start with their quality as assessment of inoculant efficacy is meaningless if it is of poor quality, if the product is contaminated or there is a low count of the focal microorganisms. Inoculant manufacturers increasingly commercialize new biofertilizers around the world. However, most of the manufacturers are not willing to improve the quality of their products, as the resultant increase in cost to the manufacturer and ultimately the farmer is a major disincentive. In addition, farmers are unlikely to judge the quality of an inoculant before they buy it (Lupwayi et al. 2000). As a consequence, farmers may purchase low-quality inoculants, obtain no or little and inconsistent effect on crops (thus limited profit), and eventually lose confidence in the benefits of inoculation (Husen et al. 2007). The importance of the wider introduction of improved regulation and controls is widely acknowledged (Jenkins and Grzywacz 2000; Husen et al. 2007; Herridge 2008) to obtain better and more consistent results in the field and remove low-quality inoculants from the markets (Bhattacharyya and Jha 2012). These efforts have mainly focused on the bacterial biofertilizers in the past, but more recent establishment of the Federation of European Mycorrhizal Fungi Producers (www.femfip.com) aims at improving quality of mycorrhizal products at least at the European level.

However, two recent publications suggest why so far biofertilizers have not performed well under field conditions. Faye et al. (2013) assessed 12 commercial AMF inoculants sourced globally in a two-step experiment under greenhouse conditions. In this study, only three inoculants significantly increased root colonization levels compared to no sterile soil. Instead of the 12 declared AMF species, 13 fungal strains were extracted from the pot culture survey, including five undeclared species, while four declared species did not produce spores. Only three inoculants increased maize growth (shoot biomass) under controlled conditions. These authors highlighted the need to pre-evaluate commercial AMF inoculants on selected crop and soils before launching large-scale field use. Herrmann et al. (2015) characterized the microbial content of 65 commercial biofertilizers manufactured in the USA, the UK, Australia, South Africa, Thailand, Kenya, and Argentina. The results showed that 64 % of the products contained one or several strains of contaminants, while only 36 % of the products could be considered as pure. Rhizobial inoculants were generally of better quality than the other PGPR-based products, and 40 % of the tested biofertilizers did not contain any of the claimed strains but only contaminants, among which human pathogen such as *Comamonas testosteroni* or *Serratia marcescens* was found. These results highlight the need for better quality control systems, to ensure that efficacious inoculants reach the end users.

Inoculant quality is driven by several factors, primarily the number and viability of the selected strain(s) and the level of contaminants that may inhibit growth of selected strains as well as provide a potential risk to the health of humans, animals, and plants. A good formulation must also provide an extended shelf life under less than optimal storage conditions (Lupwayi et al. 2000; Catroux et al. 2001), and proper labeling must give sufficient information about the nature of the product, the shelf life, and instructions for its use (Hungria et al. 2003; Husen et al. 2007). However, shelf life, expiry date, and storage information are generally not clearly stated and difficult to obtain (Gemell et al. 2005; Husen et al. 2007), and this clearly makes the establishment of a functional and reliable quality control system difficult.

To achieve the goal of setting up a quality control system worldwide, standards must be defined. As raised by Herrmann and Lesueur (2013), to date, no international agreement of standards for inoculant quality is in place. Existing regulations generally concern rhizobial inoculants only, vary from country to country, and may or may not be enforced. Where specified, standards only concern the minimum number of viable rhizobia to be inoculated per seed or per unit of weight of inoculant and the maximum level of contaminants tolerated in the final products (Lupwayi et al. 2000). The minimum standards for rhizobia required to sustain adequate nodulation have generally been presented as a number to be applied per seed and vary depending on seed size. While there are no global standards, there is general agreement in the published literature that numbers should be of 500 cfu seed$^{-1}$ for legumes with very small seeds (e.g., white clover), $10^3$ cfu seed$^{-1}$ for small seeds (e.g., subterranean clover and lucerne), $10^4$ cfu seed$^{-1}$ for medium seeds (e.g., lentil), and $10^5$ cfu seed$^{-1}$ for large seeds (e.g., soybean and faba bean) (Lupwayi et al. 2000;

**Table 4.3** Summary of the main commercial legume inoculant formulations and their properties in relation to application and efficacy

| Product | Properties relating to survival and efficacy |
|---|---|
| Moist peat: still major inoculant carrier for legumes in Australia and Canada | Long shelf life – cells undergo physiological and morphological changes that improve survival on seed |
| Liquids most common in the USA and South America (soybean) – applied to seed, large market for pre-inoculated soybean seed (the USA) | Shelf life and survival on seed of liquid formulations vary according to formulation and strain |
| Other formulations developed to provide flexibility in application: granules and freeze-dried products | Clay and peat based granules can be applied directly to the seedbed. Use of granules can be an advantage where survival on seed is a limiting factor (Brockwell et al. 1980) |
| | Water injection using freeze-dried inoculants eliminates issues of blockages in filters and spray nozzles |
| Formulations developed to improve survival on seed – dry clay powder | Dry clay powder (the USA) developed for pre-inoculated seed production (Dormal®) – claims of better survival on seed |

Herridge et al. 2002). There is evidence that increasing numbers above these levels can further increase nodulation, N fixation, and yield in both competitive and noncompetitive soil environments (Denton et al. 2013). It is interesting to point out that legume yield is not decreased when rhizobia are applied in excess, and therefore there are no conceivable maximum doses for rhizobial inoculants. On the other hand, the mode of application of the rhizobial inoculants can significantly affect the response of the plant even when the amount of rhizobia applied remains constant (Odee et al. 2002; Diouf et al. 2003).

There are no available standards for PGPR and it may explain the results described by Herrmann et al. (2015) on the quality of this category of biofertilizers. The authors of reviews on field experiments with *Azospirillum* (listed in Table 4.3) agreed that high numbers (i.e., $10^7$–$10^8$ cfu seed$^{-1}$) of *Azospirillum* are required to produce positive yield responses in cereals (Okon and Labandera–Gonzalez 1994; Dobbelaere et al. 2001). However, positive responses were observed in Argentina after application of liquid inoculants at a rate of $10^5$ cfu seed$^{-1}$ (Díaz–Zorita and Fernández–Canigia 2009). Inoculants prepared using dried alginate microbeads can provide a cell concentration of up to $10^9$ cfu seed$^{-1}$ and deliver an inoculum potential of $10^5$ cfu seed$^{-1}$ (Bashan et al. 2002, 2014). Peat cultures of *Azospirillum* have been produced commercially (i.e., Azogreen, Liphatech, France) and applied in field experiments, but survival of *Azospirillum* in peat is relatively poor as viable cell numbers can drop from $10^9$ to $10^7$ cfu seed$^{-1}$ in a matter of months (Okon and Itzigsohn 1995). The relationships between minimum cfu seed$^{-1}$, selected strain, plant genotype, inoculant formulation, and positive plant growth responses are still unclear from the published information on field application of *Azospirillum*.

Likewise, there are no broadly established standards for application of mycorrhizal inoculants. As mycorrhizal fungi are present in virtually all soils, introduced strains of mycorrhizal fungi must always compete with indigenous communities unless these have previously been removed by soil sterilization such as in intensive glasshouse production (Jansa et al. 2006). However, results described by Faye et al. (2013) strongly suggest a change in the current production system in order to significantly improve the quality of the biofertilizers and to deliver better performing AMF biofertilizers. However, the control of the purity and the load of infective propagules of AMF biofertilizers are technically more difficult than the control of bacterial biofertilizers. Some fungal isolates abundant (mainly) in tropical soils are also known to establish extensive mycelium networks, but hardly any dormant structures such as vesicles or spores, which makes them very susceptible to substrate disturbance or inoculant storage (Boddington and Dodd 2000; Jansa et al. 2003). A recent study (Ehinger et al. 2012) showed that different variants of the same AMF species with novel phenotypes could be selected among clonal spore lines generated from one initial single AMF spore. This highlights the complicated dynamic nature of AMF genetics, which needs to be better understood and exploited for the improvement of the AMF inoculant efficiency and to guarantee long-term stability of the product quality.

## 4.6 Future Research Needs

It is reasonable to expect that the need for BNF in crop production will at least double by 2050 if we are to feed a growing global population, and the expanded use of leguminous species has the potential to ameliorate a suite of problems facing agriculture around the world. BNF has the potential to reduce the application of manufactured nitrogenous fertilizer by approximately 160 million tons per annum, equating to a reduction of 270 million tons of coal or equivalent energy consumed in the production process. Given current concerns about climate change caused by greenhouse gas emissions, this represents a valuable improvement. In addition, N is likely to become increasingly expensive to produce as the cost of fossil fuels used in manufacturing and transportation processes escalates. Improvements in symbiotic BNF therefore promise significant economic benefits to farmers and consumers. Reduced reliance on external inputs such as synthetic N can improve ecosystem health and enhance the sustainability of agricultural systems ranging from industrial to subsistence farming.

Although field experiments with nonsymbiotic biofertilizers show real potential, it is still unclear as to whether these products will reliably substitute chemical fertilizer inputs to any great extent. Little is understood of the conditions under which these inoculants may work, and the lack of information about their quality, application practices, and potential in published studies makes it difficult to determine if these are optimized for any given conditions. What is very much needed, especially in the area of AMF inoculants, is the assessment of inoculum potential required for a given farming system, soil, and climatic conditions. This has been acknowledged for some time and algorithms to solve these issues were developed decades ago as illustrated in Fig. 4.3. There is also a pressing need for a more universal quality control system, providing the same regulations and standards to be used by manufacturers during formulation and production. The availability of good quality biofertilizers for farmers in countries where crops play a key role in the economy and food security must be a priority. In the absence of high-quality and efficacious biofertilizers, the dependence on the use of mineral fertilizers will not decrease. Moreover, the confidence of the end users in biofertilizers may be completely undermined if an inoculant does not clearly result in enhanced yield and/or reduced input requirements.

A real effort is required to develop new carriers and new formulations, providing "easy-to-use" products, more adapted to the farmers' requirements, together with improved quality and shelf life under less than optimal conditions. These new products are likely to contain additives that promote the survival of microorganisms during distribution and application as well as enhancing plant growth-promoting functional characteristics (Herrmann and Lesueur 2013; Bashan et al. 2014). One important aspect that is often overlooked is the fact that commercial inoculants, because they contain living microorganisms, are often dispersing potentially invasive and/or pathogenic organisms in their nonnative range (Schwartz et al. 2006). This practice has an inherent risk of environmental damage, but very little effort (at least to our knowledge) has so far been undertaken to quantify these risks and compare them rigorously with the economic and social benefits of inoculant application.

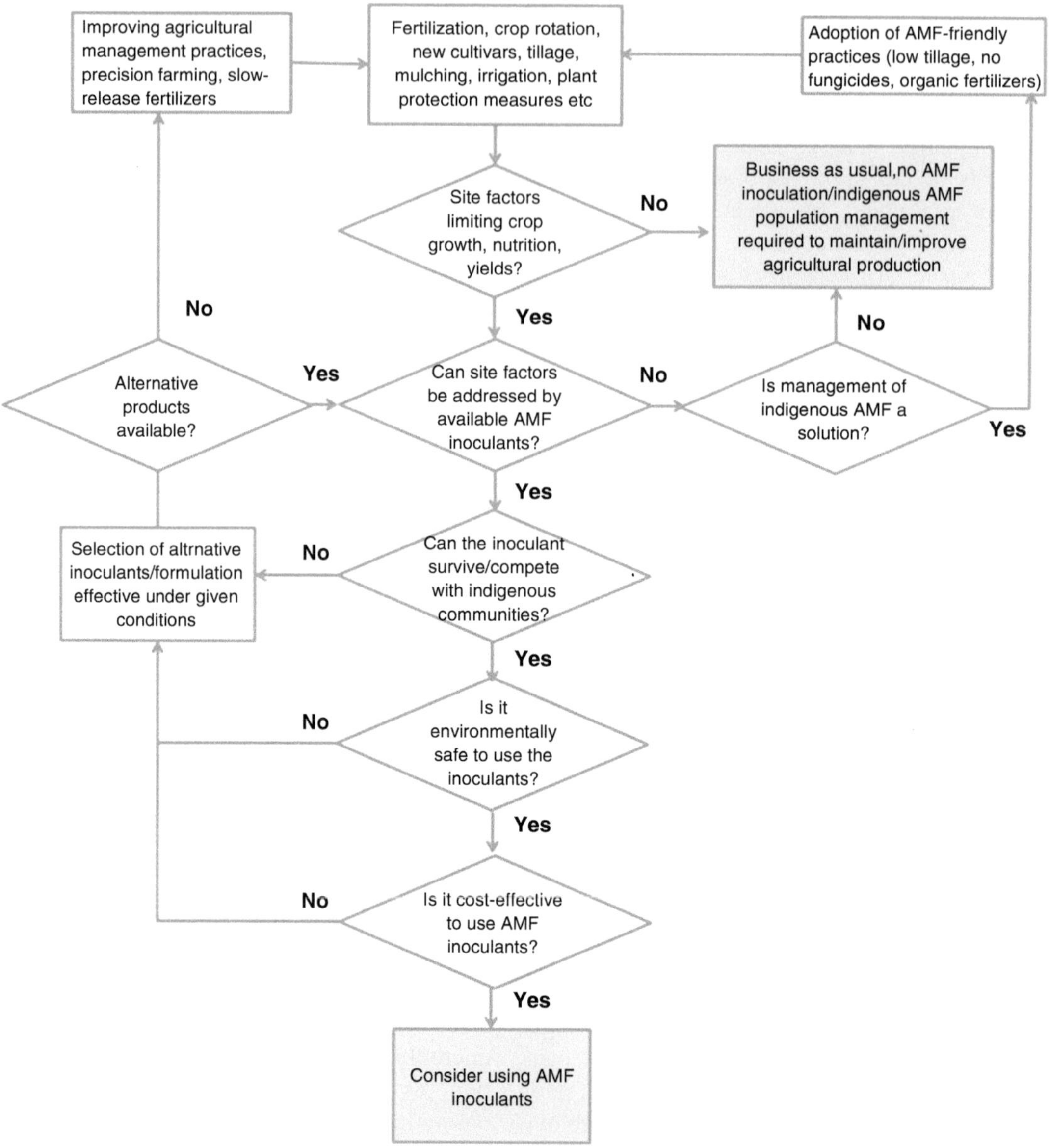


**Fig. 4.3** Scheme of a site-specific decision strategy on the use of arbuscular mycorrhizal fungal (AMF) inoculants in agricultural production. Combined and modified from Dodd and Thomson (1994), Findlay and Kendle (2001) and Schwartz et al. (2006)

## 4.7 Conclusion

Biofertilizers can significantly improve plant growth. However, in the field, beneficial impacts of biofertilizers are unfortunately rare. The way forward for changing that is to strengthen collaboration of scientists with biofertilizer manufacturers because it would be of great benefit at expediting the development of new and efficacious biofertilizers. While scientists are mainly looking at the development of new products at the laboratory scale, the private sector can assist in addressing the issue of industrial scale-up and economic feasibility of the newly developed biofertilizers. That way, all the required parameters (such as strain selection, culture conditions, specifications for different crops and target markets, as well as logistical considerations that could influence the final product) would be considered in the development of efficacious commercially viable new inoculants. Clearly, alongside the development and production aspects of biofertilizers, one of the most promising ways to increase biofertilizer efficacy is to introduce robust quality control systems. End users may then buy these products with confidence and compare their usefulness and cost–benefit ratios with traditional mineral fertilizer inputs and/or other management interventions in their specific farming system, edaphic, and environmental conditions.

**Acknowledgments** JJ was supported by the Fellowship of J E Purkyně, Czech Science Foundation (P504121665), Ministry of Education, Youth and Sports (LK11224), and the long-term development program RVO 61388971. LH and DL were partially supported by the "COMPRO" project funded by Bill and Melinda Gates Foundation (2009–2011).

The authors are indebted to Prof David Herridge for his useful review of the Ms and his important comments and suggestions to improve it.

## References

Abaidoo RC, Keyser HH, Singleton PW, Dashiell KE, Sanginga N (2007) Population size, distribution and symbiotic characteristics of indigenous *Bradyrhizobium* spp. that nodulate TGx soybean genotypes in Africa. Appl Soil Ecol 35:57–67

Adesemoye AO, Kloepper JW (2009) Plant–microbes interactions in enhanced fertilizer–use efficiency. Appl Microbiol Biotechnol 85:1–12

Althabegoiti MJ, Lopez–Garcia SL, Piccinetti C, Mongiardini EJ, Perez–Gimenez J, Quelas JI, Perticari A, Lodeiro AR (2008) Strain selection for improvement of *Bradyrhizobium japonicum* competitiveness for nodulation of soybean. FEMS Microbiol Lett 282:115–123

Alves BJ, Boddey RM, Urquiaga S (2003) The success of BNF in soybean in Brazil. Plant Soil 252:1–9

Andrade DS, Murphy PJ, Giller KE (2002) The diversity of *Phaseolus*–nodulating rhizobial populations is altered by liming of acid soils planted with *Phaseolus vulgaris* L. in Brazil. Appl Environ Microbiol 68:4025–4034

Andrews M, Cripps MG, Edwards GR (2012) The potential of beneficial microorganisms in agricultural systems. Ann Appl Biol 160:1–5

Arora NK, Khare E, Maheshwari DK (2011) Plant growth promoting rhizobacteria: constraints in bioformulation, commercialization, and future strategies. In: Maheshwari DK (ed) Plant growth and health promoting bacteria, vol 18, Microbiology monographs. Springer, Berlin

Avis TJ, Gravel V, Antoun H, Tweddell RJ (2008) Multifaceted beneficial effects of rhizosphere microorganisms on plant health and productivity. Soil Biol Biochem 40:1733–1740

Baar J (2008) From production to application of arbuscular mycorrhizal fungi in agricultural systems: requirements and needs. In: Varma A (ed) Mycorrhiza. Springer, Berlin

Babalola OO (2010) Beneficial bacteria of agricultural importance. Biotechnol Lett 32:1559–1570

Baca BE, Elmerich C (2007) Microbial production of plant hormones. In: Elmerich C, Newton WE (eds) Associative and endophytic nitrogen–fixing bacteria and cyanobacterial associations. Springer, Dordrecht

Bashan Y, de Bashan LE (2005) Plant growth–promoting. Encycl Soils Environ 1:103–115

Bashan Y, Hernandez JP, Leyva LA, Bacilio M (2002) Alginate microbeads as inoculant carriers for plant growth–promoting bacteria. Biol Fertil Soils 35:359–368

Bashan Y, De–Bashan LE, Prabhu SR, Hernandez JP (2014) Advances in plant growth–promoting bacterial inoculant technology: formulations and practical perspectives (1998–2013). Plant Soil 378:1–33

Berendsen RL, Pieterse CMJ, Bakker PAHM (2012) The rhizosphere microbiome and plant health. Trends Plant Sci 17:478–486

Berg G (2009) Plant–microbe interactions promoting plant growth and health: perspectives for controlled use of microorganisms in agriculture. App Microbiol Biotechnol 84:11–18

Bhattacharyya PN, Jha DK (2012) Plant growth–promoting rhizobacteria (PGPR): emergence in agriculture. World J Microbiol Biotechnol 28:1327–1350

Biró B, Köves–Péchy K, Vörös I, Takács T, Eggenberger P, Strasser RJ (2000) Interrelations between *Azospirillum* and *Rhizobium* nitrogen–fixers and arbuscular mycorrhizal fungi in the rhizosphere of alfalfa in sterile, AMF–free or normal soil conditions. Appl Soil Ecol 15:159–168

Boddington CL, Dodd JC (2000) The effect of agricultural practices on the development of indigenous arbuscular mycorrhizal fungi I. Field studies in an Indonesian ultisol. Plant Soil 218:137–144

Boerstler B, Thiéry O, Sýkorová Z, Berner A, Redecker D (2010) Diversity of mitochondrial large subunit rDNA haplotypes of *Glomus intraradices* in two agricultural field experiments and two semi-natural grasslands. Mol Ecol 19:1497–1511

Brockwell J, Bottomley PJ (1995) Recent advances in inoculant technology and prospects for the future. Soil Biol Biochem 27:683–697

Brockwell J, Gault RR, Chase DL, Hely FW, Zorin M, Corbin EJ (1980) An a appraisal of practical alternatives to legume seed inoculation: field experiments on seed bed inoculation with solid and liquid inoculants. Aust J Agric Res 31:47–60

Bünemann EK, Oberson A, Frossard E (2011) Phosphorus in action: biological processes in soil phosphorus cycling. Springer–Verlag, Berlin

Canfield DE, Glazer AN, Falkowski PG (2010) The evolution and future of Earth's nitrogen cycle. Science 330:192–196

Cardoso IM, Kuyper TW (2006) Mycorrhizas and tropical soil fertility. Agric Ecosyst Environ 116:72–84

Catroux G, Hartmann A, Revellin C (2001) Trends in rhizobial inoculant production and use. Plant Soil 230:21–30

Choudhary DK, Sharma KP, Gaur RK (2011) Biotechnological perspectives of microbes in agro–ecosystems. Biotechnol Lett 33:1905–1910

Compant S, Duffy B, Nowak J, Clément C, Barka EA (2005) Use of plant growth–promoting bacteria for biocontrol of plant diseases: principles, mechanisms of action, and future prospects. Appl Environ Microbiol 71:4951–4959

Compant S, Clément C, Sessitsch A (2010) Plant growth–promoting bacteria in the rhizo–and endosphere of plants: their role, colonization, mechanisms involved and prospects for utilization. Soil Biol Biochem 42:669–678

Cordell D, Drangert JO, White S (2009) The story of phosphorus: global food security and food for thought. Glob Environ Change 19:292–305

Couillerot O, Ramírez–Trujillo A, Walker V, Von Felten A, Jansa J, Maurhofer M, Défago G, Prigent–Combaret C, Comte G, Caballero–Mellado J, Moënne–Loccoz Y (2013) Comparison of prominent *Azospirillum* strains in *Azospirillum–Pseudomonas–Glomus* consortia for promotion of maize growth. Appl Microbiol Biotechnol 97:4639–4649

da Silva MF, de Souza Antônio C, de Oliveira PJ, Xavier GR, Rumjanek NG, de Barros Soares LH, Reis VM (2012) Survival of endophytic bacteria in polymer–based inoculants and efficiency of their application to sugarcane. Plant Soil 356:231–243

Dakora FD, Chimpango SBM, Valentine AJ, Elmerich C, Newton WE (2008) Biological nitrogen fixation: towards poverty alleviation through sustainable agriculture. In: Proceedings of the 15th international nitrogen fixation congress and the 12th international conference of the African association for biological nitrogen fixation. Springer, Berlin

Dalpé Y, Monreal M (2004) Arbuscular mycorrhiza inoculum to support sustainable cropping systems. Crop Manag Online. doi:10.1094/CM-2004-0301-09-RV

De–Bashan LE, Hernandez JP, Bashan Y (2012) The potential contribution of plant growth–promoting bacteria to reduce environmental degradation– a comprehensive evaluation. Appl Soil Ecol 61:171–189

Deaker R, Roughley RJ, Kennedy IR (2004) Legume seed inoculation technology – a review. Soil Biol Biochem 36:1275–1288

Deaker R, Kecskés ML, Rose MT, Khanok–on A, Ganisan K, Tran TKC, Vu TN, Phan TC, Nguyen TH, Kennedy IR (2011) Practical methods for the quality control of inoculant biofertilisers, ACIAR monograph no 147. Australian Centre for International Agricultural Research, Canberra

Denton MD, Pearce DJ, Peoples MB (2013) Nitrogen contributions from faba bean (*Vicia faba* L.) reliant on soil rhizobia or inoculation. Plant Soil 365:363–374

Díaz–Zorita M, Fernández–Canigia MV (2009) Field performance of a liquid formulation of *Azospirillum brasilense* on dryland wheat productivity. Eur J Soil Biol 45:3–11

Diouf D, Forestier S, Neyra M, Lesueur D (2003) Optimisation of inoculation of *Leucaena leucocephala* and *Acacia mangium* with rhizobium under greenhouse conditions. Ann For Sci 60:379–384

Dobbelaere S, Croonenborghs A, Thys A, Ptacek D, Vanderleyden J, Dutto P, Labandera–Gonzalez C, Caballero–Mellado J, Aguirre JF, Kapulnik Y (2001) Responses of agronomically important crops to inoculation with *Azospirillum*. Funct Plant Biol 28:871–879

Dodd JC, Thomson BD (1994) The screening and selection of inoculant arbuscular-mycorrhizal and ecotomycorrhizal fungi. Plant Soil 159:149–158

Douds DD Jr, Nagahashi G, Pfeffer PE, Kayser WM, Reider C (2005) On–farm production and utilization of arbuscular mycorrhizal fungus inoculum. Can J Plant Sci 85:15–21

Drigo B, Pijl AS, Duyts H, Kielak AM, Gamper HA, Houtekamer MJ, Boschker HTS, Bodelier PLE, Whiteley AS, van Veen JA (2010) Shifting carbon flow from roots into associated microbial communities in response to elevated atmospheric CO2. Proc Natl Acad Sci 107:10938–10942

Dumas M, Frossard E, Scholz RW (2011) Modeling biogeochemical processes of phosphorus for global food supply. Chemosphere 84:798–805

Dutta S, Podile AR (2010) Plant growth promoting rhizobacteria (PGPR): the bugs to debug the root zone. Crit Rev Microbiol 36:232–244

Ehinger MO, Croll D, Koch AM, Sanders IR (2012) Significant genetic and phenotypic changes arising from clonal growth of a single spore of an arbuscular mycorrhizal fungus over multiple generations. New Phytol 196:853–861

Faye A, Dalpé Y, Ndung'u–Magiroi K, Jefwa J, Ndoye ID, Lesueur D (2013) Evaluation of commercial arbuscular mycorrhizal inoculants on maize in Kenya. Can J Plant Sci 93:1201–1208

Ferreira AS, Pires RR, Rabelo PG, Oliveira RC, Luz JMQ, Brito CH (2013) Implications of *Azospirillum brasilense* inoculation and nutrient addition on maize in soils of the Brazilian Cerrado under greenhouse and field conditions. Appl Soil Ecol 72:103–108

Findlay CM, Kendle AD (2001) Towards a mycorrhizal application decision model for landscape management. Landsc Urban Plan 56:149–160

Frossard E, Bünemann E, Jansa J, Oberson A, Feller C (2009) Concepts and practices of nutrient management in agro–ecosystems: can we draw lessons from history to design future sustainable agricultural production systems. Die Bodenkultur 60:43–60

Gahan J, Schmalenberger A (2014) The role of bacteria and mycorrhiza in plant sulfur supply. Front Plant Sci 5:1–7

Gao X, Hoffland E, Stomph TJ, Grant CA, Zou C, Zhang F (2012) Improving zinc bioavailability in transition from flooded to aerobic rice. A review. Agron Sustain Dev 32:465–478

Gemell LG, Hartley EJ, Herridge DF (2005) Point–of–sale evaluation of preinoculated and custom–inoculated pasture legume seed. Anim Prod Sci 45:161–169

Gianinazzi S, Vosátka M (2004) Inoculum of arbuscular mycorrhizal fungi for production systems: science meets business. Can J Bot 82:1264 1271

Giri B, Giang PH, Kumari R, Prasad R, Varma A (2005) Microbial diversity in soils. In: Buscot F, Varma A (eds) Microorganisms in soils: roles in genesis and functions. Springer, Berlin

Glick BR (2012) Plant growth–promoting bacteria: mechanisms and applications. Scientifica article ID 963401

Halpern M, Bar–Tal A, Ofek M, Minz D, Muller T, Yermiyahu U (2015) The use of biostimulants for enhancing nutrient uptake. In: Sparks DL (ed) Advances in agronomy, vol 129. Elsevier, Boston, pp 141–174

Harman GE, Howell CR, Viterbo A, Chet I, Lorito M (2004) *Trichoderma* species–opportunistic, avirulent plant symbionts. Nat Rev Microbiol 2:43–56

Hayat R, Ali S, Amara U, Khalid R, Ahmed I (2010) Soil beneficial bacteria and their role in plant growth promotion: a review. Ann Microbiol 60:579–598

Herridge DF (2008) Inoculation technology for legumes. In: Dilworth MJ, James EK, Sprent JI, Newton WE (eds) Nitrogen–fixing leguminous symbioses. Springer, Dordrecht

Herridge DF, Gemell G, Hartley E (2002) Legume inoculants and quality control. In: Herridge DF (ed) Inoculants and nitrogen fixation of legumes in Vietnam. ACIAR, Canberra

Herridge DF, Peoples MB, Boddey RM (2008) Global inputs of biological nitrogen fixation in agricultural systems. Plant Soil 311:1–18

Herrmann L, Lesueur D (2013) Challenges of formulation and quality of biofertilizers for successful inoculation. Appl Microbiol Biotechnol 97:8859–8873

Herrmann L, Atieno M, Brau L, Lesueur D (2015) Microbial quality of commercial inoculants to increase BNF and nutrient use efficiency. In: de Bruijn FJ (ed) Molecular microbial ecology of the rhizosphere. Wiley – Blackwell, Hoboken

Hinsinger P, Bengough AG, Vetterlein D, Young IM (2009) Rhizosphere: biophysics, biogeochemistry and ecological relevance. Plant Soil 321:117–152

Hinsinger P, Herrmann L, Lesueur D, Robin A, Trap J, Waithaisong K, Plassard C (2015) Impact of roots, microorganisms and microfauna on the fate of soil P in the rhizosphere. In: Plaxton W, Lambers H (eds) Annual plant reviews, vol 48, Phosphorus metabolism in plants. Wiley, Chichester

Howell CR (2003) Mechanisms employed by *Trichoderma* species in the biological control of plant diseases: the history and evolution of current concepts. Plant Dis 87:4–10

Hungria M, Campo RJ, Mendes I (2003) Benefits of inoculation of the common bean (*Phaseolus vulgaris*) crop with efficient and competitive *Rhizobium tropici* strains. Biol Fertil Soils 39:88–93

Hungria M, Campo RJ, Mendes IC, Graham PH (2006) Contribution of biological nitrogen fixation to the N nutrition of grain crops in the tropics: the success of soybean (*Glycine max* L. Merr.) in South America. In: Singh RP, Shankar N, Jaiwal PK (eds) Nitrogen nutrition and sustainable plant productivity. Studium Press, Houston

Husen E, Simanungkalit RDM, Saraswati R (2007) Characterization and quality assessment of Indonesian commercial biofertilizers. Indones J Agric Sci 8:31–38

IJdo M, Cranenbrouck S, Declerck S (2011) Methods for large–scale production of AM fungi: past, present, and future. Mycorrhiza 21:1–16

Jansa J, Mozafar A, Kuhn G, Anken T, Ruh R, Sanders IR, Frossard E (2003) Soil tillage affects the community structure of mycorrhizal fungi in maize roots. Ecol Appl 13:1164–1176

Jansa J, Wiemken A, Frossard E (2006) The effects of agricultural practices on arbuscular mycorrhizal fungi. In: Frossard E, Blum WEH, Warkentin BP (eds) Function of soils for human societies and the environment. Geological Society, London

Jansa J, Bationo A, Frossard E, Rao IM (2011) Options for improving plant nutrition to increase common bean productivity in Africa. In: Bationo A, Waswa B, Okeyo JM, Maina F, Kihara J, Mokwunye U (eds) Fighting poverty in sub–Saharan Africa: the multiple roles of legumes in integrated soil fertility management. Springer, Heidelberg

Jefwa JM, Pypers P, Jemo M, Thuita M, Mutegi E, Laditi MA, Faye A, Kavoo A, Munyahali W, Herrmann L, Atieno M, Okalebo JR, Yusuf A, Ibrahim A, Ndung'u–Magiroi KW, Asrat A, Muletta D, Ncho C, Kamaa M, Lesueur D (2014) Do commercial biological and chemical products increase crop yields and economic returns under smallholder farmer conditions? In: Vanlauwe B, van Asten P, Blomme G (eds) Challenges and opportunities for agricultural Intensification of the humid highland systems of sub–Saharan Africa. Springer International Publishing, Cham

Jenkins NE, Grzywacz D (2000) Quality control of fungal and viral biocontrol agents–assurance of product performance. Biocontrol Sci Technol 10:753–777

Johnson NC, Wilson GWT, Wilson JA, Miller RM, Bowker MA (2015) Mycorrhizal phenotypes and the law of the minimum. New Phytol 205:1473–1484

Kaschuk G, Leffelaar PA, Giller KE, Alberton O, Hungria M, Kuyper TW (2010) Responses of legumes to rhizobia and arbuscular mycorrhizal fungi: a meta–analysis of potential photosynthate limitation of symbioses. Soil Biol Biochem 42:125–127

Kaur G, Reddy MS (2014) Influence of P–solubilizing bacteria on crop yield and soil fertility at multilocational sites. Eur J Soil Biol 61:35–40

Kavoo–Mwangi AM, Kahangi EM, Ateka E, Onguso J, Mukhongo RW, Mwangi EK, Jefwa JM (2013) Growth effects of microorganisms based commercial products inoculated to tissue cultured banana cultivated in three different soils in Kenya. Appl Soil Ecol 64:152–162

Khalid A, Arshad M, Zahir ZA (2006) Phytohormones: microbial production and applications. In: Ball AS, Fernandes E, Herren H, Husson O, Laing M, Palm C, Pretty J, Sanchez P, Sanginga N, Thies J (eds) Biological approaches to sustainable soil systems. CRC Press, Taylor & Francis, New York

Khan MS, Zaidi A, Wani PA (2007) Role of phosphate–-solubilizing microorganisms in sustainable agriculture – a review. Agron Sustain Dev 27:29–43

Kiers ET, Duhamel M, Beesetty Y, Mensah JA, Franken O, Verbruggen E, Fellbaum CR, Kowalchuk GA, Hart MM, Bago A, Palmer TM, West SA, Vandenkoornhyuse P, Jansa J, Bücking H (2011) Reciprocal rewards stabilize cooperation in the mycorrhizal symbiosis. Science 333:880–882

Kloepper JW, Lifshitz R, Zablotowicz RM (1989) Free–living bacterial inocula for enhancing crop productivity. Trends Biotechnol 7:39–44

Krak K, Janoušková M, Caklová P, Vosátka M, Štorchová H (2012) Intraradical dynamics of two coexisting isolates of the arbuscular mycorrhizal fungus *Glomus intraradices* sensu lato as estimated by Real–Time PCR of mitochondrial DNA. Appl Environ Microbiol 78:3630–3637

Krey T, Vassilev N, Baum C, Eichler–Löbermann B (2013) Effects of long–term phosphorus application and plant–growth promoting rhizobacteria on maize phosphorus nutrition under field conditions. Eur J Soil Biol 55:124–130

Kueneman EA, Root WR, Dashiell KE, Hohenberg J (1984) Breeding soybean for the tropics capable of nodulating effectively with indigenous *Rhizobium* spp. Plant Soil 82:387–396

Lamont BB, Pérez–Fernández M, Rodríguez–Sánchez J (2014) Soil bacteria hold the key to root cluster formation. New Phytol 3:1156–1162

Lang FB, Hijri M (2009) The complete *Glomus intraradices* mitochondrial genome sequence – a milestone in mycorrhizal research. New Phytol 183:3–6

Leake JR, Ostle NJ, Rangel–Castro JI, Johnson D (2006) Carbon fluxes from plants through soil organisms determined by field $^{13}CO_2$ pulse–labelling in an upland grassland. Appl Soil Ecol 33:152–175

Lehmann A, Barto EK, Powell JR, Rillig MC (2012) Mycorrhizal responsiveness trends in annual crop plants and their wild relatives–a meta–analysis on studies from 1981 to 2010. Plant Soil 355:231–250

Lendenmann M, Thonar C, Barnard RL, Salmon Y, Werner RA, Frossard E, Jansa J (2011) Symbiont identity matters: carbon and phosphorus fluxes between Medicago truncatula and different arbuscular mycorrhizal fungi. Mycorrhiza 21:689–702

Lesueur D, Duponnois R (2005) Relations between rhizobial nodulation and root colonization of *Acacia crassicarpa* provenances by an arbuscular mycorrhizal fungus, *Glomus intraradices* Schenk and Smith or an ectomycorrhizal fungus, *Pisolithus tinctorius* Coker & Couch. Ann For Sci 62:467–474

Lesueur D, Ingleby K, Odee D, Chamberlain J, Wilson J, Manga TT, Sarrailh JM, Pottinger A (2001) Improvement of forage production in *Calliandra calothyrsus*: methodology for the identification of an effective inoculum containing *Rhizobium* strains and arbuscular mycorrhizal isolates. J Biotechnol 91:269–282

Lugtenberg B, Kamilova F (2009) Plant–growth–promoting rhizobacteria. Annu Rev Microbiol 63:541–556

Lupwayi NZ, Olsen PE, Sande ES, Keyser HH, Collins MM, Singleton PW, Rice WA (2000) Inoculant quality and its evaluation. Field Crops Res 65:259–270

Malusa E, Sas–Paszt L, Ciesielska J (2012) Technologies for beneficial microorganisms inocula used as biofertilizers. Sci World J. ID. 491206–doi:10.1100/2012/491206

Marks BB, Megías M, Nogueira MA, Hungria M (2013) Biotechnological potential of rhizobial metabolites to enhance the performance of *Bradyrhizobium* spp. and *Azospirillum brasilense* inoculants with soybean and maize. AMB Express 3:1–10

McInnes A, Haq K (2007) Contributions of rhizobia to soil nitrogen fertility. In: Abbott LK, Murphy DV (eds) Soil biological fertility. Springer, Dordrecht

Melchiorre M, De Luca MJ, Anta GG, Suarez P, Lopez C, Lascano R, Racca RW (2011) Evaluation of bradyrhizobia strains isolated from field–grown soybean plants in Argentina as improved inoculants. Biol Fertil Soils 47:81–89

Miransari M (2011) Soil microbes and plant fertilization. App Microbiol Biotechnol 92:875–885

Morgan JAW, Bending GD, White PJ (2005) Biological costs and benefits to plant–microbe interactions in the rhizosphere. J Exp Bot 56:1729–1739

Mortimer PE, Le Roux MR, Pérez–Fernández MA, Benedito VA, Kleinert A, Xu J, Valentine AJ (2013) The dual symbiosis between arbuscular mycorrhiza and nitrogen fixing bacteria benefits the growth and nutrition of the woody invasive legume *Acacia cyclops* under nutrient limiting conditions. Plant Soil 366:229–241

Mpepereki S, Javaheri F, Davis P, Giller KE (2000) Soyabeans and sustainable agriculture. Promiscuous

soyabeans in southern Africa. Field Crops Res 65:137–149
Mummey DL, Antunes PM, Rillig MC (2009) Arbuscular mycorrhizal fungi pre–inoculant identity determines community composition in roots. Soil Biol Biochem 41:1173–1179
Musiyiwa K, Mpepereki S, Giller KE (2005) Symbiotic effectiveness and host ranges of indigenous rhizobia nodulating promiscuous soyabean varieties in Zimbabwean soils. Soil Biol Biochem 37:1169–1176
Nobbe F, Hiltner L (1896) Inoculation of the soil for cultivating leguminous plants US Patent n°570813
Odee D, Indieka S, Lesueur D (2002) Evaluation of inoculation procedures for *Calliandra calothyrs*us Meisn. grown in tree nurseries. Biol Fertil Soils 36:124–128
Okon Y, Itzigsohn R (1995) The development of *Azospirillum* as a commercial inoculant for improving crop yields. Biotechnol Adv 13:415–424
Okon Y, Labandera–Gonzalez CA (1994) Agronomic applications of *Azospirillum*: an evaluation of 20 years worldwide field inoculation. Soil Biol Biochem 26:1591–1601
Owen D, Williams AP, Griffith GW, Withers PJA (2015) Use of commercial bio–inoculants to increase agricultural production through improved phosphorus acquisition. Appl Soil Ecol 86:41–54
Palacios OA, Bashan Y, De–Bashan LE (2014) Proven and potential involvement of vitamins in interactions of plants with plant growth–promoting bacteria–an overview. Biol Fertil Soils 50:415–432
Pereg L, McMillan M (2015) Scoping the potential uses of beneficial microorganisms for increasing productivity in cotton cropping systems. Soil Biol Biochem 80:349–358
Pérez–Montaño F, Alías–Villegas C, Bellogín RA, del Cerro P, Espuny MR, Jiménez–Guerrero I, López–Baena FJ, Ollero FJ, Cubo T (2014) Plant growth promotion in cereal and leguminous agricultural important plants: from microorganism capacities to crop production. Microbiol Res 169:325–336
Phillips PWB (2004) An economic assessment of the global inoculant industry. Crop Management 3:Online
Richardson AE, Barea JM, McNeill AM, Prigent–Combaret C (2009) Acquisition of phosphorus and nitrogen in the rhizosphere and plant growth promotion by microorganisms. Plant Soil 321:305–339
Rockström J, Steffen W, Noone K, Persson Å, Chapin FS, Lambin EF, Lenton TM, Scheffer M, Folke C, Schellnhuber HJ, Nykvist B, de Wit CA, Hughes T, van der Leeuw S, Rodhe H, Sörlin S, Snyder PK, Costanza R, Svedin U, Falkenmark M, Karlberg L, Corell RW, Fabry VJ, Hansen J, Walker B, Liverman D, Richardson K, Crutzen P, Foley JA (2009) A safe operating space for humanity. Nature 461:472–475
Rodríguez–Navarro DN, Oliver IM, Contreras MA, Ruiz–Sainz JE (2011) Soybean interactions with soil microbes, agronomical and molecular aspects. Agron Sustain Dev 31:173–190
Roesch LFW, Fulthorpe RR, Riva A, Casella G, Hadwin AKM, Kent AD, Daroub SH, Camargo FAO, Farmerie WG, Triplett EW (2007) Pyrosequencing enumerates and contrasts soil microbial diversity. ISME J 1:283–290
Rooney DC, Killham K, Bending GD, Baggs E, Weih M, Hodge A (2009) Mycorrhizas and biomass crops: opportunities for future sustainable development. Trends Plant Sci 14:542–549
Saharan BS, Nehra V (2011) Plant growth promoting rhizobacteria: a critical review. Life Sci Med Res 21:1–30
Saikia SP, Dutta SP, Oswami A, Hau BS, Anjilal PB (2010) Role of *Azospirillum* in the improvement of legumes. In: Khan MS, Musarrat J, Zaidi A (eds) Microbes for legume improvement. Springer, Vienna
Saikia SP, Bora D, Goswami A, Mudoi KD, Gogoi A (2012) A review on the role of *Azospirillum* in the yield improvement of non leguminous crops. Afr J Microbiol Res 6:1085–1102
Sanon A, Andrianjaka ZN, Prin Y, Bally R, Thioulouse J, Comte G, Duponnois R (2009) Rhizosphere microbiota interfers with plant–plant interactions. Plant Soil 321:259–278
Schröder JJ, Cordell D, Smit AL, Rosemarin A (2010) Sustainable use of phosphorus: EU tender ENV. B1/ETU/2009/0025. Report no.357. Plant Research International, Wageningen
Schwartz MW, Hoeksema JD, Gehring CA, Johnson NC, Klironomos JN, Abbott LK, Pringle A (2006) The promise and the potential consequences of the global transport of mycorrhizal fungal inoculum. Ecol Lett 9:501–515
Sharma RC, Sarkar S, Das D, Banik P (2012) Impact assessment of arbuscular mycorrhiza, *Azospirillum* and chemical fertilizer application on soil health and ecology. Commun Soil Sci Plant Anal 44:1116–1126
Sindhu SS, Dua S, Verma MK, Khandelwal A (2010) Growth promotion of legumes by inoculation of rhizosphere bacteria. In: Khan MS, Musarrat J, Zaidi A (eds) Microbes for legume improvement. Springer, Vienna
Singh JS, Pandey VC, Singh DP (2011) Efficient soil microorganisms: a new dimension for sustainable agriculture and environmental development. Agric Ecosyst Environ 140:339–353
Skorupska A, Wielbo J, Kidaj D, Marek–Kozaczuk M (2010) Enhancing *Rhizobium*–legume symbiosis using signaling factors. In: Khan MS (ed) Microbes for legume improvement. Springer, Wien
Smith SE, Smith FA (2012) Fresh perspectives on the roles of arbuscular mycorrhizal fungi in plant nutrition and growth. Mycologia 104:1–13
Smith SE, Facelli E, Pope S, Smith FA (2010) Plant performance in stressful environments: interpreting new and established knowledge of the roles of arbuscular mycorrhizas. Plant Soil 326:3–20
Sýkorová Z, Börstler B, Zvolenská S, Fehrer J, Gryndler M, Vosátka M, Redecker D (2012)

Long–term tracing of *Rhizophagus irregularis* isolate BEG140 inoculated on *Phalaris arundinacea* in a coal mine spoil bank, using mitochondrial large subunit rDNA markers. Mycorrhiza 22:69–80

Thonar C, Erb A, Jansa J (2012) Real–time PCR to quantify composition of arbuscular mycorrhizal fungal communities–marker design, verification, calibration and field validation. Mol Ecol Resour 12:219–232

Thuita M, Pypers P, Herrmann L, Okalebo RJ, Othieno C, Muema E, Lesueur D (2012) Commercial rhizobial inoculants significantly enhance growth and nitrogen fixation of a promiscuous soybean variety in Kenyan soils. Biol Fertil Soils 48:87–96

Tikhonovich IA, Provorov NA (2011) Microbiology is the basis of sustainable agriculture: an opinion. Ann Appl Biol 159:155–168

Torres AR, Kaschuk G, Saridakis GP, Hungria M (2012) Genetic variability in *Bradyrhizobium japonicum* strains nodulating soybean [*Glycine max* (L.) Merrill]. World J Microbiol Biotechnol 28:1831–1835

Unkovich M, Baldock J (2008) Measurement of asymbiotic $N_2$ fixation in Australian agriculture. Soil Biol Biochem 40:2915–2921

Vacheron J, Desbrosses G, Bouffaud ML, Touraine B, Moënne-Loccoz Y, Muller D, Legendre L, Wisniewski-Dyé F, Prigent-Combaret C (2013) Plant growth–promoting rhizobacteria and root system functioning. Front Plant Sci 4:1–19

van der Heijden MGA, Martin FM, Selosse MA, Sanders IR (2015) Mycorrhizal ecology and evolution: the past, the present, and the future. New Phytol doi. doi:10.1111/nph.13288

Van Dommelen A, Vanderleyden J (2007) Associative nitrogen fixation. In: Bothe H, Ferguson SJ, Newton WE (eds) Biology of the nitrogen cycle. Elsevier, Amsterdam

Vargas L, de Carvalho TLG, Ferreira PCG, Baldani VLD, Baldani JI, Hemerly AS (2012) Early responses of rice (*Oryza sativa* L.) seedlings to inoculation with beneficial diazotrophic bacteria are dependent on plant and bacterial genotypes. Plant Soil 356:127–137

Verbruggen E, Heijden MGA, Rillig MC, Kiers ET (2013) Mycorrhizal fungal establishment in agricultural soils: factors determining inoculation success. New Phytol 197:1104–1109

Veresoglou SD, Menexes G (2010) Impact of inoculation with *Azospirillum* spp. on growth properties and seed yield of wheat: a meta–analysis of studies in the ISI Web of Science from 1981 to 2008. Plant Soil 337:469–480

Vessey JK (2003) Plant growth promoting rhizobacteria as biofertilizers. Plant Soil 255:571–586

Vosátka M, Albrechtová J, Patten R (2008) The international market development for mycorrhizal technology. In: Varma A (ed) Mycorrhiza. Springer, Berlin

Wakelin SA, Warren RA, Ryder MH (2005) Effect of soil properties on growth promotion of wheat by *Penicillium radicum*. Soil Res 42:897–904

Walker V, Couillerot O, Von Felten A, Bellvert F, Jansa J, Maurhofer M, Bally R, Moënne–Loccoz Y, Comte G (2012) Variation of secondary metabolite levels in maize seedling roots induced by inoculation with *Azospirillum*, *Pseudomonas* and *Glomus* consortium under field conditions. Plant Soil 356:151–163

Wasike VW, Lesueur D, Wachira FN, Mungai NW, Mumera LM, Sanginga N, Mburu H, Mugadi D, Wango P, Vanlauwe B (2009) Genetic diversity of indigenous *Bradyrhizobium* nodulating promiscuous soybean [*Glycine max* (L) Merr.] varieties in Kenya: impact of phosphorus and lime fertilization in two contrasting sites. Plant Soil 322:151–163

Werner D, Newton WE (2005) Nitrogen fixation in agriculture, forestry, ecology, and the environment. Springer, Dordrecht www.fao.org/nr/solaw/solaw–home/en

Yang J, Kloepper JW, Ryu CM (2009) Rhizosphere bacteria help plants tolerate abiotic stress. Trends Plant Sci 14:1–4

# Effect of Bioinoculants on the Quality of Crops

5

Elisa Bona, Guido Lingua, and Valeria Todeschini

## Contents




**Abstract**

Soil microorganisms like arbuscular mycorrhizal fungi (AMF) and plant growth-promoting rhizobacteria (PGPR) are known to positively affect plant health, growth, and nutrition. Their use was often suggested in order to reduce the input of chemicals in agriculture. Recently, it has been observed that AMF and PGPR can also affect some quality features of various crops. In this chapter we review the literature concerning the effects of soil microbes, focusing on AMF and PGPR as the most common plant-associated microorganisms, on the quality of crops. Such effects were considered according to commercial and agronomic plant categories. Current limitations of the available information and some possible future developments have been indicated.


## 5.1 Introduction

Several soil microorganisms, belonging to very different taxa, have shown the ability to improve plant growth and health, especially under conditions of low nutrient availability. The involved microbes include phosphate-solubilizing, indole acetic acid (IAA)-releasing, 1-aminocyclopropane 1-carboxylate (ACC) deaminase-producing bacteria, actinomycetes, and a number of fungal species, mycorrhizal or


E. Bona • G. Lingua (✉) • V. Todeschini
Dipartimento di Scienze e Innovazione Tecnologica, Università del Piemonte Orientale "A. Avogadro", viale T. Michel 11, 15121 Alessandria, Italy
e-mail: guido.lingua@uniupo.it




N.K. Arora et al. (eds.), *Bioformulations: for Sustainable Agriculture*,
DOI 10.1007/978-81-322-2779-3_5

not. The establishment of relationships between such microorganisms and plants results in improved plant nutrition and better tolerance of biotic and abiotic stresses. Hence, the use of inocula based on these microbes was proposed to reduce the impact of chemicals and to support sustainable agriculture.

In recent years, some of the above mentioned microorganisms have also shown the ability to affect the quality of plant products, including some fruits, vegetables, and secondary metabolites (e.g., essential oils) (Kapoor et al. 2002a; Copetta et al. 2006; Khaosaad et al. 2006; Zubek et al. 2012; Bona et al. 2015). Providing a definition for the "quality" of crops is a difficult task. It can depend on area-specific criteria, and it is often connected to economical and commercial factors or transformation processes. Anything concerning the improvement of desirable traits (like taste, smell, and look) and the healthiness of the products (e.g., increased concentration of healthy molecules, as antioxidants) can however gain a quite large consensus. The purpose of this work is, therefore, the revision of the existing literature concerning the improvement of some traits of various crops, related to the quality of products, in plants inoculated with AMF or PGPR. Other organisms, like *Trichoderma* species, can exert positive effects on plants; however, we limited this study to AMF and PGPR because they are the most widespread plant-associated microorganisms. AMF are obligate biotrophs classified into the *Glomeromycota*; they form a symbiosis with the roots of most land plants, improving plant nutrition and growth (Smith and Read 2008). PGPR are a very heterogeneous group of bacteria that can exert their beneficial role directly or indirectly; direct promotion of plant growth is usually accomplished via increased uptake of nutrients (especially nitrogen, phosphate, and iron) or interfering with the plant hormone balance, modulating the concentration of substances like auxins, cytokinin, and ethylene. Indirect beneficial effects are mostly observed when PGPR inhibit or circumscribe the noxious effects of plant pathogens (Glick 2012).

Although the above definition of "quality" is anthropocentric, the topic is of interest because it can help to:

- Broaden the possible applications for bioinoculants, providing suggestions and directions about their utility and use in agriculture, with the perspective of further economical return and lower environmental impact.
- Support reliable enterprises that commercialize inocula.
- Provide new perspectives for the definition of typical products, relying on specific interactions with microbes characteristic or dominant under certain conditions or in certain areas.
- Better understand the specificity of some interactions.
- Recognize what biological and agronomic aspects require further investigation.

The subjects of increased yield, improved P nutrition, better stress tolerance, and reduced fertilization have extensively been treated elsewhere (Gianinazzi et al. 2002; Smith and Read 2008) and will not be included in this essay. When appropriate, however, reference to the above mentioned topics will be introduced.

For the purposes of the present work, the various types of crops have been organized into categories which reflect similarities in the way of growing, in the used part of the plant, in the kind of metabolic pathway involved, and so on. The idea is that plants with similar features might benefit from the same kind of inocula. Such a criterion is of course arbitrary, but we expect that this would help in the formulation of new inoculants specifically addressed to some kind of cultivation and product.

Before starting the examination of the various crop categories, a disclaimer is necessary: we cannot make any claim of completeness. "Product quality" is almost never included in the keywords of a paper; therefore, the bibliographic search for this study went through the manual screening and sorting of (literally) thousands of papers resulting from searches based on the name of the plant species and microbial group. We apologize for any paper that has skipped our attention. In addition, for reasons of simplicity, AMF have been indicated with the names used in the original articles, in spite of the relatively recent taxonomic revision (Schüßler and Walker 2010).

As described before, we chose eight main categories on the basis of the commercial use: (1) fresh fruits, (2) nuts and oilseeds, (3) root and bulb vegetables, (4) fruiting vegetables, (5) cabbages and leaf and stem vegetables, (6) fresh and grain legumes, (7) cereals, and (8) aromatic plants, spices, tea, coffee, herbal infusion, and cocoa.

## 5.2 Fresh Fruits (Table 5.1)

Fresh fruits include citrus, Maloideae, Prunoideae, berries, grape wine, small and large fruits with inedible peel, and fruits with edible peel. They are a nutritional source of sugars, vitamins (mainly C and E), carotenoids, flavonoids, and phenol compounds collectively called phytochemicals, with antioxidant activity. The consumption of fresh fruits and vegetables in the diet is important for the prevention of several diseases (chronic, cancer, and cardiovascular diseases) (Giovannetti et al. 2012; Sbrana et al. 2014). Both abiotic (soil chemical-physical properties, growth conditions, postharvest storage) and biotic factors (plant species or cultivar and interaction with soil microorganisms, mainly AMF and PGPR) can modulate the content of antioxidant compounds. Different studies highlighted the beneficial effects of these symbiotic microorganisms on plant physiology, but only a small part of them included the assessment of the quality of plant products (leaves, roots, fruits, or vegetables). Concerning the "citrus fruit" category, Nautiyal et al. (2008) demonstrated that the rhizospheric bacterium *Bacillus lentimorbus* induces the modulation of phenolic compounds and dietary antioxidants in *Citrus sinensis*.

"Berry fruits" (blackberry, blueberry, mulberry, cranberry, strawberry, etc.), mostly belonging to the family Rosaceae, are consumed as fresh fruits and are used to produce fruit juice or jam. They are a good source of vitamins C (ascorbic acid) and B9 (folic acid), fibers, minerals, and antioxidant compounds (anthocyanins, phenolic acids, and flavonols). Many epidemiological studies reported that anthocyanin consumption decreases the risk of obesity, coronary heart disease, and various types of cancer (Karlsen et al. 2007; Krikorian et al. 2010; Kaume et al. 2012).

Blackberry (*Rubus* sp.) is an aggregate fruit, composed of small drupelets, especially rich in polyphenols included flavonoids (anthocyanins) with medicinal properties (Kaume et al. 2012; Ramos-Solano et al. 2014). In a field experiment, Ramos-Solano and coworkers demonstrated that blackberry inoculation with *Pseudomonas fluorescens* N21.4 (CECT7620) affected secondary metabolism increasing fruit production and quality in terms of flavonoids and other phenol compounds (Ramos-Solano et al. 2014). Similar results were reported for fruits of *Fragaria x ananassa* Duch (strawberry), by Castellanos-Morales et al. (2010) in plants inoculated with the AM fungus *Glomus intraradices* (Castellanos-Morales et al. 2010), by Palencia et al. (2013) in plant inoculated with *Glomus mosseae* and *G. intraradices*, and by Lingua et al. (2013) and Bona et al. (2015) in plants of the same species inoculated with *P. fluorescens* Pf4 and *Pseudomonas* sp. 5Vm1K alone or in combination with a commercial inoculum consisting of a mix of AMF (Mybasol s.r.l.). Moreover, it is noteworthy that in this last study, bioinoculants affect sugar metabolism and vitamin concentration (vitamin C and B9). Sweeter fruits are produced in PGPR-inoculated plants than in controls (Pešaković et al. 2013), and this increase of the sweetness index is especially due to sucrose concentration, as shown by Bona et al. (2015). Strawberry flavor is due to the balance of organic acids and soluble sugars (Kallio et al. 2000). Citric and malic acid are the most important acids in strawberry fruits, while glucose, fructose, and sucrose are the principal sugars, representing about the 99 % of the total carbohydrate content (Perez et al. 1997;

**Table 5.1** Fresh fruits

| Reference | Plant family | Plant species | Rhizosphere microbe | n. fruits | Maturation time | Fruit size | Acidity | Total solids | Sugars | Organic acids | Vitamins | Minerals | Flavonoid | Sweetness index | Phenol compounds |
|---|---|---|---|---|---|---|---|---|---|---|---|---|---|---|---|
| Attia et al. (2009) | Musaceae | *Maghrabi banana* | PGPR (*P. fluorescens*; *B. megaterium*) | | | | ✔ | ✔ | | | | ✔ | | | |
| Bona et al. (2015) | Rosaceae | *Fragaria x ananassa* | PGPR (*P. fluorescens* (Pf4) and *Pseudomonas* sp. (5Vm1K)); AM fungi (commercial inoculum Mybasol s.r.l. (*R. intraradices*, *G. aggregatum*, *G. viscosum*, *C. etunicatum*, *C. claroideum*)) | ✔ | ✔ | ✔ | ✔ | | ✔ | ✔ | ✔ | | | ✔ | |
| Castellanos-Morales et al. (2010) | Rosaceae | *Fragaria x ananassa* Duch | AM fungi (*G. intraradices*) | | | | | | | | | ✔ | ✔ | | ✔ |
| Esitken et al. (2006) | Rosaceae | *Prunus avium* L. (sweet cherry) | PGPR (*Pseudomonas* sp., *Bacillus* sp.) | | | | | | | | | ✔ | | | |
| Lingua et al. (2013) | Rosaceae | *Fragaria x ananassa* | PGPR (*P. fluorescens* (Pf4) and *Pseudomonas* sp. (5Vm1K)); AM fungi (commercial inoculum Mybasol s.r.l. (*R. intraradices*, *G. aggregatum*, *G. viscosum*, *C. etunicatum*, *C. claroideum*)) | | | | | | | | | | ✔ | | |
| Mamatha et al. (2002) | Moraceae, Caricaceae | Mulberry (*Morus* sp.); papaya var. Solo (*Carica papaya* L.) | PGPR (*B. coagulans*); AM fungi (*G. fasciculatum*); mixed culture of *G. mosseae* and *G. caledonium* | ✔ | | ✔ | | | | | | | | | |
| Nautiyal et al. (2008) | Rutaceae | *C. sinensis* | PGPR (*B. lentimorbus* B-30488) | | | | | | | | | | | | ✔ |

| | | | | | | | | | | | | | | | |
|---|---|---|---|---|---|---|---|---|---|---|---|---|---|---|---|
| Orhan et al. (2006) | Rosaceae | *Rubus idaeus* L. (raspberry) | PGPR (two *Bacillus* strains) | ✔ | | | ✔ | | | | | | | | |
| Palencia et al. (2013) | Rosaceae | *Fragaria* × *ananassa* Duch. "Splendor" | AM fungi (*G. mosseae* and *G. intraradices*) | | | | ✔ | ✔ | | | | | | | |
| Pešaković et al. (2013) | Rosaceae | *Fragaria x ananassa* | PGPR (*K. planticola*); liquid inoculum (*A. chroococcum*, *A. vinelandii*, *Derxia* sp., *B. megaterium*, *B. licheniformis*, and *B. subtilis*) | | ✔ | ✔ | ✔ | ✔ | ✔ | | | | | | |
| Ramos-Solano et al. (2014) | Rosaceae | *Rubus* sp. var. Lochness (blackberries) | PGPR (*P. fluorescens* N21.4 (Spanish Type Culture Collection accession number CECT 7620)) | | | | | ✔ | | | | | ✔ | | ✔ |
| Sinclair et al. (2014) | Rosaceae | *Fragaria* × *ananassa* Duch. cultivars Albion, Charlotte, and Seascape | AM fungi (*F. mosseae*; *F. caledonius*; *R. irregularis*) | | | | ✔ | | | | | | | | |
| Verginer et al. (2010) | Vitaceae | *V. vinifera* L. | PGPR (*Paenibacillus* sp.). | | | | | | | ✔ | | | ✔ | | |

Summary of the main parameters influenced by the interaction between plants, belonging to the "fresh fruits" category, and microorganisms (PGPR or AMF). The term "fruit size" includes weight, size, and biomass of the crop fruit. References are listed in alphabetical order

Castellanos-Morales et al. 2010). Besides their impact on flavor, acids can affect the gelling properties of pectin (Castellanos-Morales et al. 2010; Sinclair et al. 2014). Vitamin C is associated with healthy properties of fruits, so this is an attractive feature for consumers. Citrus fruits have the greatest reputation for a high content of vitamin C, but there are some other fruits, such as strawberries, that have a higher content of ascorbic acid (Perez et al. 1997; Bona et al. 2015).

Finally, as far as berries are concerned, Orhan et al. (2006) reported an increase of berry number in raspberry plants inoculated with *Bacillus* M3 alone or in combination with *Bacillus* OSU-142.

AMF and PGPR are able to modulate fruit quality also in woody plants such as mulberry and papaya. Mamatha et al. (2002) reported the results obtained from the inoculation of 10-year-old mulberry plants with *Glomus fasciculatum* and 1.5-year-old papaya plants (var. Solo) with a mixed culture of *G. mosseae* and *Glomus caledonium*, with or without *Bacillus coagulans*. Increased yield and P uptake are the most relevant results. Sweet cherry is one of the most important fruit crops grown in Anatolia, and Turkey is the biggest sweet cherry-producing country in the world (Esitken et al. 2006). Floral and foliar applications of *Pseudomonas* BA-8 and *Bacillus* OSU-142 in sweet cherry plants enhance fruit size, compared with the control (Esitken et al. 2006).

In some crops, volatile compounds are important for the flavor of the final products, as in the case of grapevine (*Vitis vinifera* L.). In fact, in this plant, compounds such as 1-propanol, 2-butanone have a strong impact on the aroma of the produced wines and give them their characteristic flavor. Verginer et al. (2010) reported the isolation and identification of different rhizospheric microorganisms (bacteria and fungi) that influenced the production of volatile compounds in fruits.

Finally, Attia and coworkers reported that banana plants inoculated with *P. fluorescens* and *Bacillus megaterium* show enhanced fruit size, acidity, total soluble solids, and micro- and macronutrient content of fruits (Attia et al. 2009).

## 5.3 Nuts and Oilseeds (Table 5.2)

In this section we have included plants with dry seeds for the production of oils and olive trees. Some cereals, like maize, might have been considered as well, but we decided to discuss each species only in one section.

Overall the amount of available information is rather limited. For instance, there are thousands of papers concerning the interaction between sunflower and AMF or PGPR, but, to the best of our knowledge, none reports information concerning the quality of seeds or oil.

In the case of peanuts (*Arachis hypogaea*), *G. mosseae* and *Rhizobium* sp. can increase the shoot and root content of phenolic compounds; the effect is highest with dual inoculation, and it is associated to the appearance of some new molecules, absent in the uninoculated plants (Devi and Reddy 2002). These results suggest that it might be worth searching for some effects on the fruits.

Flax (*Linum usitatissimum*) can be grown in different varieties, for the production of fiber or oil. Inoculation with *G. mosseae* or *G. intraradices*, or their mixture, results in increased fiber production. In oil flax, the same two AMF slightly decrease the amount of non-saturated fatty acids and increase that of saturated fatty acids. These two AMF also show competition against *Sinorhizobium* sp., decreasing its abundance when co-inoculated. As a consequence, in this condition, the effect of co-inoculation is not synergistic (Rydlová et al. 2011).

AMF can also colonize various woody (shrubs or trees) species, including hazelnut (*Corylus avellana*), walnut (*Juglans regia*), macadamia (*Macadamia tetraphylla*), and olive (*Olea europaea*), whose fruits or seeds are used for the production of oil. So far, the effects of colonization on fruit quality have not been investigated, but it was shown that *G. mosseae* or *G. intraradices* BEG 72 can dramatically improve the survival of micropropagated walnut plants if inoculated in the post-acclimatization

**Table 5.2** Nuts and oilseeds

| Reference | Plant family | Plant species | Rhizosphere microbe | n. fruits | Fruit size | Minerals | Phenol compounds | Oil content |
|---|---|---|---|---|---|---|---|---|
| Devi and Reddy (2002) | Fabaceae | *Arachis hypogaea* L. | PGPR (*Rhizobium* sp.); AM fungi (*G. mosseae*) | | | | ✓ | |
| Estaun et al. (2003) | Oleaceae | *O. europaea* L. | AM fungi (*G. intraradices*, *G. mosseae*) | ✓ | ✓ | | | |
| Rydlova et al. (2011) | Linaceae | *L. usitatissimum* L. (oil and fiber flax) | PGPR (*Sinorhizobium* sp.); AM fungi (*G. mosseae*, *G. intraradices*, *G. mosseae* + *G. intraradices*) | | | | | ✓ |

Summary of the main parameters influenced by the interaction between plants, belonging to the "nuts and oilseeds" category, and microorganisms (PGPR or AMF). The term "fruit size" includes weight, size, and biomass of the crop fruit. References are listed in alphabetical order

phase (Dolcet-Sanjuan et al. 1996). In the case of micropropagated hazelnut plants, colonization with AMF improves growth rather than survival (Mirabelli et al. 2009). Drought tolerance of Macadamias improves when mycorrhized (Yooyongwech et al. 2013). Similarly, we could not find any information concerning the effect of PGPR on the quality of *O. europaea* fruits; actually, the only report concerns the rooting induction of some auxin-producing bacteria on olive cuttings (Montero-Calasanz et al. 2013). Olive trees can form arbuscular mycorrhiza (Citernesi et al. 1998) although species-specific responses have been reported: three different cultivars of olive plants have been found to be colonized by a *Gigaspora* species, but only one of the three by a *Glomus* species (Chatzistathis et al. 2013). The use of inocula has been evaluated in order to increase the success of transplantation (Palenzuela et al. 2002), to improve water-deficit tolerance (Caravaca et al. 2003), and to protect plants against parasites (Castillo et al. 2006). Inoculation with *G. mosseae* or *G. intraradices* improves plant growth and fruit yield compared to non-inoculated plants, with more pronounced effects for the latter fungus; however, under field conditions, the inocula effects tend to diminish with time, due to the presence of naturally occurring propagules which trigger the colonization of uninoculated plants (Estaun et al. 2003). The phospholipid and glycolipid content of leaves increases in olive plants inoculated with *G. intraradices* DAOM 197198 (Mechri et al. 2014); unfortunately, no data are yet available on the lipid content of fruits.

## 5.4 Root and Bulb Vegetables (Table 5.3)

Potatoes, sweet potatoes, carrots, radishes, turnips, onions, green onions, garlics, and shallots belong to this category. Data present in the literature on the microorganism-plant interaction in food quality mostly concern the bulb vegetables, with particular attention to the family Alliaceae.

**Table 5.3** Root and bulb vegetables

| Reference | Plant family | Plant species | Rhizosphere microbe | n. bulbs/ roots | Bulb size | Firmness | Total solids | Organic acids | Vitamins | Minerals | Carotenoids | Flavonoid | Phenol compounds | Alliin |
|---|---|---|---|---|---|---|---|---|---|---|---|---|---|---|
| Abdel-Razzak and El-Sharkawy (2013) | Alliaceae | *A. sativum* | PGPR (commercial inoculum Halex-2 (*Azospirillum sp.*, *Azotobacter sp.*, *Klebsiella sp.*)) | ✔ | ✔ | | | | | | | | | |
| Albrechtova et al. (2012) | Alliaceae | *Allium cepa* L. cv. Alice | AM fungi (commercial inoculum Symbivit (Symbiom s.r.o) (*G. intraradices* BEG 140, *G. mosseae* BEG 95, *G. etunicatum* BEG 92, *G. claroideum* BEG 96, *G. microaggregatum* BEG 56, *G. geosporum* BEG 199. Single inoculation *G. intraradices* BEG 140)) | | ✔ | | | | | ✔ | | | | |
| Borde et al. (2009) | Alliaceae | *A. sativum* | AM fungi (*G. fasciculatum*) | | ✔ | | | | | | | | | ✔ |
| Charron et al. (2001) | Alliaceae | *Allium cepa* L. | AM fungi (*G. intraradices*, *G. versiforme*) | | | ✔ | | | | | | | | |
| Mogren et al. (2008) | Alliaceae | *A. cepa* L. | AM fungi (*G. intraradices* Schenck and Smith (BEG 87)) | | | | | | | | | ✔ | | |
| Nautiyal et al. (2008) | Apiaceae | *Daucus carota* | PGPR (*B. lentimorbus* B-30488) | | | | | | | | | | ✔ | |
| Perner et al. (2008) | Alliaceae | *A. cepa* L. | AM fungi (TerraVital Hortimix comprising *G. mosseae*, *G. intraradices*, *G. claroideum*, and *G. microaggregatum*) | | | | | | | | | ✔ | | ✔ |
| Wang et al. (2011) | Apiaceae Alliaceae | *Daucus carota*, *Allium fistulosum* L. | AM fungi (*G. intraradices* BEG 141, *G. mosseae* BEG 167) | | ✔ | | | | | | | | | |

Summary of the main parameters influenced by the interaction between plants, belonging to the "root and bulb vegetables" category, and microorganisms (PGPR or AMF). The term "fruit size" includes weight, size, and biomass of the crop fruit. References are listed in alphabetical order

For instance, Abdel-Razzak and El-Sharkawy (2013) found an increase of bulb and clove number in *Allium sativum* plants treated with a commercial inoculum Halex-2 (consisting of growth-promoting nonsymbiotic N-fixing bacteria *Azospirillum* sp., *Azotobacter* sp., and *Klebsiella* sp.) (Abdel-Razzak and El-Sharkawy 2013) and with AMF inoculation (Borde et al. 2009). With onion, garlic is one of the most cultivated species of bulb vegetables. It is used as a spice and condiment and has high medicinal (e.g., it contrasts high blood pressure and arteriosclerosis), antioxidant (Chung 2006), antimicrobial, and antiviral properties (Harris et al. 2001) due to the presence of allicin (Borde et al. 2009). This organo-sulfur compound is produced from alliin by the enzyme alliinase (Corzo-Martinez et al. 2007), and it is responsible of the bulb flavor and aroma. A significant increase of alliin and alliinase content was shown in garlic plants, grown in field and inoculated with the AMF *G. fasciculatum* (Borde et al. 2009).

*Allium cepa* production is common worldwide and according to estimates it is over 250,000 tonnes (FAOSTAT 2010). The beneficial effects derived from onion consumption depend on two groups of chemical compounds: the flavor precursors alkenyl sulfoxides and flavonoids (Griffiths et al. 2002; Mogren et al. 2008). Yellow onions contain high concentration of flavonols, and the main flavonol glycosides have been shown to be quercetin 4-′-monoglucoside and quercetin 3,4′-diglucoside (Price and Rhodes 1997). It was reported that these compounds are modulated by AMF as shown by Albrechtova et al. (2012) and Perner et al. (2008). However, it should be underlined that different fungi induce different response in the same plants. In fact, a mix of *Glomus* sp. induces the best total antioxidant capacity in *A. cepa* cv. Alice bulbs if compared with *G. intraradices* BEG 140 alone that, on the contrary, is most effective in increasing mineral (Mg and K) accumulation (Albrechtova et al. 2012). Similarly, Charron et al. (2001) demonstrated that *Glomus versiforme* induces the production of firmer bulbs than *G. intraradices* (Charron et al. 2001).

Inoculation with *G. intraradices* BEG 141 leads to increased biomass either in carrot or in green onion (Wang et al. 2011). To the best of our knowledge, no data concerning PGPR effects on bulb quality are available in the literature. Within "root and bulb vegetables" category, the only paper about carrot quality reported an increase of phenol compounds in plants inoculated with *B. lentimorbus* B-30488 (Nautiyal et al. 2008).

## 5.5 Fruiting Vegetables (Table 5.4)

Solanaceous species, like tomato and pepper, establish root symbiosis with different AMF. Tomato, in particular, has become a model species for studies on plant interactions with AMF and PGPR, and effects of the microorganisms on the quality of tomatoes have been demonstrated. In particular, Zouari et al. (2014) demonstrated that AM symbiosis with *Funneliformis mosseae* induces the upregulation of genes involved in photosynthesis, stress response, transport, amino acid synthesis, and carbohydrate metabolism and the downregulation of ethylene response genes in tomatoes. Also Salvioli et al. (2012) previously demonstrated the modulation of gene transcription and of amino acid composition in fruits of plants inoculated with AMF. Nzanza et al. (2012) reported increase in the number of marketable fruits and of the concentration of nutrients such as minerals (P, K, Ca, Mg), vitamin C, and total flavonoids in tomatoes of plants inoculated with *Trichoderma harzianum* and *G. mosseae*. Similar results were reported by Giovannetti et al. (2012) (enhancement of lycopene concentration in tomatoes of plants inoculated with *G. intraradices*), by García-Fraile et al. (2012) (enhancement of fruits size and mineral concentrations in plant inoculated with *Rhizobium* sp.), by Sirichaiwetchakul et al. (2011) who showed that AMF (*Glomus* sp., *G. mosseae*, *Acaulospora* sp., *Entrophospora schenckii*, *Scutellospora fulgida*) raise the amount of total soluble solid and ascorbic acid in tomatoes, by Lenin

**Table 5.4** Fruiting vegetables

| Reference | Plant family | Plant species | Rhizosphere microbe | Allergens and metabolism | Flowers | n. fruits | Maturation time | Fruit size | Acidity | Total solids | Sugars | Organic acids | Minerals | Carotenoids | Flavonoid | Chlorophylls | Xanthophylls |
|---|---|---|---|---|---|---|---|---|---|---|---|---|---|---|---|---|---|
| Castillo et al. (2009) | Solanaceae | *C. annuum* L. "Cacho de cabra" | AM fungi (*G. intraradices* (commercial; IC); *G. claroideum* (native; IN)) | | | | ✓ | | ✓ | ✓ | | ✓ | | | | | |
| Del Amor et al. (2008) | Solanaceae | *C. annuum* L. (sweet pepper) | PGPR (Biopron (*A. brasilense*, *P. dispersa*)) | | | | | | | | | ✓ | | | | | |
| Garcia-Fraile et al. (2012) | Solanaceae | *Solanum lycopersicum*, *C. annuum* | PGPR (*R. leguminosarum*) | | ✓ | ✓ | | ✓ | | | | | ✓ | | | | |
| Giovannetti et al. (2012) | Solanaceae | *S. lycopersicum* L. var. Moneymaker | AM fungi (*G. intraradices* (IMA6)) | | | | | | | | | | ✓ | ✓ | | | |
| Kaya et al. (2009) | Solanaceae | *C. annuum* L. | AM fungi (*G. clarum*) | | | | | | | | | | ✓ | | | | |
| Lee et al. (2011) | Solanaceae | *C. annuum* L. | PGPR (*M. oryzae*) | | | | | ✓ | | | | | | | | | |
| Lenin et al. (2010) | Solanaceae | *S. lycopersicum* L., *S. melongena* L., *C. annuum* L. | AM fungi (*G. fasciculatum*) | | | | | | | | | | ✓ | | | ✓ | |
| Lu et al. (2006) | Cucurbitaceae | *C. sativus* | AM fungi (*G. versiforme*, *G. mosseae*, *G. intraradices*, and their mixtures) | | | ✓ | | | | | ✓ | | | | | | |
| Mandyal et al. (2012) | Solanaceae | *C. annuum* L. | PGPR (*Bacillus* sp.) | | | ✓ | | | | | | | ✓ | | | | |
| Mena-Violante et al. (2006) | Solanaceae | *C. annuum* L. cv. San Luis | AM fungi (*G. fasciculatum*, Los Tuxtla (AMFT), AMFD) | | | | | ✓ | | | | | | ✓ | | | ✓ |
| Nzanza et al. (2012) | Solanaceae | *Solanum lycopersicum* L. | AM fungi (commercial inoculum *T. harzianum* DB 103; Biocult© *G. mosseae*) | | | ✓ | | ✓ | | | | | ✓ | ✓ | ✓ | | |
| Ortas (2010) | Cucurbitaceae | *C. sativus* | AM fungi (*G. mosseae*, *G. etunicatum*, *G. clarum*, *G. caledonium*) | | | ✓ | | | | | | | ✓ | | | | |
| Salvioli et al. (2012) | Solanaceae | *S. lycopersicum* L. | AM fungi (*G. mosseae*) | ✓ | ✓ | ✓ | | | | | | | | | | | |
| Schwarz et al. (2011) | Solanaceae | *S. lycopersicum* L. | AM fungi (*G. mosseae*) | ✓ | | | | | | | | | | | | | |

| | | | | | | | | | | | | | | | | |
|---|---|---|---|---|---|---|---|---|---|---|---|---|---|---|---|---|
| Sirichaiwetchakul et al. (2011) | Solanaceae | *S. lycopersicum* L. (cherry tomatoes) | AM fungi (*Glomus* sp., *G. mosseae*, *Acaulospora* sp., *E. schenckii*, *S. fulgida*) | | ✓ | ✓ | | ✓ | | ✓ | ✓ | | | | | |
| Ulrichs et al. (2008) | Solanaceae | *S. lycopersicum* L. | AM fungi (*Glomus* sp.) | | | | | | | | | | ✓ | | | |
| Zouari et al. (2014) | Solanaceae | *S. lycopersicum* L. | AM fungi (*F. mosseae*) | ✓ | | | | | | | | | | | | |

Summary of the main parameters influenced by the interaction between plants, belonging to the "fruiting vegetables" category, and microorganisms (PGPR or AMF)

et al. (2010) (increase of chlorophylls, proteins, and minerals and reduction of starch and sugars in tomatoes of plants inoculated with *G. fasciculatum*), and by Ulrichs et al. (2008) (increment of lycopene, β-carotene, and phenolic contents in fruits of AM plants).

Finally, Schwarz et al. (2011) reported a modulation of different proteins potentially allergenics in tomato fruits, but they explain that this modulation is not sufficient to increase allergenic skin reactivity in a group of patients being allergic to tomato.

The eggplant or aubergine (*Solanum melongena* L.), belonging to the family Solanaceae, is cultivated worldwide for its fruits. Only one work reports the use of AMF (*G. fasciculatum*) to improve eggplant quality (Lenin et al. 2010), describing the accumulation of minerals, proteins, and chlorophylls in inoculated plants.

*Capsicum annuum*, with cultivated varieties including bell, sweet chili, and paprika peppers, is a perennial herbaceous plant belonging to the family Solanaceae, which originated in Central and South America and the Caribbean.

Different papers reported data concerning rhizospheric microorganisms (mostly PGPR) producing effects on fruit quality of *C. annuum*. In particular, Mena-Violante et al. (2006) and Kaya et al. (2009) reported results of AMF inoculation on fruits growth of chile ancho (*C. annuum* L. cv. San Luis) (Mena-Violante et al. 2006) and of pepper fruits (Kaya et al. 2009). Moreover, Castillo et al. (2009) reported that inoculation of *C. annuum* plants with *G. intraradices* and *Glomus claroideum* induces a precocity in fruit production (49 days earlier) and a 177 % higher fresh weight than fruits of uninoculated plants. Concerning PGPR effects on pepper fruits, del Amor et al. (2008) investigated the influence of a commercial product (Biopron), consisting of the bacteria *Azospirillum brasilense* and *Pantoea dispersa* on sweet pepper fruits, and reported an increased concentration of citric, ascorbic, and succinic acids in green fruits compared to the fruits of uninoculated plants. Significant effects on fruit growth were reported by Lee et al. (2011) in red pepper inoculated with *Methylobacterium oryzae* CBMB20. The same result was reported by García-Fraile et al. (2012) in pepper var. "verde italiano" inoculated with *Rhizobium* strains and by Mandyal et al. (2012) in sweet pepper inoculated with *Bacillus* sp.

*Cucumis sativus*, the garden cucumber, is a widely cultivated plant in the gourd family (Cucurbitaceae), which includes squash, and in the same genus are muskmelon and cantaloupe. Cucumber originated in India, where it appears to have been cultivated for more than 3000 years, and then spread to China. Hundreds of cultivars of varying size and color are now grown in warm areas worldwide, commercially and in home gardens. Scanty data are available about the use of PGPR and AMF in cucumber cultivation. The effects of different AMF on the quality of cucumber fruits have been reported in a couple of papers. They included the increase of P and Zn concentration (after inoculation with *G. mosseae*, *G. etunicatum*, *G. clarum*, *G. caledonium*, and a mixture of them, Ortas 2010) and the enhancement of soluble sugar, amino acid, and soluble protein concentrations in fruits of inoculated (with *G. versiforme*, *G. mosseae*, and *G. intraradices*) plants of *C. sativus* (Lu et al. 2006).

## 5.6 Cabbages and Leaf and Stem Vegetables (Table 5.5)

Cabbages are a popular, nutritious vegetable crop, belonging to the family Brassicaceae, which also includes brussels sprouts, cauliflower, broccoli, turnips, etc. They are very rich in vitamin C, vitamin A, and vitamin E which, together with cysteine, glucosinolate, and sulforaphane, are antioxidant components and whose ingestion protects against free radicals and the diseases that they cause, such as different types of cancer (Rosen et al. 2005). To our knowledge, there are few data concerning the use of bioinoculants (only PGPR because most Brassicaceae do not form mycorrhizae) in cabbage cultivation and the effects that these microorganisms could have on crop quality. Verma and Maurya (2013) reported the effects of *P. fluorescens* on cabbage cv. Golden and Yildirim (2011) who investigated the influence of *Bacillus cereus* ($N_2$ fixing), *Brevibacillus reuszeri* (P solubilizing), and *Rhizobium rubi* (both $N_2$ fixing and P solubilizing) on broccoli. In both studies, size and mineral concentration are enhanced following the inoculation with PGPR. Moreover, Yildirim (2011) demonstrated that heads of cabbages inoculated with the tree strains tested have higher chlorophyll content than controls (14.7 %, 14.0 %, and 13.7 %, respectively, in plants inoculated with *B. cereus*, *R. rubi*, or *B. reuszeri*).

Lettuce (*Lactuca sativa* L.) is one of the most used vegetables in the human diet. So, the total production of lettuce and chicory in the European Union in 2010 achieved 3,000,000 t (FAO Statistic Division), with Italy and Spain as main producers. This vegetable exhibits healthy properties mainly due to anthocyanins, carotenoids, and phenolics, with high fiber content and useful amounts of some minerals in its tissues (Azcón et al. 2003; Baslam et al. 2011a). These characteristics are modulated by both the use of fertilizers and the inoculation with AMF. In particular, the association of lettuce with AMF (a commercial inoculum consisting in a mixture of *G. intraradices* and *G. mosseae*) results in a greater quantity of anthocyanins in plants grown with different fertilization (Hewitt solution and/or water-insoluble fraction of a "rhizosphere-controlled fertilizer") (Baslam et al. 2011a, b, 2013). To our knowledge, no results are reported in the literature about the effects of PGPR on lettuce.

Fennel (*Foeniculum vulgare* Mill.) belongs to the family Apiaceae. Besides its use as a vegetable, the pharmacopoeial use is focused on the fruits and the important ingredient is the oil (Mahmoud and Hassan 2013). Fennel is composed of a white or pale green bulb from which closely superimposed stalks are arranged. The stalks are topped with feathery green leaves near which flowers grow and produce fennel seeds. The bulb, stalk, leaves, and seeds are edible. Fennel is also considered as a spice due to

**Table 5.5** Cabbages, leaf, and stem vegetables

| Reference | Plant family | Plant species | Rhizosphere microbe | Fruit size | Sugars | Organic acids | Minerals | Carotenoids | Flavonoid | Chlorophylls | Phenol compounds |
|---|---|---|---|---|---|---|---|---|---|---|---|
| Azcón et al. (2003) | Compositae | *L. sativa* L. | AM fungi (*G. mosseae*) | | | | ✔ | | | | |
| Baslam et al. (2011a) | Compositae | *L. sativa* L. var. capitata, *L. sativa* var. longifolia | AM fungi (*G. fasciculatum*, AEGIS endo granulo) | | | | ✔ | ✔ | ✔ | | ✔ |
| Baslam et al. (2011b) | Compositae | *L. sativa* L. var. Capitata | AM fungi (AEGIS endo granulo: commercial inoculum containing *G. intraradices* and *G. mosseae*) | | ✔ | ✔ | ✔ | ✔ | ✔ | ✔ | ✔ |
| Baslam et al. (2013) | Compositae | *L. sativa* Cogollos de Tudela (CT) (var. longifolia); Batavia Rubia Munguía (BRM) (var. Capitata); Maravilla de Verano (MV) (var. Capitata) | AM fungi (*G. fasciculatum*; commercial inoculum AEGIS Endo Gránulo (*R. intraradices* *F. mosseae*)) | ✔ | | | ✔ | ✔ | ✔ | ✔ | |
| Ceccarelli et al. (2010) | Asteraceae | *C. cardunculus* L. var. *scolymus* | AM fungi (*G. intraradices*, *G. mosseae*) | | | | | | | | ✔ |
| Mahmoud and Hassan (2013) | Apiaceae | *F. vulgare* | Biofertilizer + compost | ✔ | | | ✔ | | | ✔ | |
| Verma and Maurya (2013) | Brassicaceae | *Brassica oleracea* var. capitata cv. Golden acre | PGPR (*P. fluorescens*) | ✔ | | | ✔ | | | | |
| Yildirim (2011) | Brassicaceae | *B. oleracea* var. italica | PGPR (*B. cereus*, *B. reuszeri*, *R. rubi*) | ✔ | | | ✔ | | | ✔ | |

Summary of the main parameters influenced by the interaction between plants, belonging to the "cabbage, leaf, and stem vegetables" category, and microorganisms (PGPR or AMF). The term "fruit size" includes weight, size, and biomass of the crop fruit. References are listed in alphabetical order terpenic compounds isolated from its fruits. It has been demonstrated that the inoculation with biofertilizers in association with compost enhances production of essential oil and chlorophylls (Mahmoud and Hassan 2013).

A special case is that of artichoke (*Cynara cardunculus* L. var. *scolymus*) because flower heads constitute the edible part. An increase in total phenolic content and antioxidant activity was reported in the leaves and flower heads after inoculation with *G. intraradices*, either alone or in mixture with *G. mosseae*. The effect was observed both in greenhouse and in field experiments (Ceccarelli et al. 2010).

## 5.7 Fresh and Grain Legumes (Table 5.6)

Common beans represent the most important source of dietary proteins from plants, especially in developing countries, where its cultivation has often been relegated to marginal areas, characterized by poor soils (Ballesteros-Almanza et al. 2010). So, the use of biofertilizers can improve soil production and restoration of soil fertility (P and N availability). Yadegari and Rahmani (2010) demonstrated that the inoculation with *P. fluorescens* P93, *Azospirillum lipoferum* S21, and *Rhizobium* strains 133 and 136 increases fruit number, fruit size, and protein concentration in bean.

Chickpea (*Cicer arietinum*) is the second most widely grown legume crop after soybean, accounting for a substantial proportion of human dietary nitrogen intake and playing a crucial role in food security in developing countries. The inoculation with PGPR *A. brasilense*, as reported in Hamaoui et al. (2001), enhances chickpea fruit number and yield.

Faba bean is a legume which is used as a green vegetable or dried, fresh, or canned in the Middle East, Mediterranean region, China, and Ethiopia. Also, it is considered as a cash crop in Egypt and Sudan (El Wakeil and El Sebai 2007). Two papers report the effects of PGPR inoculation on faba bean quality: in the first, Hewedy

**Table 5.6** Fresh and grain legumes

| Reference | Plant family | Plant species | Rhizosphere microbe | n. fruits | Fruit size | Protein | Total solids | Sugar | Vitamins | Minerals |
|---|---|---|---|---|---|---|---|---|---|---|
| Abbas (2013) | Fabaceae | *Vicia faba* cv. Giza 3 | PGPR (*B. licheniformis*) | | ✓ | | | ✓ | ✓ | |
| Hamaoui et al. (2001) | Fabaceae | *C. arietinum* | PGPR (*A. brasilense*) | ✓ | ✓ | | | | | |
| Hewedy (2011) | Fabaceae | *Vicia faba* cv. Giza 843 | PGPR (*S. chibaensis*) | ✓ | ✓ | | | | ✓ | |
| Kumari et al. (2010) | Fabaceae | *Pisum sativum* L. | PGPR (*Rhizobium* sp.) | | | ✓ | | | | ✓ |
| Yadegari and Rahmani (2010) | Fabaceae | *Phaseolus vulgaris* | PGPR (*P. fluorescens* P93, *A. lipoferum* S21, *Rhizobium strain* 133 and 136) | ✓ | ✓ | ✓ | | | | |

Summary of the main parameters influenced by the interaction between plants, belonging to the "fresh and grain legumes" category, and microorganisms (PGPR or AMF). The term "fruit size" includes weight, size, and biomass of the crop fruit. References are listed in alphabetical order

(2011) showed that inoculation of faba bean with *Streptomyces chibaensis* induces an increased accumulation of minerals (N, P, K), proteins, and vitamins (ascorbic acid, riboflavin, and thiamin) and a yield enhancement (Hewedy 2011). In the second, Abbas (2013) described the influence of biostimulants (bacteria and yeast) on the growth and biochemical composition of faba bean. In particular, this work demonstrated that, besides increased biomass production, biostimulants (*Bacillus licheniformis* and, interestingly, also yeast) significantly increase pigment, carotenoid, and total carbohydrate content (Abbas 2013).

Finally, only one work reported the effects of rhizobial inoculation on the quality of pea: Kumari et al. (2010) reported that seed inoculation with biofertilizer (*Rhizobium* and other PGPR) induces an increase in the number and size of peas.

To our knowledge, no data are available on the effects of AMF on bean quality. This is an important point because AMF could be very relevant for the exploitation of marginal areas.

## 5.8 Cereals (Table 5.7)

According to FAO, cereals are defined as a group of species generally, but not exclusively, belonging to the gramineous family (i.e., Poaceae) that produce dry seeds rich in starch. Therefore, this word does not define a botanical taxon. This definition excludes Poaceae used as fodder, but includes plants of other families that produce grain seeds with high starch content. The Poaceae cereals (as wheat, rice, maize, sorghum, oats, barley, rye, millet, etc.) are sometimes considered "true cereals," while species belonging to other families, like buckwheat, quinoa, and amaranthus, are referred to as "pseudo-cereals." Because of their high content of starch, cereals are the staple food for mankind. Rice is the main cereal in Asia, wheat in Europe, and maize in South America; however, the most largely cultivated species is maize, because of its multiple use, as a food and as a source of bioplastic and biofuel. However, several cereal seeds also contain proteins and oils. The concentration of these two groups of substances can be important for their features, for instance, durum wheat is considered of better quality when richer in proteins, and oils can be extracted for use as well (as it happens for maize and rice). Concentration of proteins and oils is often regarded as a measure of cereals quality.

Cereals have been selected for disease resistance and high yield traits and not for their ability to form mycorrhizae. However, colonization has been reported, either in field or controlled conditions, for most species of this group (Raju et al. 1990; Braunberger et al. 1991; Secilia and Bagyaraj 1994; Hawkins and George 1997; Zhu et al. 2003; Likar et al. 2008; Criado et al. 2015), with the exception of quinoa (*Chenopodium quinoa*), which shows negligible AM root colonization and no arbuscule formation at all (Urcelay et al. 2011). This is not strange, since it belongs to the family Chenopodiaceae, known to include mostly non-mycorrhizal species.

Cereals inoculated with soil microorganisms (AMF, bacteria or their combination) have been tested for resistance to drought (Sajedi et al. 2010), for response to various levels of P availability (Braunberger et al. 1991; Al-Karaki and Clark 1999), to high temperatures, salinity, or pathogens. Results often showed better P nutrition, higher biomass and grain yield (also related to a higher number of tillers – Khan and Zaidi 2007; Midrarullah and Mirza 2014), increased concentration of Zn in grains (Sharif et al. 2011; Subramanian et al. 2013; Berta et al. 2014; Abaid-Ullah et al. 2015), and lower concentrations of some potentially phytotoxic elements, like Mn or Al (Karagiannidis and Hadjisavva-Zinoviadi 1998; Cornejo et al. 2008), in comparison to uninoculated plants.

Al-Karaki and Clark (1999) were the first to investigate parameters related with quality in mycorrhizal plants of durum wheat grown at different P concentrations. The grain protein and lipid concentration are not increased by AM colonization, but their total content per plant is increased due to enhanced yield (via

**Table 5.7** Cereals

| Reference | Plant family | Plant species | Rhizosphere microbe | n. fruits | Fruit size | Starch | Minerals | Lipid | Protein | Carotenoids |
|---|---|---|---|---|---|---|---|---|---|---|
| Abaid-Ullah et al. (2015) | Poaceae | *T. aestivum* L. | PGPR (*S. liquefaciens*, *S. marcescens*, *B. thuringiensis*) | | ✔ | | ✔ | | | |
| Al-Karaki and Clark (1999) | Poaceae | *T. durum* L. | AM fungi (*G. mosseae*) | | | | ✔ | ✔ | ✔ | |
| Behara and Rautaray (2010) | Poaceae | *T. turgidum* var. durum | PGPR (*Azotobacter* sp., *A. brasilense*, *P. vulgaris*, *Kurthia* sp., *K. planticola*, *B. subtilis*); AM fungi (*G. fasciculatum*) | | | | | | | ✔ |
| Behn (2008) | Poaceae | *T. aestivum* L. | PGPR (*P. fluorescens* RA56); AM fungi (*G. mosseae*) | ✔ | ✔ | | | | | |
| Berta et al. (2014) | Poaceae | *Zea mays* | PGPR (*P. fluorescens* Pf4); AM fungi (commercial inoculum Mybasol s.r.l. (*R. intraradices*, *G. aggregatum*, *G. viscosum*, *G. etunicatum*, *G. claroideum*)) | | ✔ | ✔ | ✔ | | ✔ | |
| Cornejo et al. (2008) | Poaceae | *T. aestivum* L. | AM fungi (*G. etunicatum* CH110) | | | | ✔ | | | |
| Criado et al. (2015) | Poaceae | *Hordeum vulgare* | AM fungi (*R. intraradices*) | | ✔ | | ✔ | | | |
| Karagiannidis and Hadjisavva-Zinoviadi (1998) | Poaceae | *Triticum turgidum* var. durum | AM fungi (*G. mosseae*) | | | | ✔ | | | |
| Khan and Zaidi (2007) | Poaceae | *T. aestivum* L. | PGPR (*A. chroococcum*, *Bacillus* sp.); AM fungi (*G. fasciculatum*) | | ✔ | | | | | |
| Kumar et al. (2011) | Poaceae | *T. aestivum* L. | AM fungi (*G. mosseae*) | | | | | | | |
| Midrarullah and Mirza (2014) | Poaceae | *Oryza sativa* | PGPR (*A. brasilense* R1, *A. lipoferum* RSWT1, and *Pseudomonas* Ky1) | ✔ | | | | | | |
| Sajedi et al. (2010) | Poaceae | *Z. mays* L. | AM fungi (*G. intraradices*) | ✔ | ✔ | | | | | |

| | | | | | | | | | | |
|---|---|---|---|---|---|---|---|---|---|---|
| Sharif et al. (2011) | Poaceae | *Z. mays*, *Sorghum bicolor* L., *Pennisetum glaucum* L., *Vigna mungo* L., *Vigna radiata* | AM fungi (*G. fasciculatum*, *G. intraradices*, *G. mosseae*, *G. aggregatum*, *A. mellea*) | ✓ | ✓ | | | | | |
| Subramanian et al. (2013) | Poaceae | *Z. mays* L. | AM fungi (*G. intraradices*) | | | | ✓ | | | |

Summary of the main parameters influenced by the interaction between plants, belonging to the "cereals" category, and microorganisms (PGPR or AMF). The term "fruit size" includes weight, size, and biomass of the spike or grain. References are listed in alphabetical order

increased number of tillers). Plants grown at high P concentration show a similar effect, and thus the mycorrhizal effect might be explained as nutritional.

More recently, three further studies have explored the consequences of inoculation with combinations of plant beneficial microorganisms on the quality of grains in cereals. Khan and Zaidi (2007) have investigated the effect of several combinations of plant growth-promoting (PGP) microorganisms, with or without the AMF *G. fasciculatum*, on the growth, yield, and quality of *Triticum aestivum*. The different PGP microorganisms had been selected because of their ability to fix nitrogen (*Azotobacter chroococcum*) or solubilize phosphate (the bacterium *Bacillus* sp. 8 and the fungus *Penicillium variabile*). While not all the possible combinations were tested, it is most interesting and valuable that the experiment was run as a field test. In terms of productivity, the various inoculants are generally able to increase, at different degrees, the plant biomass (often because of the increased number of tillers), straw production, and grain yield. For what concerns quality, single inoculations are generally not able to increase the grain protein content compared to the control, with the remarkable exception of *A. chroococcum* which leads to a grain protein content that is 2.4 times higher than that of the uninoculated controls. Overall, the multiple combinations (inocula consisting of three or four microbes) were consistently resulting in better performance for grain protein content (about twice the control). While the positive effect of a N-fixing bacterium such as *A. chroococcum* on protein accumulation is not surprising, the effect of this microbe is also related to its ability to produce various growth-promoting molecules, affecting wheat growth and development. The effects observed for combined microorganisms are often larger than the sum of the single increases, suggesting a synergistic effect. It must also be observed that *P. variabile* mostly exerts a negative effect on the various parameters considered (in terms of growth, yield, and grain protein content), alone or in combination with other microbes.

The previous study concerned one AMF and three other PGP microorganisms; on the contrary, the effects of a mycorrhizal inoculum containing five fungal species (*R. intraradices*, *G. aggregatum*, *G. viscosum*, *G. etunicatum*, and *G. claroideum*) and/or of a single bacterial strain (*P. fluorescens* strain Pf4, a PGPB able to produce siderophores and IAA and to solubilize phosphate) were assessed in maize by Berta et al. (2014). Beyond improving plant yield in several ways (increasing seed size, number of seeds per spike, and spike length and width), they are also able to differently influence the storage compounds in the grains. Bacteria increase starch content (and decrease the protein content), while AMF induce a higher accumulation of proteins (either soluble or unsoluble) and reduce the amount of digestible starch. Each inoculum (AM mix, Pf4, or their co-inoculation) also results in higher concentration of iron and zinc in the seeds. Results concerning proteins are of special interest, since the main maize allergens are found to the water-soluble protein group (Pastorello et al. 2009).

The third study concerns *Hordeum vulgare*. The commercial quality of malting barley after inoculation with *R. intraradices* was evaluated by Criado et al. (2015).

For malting barley, grain size and N concentration (ideally between 1.6 % and 1.8 %) are some of the most important features. The AM symbiosis does not affect the grain size in a significant way, but it increases N concentration in seeds, while in non-mycorrhizal plants the grain N concentration does not meet the standards required by industry.

In addition, a short note by Kumar et al. (2011) describes the effects of wheat inoculation with *G. mosseae* and seed priming (soaking overnight before sawing) on yield and grain quality. The authors claim that "The fungal inoculation improved the quality of the grain, increasing protein content and wet gluten. The study revealed that the agricultural practices followed in the region are adversely affecting the colonization of indigenous mycorrhizae" but the data included in the paper suggest for a possible effect by seed priming and not by inoculation.

Rice is a special case among cereals, because it is generally grown on (temporarily) flooded soils. Early studies described many wetland plants as non-mycorrhizal (e.g., Khan 1974; Anderson et al. 1984). However, the AM colonization of several hydrophytic species has been reported (Brown and Bledsoe 1996; Turner et al. 2000; Bohrer et al. 2004), and rice plants form arbuscular mycorrhiza; yield of AM rice plants improves even when they are flooded (Solaiman and Hirata 1997). Arsenic accumulation is a major problem in rice cultivation because of the toxic and carcinogenic effects of this element. Since arsenate can use phosphate transporters, the role of AMF is of special interest in this context. The concentration of As, in mycorrhizal plants, seems to vary according to the different plant cultivars and fungal species (Li et al. 2011) or their combinations (Chan et al. 2013). This fact opens the possibility to select the best formulations of fungal specie (singly or in a mixture) to be inoculated in chosen rice varieties and, therefore, improve rice quality in terms of healthiness, lowering As content.

Some studies describe the lack of effects on quality. Behn (2008) reports the testing of *G. mosseae* and *P. fluorescens* strain RA56, alone or in combination, for protective effects against *Gaeumannomyces graminis* in *T. aestivum*. While both the microorganisms (alone or in combination) are able to protect the plant from the pathogen, they fail to affect the grain quality, measured as protein content; this is true with and without the pathogen. In another paper, the effect of various microbes (including AMF and phosphate-solubilizing and N-fixing bacteria) was tested on *Triticum turgidum* at 50 % fertilization for N, P, and K and compared with 100 % fertilization and unfertilized controls. The bacteria (alone or in combination) was unable to improve any of the quality parameters (β-carotene concentration, yellow berry, protein content, hectoliter weight) when compared with uncolonized plants at 100 % or 50 % fertilization; AM plants were the worst performers, but it must be noted that the fungal species composition of the inoculums is not indicated and the level of colonization was not assessed (Behara and Rautaray 2010).

Overall, the above mentioned results clearly indicate a role for microorganisms in determining the seed quality of cereals, with possible positive effects on the grain concentration of mineral nutrients (like Cu, Fe, or Zn), lipids, and proteins. In addition, inoculated plants often show a significantly higher yield and larger seeds. Inoculation with more than one microorganism frequently results in better results than single inoculations.

## 5.9 Aromatic Plants, Spices, Tea, Coffee, Herbal Infusion, and Cocoa (Table 5.8)

Several plants mainly belonging to the families Lamiaceae, Asteraceae, and Apiaceae are included in this category. Some of them are known as aromatic plants, while others as spices or drugs. They are cultivated worldwide for pharmaceutical, herbal, cosmetic, and culinary uses. A number of studies demonstrate that the associations of PGPR and AMF with aromatic and officinal plants positively affect the quantitative and/or qualitative profile of the secondary metabolites (e.g., essential oils and alkaloids), in addition to improved growth parameters, biomass and yield (Khaosaad et al. 2006; Copetta et al. 2007; Banchio et al. 2008; Arango et al. 2012; Roshanpour et al. 2014; Silva et al. 2014).

The genus *Ocimum* comprises more than 150 species and is one of the most representative genera of the family Lamiaceae (Evans 1996). *Ocimum basilicum* L. (sweet basil) is cultivated in several regions all over the world. It is an annual herb used as food flavoring, with antiviral/antimicrobial activities and good medicinal properties, mostly due to its essential oils, stored in leaf glandular hairs (in particular eugenol and linalool) (Simon et al. 1990; Jirovetz et al. 2003; Toussaint et al. 2008; Roshanpour et al. 2014). Because of all the above mentioned qualities, its cultivation is of great economic interest.

Copetta and coworkers (2007) tested the effects of three species of AMF (*G. mosseae*, *Gigaspora margarita*, and *G. rosea*) on sweet basil plants (*O. basilicum* var. Genovese) showing that these fungi differently affect the essential oil production: *G. mosseae* does not alter the concentration of the analyzed compounds relative to control plants, while *G. rosea* and *G. margarita* significantly increase bornyl acetate and methyl eugenol, respectively; both the *Gigaspora* species enhance δ-cadinene content. A modulation of the essential oil profile due to the AMF inoculation was also observed by Toussaint et al. (2008) in plants of the same variety in the presence of *G. mosseae* and by Zolfaghari et al. (2012) in *O. basilicum* L. inoculated with *G. mosseae*, *G. intraradices*, or *G. fasciculatum* (with the best results by *G. mosseae* and *G. fasciculatum*). Not only AMF but also PGPR influence the synthesis of basil essential oils. In fact, Banchio et al. (2009) found an increase of R-terpineol and eugenol in leaves of plants exposed either to airborne volatiles from the beneficial soil bacterium *Bacillus subtilis* (GB03) or to root inoculation with the same microorganism. Integrated application of three different bacteria, *A. chroococcum*, *A. lipoferum*, and *Bacillus circulans*, improves the essential oil yield and shoot biomass of basil plants (Roshanpour et al. 2014). Similar results on shoot parameters were reported by Maleki et al. (2013) in plants co-inoculated with AMF (*G. intraradices*) and PGPR (*A. lipoferum* and *A. chroococcum*).

Genus *Salvia* is another important and representative genus of the family Lamiaceae. *Salvia officinalis* L. is used either in food industry (as herb or spice) or in the pharmaceutical sector (against inflammations and infections of the mouth and throat). As basil, sage antioxidant activity is due in great part to the presence in leaves of essential oils (Deans and Simpson 2000) and of two polyphenol compounds, rosmarinic and carnosic acid (Cuvelier et al. 1996). It was shown that besides the AM symbiosis, the nutrient supply, especially P, affects secondary metabolism in sage (Nell et al. 2009; Geneva et al. 2010). Nell

**Table 5.8** Aromatic plants, species, tea, coffee, herbal infusion, and cocoa

| Reference | Plant family | Plant species | Rhizosphere microbe | Biomass | Minerals | Organic acids | Carotenoids | Flavonoid | Chlorophylls | Xanthophylls | Phenol compounds | Terpenoids | Alkaloids | Essential oil |
|---|---|---|---|---|---|---|---|---|---|---|---|---|---|---|
| Al-Fraihat et al. (2011) | Lamiaceae | *Majorana hortensis*, L. (or *Origanum majorana* L.) | PGPR (The nitrobein: *Azotobacter*, a nitrogen-fixing bacteria; Halex-2, a mixture of N-fixing bacteria of the genera *Azotobacter*, *Azospirillum*, and *Klebsiella*) | | | | | | | | | | | ✔ |
| Arango et al. (2012) | Lamiaceae | *Mentha piperita* | AM fungi (*G. mosseae* isolate SB1 (Spegazzini Institute Collection, UNLP); *G. intraradices*, isolate A4; *G. intraradices*, isolate B1 Glomeromycota (In Vitro Bank, BGI, Buenos Aires, Argentina)) | | ✔ | | | | | | | | | ✔ |
| Awasthi et al. (2011) | Asteraceae | *Artemisia annua* | PGPR (*B. subtilis*, *Stenotrophomonas* sp.); AM fungi (*G. mosseae*, *G. aggregatum*, *G. fasciculatum*, *G. intraradices*) | ✔ | | | | | | | | | ✔ | |
| Bagyalakshmi et al. (2012) | Theaceae | *Camellia sinensis* | PGPR (*P. putida* strain KSB (potassium-solubilizing bacteria)) | | ✔ | | ✔ | | ✔ | | ✔ | | ✔ | |
| Banchio et al. (2008) | Lamiaceae | *Origanum majorana* L. | PGPR (*P. fluorescens* WCS417r; *B. subtilis* 09; *S. meliloti* Rm1021; *Bradyrhizobium* sp. USDA 4438) | ✔ | | | | | | | | | | |
| Banchio et al. (2009) | Lamiaceae | *Ocimum basilicum* | PGPR (*B. subtilis* strain GB03) | | | | | | | | | | | ✔ |
| Bharti et al. (2013) | Lamiaceae | *Mentha arvensis* cv. Kosi | AM fungi (*G. mosseae*, *G. aggregatum*, *G. fasciculatum*, *G. intraradices*) | ✔ | | | ✔ | | ✔ | | | | | ✔ |
| Binet et al. (2011) | Asteraceae | *Artemisia umbelliformis* | AM fungi (*G. tenue*, *G. intraradices*, *G. claroideum/etunicatum*, *Acaulospora* sp. (isolated from soil)) | | | | | | | | | | | ✔ |
| Chaudhary et al. (2008) | Asteraceae | *Artemisia annua* L. | AM fungi (*G. macrocarpum*, *G. fasciculatum*) | | ✔ | | | | | | | | | |
| Chulan (1991) | Sterculiaceae | *Theobroma cacao* L. | AM fungi (*S. calospora*) | | | | | | | | | | | |

| | | | | | | | | | | | | | | |
|---|---|---|---|---|---|---|---|---|---|---|---|---|---|---|
| Copetta et al. (2006, 2007) | Lamiaceae | *Ocimum basilicum* | AM fungi (*G. mosseae*, *G. margarita*, *G. rosea*) | | | | | | | | | | | ✔ |
| Darzi et al. (2012) | Apiaceae | *Anethum graveolens* | PGPR (mixture of *A. chroococcum* and *A. lipoferum*) | | | | | | | | ✔ | | | ✔ |
| Farahani et al. (2008) | Apiaceae | *Coriandrum sativum* | AM fungi (*G. hoi*) | ✔ | | | | | | | | | | ✔ |
| Freitas et al. (2004) | Lamiaceae | *Mentha arvensis* L. | AM fungi (*G. clarum*, *G. etunicatum*, *G. margarita*, *A. scrobiculata*) | ✔ | | | | | | | | | | |
| Geneva et al. (2010) | Lamiaceae | *Salvia officinalis* L. | AM fungi (*G. intraradices*) | | | ✔ | | ✔ | | | ✔ | | | ✔ |
| Gupta et al. (2002) | Lamiaceae | *Mentha arvensis* L. (cv. Kalka, Shivalik, and Gomati) | AM fungi (*G. fasciculatum*) | ✔ | ✔ | | | | | | | | | ✔ |
| Jurkiewicz et al. (2010) | Asteraceae | *Arnica montana* | AM fungi (*G. intraradices* or inoculum composed of several *Glomus* strains) | | | | | | | | ✔ | | | |
| Kapoor et al. (2002a) | Apiaceae | *C. sativum* | AM fungi (*G. macrocarpum*, *G. fasciculatum*) | | ✔ | | | | | | | | | ✔ |
| Kapoor et al. (2002b) | Apiaceae | *Anethum graveolens* L. | AM fungi (*G. macrocarpum*, G. *fasciculatum*) | | | | | | | | | | | ✔ |
| Kapoor et al. (2004) | Apiaceae | *F. vulgare* Mill | AM fungi (*G. macrocarpum*, *G. fasciculatum*) | | ✔ | | | | | | | | | ✔ |
| Kapoor et al. (2007) | Asteraceae | *Artemisia annua* L. | AM fungi (*G. macrocarpum*, *G. fasciculatum*) | | | | | | | | | | | ✔ |
| Karagiannidis et al. (2011) | | Oregano and mint (collected from Greek mountain) | AM fungi (two strains *G. etunicatum* and one *G. lamellosum* (collected from field)) | | ✔ | | | | | | | | | ✔ |
| Khaosaad et al. (2006) | Lamiaceae | *Origanum vulgare* ssp. | AM fungi (*G. mosseae*) | ✔ | | | | | | | | | | ✔ |
| Maleki et al. (2013) | Lamiaceae | *Ocimum basilicum* L. | PGPR (*A. chroococcum*, *A. lipoferum*); AM fungi (*G. intraradices*) | ✔ | | | | | | | | | | |
| Morone-Fortunato and Avato (2008) | Lamiaceae | *Origanum vulgare* | AM fungi (*G. viscosum*) | ✔ | | | | | | | | | | ✔ |

(continued)

**Table 5.8** (continued)

| Reference | Plant family | Plant species | Rhizosphere microbe | Biomass | Minerals | Organic acids | Carotenoids | Flavonoid | Chlorophylls | Xanthophylls | Phenol compounds | Terpenoids | Alkaloids | Essential oil |
|---|---|---|---|---|---|---|---|---|---|---|---|---|---|---|
| Nell et al. (2009) | Lamiaceae | *Salvia officinalis* L. | AM fungi (*G. mosseae* (BEG 12) and *G. intraradice* (BB-E)-BIORIZE/AGRAUXINE; commercial inoculum "Symbivit®" consisting of *G. mosseae*, *G. intraradices*, *G. cladoideum*, *G. microagregatum*, *G. caledonium* and *G. etunicatum*) | | ✔ | | | | | | | | | ✔ |
| Nell et al. (2010) | Valerianaceae | *Valeriana officinalis* L. | AM fungi (*G. mosseae* and *G. intraradices* (BB-E), BIORIZE/AGRAUXINE; commercial inoculum "Symbivit®" consisting of *G. mosseae*, *G. intraradices*, *G. cladoideum*, *G. microagregatum*, *G. caledonium*, and *G. etunicatum*) | ✔ | | | | | | | | ✔ | | |
| Oliveira et al. (2014) | Passifloraceae | *Passiflora alata* Curtis | AM fungi (*G. albida*) | | | | | ✔ | | | ✔ | | | |
| Roshanpour et al. (2014) | Lamiaceae | *Ocimum basilicum* | PGPR (*A. chroococcum*, *A. lipoferum*, *B. acillus circulans*) | ✔ | | | | | | | | | | ✔ |
| Silva et al. (2014) | Lamiaceae | *Mentha x piperita* L. var. citrata | AM fungi (*A. morrowiae*, *R. clarus*, *S. calospora*) | ✔ | ✔ | | | | | | | | | ✔ |
| Singh et al. (2010) | Theaceae | *Camellia sinensis* (L.) O. Kuntze | AM fungi (consortia containing: *A. spinosa*, A. sp. 1, *G. aggregatum*, *G. ambisporum*, *G. clavisporum*, *G. geosporum*, *G. mosseae*, *G. pustulatum*, and *Glomus* sp. 7.; consortia containing *A. foveta*, *A. scrobiculata*, *G. ambisporum/heterosporum*, *G. clarum*, *G. fasciculatum*, *G. heterosporum*, *G. intraradices*, *Glomus* sp. 6, *Glomus* sp. 8, *Glomus* sp. 9, and *Glomus* sp. 10) | ✔ | | | | | | | ✔ | | ✔ | |

| | | | | | | | | | | | | | | |
|---|---|---|---|---|---|---|---|---|---|---|---|---|---|---|
| Toussaint et al. (2008) | Lamiaceae cv. Genovese | *Ocimum basilicum* | AM fungi (*G. mosseae*) | | | | | | ✔ | | | ✔ | | |
| Zitterl-Eglseer et al. (2015) | Apiaceae | *Angelica archangelica* L. | AM fungi (*G. mosseae* and *G. intraradices* (BB-E), BIORIZE/AGRAUXINE; commercial inoculum Symbivit containing *G. mosseae* (BEG 99), *G. intraradices* (BEG 98), *G. claroideum* (BEG 93), *G. microaggregatum* (BEG 56), *G. caledonium* (BEG 97), and *G. etunicatum* (BEG 92)) | ✔ | ✔ | | | | | | | ✔ | | |
| Zolfaghari et al. (2012) | Lamiaceae | *Ocimum basilicum* | AM fungi (*G. mosseae*, *G. fasciculatum*) | ✔ | | | | | | | | | | ✔ |
| Zubek et al. (2012) | Clusiaceae | *Hypericum perforatum* | AM fungi (*R. intraradices* (BEG 140); *F. mosseae*; mix: *F. constrictum* (262-5 Walker), *F. geosporum* (UNIJAG.PL.12-2), *F. mosseae*, *R. intraradices* (BEG 140)) | ✔ | | | | | | | | | ✔ | |

Summary of the main parameters influenced by the interaction between plants, belonging to the "aromatic plants, spices, tea coffee, herbal infusion, and cocoa" category, and microorganisms (PGPR or AMF). The term "biomass" includes weight, size, and biomass of the shoots. References are listed in alphabetical order et al. (2009), testing three different fungal inocula and two levels of phosphate, observed that P supply, but not AM inoculation, increases total phenolic and rosmarinic acid concentrations in leaves of *S. officinalis*. However, Geneva et al. (2010) observed that AM colonization combined with foliar fertilization (N, P, K) improves dry biomass production and enhances the content of antioxidant metabolites, such as ascorbate and reduced glutathione. Plants inoculated with *G. intraradices* also show an increase of essential oil content, in particular bornyl acetate, 1,8-cineole, and α- and β-thujones, while foliar fertilization increased bornyl acetate and camphor. Similar results were obtained by Karagiannidis et al. (2011) in AM-oregano plants grown at different P levels.

However, in some cases the improvement of essential oil production is not directly dependent on mineral nutrition (especially P) as observed by Khaosaad et al. (2006) on three cultivars of *Origanum vulgare* L. inoculated with *G. mosseae*. This is in accordance with results obtained by Copetta et al. (2006) in sweet basil plants. In such cases other factors as plant genotype, fungal species, and their interaction should be considered. An analysis of the components of essential oils was conducted in micropropagated oregano plants, inoculated or not with *G. viscosum*, at various times after transplanting, showing a combined effect of time and inoculation on the molecular composition of the oil (Morone-Fortunato and Avato 2008).

*Majorana hortensis* Moench., also known as *Origanum majorana* L., is a perennial herbaceous plant from the Mediterranean area. It has a peculiar, spicy smell and taste due to its isoprenoid compounds of which pinene, limonene, linalool, terpinene, and eugenol are the most abundant (Al-Fraihat et al. 2011). The symbiotic interaction between *O. majorana* plants and four different PGPR results in the increase of essential oil content, depending on the rhizobacterial species, without alteration of oil composition (Banchio et al. 2008). The best results were observed in presence of *Bradyrhizobium* sp. and *P. fluorescens*. According to these findings, Al-Fraihat and coworkers (2011) reported an enhancement of essential oil content and yield per plant and per hectare, a larger biomass production, and an improvement of mineral accumulation in marjoram plants treated with two commercial inocula "nitrobein" and "Halex-2" containing one or more N-fixing bacteria, respectively. We couldn't find data on the effects of AMF on marjoram plant quality.

Different species of the genus *Mentha* are used in pharmaceutical, food, cosmetics, and beverages (Verma et al. 2010). Their typical flavor and aroma are due to the presence of essential oils, a natural source of menthol, menthone, isomenthone, menthofuran, carvone, linalool, and linalyl acetate (Verma et al. 2010; Bharti et al. 2013). Freitas et al. (2004), testing the effect of four AMF on plants of *Mentha arvensis* L. grown at different P levels, found a higher essential oil content (up to 89 %) in AM plants than in uninoculated ones. Similar results on the production of secondary metabolites were obtained by Gupta et al. (2002) on three cultivars of *M. arvensis* inoculated with *G. fasciculatum*. These results underline that not only different fungi induce different effects on the same cultivar but also that different cultivars produce different metabolites in the presence of the same fungus. In addition to this, as mentioned above, another factor should be considered: mineral nutrition (in particular N, P, K). When no phosphorus is added, AMF inoculation increases biomass production, volatile compound, and menthol content in *Mentha viridis* (Karagiannidis et al. 2011). No increment in essential oil and menthol content occurs at high P level. An enhancement of secondary metabolite production in *Mentha piperita* L. (peppermint) plants, inoculated with AMF in relation to nutrient availability and uptake, was reported also by Silva et al. (2014) and by Arango et al. (2012).

In addition to the plant belonging to the family Lamiaceae, others from the families Asteraceae, Apiaceae, Apocynaceae, Clusiaceae, Geraniaceae, Passifloraceae, and Valerianaceae produce many secondary metabolites of commercial interest, including terpenes, phenols, and alkaloids, whose biosynthesis is influenced by the presence of microorganisms in the soil or in the growing medium. In particular, AMF increase the production of sesquiterpene lactones in *Arnica montana* (Jurkiewicz et al. 2010), geraniol and linalool in coriander (*Coriandrum sativum L.* – Kapoor et al. 2002a; Farahani et al. 2008), anethol in fennel (*F. vulgare* Mill. – Kapoor et al. 2004), and limonene and carvone in *Anethum graveolens* L. (Kapoor et al. 2002b). Nitrogen-fixing bacteria affect essential oil and carvone content in dill (Darzi et al. 2012).

Moreover, Kapoor et al. (2007), Chaudhary et al. (2008), Awasthi et al. (2011), and Binet et al. (2011) investigated the effects of AMF or PGPR on *Artemisia* plants showing a general improvement of plant mineral nutrition and a stimulation of bioactive compounds production. Mycorrhiza boost artemisinin content increasing the density of glandular trichome in leaves (Kapoor et al. 2007). This is in accordance with a previous study on *Ocimum basilicum* plants inoculated with three different AMF (Copetta et al. 2006).

*Hypericum perforatum* produces two important molecules including hypericin and pseudohypericin, known to have antiviral activity and inhibitory effects toward the acquired immune deficiency syndrome (AIDS). Zubek et al. (2012) observed that hypericin and pseudohypericin contents significantly increase in the presence of AM symbiosis.

Plants of *Passiflora alata* inoculated with *Gigaspora albida* produce a higher amount of biomolecules in leaves, especially flavonoids, in respect of control plants (Oliveira et al. 2014).

Similar results are reported also in some officinal plants including *Valeriana officinalis* (Nell et al. 2010) and *Angelica archangelica* (Zitterl-Eglseer et al. 2015) largely employed in herbal uses.

*Theobroma cacao* L. is a small tree of the family Sterculiaceae, native of South America. Its berries contain a variable number of seeds, depending on the plant cultivar. It is known either as drug, for the alkaloid content, mainly caffeine and theobromine, or as oleaginous plant since large amounts of cocoa butter are extracted from its seeds. For these qualities *T. cacao* is used in confectionery, pharmaceutical, and cosmetic industry. There are many studies concerning the extraction methods of alkaloids or cocoa butter and their different uses in industry (Li and Hartland 1996; Schulz 2004), but no studies on the effects of microbe inoculation on cocoa quality are found in the literature except one, in which an increase of nutrient uptake (N, P, K) in *Theobroma* plants inoculated with the AMF *Scutellospora calospora* was reported (Chulan 1991).

Tea (*Camellia sinensis*), belonging to the family Theaceae, is an important crop cultivated in different areas of the world. Chinese people (from which the name "sinensis" was derived) were the first to use tea as medicinal drink (Eden 1958). There are different varieties of *C. sinensis*. Their quality depends on many factors such as organic and inorganic composition of harvested shoots, plant mineral nutrition, and microbe interaction. Changes in the active compounds, flavonoids, amino acids, vitamins (C, E, K), caffeine, and polysaccharides are responsible for the different taste, flavor, and color (Mondal et al. 2004; Bagyalakshmi et al. 2012). It is noteworthy that the vitamin C content in tea leaves is the same one found in lemon (Mondal et al. 2004). Tea consumption has beneficial effects on human health because of its antioxidant and antimicrobial activity, besides protecting cell membranes from oxidative damages, preventing coronary heart diseases and normalizing blood pressure.

In literature, scanty data on the effects of AMF or PGPR inoculations on tea quality are reported. Singh et al. (2010) tested the effect of two different mixed fungal consortia, one containing indigenous populations of fungi and another containing cultivated AMF. The results showed an increase of amino acids, total protein, total polyphenols, sugar, and caffeine content in AM plants with better response in tea plants inoculated with native AMF.

Moreover, in a field experiment, conducted to investigate the effect of *Pseudomonas putida* (an indigenous potassium-solubilizing bacteria) on tea quality, Bagyalakshmi et al. (2012) observed an improvement of the content of theaflavin, thearubigin, and caffeine and highly polymerized substances, total liquor color, briskness, and color and flavor indexes in plants inoculated with bacteria.

Finally, the effect of AMF on the quality of plants used in traditional medicine of China has been reviewed by Zeng et al. (2013).

## 5.10 Conclusion

In this chapter we have summarized the effects of soil microbes on different plant parts, concerning storage (roots, bulbs, tubers), vegetative (leaves), or reproductive (fruits) structures. Modifications concerned the accumulation of mineral nutrients, sugars, lipids, and a number of chemically diverse secondary metabolites. In some cases, the reported variations can be explained by the nutritional improvement operated by the microorganisms in favor of the plant, but this is not the rule.

The number of AMF species used in the described experiments is rather small even considering the limited number of Glomeromycota. It is clear that widening the range of tested AM species could only be positive for a better understanding of the effects induced in plants. Also the number of plants species is not extremely large (about 30). This is most likely related to the overwhelming importance of a relatively small number of plant species for agriculture, human and animal nutrition, and industry.

In spite of these limitations, some facts do emerge:

- Soil microbes can affect the quality of crops (often increasing yield as well).
- Even if most experiments have been carried under controlled conditions and on a reduced scale, the relatively rare tests run in the field show that a practical exploitation is feasible and can result in better products in real life agriculture.
- Combinations of different microorganisms often perform better than single inoculants; combinations can sometimes act synergistically, some other addictively.
- Different microorganisms can give different results on the same plant species (for instance, promoting the accumulation of a specific secondary metabolite). This opens to the possibility of using selected isolates or strains according to the environmental conditions or according to the kind of result or product that is expected.

Therefore, a new role and exploitation of AMF and PGPR, as enhancers of crop quality, can be described, in addition to increasing yield, providing valuable ecosystem services and protecting plants from pathogens. The selection of new crop cultivars should consider the ability of plant to interact with the beneficial soil microbes.

**Acknowledgments** To the memory of my friend Horst Vierheilig, who first prompted me to write this review, sitting by a swimming pool in Marrakech, in 2010. It was originally meant to be submitted to a journal. Then, several things happened, delaying, hampering, and stopping this work. In 2014 I was invited by Raffaella Balestrini to submit a chapter for the present book, and I immediately thought about this topic. Once more, I would not have been able to complete this chapter if it had not been for my fellow authors Elisa and Valeria.

## References

Abaid-Ullah M, Hassan M, Jamil M, Brader G, Kausar M, Shah N, Sessitsch A, Yusuf Hafeez F (2015) Plant growth promoting rhizobacteria: an alternate way to improve yield and quality of wheat (*Triticum aestivum*). Int J Agric Biol 17:51–60

Abbas S (2013) The influence of biostimulants on the growth and on the biochemical composition of *Vicia faba* cv. Giza 3 beans. Rom Biotechnol Lett 18:8061–8068

Abdel-Razzak H, El-Sharkawy G (2013) Effect of biofertilizer and humic applications on growth, yield, quality and storability of two garlic (*Allium sativum* L.) cultivars. Asian J Crop Sci 5:48–64

Albrechtova J, Latr A, Nedorost L, Nedorost L, Pokluda R, Posta K, Vosatka M (2012) Dual inoculation with mycorrhizal and saprotrophic fungi applicable in sustainable cultivation improves the yield and nutritive value of onion. Sci World J 2012:374091

Al-Fraihat AH, Al-Dalain SYA, Al-Rawashdeh ZB (2011) Effect of organic and biofertilizers on growth, herb yield and volatile oil of marjoram plant grown in Ajloun region, Jordan. J Med Plants Res 5:2822–2833

Al-Karaki GN, Clark RB (1999) Mycorrhizal influence on protein and lipid of durum wheat grown at different soil phosphorus levels. Mycorrhiza 9:97–101

Anderson R, Liberta A, Dickman L (1984) Interaction of vascular plants and vesicular-arbuscular mycorrhizal fungi across a soil moisture-nutrient gradient. Oecologia 64:111–117

Arango M, Ruscitti M, Ronco M, Beltrano J (2012) Mycorrhizal fungi inoculation and phosphorus fertilizer on growth, essential oil production and nutrient uptake in peppermint (*Mentha piperita* L.). Bras J Med Plants 14:692–699

Attia M, Ahmed M, El-Sonbaty M (2009) Use of biotechnologies to increase growth, productivity and fruit quality of *Maghrabi banana* under different rates of phosphorus. World J Agric Sci 5:211–220

Awasthi A, Bharti N, Nair N, Singh R, Shukla AK, Gupta MM, Darokar MP, Kalra A (2011) Synergistic effect of *Glomus mosseae* and nitrogen fixing *Bacillus subtilis* strain Daz26 on artemisinin content in *Artemisia annua* L. Appl Soil Ecol 49:125–130

Azcón R, Ambrosano E, Charest C (2003) Nutrient acquisition in mycorrhizal lettuce plants under different phosphorus and nitrogen concentration. Plant Sci 165:1137–1145

Bagyalakshmi B, Ponmurugan P, Marimuthu S (2012) Influence of potassium solubilizing bacteria on crop productivity and quality of tea (*Camellia sinensis*). Afr J Agric Res 7:4250–4259

Ballesteros-Almanza L, Altamirano-Hernandez J, Santoyo G, Sanchez-Yañez JM, Valencia-Cantero E, Macias-Rodriguez L, Lopez-Bucio J, Cardenas-Navarro R, Farias-Rodriguez R (2010) Effect of co-inoculation with mycorrhiza and rhizobia on the nodule trehalose content of different bean genotypes. Open Microbiol J 4:83–92

Banchio E, Bogino PC, Zygadlo J, Giordano W (2008) Plant growth promoting rhizobacteria improve growth and essential oil yield in *Origanum majorana* L. Biochem Syst Ecol 36:766–771

Banchio E, Xie X, Zhang H, Paré P (2009) Soil bacteria elevate essential oil accumulation and emissions in sweet basil. J Agric Food Chem 57:653–657

Baslam M, Pascual I, Sánchez-Díaz M, Erro J, García-Mina JM, Goicoechea N (2011a) Improvement of nutritional quality of greenhouse-grown lettuce by arbuscular mycorrhizal fungi is conditioned by the source of phosphorus nutrition. J Agric Food Chem 59:11129–11140

Baslam M, Garmendia I, Goicoechea N (2011b) Arbuscular mycorrhizal fungi (AMF) improved growth and nutritional quality of greenhouse-grown lettuce. J Agric Food Chem 59:5504–5515

Baslam M, Esteban R, García-Plazaola JI, Goicoechea N (2013) Effectiveness of arbuscular mycorrhizal fungi (AMF) for inducing the accumulation of major carotenoids, chlorophylls and tocopherol in green and red leaf lettuces. Appl Microbiol Biotechnol 97:3119–3128

Behara U, Rautaray S (2010) Effect of biofertilizers and chemical fertilizers on productivity and quality parameters of durum wheat (*Triticum turgidum*) on a Vertisol of Central India. Arch Agron Soil Sci 56:65–72

Behn O (2008) Influence of *Pseudomonas fluorescens* and arbuscular mycorrhiza on the growth, yield, quality and resistance of wheat infected with *Gaeumannomyces graminis*. J Plant Dis Prot 115:4–8

Berta G, Copetta A, Gamalero E, Bona E, Cesaro P, Scarafoni A, D'Agostino G (2014) Maize development and grain quality are differentially affected by mycorrhizal fungi and a growth-promoting pseudomonad in the field. Mycorrhiza 24:161–170

Bharti N, Baghel S, Barnawal D, Yadav A, Kalra A (2013) The greater effectiveness of *Glomus mosseae* and *Glomus intraradices* in improving productivity, oil content and tolerance of salt stressed menthol mint (*Mentha arvensis*). J Sci Food Agric 93:2154–2161

Binet MN, Van Tuinen D, Deprêtre N, Koszela N, Chambon C, Gianinazzi S (2011) Arbuscular mycorrhizal fungi associated with *Artemisia umbelliformis* Lam, an endangered aromatic species in Southern French Alps, influence plant P and essential oil contents. Mycorrhiza 21:523–535

Bohrer K, Friese C, Amon J (2004) Seasonal dynamics of arbuscular mycorrhizal fungi in differing wetland habitats. Mycorrhiza 14:329–337

Bona E, Lingua G, Manassero P, Cantamessa S, Marsano F, Todeschini V, Copetta A, D'Agostino G, Massa N, Avidano L, Gamalero E, Berta G (2015) AM fungi and PGP pseudomonads increase flowering, fruit production, and vitamin content in strawberry grown at low nitrogen and phosphorus levels. Mycorrhiza 25:181–193

Borde M, Dudhane M, Jite PK (2009) Role of bioinoculant (AM fungi) increasing in growth, flavor content and yield in *Allium sativum* L. under field condition. Not Bot Hortic Agrobot Cluj 37:124–128

Braunberger P, Miller MH, Peterson RL (1991) Effect of phosphorus nutrition on morphological characteristics of vesicular arbuscular mycorrhizal colonization of maize. New Phytol 119:107–113

Brown M, Bledsoe C (1996) Spatial and temporal dynamics *Jaumea carnosa*, a tidal saltmarsh halophyte. J Ecol 84:703–715

Caravaca F, Diaz E, Barea J, Azcón-Aguilar C, Roldán A (2003) Photosynthetic and transpiration rates of *Olea europaea* subsp *sylvestris* and *Rhamnus lycioides* as affected by water deficit and mycorrhiza. Biol Plant 46:637–639

Castellanos-Morales V, Villegas J, Wendelin S, Vierheilig H, Eder R, Cárdenas-Navarro R (2010) Root colonisation by the arbuscular mycorrhizal fungus *Glomus intraradices* alters the quality of strawberry fruits (*Fragaria x ananassa* Duch.) at different nitrogen levels. J Sci Food Agric 90:1774–1782

Castillo P, Nico A, Azcon-Aguilar C, Del Río RC, Calvet C, Jiménez-Díaz RM (2006) Protection of olive planting stocks against parasitism of root-knot nematodes by arbuscular mycorrhizal fungi. Plant Pathol 55:705–713

Castillo RC, Sotomayor S, Ortiz OC, Leonelli CG, Borie BF, Rubio HR (2009) Effect of arbuscular mycorrhizal fungi on an ecological crop of chili peppers (*Capsicum annuum* L.). Chil J Agric Res 69:79–87

Ceccarelli N, Curadi M, Martelloni L, Martelloni L, Sbrana C, Picciarelli P, Giovannetti M (2010) Mycorrhizal colonization impacts on phenolic content and antioxidant properties of artichoke leaves and flower heads two years after field transplant. Plant Soil 335:311–323

Chan WF, Li H, Wu FY, Wu SC, Wong MH (2013) Arsenic uptake in upland rice inoculated with a combination or single arbuscular mycorrhizal fungi. J Hazard Mater 262:1116–1122

Charron G, Furlan V, Bernier-Cardou M, Doyon G (2001) Response of onion plants to arbuscular mycorrhizae. Mycorrhiza 11:187–197

Chatzistathis T, Orfanoudakis M, Alifragis D, Therios I (2013) Colonization of Greek olive cultivars' root system by arbuscular mycorrhizal fungus: root morphology, growth, and mineral nutrition of olive plants. Sci Agric 70:185–194

Chaudhary V, Kapoor R, Bhatnagar AK (2008) Effectiveness of two arbuscular mycorrhizal fungi on concentrations of essential oil and artemisinin in three accessions of *Artemisia annua* L. Appl Soil Ecol 40:174–181

Chulan HA (1991) Effect of fertilizer and endomycorrhizal inoculum on growth and nutrient-uptake of cocoa (*Theobroma cacao* L.) seedlings. Biol Fertil Soils 11:250–254

Chung L (2006) The antioxidant properties of garlic compounds: allyl cysteine, alliin, allicin, and allyl disulfide. J Med Food 9:205–213

Citernesi A, Vitagliano C, Giovannetti M (1998) Plant growth and root system morphology of *Olea europaea* L. rooted cuttings as influenced by arbuscular mycorrhizas. J Hortic Sci Biotechnol 73:647–654

Copetta A, Lingua G, Berta G (2006) Effects of three AM fungi on growth, distribution of glandular hairs, and essential oil production in *Ocimum basilicum* L. var. Genovese. Mycorrhiza 16:485–494

Copetta A, Lingua G, Bardi L, Masoero G, Berta G (2007) Influence of arbuscular mycorrhizal fungi on growth and essential oil composition in *Ocimum basilicum* var. Genovese. Caryologia 60:106–110

Cornejo P, Castillo C, Azcon R, Borie F (2008) Mycorrhizal effectiveness on wheat nutrient acquisition in an acidic soil from southern chile as affected by nitrogen sources. J Plant Nutr 31:1555–1569

Corzo-Martinez M, Corzo N, Villamiel M (2007) Biological properties of onion and garlic. Trends Food Sci Technol 18:609–625

Criado M, Gutierrez Boem F, Roberts I, Caputo C (2015) Post-anthesis N and P dynamics and its impact on grain yield and quality in mycorrhizal barley plants. Mycorrhiza 25:229–235

Cuvelier M, Richard H, Berset C (1996) Antioxidative activity and phenolic composition of pilot-plant and commercial extracts of sage and rosemary. J Am Oil Chem Soc 73:645–652

Darzi M, Hadi M, Rejali F (2012) Effects of the application of vermicompost and nitrogen fixing bacteria on quantity and quality of the essential oil in dill (*Anethum graveolens*). J Med Plants Res 6:3793–3799

Deans S, Simpson E (2000) Antioxidants from *Salvia officinalis*. In: Kintzios, medicinal and aromatic plants – industrial profiles, sage the genus Salvia, vol 14, Harwood Academic Publishers, pp 185–192

Del Amor FM, Serrano-Martínez A, Fortea MI, Leguac P, Núñez-Delicado E (2008) The effect of plant-associative bacteria (*Azospirillum* and *Pantoea*) on the fruit quality of sweet pepper under limited nitrogen supply. Sci Hortic 117:191–196

Devi MC, Reddy MN (2002) Phenolic acid metabolism of groundnut (*Arachis hypogaea* L.) plants inoculated with VAM fungus and Rhizobium. Plant Growth Regul 37:151–156

Dolcet-Sanjuan R, Claveria E, Camprubí A, Estaun V, Calvet C (1996) Micropropagation of walnut trees (*Juglans regia* L.) and response to arbuscular mycorrhizal inoculation. Agronomie 16:639–645

Eden T (1958) The development of tea culture. In: Eden T (ed) Tea. Longman, London, pp 1–4

El Wakeil N, El Sebai T (2007) Role of biofertilizer on faba bean growth, yield, and its effect on bean aphid and the associated predators. Res J Agric Biol Sci 3:800–880

Esitken A, Pirlak L, Turan M, Sahin F (2006) Effects of floral and foliar application of plant growth promoting rhizobacteria (PGPR) on yield, growth and nutrition of sweet cherry. Sci Hortic 110:324–327

Estaun V, Camprubi A, Calvet C, Pinochet J (2003) Nursery and field response of olive trees inoculated with two arbuscular mycorrhizal fungi, *Glomus intraradices* and *Glomus mosseae*. J Am Soc Hortic Sci 128:767–775

Evans W (1996) Trease and Evans' pharmacognosy. WB Sounders Company, London, p 48

FAOSTAT (2010) http://faostat.fao.org

Farahani HA, Lebaschi MH, Hamidi A (2008) Effects of arbuscular mycorrhizal fungi, phosphorus and water stress on quantity and quality characteristics of coriander. J Adv Nat Appl Sci 2:55–59

Freitas MSM, Martins MA, Vieira IJC (2004) Produção e qualidade de óleos essenciais de *Mentha arvensis* em resposta à inoculação de fungos micorrízicos arbusculares. Pesqui Agrop Bras 39:887–894

García-Fraile P, Carro L, Robledo M, Ramírez-Bahena MH, Flores-Félix JD, Fernández MT, Mateos PF, Rivas R, Igual JM, Martínez-Molina E, Peix A, Velázquez E (2012) *Rhizobium* promotes non-legumes growth and quality in several production steps: towards a biofertilization of edible raw vegetables healthy for humans. PLoS One 7:e38122. doi:10.1371/journal.pone.0038122

Geneva MP, Stancheva IV, Boychinova MM, Mincheva NH, Yonova PA (2010) Effects of foliar fertilization and arbuscular mycorrhizal colonization on *Salvia officinalis* L. growth, antioxidant capacity, and essential oil composition. J Sci Food Agric 90:696–702

Gianinazzi S, Schüepp H, Barea JM, Haselwandter K (2002) Mycorrhiza technology in agriculture: from genes to bioproducts. Birkhäuser Verlag, Basel

Giovannetti M, Avio L, Barale R, Boychinova MM, Mincheva NH, Yonova PA (2012) Nutraceutical value and safety of tomato fruits produced by mycorrhizal plants. Br J Nutr 107:242–251

Glick BR (2012) Plant growth-promoting bacteria: mechanisms and applications. Scientifica (Cairo) 2012:963401. doi:10.6064/2012/963401

Griffiths G, Trueman L, Crowthder T, Thomas B, Smith B (2002) Onions- a global benefit to health. Phytother Res 16:603–615

Gupta ML, Prasad A, Ram M, Kumar S (2002) Effect of the vesicular arbuscular mycorrhizal (VAM) fungus *Glomus fasciculatum* on the essential oil yield related characters and nutrient acquisition in the crops of different cultivars of menthol mint (*Mentha arvensis*) under field conditions. Bioresour Technol 81:77–79

Hamaoui B, Abbadi J, Burdman S, Rashid A, Sarig S, Okon Y (2001) Effects of inoculation with *Azospirillum brasilense* on chickpeas (*Cicer arietinum*) and faba beans (*Vicia faba*) under different growth conditions. Agronomie 21:553–560

Harris J, Cottrell S, Plummer S, Lloyd D (2001) Antimicrobial properties of *Allium sativum* (garlic). Appl Microbiol Biotechnol 57:282–286

Hawkins H, George E (1997) Hydroponic culture of the mycorrhizal fungus *Glomus mosseae* with *Linum usitatissimum* L., *Sorghum bicolor* L. and *Triticum aestivum* L. Plant Soil 196:143–149

Hewedy M (2011) Associative effect of the rhizobacteria *Streptomyces chibaensis* and commercial biofertilizers on the growth, yield and nutritional value of *Vicia faba*. J Am Sci 7:552–559

Jirovetz L, Bichbauer G, Shafi M, Kaniampady M (2003) Chemotaxonomical analysis of the essential oil aroma compounds of four different *Ocimum* species from Southern India. Eur Food Res Technol 217:120–124

Jurkiewicz A, Ryszka P, Anielska T, Waligórski P, Białońska D, Góralska K, Tsimilli-Michael M, Turnau K (2010) Optimization of culture conditions of *Arnica montana* L.: effects of mycorrhizal fungi and competing plants. Mycorrhiza 20:293–306

Kallio H, Hakala M, Pelkkikangas A, Lapvetelainen A (2000) Sugars and acids of strawberry varieties. Eur Food Res Technol 212:81–85

Kapoor R, Giri B, Mukerji KG (2002a) Mycorrhization of coriander (*Coriandrum sativum* L) to enhance the concentration and quality of essential oil. J Sci Food Agric 82:339–342

Kapoor R, Giri B, Mukerji KG (2002b) Glomus macrocarpum: a potential bioinoculant to improve essential oil quality and concentration in Dill (Anethum graveolens L.) and Carum (Trachyspermum ammi (Linn.) Sprague). World J Microbiol Biotechnol 18:459–463

Kapoor R, Giri B, Mukerji KG (2004) Improved growth and essential oil yield and quality in *Foeniculum vulgare* mill on mycorrhizal inoculation supplemented with P-fertilizer. Bioresour Technol 93:307–311

Kapoor R, Chaudhary V, Bhatnagar AK (2007) Effects of arbuscular mycorrhiza and phosphorus application on artemisinin concentration in *Artemisia annua* L. Mycorrhiza 17:581–587

Karagiannidis N, Hadjisavva-Zinoviadi S (1998) The mycorrhizal fungus *Glomus mosseae* enhances growth, yield and chemical composition of a durum wheat variety in 10 different soils. Nutr Cycle Agroecosyst 52:1–7

Karagiannidis N, Thomidis T, Panou-Filotheou E (2011) Effects of *Glomus lamellosum* on growth, essential oil production and nutrients uptake in selected medicinal plants. J Agric Sci 4:137–144

Karlsen A, Retterstol L, Laake P, Paur I, Bøhn SK, Sandvik L, Blomhoff R (2007) Anthocyanins inhibit nuclear factor-k B activation in monocytes and reduce plasma concentrations of pro-inflammatory mediators in healthy adults. J Nutr 137:1951–1954

Kaume L, Howard LR, Devareddy L (2012) The blackberry fruit: a review on its composition and chemistry, metabolism and bioavailability, and health benefits. J Agric Food Chem 60:5716–5727

Kaya C, Ashraf M, Sonmez O, Aydemira S, Levent Tuna A, Cullu MA (2009) The influence of arbuscular mycorrhizal colonisation on key growth parameters and fruit yield of pepper plants grown at high salinity. Sci Hortic 121:1–6

Khan A (1974) Occurrence of mycorrhizas in halophytes, hidrophytes and xerophytes, and of endogone spores in adjacent soils. J Gen Microbiol 81:7–14

Khan M, Zaidi A (2007) Synergistic effects of the inoculation with plant growth-promoting rhizobacteria and an arbuscular mycorrhizal fungus on the performance of wheat. Turk J Agric 31:355–362

Khaosaad T, Vierheilig H, Nell M, Zitterl-Eglseer K, Novak J (2006) Arbuscular mycorrhiza alter the concentration of essential oils in oregano (*Origanum* sp., Lamiaceae). Mycorrhiza 16:443–446

Krikorian R, Shidler M, Nash TA, Kalt W, Vinqvist-Tymchuk MR, Shukitt-Hale B, Joseph JA (2010) Blueberry supplementation improves memory in older adults. J Agric Food Chem 58:3996–4000

Kumar A, Sharma K, Gera R (2011) Arbuscular mycorrhizae (*Glomus mosseae*) symbiosis for increasing the yield and quality of wheat (*Triticum aestivum*). Indian J Agric Sci 81:478–480

Kumari A, Singh O, Kumar R, Singh AK, Singh R (2010) Effect of integrated nutrient management on yield and quality of dwarf pea (*Pisum sativum* L.). Veg Sci 37:149–152

Lee M, Chauhan P, Yim WJ, Lee GJ, Kim YS, Park KW, Sa TM (2011) Foliar colonization and growth promotion of red pepper (*Capsicum annuum* L.) by *Methylobacterium oryzae* CBMB20. J Appl Biol Chem 54:120–125

Lenin M, Selvakumar G, Thamizhiniyan P, Rajendiran R (2010) Growth and biochemical changes of vegetable seedlings induced by arbuscular mycorrhizal fungus. J Exp Sci 1:27–31

Li S, Hartland S (1996) A new industrial process for extracting cocoa butter and xanthines with supercritical carbon dioxide. J Am Oil Chem Soc 73:423–429

Li H, Ye ZH, Chan WF, Chen XW, Wu FY, Wu SC, Wong MH (2011) Can arbuscular mycorrhizal fungi improve grain yield, as uptake and tolerance of rice grown under aerobic conditions? Environ Pollut 159:2537–2545

Likar M, Bukovnik U, Kreft I, Chrungoo NK, Regvar M (2008) Mycorrhizal status and diversity of fungal endophytes in roots of common buckwheat (*Fagopyrum esculentum*) and tartary buckwheat (*F. tataricum*). Mycorrhiza 18:309–315

Lingua G, Bona E, Manassero P, Marsano F, Todeschini V, Cantamessa C, Copetta A, D'Agostino G, Gamalero E, Berta G (2013) Arbuscular mycorrhizal fungi and plant growth-promoting pseudomonads increases anthocyanin concentration in strawberry fruits (*Fragaria x ananassa* var. Selva) in conditions of reduced fertilization. Int J Mol Sci 14:16207–16225

Lu G, Chen G, Qi G, Gao Z (2006) Effects of arbuscular mycorrhizal fungi on the growth and fruit quality of plastic greenhouse *Cucumis sativus* L. J Appl Ecol 17:2352–2356

Mahmoud AWM, Hassan AZA (2013) Response of fennel plants to organic biofertilizer in replacement of chemical fertilization. Topclass J Agric Res 1:29–35

Maleki V, Ardakani MR, Rejali F, Taherpour AA (2013) Physiological responses of sweet basil (*Ocimum basilicum* L.) to triple inoculation with *Azotobacter*, *Azospirillum*, *Glomus intraradices* and foliar application of citric acid. Ann Biol Res 4:62–71

Mamatha G, Bagyaraj DJ, Jaganath S (2002) Inoculation of field-established mulberry and papaya with arbuscular mycorrhizal fungi and a mycorrhiza helper bacterium. Mycorrhiza 12:313–316

Mandyal P, Kaushal R, Sharma K, Kaushal M (2012) Evaluation of native PGPR isolates in bell pepper for enhanced growth, yield and fruit quality. Int J Farm Sci 2:28–35

Mechri B, Attia F, Tekaya M, Cheheb H, Hammami M (2014) Colonization of olive trees (*Olea europaea* L.) with the arbuscular mycorrhizal fungus *Glomus* sp modified the glycolipids biosynthesis and resulted in accumulation of unsaturated fatty acids. J Plant Physiol 171:1217–1220

Mena-Violante HG, Ocampo-Jiménez O, Dendooven L, Martínez-Soto G, González-Castañeda J, Davies FT Jr, Olalde-Portugal V (2006) Arbuscular mycorrhizal fungi enhance fruit growth and quality of chile ancho (*Capsicum annuum* L. cv San Luis) plants exposed to drought. Mycorrhiza 16:261–267

Midrarullah AB, Mirza M (2014) Response of rice to inoculation with plant growth promoting rhizobacteria in control lab environment and field experiment. Pak J Bot 46:1121–1124

Mirabelli C, Pierandrei F, Rea E (2009) Effect of arbuscular mycorrhizal fungi on micropropagated hazelnut (*Corylus avellana* L.) plants. Acta Hortic 812:467–472

Mogren L, Caspersen S, Olsson M, Gertsson U (2008) Organically fertilized onions (*Allium cepa* L.): effects of the fertilizer placement method on quercetin content and soil nitrogen dynamics. J Agric Food Chem 56:361–367

Mondal TK, Bhattacharya a, Laxmikumaran M, Singh Ahuja P (2004) Recent advances of tea (*Camellia sinensis*) biotechnology. Plant Cell Tissue Organ Cult 76:195–254

Montero-Calasanz MC, Santamaria C, Albareda M, Daza A, Duan J, Glick BR, Camacho M (2013) Alternative rooting induction of semi-hardwood olive cuttings by several auxin-producing bacteria for organic agriculture systems. Span J Agric Res 11:146–154

Morone-Fortunato I, Avato P (2008) Plant development and synthesis of essential oils in micropropagated and mycorrhiza inoculated plants of *Origanum vulgare* L. ssp. hirtum (Link) Ietswaart. Plant Cell Tissue Organ Cult 93:139–149

Nautiyal C, Govindarajan R, Lavania M, Pushpangadan P (2008) Novel mechanism of modulating natural antioxidants in functional foods: involvement of plant growth promoting rhizobacteria NRRL B-30488. J Agric Food Chem 56:4474–4481

Nell M, Vötsch M, Vierheilig H, Steinkellner S, Zitterl-Eglseer K, Franz C, Novaka J (2009) Effect of phosphorus uptake on growth and secondary metabolites of garden sage (*Salvia officinalis* L.). J Sci Food Agric 89:1090–1096

Nell M, Wawrosch C, Steinkellner S, Vierheilig H, Kopp B, Lössl A, Franz C, Novak J, Zitterl-Eglseer K (2010) Root colonization by symbiotic arbuscular mycorrhizal fungi increases sesquiterpenic acid concentrations in *Valeriana officinalis* L. Planta Med 76:393–398

Nzanza B, Marais D, Soundy P (2012) Yield and nutrient content of tomato (*Solanum lycopersicum* L.) as influenced by *Trichoderma harzianum* and *Glomus mosseae* inoculation. Sci Hortic 144:55–59

Oliveira MS, Campos MAS, Silva FSB (2014) Arbuscular mycorrhizal fungi and vermicompost to maximize the production of foliar biomolecules in *Passiflora alata* Curtis seedlings. J Sci Food Agric. doi:10.1002/jsfa.6767

Orhan E, Esitken A, Ercisli S, Turan M, Sahin F (2006) Effects of plant growth promoting rhizobacteria (PGPR) on yield, growth and nutrient contents in organically growing raspberry. Sci Hortic 111:38–43

Ortas I (2010) Effect of mycorrhiza application on plant growth and nutrient uptake in cucumber production under field conditions. Span J Agric Res 8:116–122

Palencia P, Martinez F, Weiland C (2013) Effect of arbuscular mycorrhizal fungi on quality of strawberry fruit in soilless growing system. Acta Hortic 1013:493–498

Palenzuela J, Azcon-Aguilar C, Figueroa D, Caravaca F, Roldán A, Barea JM (2002) Effects of mycorrhizal inoculation of shrubs from Mediterranean ecosystems and composted residue application on transplant performance and mycorrhizal developments in a desertified soil. Biol Fertil Soils 36:170–175

Pastorello E, Farioli L, Pravettoni V, Scibilia J, Conti A, Fortunato D, Borgonovo L, Bonomi S, Primavesi L, Ballmer-Weber B (2009) Maize food allergy: lipid-transfer proteins, endochitinases, and alphazein precursor are relevant maize allergens in double-blind placebo controlled maize-challenge-positive patients. Anal Bioanal Chem 395:93–102

Perez G, Olias R, Espada J, Olías JM, Sanz C (1997) Rapid determination of sugars, nonvolatile acids, and ascorbic acid in strawberry and other fruits. J Agric Food Chem 45:3545–3549

Perner H, Rohn S, Driemel G, Batt N, Schwarz D, Kroh LW, George E (2008) Effect of nitrogen species supply and mycorrhizal colonization on organo sulfur and phenolic compounds in onions. J Agric Food Chem 56:3538–3545

Pešaković M, Karaklajić-Stajić Ž, Milenković S, Mitrović O (2013) Biofertilizer affecting yield related characteristics of strawberry (*Fragaria×ananassa* Duch.) and soil micro-organisms. Sci Hortic 150:238–243

Price K, Rhodes M (1997) Analysis of the major flavonol glycosides present in four varieties of onion (*Allium cepa*) and changes in composition resulting from autolysis. J Sci Food Agric 74:331–339

Raju P, Clark R, Ellis J, Maranville J (1990) Effects of species of VA-mycorrhizal fungi on growth and mineral uptake of sorghum at different temperatures. Plant Soil 121:165–170

Ramos-Solano B, Garcia-Villaraco A, Gutierrez-Mañero FJ, Lucas JA, Bonilla A, Garcia-Seco D (2014) Annual changes in bioactive contents and production in field-grown blackberry after inoculation with *Pseudomonas fluorescens*. Plant Physiol Biochem 74:1–8

Rosen CJ, Fritz VA, Gardner GM, Hecht SS, Carmella SG, Kenney PM (2005) Cabbage yield and glucosinolate concentrations as affected by nitrogen and sulfur fertility. HortScience 40:1493–1498

Roshanpour N, Darzi MT, Haj M, Hadi S (2014) Effects of plant growth promoter bacteria on biomass and yield of basil (*Ocimum basilicum* L.). Int J Adv Biol Biomed Res 2:2077–2085

Rydlová J, Püschel D, Sudová R, Gryndler M, Mikanová O, Vosátka M (2011) Interaction of arbuscular mycorrhizal fungi and rhizobia: Effects on flax yield in spoil-bank clay. J Plant Nutr Soil Sci 174:128–134

Sajedi N a, Ardakani MR, Rejali F, Mohabbati F, Miransari M (2010) Yield and yield components of hybrid corn (*Zea mays* L.) as affected by mycorrhizal symbiosis and zinc sulfate under drought stress. Physiol Mol Biol Plants 16:343–351

Salvioli A, Zouari I, Chalot M, Bonfante P (2012) The arbuscular mycorrhizal status has an impact on the transcriptome profile and amino acid composition of tomato fruit. BMC Plant Biol 12:44

Sbrana C, Avio L, Giovannetti M (2014) Beneficial mycorrhizal symbionts affecting the production of health-promoting phytochemicals. Electrophoresis 35:1535–1546

Schüßler A, Walker C (2010) The Glomeromycota: a species list with new families and genera. The Royal Botanic Garden/Botanische Staatssammlung Munich/Oregon State University, Edinburgh/Munich/Oregon. ISBN-13: 978-1466388048; ISBN-10: 1466388048

Schulz H (2004) Application in analysis of coffee, tea, cocoa, tobacco, spices, medicinal and aromatic plants, and related products. In: Roberts C, Workman J Jr, Reeves J (eds) Near-infrared spectroscopy in agriculture, Agronomy Monograph No. 44. American Society of Agronomy/Crop Science of America/Soil Science Society of America, Madison, pp 345–376

Schwarz D, Welter S, George E (2011) Impact of arbuscular mycorrhizal fungi on the allergenic potential of tomato. Mycorrhiza 21:341–349

Secilia J, Bagyaraj D (1994) Selection of efficient vesicular–arbuscular mycorrhizal fungi for wetland rice-a preliminary screen. Mycorrhiza 4:265–268

Sharif M, Sarir M, Muhammad D, Shafi M (2011) Response to different crops to arbuscular mycorrhiza fungal inoculation in phosphorus deficient soil. Commun Soil Sci Plant Anal 42:2299–2309

Silva V, Alves P, de Oliveira R, de Jesus R, do Bonfim CL, Gross E (2014) Influence of arbuscular mycorrhizal fungi on growth, mineral composition and production of essential oil in *Mentha x piperita* L. var. citrata (Ehrh.) Briq. under two phosphorus levels. J Med Plants Res 8:1321–1332

Simon J, Quinn J, Murray R (1990) Basil: a source of essential oils. In: Janick J, Simon J (eds) Advances in new crops. Timber Press, Portland, pp 484–489

Sinclair G, Charest C, Dalpé Y, Khanizadeh S (2014) Influence of colonization by arbuscular mycorrhizal fungi on three strawberry cultivars under salty conditions. Agric Food Sci 23:146–158

Singh S, Pandey A, Kumar B, Palni LMS (2010) Enhancement in growth and quality parameters of tea [*Camellia sinensis* (L.) O. Kuntze] through inoculation with arbuscular mycorrhizal fungi in an acid soil. Biol Fertil Soils 46:427–433

Sirichaiwetchakul S, Sirithorn P, Manakasem Y (2011) Arbuscular mycorrhizal fungi on growth, fruit yield, and quality of cherry tomato under glasshouse conditions. Suranaree J Sci Technol 18:273–280

Smith SE, Read DJ (2008) Mycorrhizal symbiosis. Academic, New York

Solaiman M, Hirata H (1997) Responses of directly seeded wetland rice to arbuscular mycorrhizal inoculation. J Plant Nutr 20:1479–1487

Subramanian K, Balakrishnan N, Senthil N (2013) Mycorrhizal symbiosis to increase the grain micronutrient content in maize. Aust J Crop Sci 7:900–910

Toussaint J, Kraml M, Nell M, Smith SE, Smith FA, Steinkellner S, Schmiderer C, Vierheilig H, Novak J (2008) Effect of *Glomus mosseae* on concentrations of rosmarinic and caffeic acids and essential oil compounds in basil inoculated with *Fusarium oxysporum* f. sp. basilici. Plant Pathol 57:1109–1116

Turner S, Amon J, Schneble R, Friese C (2000) Mycorrhizal fungi associated with plants in ground-water fed wetlands. Wetlands 20:200–204

Ulrichs C, Fischer G, Büttner C, Mewis I (2008) Comparison of lycopene, b -carotene and phenolic contents of tomato using conventional and ecological horticultural practices, and arbuscular mycorrhizal fungi (AMF). Agron Colomb 26:40–46

Urcelay C, Acho J, Joffre R (2011) Fungal root symbionts and their relationship with fine root proportion in native plants from the Bolivian Andean highlands above 3,700 m elevation. Mycorrhiza 21:323–330

Verginer M, Leitner E, Berg G (2010) Production of volatile metabolites by grape-associated microorganisms. J Agric Food Chem 58:8344–8350

Verma R, Maurya BR (2013) Effect of bio-organics and fertilizers on yield and nutrient uptake by cabbage. Ann Plant Soil Res 15:35–38

Verma RS, Rahman L, Verma RK, Chauhan A, Yadav AK, Singh A (2010) Essential oil composition of menthol mint (*Mentha arvensis*) and peppermint (*Mentha piperita*) cultivars at different stages of plant growth from Kumaon region of Western Himalaya. Open Access J Med Aromat Plants 1:13–18

Wang FY, Tong RJ, Shi ZY, Xu XF, He XH (2011) Inoculations with arbuscular mycorrhizal fungi increase vegetable yields and decrease phoxim concentrations in carrot and green onion and their soils. PLoS ONE 6:e16949. doi:10.1371/journal.pone.0016949

Yadegari M, Rahmani H (2010) Evaluation of bean (*Phaseolus vulgaris*) seeds inoculation with *Rhizobium phaseoli* and Plant Growth Promoting Rhizobacteria (PGPR) on yield and yield components. Afr J Agric Res 5:792–799

Yildirim E (2011) Promotion of broccoli by plant growth promoting rhizobacteria. Hortic Sci 46:932–936

Yooyongwech S, Phaukinsang N, Cha-um S, Supaibulwatana K (2013) Arbuscular mycorrhiza improved growth performance in *Macadamia tetraphylla* L. grown under water deficit stress involves soluble sugar and proline accumulation. Plant Growth Regul 69:285–293

Zeng Y, Guo LP, Chen BD, Hao ZP, Wang JY, Huang LQ, Yang G, Cui XM, Yang L, Wu ZX, Chen ML, Zhang Y (2013) Arbuscular mycorrhizal symbiosis and active ingredients of medicinal plants: current research status and prospectives. Mycorrhiza 23:253–266

Zhu Y, Smith F, Smith S (2003) Phosphorus efficiencies and responses of barley (*Hordeum vulgare* L.) to arbuscular mycorrhizal fungi grown in highly calcareous soil. Mycorrhiza 13:93–100

Zitterl-Eglseer K, Nell M, Lamien-Meda A, Steinkellner S, Wawrosch C, Kopp B, Zitterl W, Vierheilig H, Novak J (2015) Effects of root colonization by symbiotic arbuscular mycorrhizal fungi on the yield of pharmacologically active compounds in *Angelica archangelica* L. Acta Physiol Plant 37:21

Zolfaghari M, Nazeri V, Sefidkon F, Rejali F (2012) Effect of arbuscular mycorrhizal fungi on plant growth and essential oil content and composition of *Ocimum basilicum* L. Iran J Plant Physiol 3:643–650

Zouari I, Salvioli A, Chialva M, Novero M, Miozzi L, Tenore GC, Bagnaresi P, Bonfante P (2014) From root to fruit: RNA-Seq analysis shows that arbuscular mycorrhizal symbiosis may affect tomato fruit metabolism. BMC Genomics 15:221

Zubek S, Mielcarek S, Turnau K (2012) Hypericin and pseudohypericin concentrations of a valuable medicinal plant *Hypericum perforatum* L. are enhanced by arbuscular mycorrhizal fungi. Mycorrhiza 22:149–156

# Part II

# Diverse Applications

# Biological Potential of Arbuscular Mycorrhizal Fungi

6

Raffaella Balestrini

## Contents




**Abstract**

Microorganisms play a key role in preserving soil fertility in forest and agro-ecosystems. The exploitation of their beneficial traits represents a promising avenue for the development of more sustainable agriculture. In this chapter, attention is focused on arbuscular mycorrhizal (AM) fungi, considering the aspects that have been highlighted through the sequencing of the *Rhizophagus irregularis* genome and on the mechanisms involved in the nutritional exchanges that take place during their interaction with plants. Examples of the use of this group of fungi in applicative projects are also reported.


## 6.1 Introduction

Modern agriculture and farming practices are mainly based on the cultivation of high-yield varieties, combined with the use of agrochemicals, i.e., fertilizers and pesticides (Schlaeppi and Bulgarelli 2014). Fertilizers provide the main nutrients required by plants and, among these nutrients, phosphorus (P) represents a particularly critical component (Roy-Bolduc and Hijri 2011). As mineral fertilizers (such as phosphate) are derived from nonrenewable natural resources and agrochemicals are often hazardous to the environment, alternative and more


R. Balestrini (✉)
Istituto per la Protezione Sostenibile delle Piante (IPSP), Consiglio Nazionale delle Ricerche (CNR), Viale Mattioli 25, 10125 Torino, Italy
e-mail: raffaella.balestrini@ipsp.cnr.it





N.K. Arora et al. (eds.), *Bioformulations: for Sustainable Agriculture*,
DOI 10.1007/978-81-322-2779-3_6

sustainable practices should be explored (Schlaeppi and Bulgarelli 2014). Microorganisms play a key role in preserving soil fertility in forest and agro-ecosystems. The exploitation of beneficial microbial traits, for the host plant or the environment, represents a promising avenue for the development of a more sustainable next-generation agriculture (Schlaeppi and Bulgarelli 2014). Together with rhizobacteria, the most important biofertilizer microorganisms are mycorrhizal fungi, and in particular arbuscular mycorrhizal (AM) fungi, which form symbioses with the roots of the most important crop species (e.g., tomato, maize, wheat, rice, and potato). Mycorrhizal fungi are usually considered biofertilizers, since their best understood function in the symbiosis is an improvement in plant mineral nutrient acquisition, in exchange for C compounds derived from the photosynthetic process, which results in positive host growth responses as well as in a better response to stress conditions (Balestrini et al. 2015; van der Heijden et al. 2015). They are considered to be essential elements for plant nutrition as their hyphae can extend for many meters and they can uptake the mineral nutrients present in the soil and transfer them to the roots. In recent years, several works have been focused on verifying the effect of AM fungi on the edible portions of plants and on human-relevant nutrients (Giovannetti et al. 2012a; Hart et al. 2014; Sbrana et al. 2014; Zouari et al. 2014; Chap. 5 by Bona et al. in this book).

The characteristics of the soil in which plants grow are the first parameters that can influence the productivity as well as the product quality of plants. The reduction (or even the disappearance) of mycorrhizal fungal propagules, as a result of certain cultural practices in agricultural systems, such as the use of chemicals on the soil or soil treatments, can be considered as an indicator of the decreased quality of the plant–soil system (Giovannetti et al. 2006). AM fungi can be considered the most important microorganisms that need to be monitored in the environmental impact assessments of the cultural practices used in agriculture. Some saprotrophic fungal species can also play a beneficial role on plants and soil, by functioning as natural biocontrol agents. *Trichoderma* spp., the most commonly applied biocontrol fungi, has been studied extensively using a variety of research approaches (Mukherjee et al. 2013). Together with other beneficial microbes, they could be involved in the disease-suppressive phenomenon of soils in which crop plants suffer less from specific soilborne pathogens than expected (Mazzola 2002). In the last few years, the metagenomes of fungal communities associating with plant roots have been revealed, and great interest has been shown in having information about natural soil microbiota (Orgiazzi et al. 2012; Schlaeppi and Bulgarelli 2014; van der Heijden et al. 2015). Here, attention is focused on AM fungi, and in particular on the aspects highlighted through the sequencing of *R. irregularis* genome (Tisserant et al. 2013; Lin et al. 2014), and on the mechanisms involved in the nutritional exchanges that take place during the interaction with the plants. Examples of the use of this group of fungi in applicative projects are also reported.

## 6.2 Looking Inside the Genome of an AM Fungus

AM fungi are obligate biotrophs that colonize the roots of plants and form highly branched structures (arbuscules) inside the cortical cells of the roots (Fig. 6.1). Although AM fungal

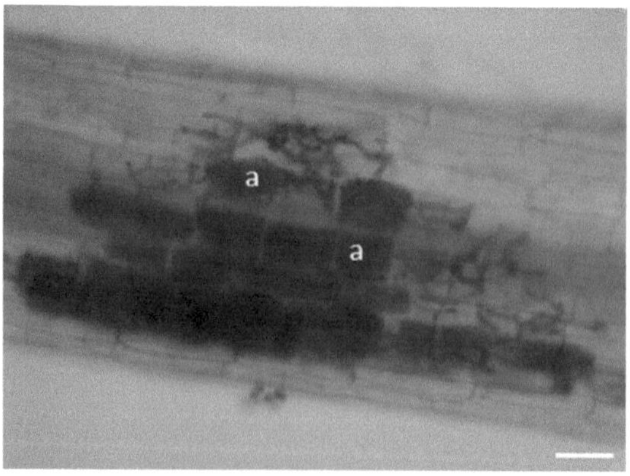


**Fig. 6.1** Tomato root colonized by an AM fungus (*Cotton blue staining*), where intraradical hyphae and arbuscules (a) are visible. Bar, 40 μm

spores can germinate in the absence of host plants (pre-symbiotic phase), they depend on the establishment of intracellular symbiosis to complete the fungal life cycle and produce the next generation of spores. The genome of the AM fungus *R. irregularis* DAOM-197198 (formerly *Glomus intraradices*) was the third mycorrhizal genome to be published (Tisserant et al. 2013; Lin et al. 2014). Interestingly, the observed low level of genome polymorphism does not support the presence of highly diverged multiple genomes in the multinucleated *R. irregularis* (Tisserant et al. 2013). A *de novo* genome sequencing of the individual nuclei of *R. irregularis* has revealed a low level of polymorphism between the nuclei (Lin et al. 2014), in agreement with the previously mentioned result. The first three sequenced mycorrhizal genomes, i.e., the ectomycorrhizal *Laccaria bicolor* (Martin et al. 2008) and *Tuber melanosporum* (Martin et al. 2010) and the AM fungus *R. irregularis*, show substantial losses in plant cell wall-degrading enzymes. Recently, new fungal genomes, including ectomycorrhizal, orchid, and ericoid species, have been sequenced with the aim of elucidating the genetic bases of the mycorrhizal lifestyle (Kohler et al. 2015). The latter work has confirmed that a convergent evolution of ectomycorrhizal (ECM) fungi involves a reduced set of genes encoding plant cell wall (PCW)-degrading enzymes, compared to their ancestral wood decayers. In spite of the intraroot habitat, PCW-degrading enzymes lack in *R. irregularis*, thus suggesting that the penetration of the root cells relies on plant enzymes (Tisserant et al. 2013). It can be hypothesized that signal molecules released by the fungus are perceived by the plant cells which, in turn, activate their own enzymes, thus allowing the penetration of the fungus through the cell wall and the intracellular development of the functional structures of the symbiosis, i.e., the arbuscules (Balestrini and Bonfante 2014; Kuo et al. 2014). Invertase or sucrose transporter genes have not been identified in its genome, thus suggesting a great dependence of the fungus on its host for carbohydrates, as well as the key enzymes that are involved in the biosynthesis of fungal toxins and thiamine (Tisserant et al. 2013). *R. irregularis* also has a relatively small repertoire of effector-like proteins (Lin et al. 2014), including five putative secreted proteins with a so-called crinkler (CRN) domain. Secreted CRN domain effectors are abundantly present in the oomycete plant pathogens of the *Phytophthora* genus, and similar proteins have been recognized in *Batrachochytrium dendrobatidis*, but not in other sequenced fungi (Lin et al. 2014). A *R. irregularis* effector protein (SP7) has already been characterized (Kloppholz et al. 2011), and genome analysis has shown the presence of 13 sequences with similarity to SP7. Alignment of these protein sequences conserved features are also present in SP7, which would seem to indicate that these proteins are good candidates for the display of effector functionality (Lin et al. 2014). Transcriptomics has revealed that a battery of mycorrhiza-induced small secreted proteins (MiSSPs) is expressed in symbiotic tissues (Tisserant et al. 2013). Starting from the genome sequence, a monosaccharide transporter (MST2), which is specific to and required for the symbiosis, has already been characterized (Helber et al. 2011), and its expression has been correlated to that of the mycorrhiza-specific plant phosphate transporter gene PT4, which is used as an indicator of a functional AM association.

Taken together, the results show that the obligate biotrophy of this fungus does not seem to be associated with a large reduction in metabolic complexity, as observed in many obligate biotrophic pathogens, and its ability to interact with the soil environment, regarding nutrient uptake, is maintained in the symbiotic fungus (Tisserant et al. 2013).

## 6.3 New Insights Into Nutrient Exchanges in AM Symbiosis

AM fungi colonize the roots and form highly branched structures inside root cortical cells, i.e., the arbuscules, which are the functional site of nutrient exchange. The formation of symbiosis with AM fungi leads to an improved nutritional status of the host plant (Smith and Smith 2012), especially under low nutrient availability, allowing the plant to access unexplored soil niches and absorb usually inaccessible nutrient sources. The fungus, in turn, receives photosynthetic carbohydrates. The extraradical AM mycelium captures water and mineral nutrients from the soil and is an important sink for host carbon. Inside the roots, carbohydrates and mineral nutrients are then exchanged across the interface between the plant and the fungus (Balestrini et al. 2015). These aspects have been studied extensively from both physiological and molecular perspectives (Harrison et al. 2002; Bucher 2007; Smith and Smith 2011). It has clearly been demonstrated that plants possess a symbiotic phosphate uptake pathway (Bucher 2007; Smith and Smith 2011). Moreover, ammonium transporters specifically expressed in arbusculated cells have also been discovered (Gomez et al. 2009; Guether et al. 2009a; Kobae et al. 2010). Evidence of the involvement of AM symbiosis in the transfer of mineral nutrients has been obtained in studies of several plant species (Nagy et al. 2005; Paszkowski et al. 2002; Guether et al. 2009b; Hogekamp et al. 2011; Casieri et al. 2013 for a review). On the fungal side, it is worth noting the fact that a phosphate transporter (Balestrini et al. 2007; Tisserant et al. 2012; Fiorilli et al. 2013) and an ammonium one (Pérez-Tienda et al. 2011) have been found as expressed in arbuscules. Transcriptomic data have also shown the upregulation of a putative amino acid transporter in the intraradical mycelium *vs* the extraradical one (Tisserant et al. 2012). Recently, a *R. irregularis* gene sequence, similar to members of the PTR2 family of fungal oligopeptide transporters (called *RiPTR2*), has been identified and characterized (Belmondo et al. 2014). Gene expression analysis has shown that *RiPTR2* is expressed in the extraradical mycelium, thus suggesting a role in the uptake of organic N from the soil. However, a higher expression has been observed in the intraradical stage, i.e., arbuscules, which would seem to suggest a role in the mobilization of organic N in mycorrhizal roots (Belmondo et al. 2014). Considering the plant side, two *Medicago truncatula* PTR genes have also been described as mycorrhiza responsive and their transcripts have been detected in arbusculated cells (Gomez et al. 2009). Overall, these results suggest that the fungus may reabsorb nutrients released at the periarbuscular interface, thus exerting control over the amount of nutrients delivered to the host (Balestrini et al. 2007; Belmondo et al. 2014). A large number of papers have been published on phosphate (Pi) and nitrogen (N) uptake and their transfer during symbiosis (Casieri et al. 2013). Recently, new information has been obtained on sulfur (S) acquisition. The *G. intraradices* fungus has been shown, in monoxenic conditions, to take up both sulfate- and S-containing amino acids and to transfer them to the plant (Allen and Shachar-Hill 2009). On the plant side, recent investigations have revealed that AM symbiosis can influence the expression of plant sulfate transporters and, as a consequence, can improve the S nutritional status of the host plant (Casieri et al. 2012; Giovannetti et al. 2014). The transcriptomic response of *M. truncatula* to different sulfate concentrations in mycorrhizal and non-mycorrhizal plants has been investigated, and it has been found that the regulation of *M. truncatula* sulfate transporters depends on the S concentration and on the mycorrhizal colonization in both the root and shoot (Casieri et al. 2012). Sieh and colleagues (2013) have analyzed the influence of mycorrhizal colonization on the metabolite composition of leaves and the expression of sulfur starvation-related genes in *M. truncatula* grown under different sulfur and phosphate fertilization treatments. Giovannetti et al. (2014) have recently identified and characterized a novel AM-dependent sulfate transporter, LjSultr1; 2, that also responds to S starvation. Its transcript accumulates specifically in the arbusculated cells of mycorrhizal roots. The authors have also shown the importance of this sulfate transporter in the uptake of sulfate in *Lotus japonicus* seedlings (4 and 10 days old), even in the absence of the mycorrhizal fungus. This would seem to suggest that a single gene, *LjSultr1;2*, mediates both direct and symbiotic pathways. Although potassium ($K^+$) is one of the most abundant elements in the soil, its availability is very low, limiting plant growth and productivity of ecosystems (Garcia and Zimmermann 2014). The contribution of AM symbiosis to the enhancement of plant $K^+$ nutrition is not clearly understood yet, and only a few data have been reported on the putative AM symbiosis contribution to $K^+$ acquisition (Garcia and Zimmermann 2014). Interestingly, it has been suggested that the improvement in $K^+$ nutrition, due to mycorrhizal symbiosis, could act on the abiotic stress tolerance of the host plant, e.g., under salinity and drought stress conditions (Garcia and Zimmermann 2014 and references therein). The upregulation of several genes involved in nutrient transport, which may be involved in Pi, S, N, and $K^+$ uptake by the plant from the fungus, has been observed in *L. japonicus* mycorrhizal roots (Guether et al. 2009b). Among them, two putative aquaporin genes, named *LjNIP1* and *LjXIP1*, have been identified. A further gene expression analysis has revealed a good correlation between the expression of *LjNIP1* and *LjPT4*, the phosphate transporter that is considered a marker gene of mycorrhizal functionality. *LjNIP1* transcripts have been localized exclusively in arbuscule-containing cells using laser microdissection (Giovannetti et al. 2012b), and it has been suggested that LjNIP1 could be considered a novel molecular marker of mycorrhizal symbiosis. On the fungal side, two functional aquaporin genes, *GintAQPF1* and *GintAQPF2*, have recently been cloned from the AM fungus *G. intraradices*. Gene expressions were analyzed during symbiosis under drought stress, and the potential transport of water via AMF to the host plants was suggested (Li et al. 2013a, b). Fungal aquaporins could play a role not only in helping AM fungi to resist drought stress but also in plant drought tolerance (Li et al. 2013b). Soil waterlogging is considered a major abiotic stress that affects the growth, development, and survival of numerous plant species, in natural as well as in agricultural and horticultural ecosystems (Parent et al. 2008). The role of AM symbiosis in the tolerance of tomato plants to flooding has recently been reported (Calvo-Polanco et al. 2014). In particular it was shown that, under flooded conditions, the hydraulic conductivity of the roots of plants inoculated with *R. irregularis* increased, and this effect was correlated to a higher expression of the plant aquaporin SlPIP1;7 and of the fungal aquaporin GintAQP1.

Looking at these results, several efforts have been made in the last few years to highlight the role of AM symbiosis in plant nutrient and water uptake (Casieri et al. 2013; Bárzana et al. 2014). The correlation of molecular and biochemical data to ecophysiological data will constitute a major advance (van der Heijden et al. 2015). This could in fact lead to an improvement in the knowledge on the impact of AM symbiosis on plant performances, under different growth conditions, and on the role of AM fungi in plant protection against biotic and abiotic stresses.

## 6.4 Exploitation of AM Fungi in Sustainable Agriculture

The use of AM fungi in agriculture has not yet been widely adopted in the typical intensive agriculture of Europe and North America, while advances have been made by developing countries, such as Cuba, India, and Mexico, where chemical fertilizers are prohibitively expensive (Roy-Bolduc and Hijri 2011). New methods for the large-scale production of AM fungi (Ijdo et al. 2011) and seed coating using AM fungi (Vosatka et al. 2012) have been developed in recent years. Recently, the use of in vitro produced AM fungi has led to significant yield increases in the globally important food security crop cassava (Ceballos et al. 2013). Roy-Bolduc and Hijri (2011) have reported that two main approaches to manage AM fungi in agriculture programs are possible: using an inoculum of selected AM fungal strains or adopting practices that can enhance indigenous AM fungi. In the latter case, they are well adapted to local conditions, although they might have a lower efficiency than selected strains. In the last few years, several researches have focused on the selection of more efficient AM fungal strains, and a crucial point is that of finding the most efficient plant/AM fungus combinations. The first step is the identification of AM fungal soil communities from several ecosystems and the study of the shifts in these communities, i.e., under different soil managements (Alguacil et al. 2008; Lumini et al. 2010; Borriello et al. 2012). Several papers have been published in the last few years on the identification of AM fungal communities using both morphological and molecular methods (Balestrini et al. 2010; Al-Yahyaìei et al. 2011; Lumini et al. 2011; Berruti et al. 2014; Borriello et al. 2015), also under abiotic stress conditions, e.g., salinity (Estrada et al. 2013a, b). A major advancement has been made, thanks to the development of "high-throughput" sequencing techniques, such as 454 sequencing platforms, which have greatly changed the analysis of environmental samples in terms of yields, cost, and speed, and has resulted in thousands of sequences being obtained in a short time (Lindahl et al. 2013; Opik et al. 2013; Balestrini et al. 2015 and references therein). The selection of a suitable host plant for the mass production of indigenous AM fungi has also been made (Abdullahi et al. 2014). An increasing number of companies produce and sell certified AM fungal inoculants for horticulture and agriculture, although their success in root colonization is not predictable (Berruti et al. 2013). Several field investigations have shown that different AM fungal species/isolates may lead to different colonization rates and different growth responses in plants. In addition, different plant genotypes can respond differently to AM fungal inoculation (An et al. 2010). Recently, the application of commercial inocula, constituted by an AM fungal isolate (*Funneliformis mosseae*), or a consortium of different fungi and bacteria in pot-cultivated *Camellia japonica*, has shown that the applied AM fungal species/isolates poorly colonized the root system, which would seem to suggest that commercial formulations should be more targeted and host specific to successfully colonize the host root and, as a consequence, potentially express their benefits (Berruti et al. 2013). AM fungal species that were found in association with camellia plants in seminatural ecosystems have been used in a parallel study on C. *japonica* "Nuccio's Pearl," and a higher level of colonization has been obtained. The identification of the AM fungal species that form the arbuscules inside the roots has also been possible using the microdissection approach (Berruti et al. 2013). Moreover, it has been reported that the tolerance of maize to salt stress is enhanced more by AM fungi isolates from saline environments than by the collection ones (Estrada et al. 2013a, b).

The appropriate application of AM fungi to improve food security, by increasing the overall yield of important crops, has been discussed in a recent perspective article by Rodriguez and Sanders (2015). The main challenges of using AM fungi to significantly improve the production of important food security crops, especially in the tropics, have been outlined, with particular attention being paid to ecological aspects related to the introducing of non-native AM fungi, which could have an impact on the existing AM fungal communities. It should be recalled that AM fungal inoculants are difficult to trace in field experiments. However, in the presence of the *R. irregularis* reference genome, the partial genome re-sequencing of multiple isolates from different geographic origins will facilitate the identification of genome-wide markers for future genomic population studies as well as the discrimination between introduced *R. irregularis* strains and native ones (Rodriguez and Sanders 2015).

## 6.5 Conclusions

In the last years, major advancements have been done in the knowledge of the biology of AM fungi and the ecology of AM symbiosis. As reported by van der Heijden and colleagues (2015), a further frontier will be to develop predictive models that help us to decide when, and under what conditions, application of a treatment based on mycorrhizal fungi is needed and advantageous.

## References


Abdullahi R, Lihan S, Edward R (2014) Isolation of indigenous arbuscular mycorrhizal fungi and selection of host plant for inoculum production. Int J Biosci 5:116–122

Alguacil MM, Lumini E, Roldan A, Salinas–Garcia JR, Bonfante P, Bianciotto V (2008) The impact of tillage practices on arbuscular mycorrhizal fungal diversity in subtropical crops. Ecol Appl 18:527–536

Allen JW, Shachar-Hill Y (2009) Sulfur transfer through an arbuscular mycorrhiza. Plant Physiol 149:549–560

Al-Yahyaìei MN, Oehl F, Vallino M, Lumini E, Redecker D, Wiemken A, Bonfante P (2011) Unique arbuscular mycorrhizal fungal communities uncovered in date palm plantations and surrounding desert habitats of Southern Arabia. Mycorrhiza 21:195–209

An GH, Kobayashi S, Enoki H, Sonobe K, Muraki M, Karasawa T et al (2010) How does arbuscular mycorrhizal colonization vary with host genotype? An example based on maize (*Zea mays*) germplasms. Plant Soil 327:441–453

Balestrini R, Bonfante P (2014) Cell wall remodeling in mycorrhizal symbiosis: a way towards biotrophism. Front Plant Sci 5:237

Balestrini R, Gomez-Ariza J, Lanfranco L, Bonfante P (2007) Laser microdissection reveals that transcripts for five plant and one fungal phosphate transporter genes are contemporaneously present in arbusculated cells. Mol Plant Microbe Interact 20:1055–1062

Balestrini R, Magurno F, Walker C, Lumini E, Bianciotto V (2010) Cohorts of arbuscular mycorrhizal fungi (AMF) in *Vitis vinifera*, a typical Mediterranean fruit crop. Environ Microbiol Rep 2:594–604

Balestrini R, Lumini E, Borriello R, Bianciotto V (2015) Plant–soil biota interactions. In: Paul EA (ed) Soil microbiology, ecology and biochemistry, 4th edn. Academic Press, Elsevier, London/San Diego/Oxford

Bárzana G, Aroca R, Bienert GP, Chaumont F, Ruiz-Lozano JM (2014) New insights into the regulation of aquaporins by the arbuscular mycorrhizal symbiosis in maize plants under drought stress and possible implications for plant performance. Mol Plant Microbe Interact 27:349–363

Belmondo S, Fiorilli V, Pérez–Tienda J, Ferrol N, Marmeisse R, Lanfranco L (2014) A dipeptide transporter from the arbuscular mycorrhizal fungus *Rhizophagus irregularis* is upregulated in the intraradical phase. Front Plant Sci 5:436

Berruti A, Borriello R, Lumini E, Scariot V, Bianciotto V, Balestrini R (2013) Application of laser microdissection to identify the mycorrhizal fungi that establish arbuscules inside root cells. Front Plant Sci 4:135

Berruti A, Borriello R, Orgiazzi A, Barbera AC, Lumini E, Bianciotto V (2014) Arbuscular mycorrhizal fungi and their value for ecosystem management. In: Grillo O (ed) The dynamic balance of the planet. Intech, Rijeka

Borriello R, Lumini E, Girlanda M, Bonfante P, Bianciotto V (2012) Effects of different management practices on arbuscular mycorrhizal fungal diversity in maize fields by a molecular approach. Biol Fert Soils 48:911–922

Borriello R, Berruti A, Lumini E, Della Beffa MT, Scariot V, Bianciotto V (2015) Edaphic factors trigger diverse AM fungal communities associated to exotic camellias in closely located Lake Maggiore (Italy) Sites. Mycorrhiza 25:253–265

Bucher M (2007) Functional biology of plant phosphate uptake at root and mycorrhiza interfaces. Tansley review. New Phytol 173:11–26

Calvo-Polanco M, Molina S, Zamarreño AM, García-Mina JM, Aroca R (2014) The symbiosis with the arbuscular mycorrhizal fungus *Rhizophagus irregularis* drives root water transport in flooded tomato plants. Plant Cell Physiol 55:1017–1029

Casieri L, Gallardo K, Wipf D (2012) Transcriptional response of *Medicago truncatula* sulfate transporters to arbuscular mycorrhizal symbiosis with and without sulphur stress. Planta 235:1431–1447

Casieri L, Ait Lahmidi N, Doidy J, Fourrey C, Migeon A, Bonneau L et al (2013) Biotrophic transportome in mutualistic plant–fungal interactions. Mycorrhiza 23:597–625

Ceballos I, Ruiz M, Fernández C, Peña R, Rodriguez A, Sanders IR (2013) The *in vitro* mass–produced model mycorrhizal fungus, *Rhizophagus irregularis*, significantly increases yields of the globally important food security crop cassava. PLoS One 8:e70633

Estrada B, Aroca R, Maathuis FJM, Barea JM, Ruiz-Lozano JM (2013a) Arbuscular mycorrhizal fungi native from a Mediterranean saline area enhance maize tolerance to salinity through improved ion homeostasis. Plant Cell Environ 36:1771–1782

Estrada B, Aroca R, Barea JM, Ruiz-Lozano JM (2013b) Native arbuscular mycorrhizal fungi isolated from a saline habitat improved maize antioxidant systems and plant tolerance to salinity. Plant Sci 201–202:42–51

Fiorilli V, Lanfranco L, Bonfante P (2013) The expression of *GintPT*, the phosphate transporter of *Rhizophagus irregularis*, depends on the symbiotic status and phosphate availability. Planta 237:1267–1277

Garcia K, Zimmermann SD (2014) The role of mycorrhizal associations in plant potassium nutrition. Front Plant Sci 5:337

Giovannetti M, Turrini A, Strani P, Sbrana C, Avio L, Pietrangeli B (2006) Mycorrhizal fungi in ecotoxicological studies: soil impact of fungicides, insecticides and herbicides. Prev Today 2:47–61

Giovannetti M, Avio L, Barale R, Ceccarelli N, Cristofani R, Iezzi A, Mignolli F, Picciarelli P, Pinto B, Reali D, Sbrana C, Scarpato R (2012a) Nutraceutical value and safety of tomato fruits produced by AM plants. Br J Nutr 107:242–251

Giovannetti M, Balestrini R, Volpe V, Guether M, Straub D, Costa A et al (2012b) Two putative–aquaporin genes are differentially expressed during arbuscular mycorrhizal symbiosis in *Lotus japonicus*. BMC Plant Biol 12:186

Giovannetti M, Tolosano M, Volpe V, Kopriva S, Bonfante P (2014) Identification and functional characterization of a sulfate transporter induced by both

sulfur starvation and mycorrhiza formation in *Lotus japonicus*. New Phytol 204:609–619
Gomez SK, Javot H, Deewatthanawong P, Torres-Jerez I, Tang Y, Blancaflor EB et al (2009) *Medicago truncatula* and *Glomus intraradices* gene expression in cortical cells harboring arbuscules in the arbuscular mycorrhizal symbiosis. BMC Plant Biol 9:10
Guether M, Neuhäuser B, Balestrini R, Dynowski M, Ludewig U, Bonfante P (2009a) A mycorrhizal–specific ammonium transporter from *Lotus japonicus* acquires nitrogen released by arbuscular mycorrhizal fungi. Plant Physiol 150:73–83
Guether M, Balestrini R, Hannah M, He J, Udvardi MK, Bonfante P (2009b) Genome–wide reprogramming of regulatory networks, transport, cell wall and membrane biogenesis during arbuscular mycorrhizal symbiosis in *Lotus japonicus*. New Phytol 182:200–212
Harrison MJ, Dewbre GR, Liu J (2002) A phosphate transporter from *Medicago truncatula* involved in the acquisition of phosphate released by arbuscular mycorrhizal fungi. Plant Cell 14:2413–2429
Hart M, Ehret DL, Krumbein A, Leung C, Murch S, Turi C, Franken P (2014) Inoculation with arbuscular mycorrhizal fungi improves the nutritional value of tomatoes. Mycorrhiza 25:359–376
Helber N, Wippel K, Sauer N, Schaarschmidt S, Hause B, Requena N (2011) A versatile monosaccharide transporter that operates in the arbuscular mycorrhizal fungus *Glomus* sp. is crucial for the symbiotic relationship with plants. Plant Cell 23:3812–3823
Hogekamp C, Arndt D, Pereira PA, Becker JD, Hohnjec N, Kuster H (2011) Laser microdissection unravels cell–type–specific transcription in arbuscular mycorrhizal roots, including CAAT–box transcription factor gene expression correlating with fungal contact and spread. Plant Physiol 157:2023–2043
IJdo M, Cranenbrouck S, Declerck S (2011) Methods for large–scale production of AM fungi: past, present, future. Mycorrhiza 21:1–16
Kloppholz S, Kuhn H, Requena N (2011) A secreted fungal effector of *Glomus intraradices* promotes symbiotic biotrophy. Curr Biol 21:1204–1209
Kobae Y, Tamura Y, Takai S, Banba M, Hata S (2010) Localized expression of arbuscular mycorrhiza–inducible ammonium transporters in soybean. Plant Cell Physiol 51:1411–1415
Kohler A, Kuo A, Nagy LG, Morin E, Barry KW, Buscot F et al (2015) Convergent losses of decay mechanisms and rapid turnover of symbiosis genes in mycorrhizal mutualists. Nat Genet 47:410–415
Kuo A, Kohler A, Martin FM, Grigoriev IV (2014) Expanding genomics of mycorrhizal symbiosis. Front Microbiol 5:582
Li T, Hu YJ, Hao ZP, Li H, Wang Y-S, Chen B-D (2013a) First cloning and characterization of two functional aquaporin genes from an arbuscular mycorrhizal fungus *Glomus intraradices*. New Phytol 197:617–630
Li T, Hu YJ, Hao ZP, Li H, Chen BD (2013b) Aquaporin genes *GintAQPF1* and *GintAQPF2* from *Glomus intraradices* contribute to plant drought tolerance. Plant Signal Behav 8:e24030
Lin K, Limpens E, Zhang Z, Ivanov S, Saunders DGO, Mu D et al (2014) Single nucleus genome sequencing reveals high similarity among nuclei of an endomycorrhizal fungus. PLoS Genet 10:e1004078
Lindahl BD, Nilsson RH, Tedersoo L, Abarenkov K, Carlsen T, Kjøller R, Kõljalg U, Pennanen T, Rosendahl S, Stenlid J, Kauserud H (2013) Fungal community analysis by high–throughput sequencing of amplified markers – a user's guide. New Phytol 199:288–299
Lumini E, Orgiazzi A, Borriello R, Bonfante P, Bianciotto V (2010) Disclosing arbuscular mycorrhizal fungal biodiversity in soil through a land–use gradient using a pyrosequencing approach. Environ Microbiol 12:2165–2179
Lumini E, Vallino M, Alguacil MM, Romani M, Bianciotto V (2011) Different farming and water regimes in Italian rice fields affect arbuscular mycorrhizal fungal soil communities. Ecol Appl 21:1696–1707
Martin F, Aerts A, Ahrén D, Brun A, Danchin EG, Duchaussoy F, Gibon J, Kohler A (2008) The genome of *Laccaria bicolor* provides insights into mycorrhizal symbiosis. Nature 452:88–92
Martin F, Kohler A, Murat C, Balestrini R, Coutinho P, Jaillon O, Montanini B, Morin E, Noel B, Percudani R, Porcel B, Rubini et al (2010) Perigord black truffle genome uncovers evolutionary origins and mechanisms of symbiosis. Nature 464:1033–1038
Mazzola M (2002) Mechanisms of natural soil suppressiveness to soilborne diseases. Antonie Van Leeuwenhoek 81:557–564
Mukherjee PK, Horwitz BA, Herrera-Estrella A, Schmoll M, Kenerley CM (2013) Trichoderma research in the genome era. Annu Rev Phytopathol 51:105–129
Nagy R, Karandashov V, Chague V, Kalinkevich K, Tamasloukht M, Xu G, Jakobsen I, Levy AA, Amrhein N, Bucher M (2005) The characterization of novel mycorrhiza–specific phosphate transporters from *Lycopersicon esculentum* and *Solanum tuberosum* uncovers functional redundancy in symbiotic phosphate transport in solanaceous species. Plant J 42:236–250
Öpik M, Zobel M, Cantero JJ, Davison J, Facelli JM, Hiiesalu I et al (2013) Global sampling of plant roots expands the described molecular diversity of arbuscular mycorrhizal fungi. Mycorrhiza 23:411–430
Orgiazzi A, Lumini E, Nilsson RH, Girlanda M, Vizzini A, Bonfante P, Bianciotto V (2012) Unravelling soil fungal communities from different Mediterranean land-use backgrounds. PLoS ONE 7(4), e34847. doi:10.1371/journal.pone.0034847
Parent C, Capelli N, Berger A, Crèvecoeur M, Dat JF (2008) An overview of plant responses to soil waterlogging. Plant Stress 2:20–27
Paszkowski U, Kroken S, Roux C, Briggs S (2002) Rice phosphate transporters include an evolutionarily divergent gene specifically activated in arbuscular mycorrhizal symbiosis. Proc Natl Acad Sci U S A 99:13324–13329

Perez–Tienda J, Testillano PS, Balestrini R, Fiorilli V, Azcon–Aguilar C, Ferrol N (2011) GintAMT2, a new member of the ammonium transporter family in the arbuscular mycorrhizal fungus *Glomus intraradices*. Fungal Genet Biol 48:1044–1055

Rodriguez A, Sanders IR (2015) The role of community and population ecology in applying mycorrhizal fungi for improved food. ISME J 9:1053–1061

Roy-Bolduc A, Hijri M (2011) The use of mycorrhizae to enhance phosphorus uptake: a way out the phosphorus crisis. J Biofertil Biopest 2:104

Sbrana C, Avio L, Giovannetti M (2014) Beneficial mycorrhizal symbionts affecting the production of health–promoting phytochemicals. Electrophoresis 35:1535–1546

Schlaeppi K, Bulgarelli D (2014) The plant microbiome at work. Mol Plant-Microbe Interact 28:212–217

Sieh D, Watanabe M, Devers EA, Brueckner F, Hoefgen R, Krajinski F (2013) The arbuscular mycorrhizal symbiosis influences sulfur starvation responses of *Medicago truncatula*. New Phytol 197:606–616

Smith FA, Smith SE (2011) What is the significance of the arbuscular mycorrhizal colonisation of many economically important crop plants? Plant Soil 348:63–79

Smith SE, Smith FA (2012) Fresh perspectives on the roles of arbuscular mycorrhizal fungi in plant nutrition and growth. Mycologia 104:1–13

Tisserant E, Kohler A, Dozolme-Seddas P, Balestrini R, Benabdellah K, Colard A et al (2012) The transcriptome of the arbuscular mycorrhizal fungus *Glomus intraradices* (DAOM 197198) reveals functional tradeoffs in an obligate symbiont. New Phytol 193:755–769

Tisserant E, Malbreil M, Kuo A, Kohler A, Symeonidi A, Balestrini R, Charron P et al (2013) The genome of an arbuscular mycorrhizal fungus provides insights into the oldest plant symbiosis. PNAS 110:20117–20122

van der Heijden MGA, Martin FM, Selosse M-A, Sanders IR (2015) Mycorrhizal ecology and evolution: the past, the present, and the future. New Phytol 205:1406–1423

Vosátka M, Látr A, Gianinazzi S, Albrechtová J (2012) Development of arbuscular mycorrhizal biotechnology and industry: current achievements and bottlenecks. Symbiosis 58:29–37

Zouari I, Salvioli A, Chialva M, Novero M, Miozzi L, Tenore GC, Bagnaresi P, Bonfante P (2014) From root to fruit: RNA-Seq analysis shows that arbuscular mycorrhizal symbiosis may affect tomato fruit metabolism. BMC Genomics 15:221

# The Use of Arbuscular Mycorrhizal Fungi in Combination with *Trichoderma* spp. in Sustainable Agriculture

7

Jose Antonio Pascual

## Contents




**Abstract**

Arbuscular mycorrhizal (AM) fungi as well as fungal biological control agent such as *Trichoderma* spp. have shown beneficial effects on plant growth and health. The study of the interactions between AM fungi, fungal antagonists and plants is of great interest. Before using them, it is in fact important to understand whether and how the two different beneficial fungal components can influence each other and/or the host plant. In this chapter, the attention is focused on the plant responses to the inoculation with AM fungi and *Trichoderma* spp., alone or in combination, with particular attention to the impact on the plant hormonal balance.


## 7.1 Interactions of AMF with Trichoderma: An Example of Microbial Interactions

The use of AM fungi (AMF) as bioinoculants can benefit plant growth and health, and their use could contribute to a reduction in the use of pesticides and other environmentally harmful agrochemical products currently required for optimal plant growth and health (Barea et al. 1997). The degree of the effect on plants depends on the plant species and the AMF (Jeffries and Barea 2001). The successful use of AMF in sustainable agriculture


J.A. Pascual (✉)
Centro de Edafología y Biología Aplicada del Segura, CEBAS-CSIC, PO BOX 4195, Campus Universitario de Espinardo. 25, 30100 Murcia, Spain
e-mail: jpascual@cebas.csic.es





N.K. Arora et al. (eds.), *Bioformulations: for Sustainable Agriculture*,
DOI 10.1007/978-81-322-2779-3_7

requires a selection of the appropriate host/fungus combination, infectivity and efficacy (Haas and Krikun 1985; Tarbell and Koske 2007). The study of interactions between some other beneficial organisms associated with plant roots is important, because such interactions could either enhance or inhibit the beneficial effects of individual species. Little is known about the interactions between AMF and beneficial saprophytic fungi, and the few studies published on this topic do not provide any conclusive findings (Green et al. 1999; Vázquez et al. 2000). Even more important, the beneficial effect that is attributed to these interactions under controlled experimental conditions may not be reflected in field experiments (Calvet et al. 1992; McAllister et al. 1997; Fracchia et al. 1998; Vázquez et al. 2000; Martinez et al. 2004). These interactions must be identified and characterised for their successful establishment in plants. AMFs have been shown to cause changes in the rhizosphere that could affect both the quality and quantity of microbial populations and to promote the activity of other microorganisms that could compete with soilborne pathogens (Filion et al. 1999; Barea et al. 2002). It has also been demonstrated that some *Trichoderma* strains may influence AMF activity (Green et al. 1999; Vazquez et al. 2000; Brimner and Boland 2003; Martincz et al. 2004).

Martinez-Medina et al. (2009) hypothesised that the combined use of AMF and *Trichoderma harzianum* (Rifai) could improve plant growth and, at the same time, help to protect against *Fusarium oxysporum* pathogen, within the context of a sustainable soil–plant system. It should be noted that the effects of the interaction between AMF and *T. harzianum* may be very different, depending on the AMF strain and the saprophytic strain (Martinez et al. 2004). Martinez-Medina et al. (2009) found no additional suppressive effect of *T. harzianum* when *Glomus mosseae* was co-inoculated, because the AMF itself caused reduction of soilborne pathogen. This suppressive effect was also shown against other microorganisms such as the co-inoculated *T. harzianum* strain (Azcón-Aguilar and Barea 1996). It has also been reported that the presence of *T. harzianum* significantly increased root colonisation by several AMF species, reaching values significantly higher than the most effective AMF (*G. mosseae*) inoculated alone, in contrast with the absence of effects observed in other studies (Camprubí et al. 1995; Vázquez et al. 2000). On the other hand, *T. harzianum* populations were generally reduced in the presence of AMF, except in the case of the co-inoculation with *Glomus intraradices*. This could be due to the modifications of root exudates or to substances released by the extramatrical mycelium (McAllister et al. 1994; Filion et al. 1999; Martínez et al. 2004). Competition between *T. harzianum* and AMF in the first establishment stages, which could be maintained in time, has been also proposed to explain these findings (McAllister et al. 1994; Green et al. 1999). Competition for host photosynthates or sites, microbial changes in the rhizosphere due to the presence of AMF species, and induction of local and systemic defence responses have been proposed (Caron 1989; Azcón-Aguilar and Bago 1994). By contrast, Filion et al. (1999) observed an improvement in germination of *T. harzianum* conidia by *G. intraradices*, suggesting a complex interaction between *T. harzianum* and the AMF. Recently, the interaction between *Trichoderma atroviride* (PKI1), two AMF species and the host plant has been studied under in vitro culture conditions. Using in vivo imaging of green fluorescent protein-tagged lines, Lace et al. (2014) showed that *T. atroviride* did not activate symbiotic-like responses in the plant cells at the contact points. Furthermore, it has been observed that *T. atroviride* strain parasitized the hyphae of the two AMF species (*Gigaspora gigantea* and *Gigaspora margarita*) through localised wall breaking and degradation, although this was not associated with an evident chitin lysis or the upregulation of two major chitinase genes. *T. atroviride* colonised extensive areas of the root epidermis, in association with localised cell death, and plant colonisation by both the fungal symbionts has also been observed when *Trichoderma* was applied to a pre-established AM symbiosis. Table 7.1 summarises the main interaction effects related to the interaction between AMF and *T. harzianum* in some crops.

**Table 7.1** Summary about the interaction between AMF, *T. harzianum* and some plants

| | Growth and nutrition | Control against *F. oxysporum* | Induction of systemic response in absence of the pathogen | Induction of systemic response in presence of the pathogen | Differences on root gene expression |
|---|---|---|---|---|---|
| *T. harzianum* | Increase melon growth and nutrition, particularly by solubilising nitrogen to be uptaken by roots (Contreras-Cornejo et al. 2009) | Increase melon resistance against plant pathogens, although it depends on the species (Yedidia et al. 2003; Martinez-Medina et al. 2014) | Increase SA and JA levels (Hanson and Howell 2004) | Response through SA, ACC and ABA due to pathogen was modulated by the presence of the *T. harzianum* strain (Martinez-Medina et al. 2013) | Induction of gene (*LOX* and *PAL*) related to systemic induced resistance (SIR) (Yedidia et al. 2003; Shoresh et al. 2005) |
| AMF | Increase melon plant growth and nutrition particularly by solubilising phosphorus to be uptaken by roots (Pozo and Azcon-Aguilar 2007) | It can increase plant resistance against plant pathogens, but it is more rarely than *T harzianum* (Pozo and Azcon-Aguilar 2007; Martinez-Medina et al. 2011) | No increase in SA and JA levels (Pozo et al. 2002) | Possible *priming* effect related to jasmonic route (Pozo et al. 2009). Modulated response through ACC and ABA by pathogen action | Induction of gene (*PAL*, *PR1* and *PR3*) related to systemic acquired resistance (SAR) (Vierheilig 2004) |
| *T. harzianum* + AMF | Increase but it is dependent on the AMF–*T. harzianum* combination. A good combination observed was for *G. mosseae*–*T. harzianum* (Martinez-Medina et al. 2009) | Reduction of the plant disease but it was not synergic between the different assayed coinoculations (Saldajeno et al. 2008; Martinez-Medina et al. 2009) | Decrease the levels of salicylic and JA in respect to the ones only inoculated with *T. harzianum* (Martinez-Medina et al. 2011; Garmendia et al. 2005) | Pattern similar to *T. harzianum*-inoculated plants (Martinez-Medina et al. 2011) | Decrease in the levels of gene expression related to the ones showed in the AMF-inoculated plants that reflect a loose on SAR (Vinale et al. 2008b) |

## 7.2 The Use of a Combination of AMF–*Trichoderma* to Improve Plant Health

It has been well documented that AMF themselves can control a whole range of plant pathogens, and their effect on plant yield and nutrient uptake depends on the type of species and even on the strain. The alleviation of damage caused by soilborne pathogens, such as *Phytophthora*, *Fusarium*, *Pythium*, *Rhizoctonia*, *Sclerotium* and *Verticillium*, has been reported in mycorrhizal plants (Bi et al. 2007). Martinez-Medina et al. (2009) showed that the percentage of plants infected by *F. oxysporum* in AMF-inoculated soils decreased significantly, compared to non-inoculated soils, but some AMF, such as *Glomus claroideum* and *G. mosseae*, did not show any suppressive effect on melon plants. If these treatments were also co-inoculated with *T. harzianum*, the percentage of infected plants would have been reduced significantly. As far as suppressing disease incidence is concerned, *T. harzianum* was seen to be more effective than AMF.

Several studies have reported on the biocontrol capacity of *Trichoderma* species (Chet 1987; Chet et al. 1997; de Meyer et al. 1998; Yedidia et al. 2003; Harman 2000; Howell et al. 2000; Soresh et al. 2005; Martínez-Medina et al. 2009). Various biocontrol mechanisms have been reported, such as mycoparasitism, antibiotic production, competition or the induction of local and systemic defence responses (Howell 2003; Yedidia et al. 2003; Harman et al. 2004). However, most studies on the reduction of disease by *Trichoderma* spp. have focused on microbial interactions rather than on host plant responses (Shoresh et al. 2005; Korolev et al. 2008). Moreover, a number of mechanisms have been proposed to explain growth enhancement by *Trichoderma* spp. (Harman et al. 2004; Benítez et al. 2004).

## 7.3 Crosstalk Between AMF–*Trichoderma* and Plants

Plants react to external abiotic and biotic alterations with a similar response to that of a plant pathogen reaction, which triggers a wide range of defence mechanisms in plants that confer protection against pathogenic invasion. Fungal interactions with phytohormonal signalling and the induction of resistance against pathogens have also been reported as growth-promoting mechanisms (Vassilev et al. 2006). A possible role of indole acetic acid (IAA) in the growth stimulation of tomato plants produced by inoculation with *Trichoderma aureoviride* was proposed by Gravel et al. (2007). Vinale et al. (2008a) observed that some secondary metabolites that are involved directly in the *Trichoderma*–plant interaction may act as auxin inducers. Phytohormone signalling pathways are currently emerging as important regulators of AM establishment (Gutjahr 2014). Mycorrhizal fungi have important physiological implications for the host plants that are mediated through a hormone response that involves an altering in the levels of cytokinins, gibberellins, ethylene, abscisic acid (ABA), jasmonic acid (JA) and auxin (Hause et al. 2002; Ludwig-Müller et al. 2002; Vierheilig 2004; Foo et al. 2013; Fernández et al. 2014). This alteration has been reported not only at a root level but also for the transportation of these hormones to the aerial hosted plant shoots (Toussaint 2007). It has been suggested that phytohormones, such as IAA, cytokinins and ethylene, may alter shoot or root growth (Tsavkelova et al. 2005). In addition, plant hormones, such as jasmonic acid (JA), salicylic acid (SA) and ethylene, may act as signalling molecules during the activation of systemic resistance in plant shoots (Abeles et al. 1992; Sticher et al. 1997; Pieterse et al. 2003). Recent studies have shown a systemic effect in the shoots of mycorrhizal roots (Liu et al. 2007; Fiorilli et al. 2009). The response of the plant differs, depending on which AMF is inoculated. *G. intraradices*-inoculated plants have shown increased shoot IAA and ACC, the ethylene precursor, while a decrease has been observed using *G. mosseae*. These findings reinforce the idea that the effects of inoculation with different fungi on the hormonal balance of plants may vary to a great extent. The analysis of the defence-related hormones, SA, ABA and JA, and the expression of marker genes of the pathways in which they are involved have recently revealed significant changes in the roots of mycorrhizal plants, and the changes seem to be depended on both the plant and AMF that are used (Fernandez et al. 2014). An important role in promoting the association with AMF has been shown for strigolactones, which can be exudated by plant roots (Akiyama et al. 2005). Furthermore, there have also been several studies related to the induction of resistance by AMF, focusing especially on the activation of plant defence mechanisms in roots (Pozo and Azcón-Aguilar 2007; Avis et al. 2008; Pozo et al. 2009).

Several reports have demonstrated that the interaction between AMFs and *Trichoderma* spp. may be beneficial for both plant growth and plant disease control (Linderman 1992; Barea et al. 1997; Saldajeno et al. 2008; Martínez-Medina et al. 2009). A synergistic effect of some saprophytic fungi on AMF spore germination and colonisation has been confirmed (Calvet et al. 1993; McAllister et al. 1996; Fracchia et al. 1998). For example, it has been reported that some *Trichoderma* strains may influence AMF activity (Calvet et al. 1992, 1993; Brimner and Boland 2003; Martinez et al. 2004; Martínez-Medina et al. 2009). Volatile and soluble exudates produced by saprophytic fungi are involved in these effects (McAllister et al. 1994, 1995; Fracchia et al. 1998). Nevertheless, the results of research on the interactions between soil saprophytic and AMF differ widely, even when the same species of saprophytic fungi are involved. For example, *T. harzianum* was found to have antagonistic, neutral and stimulating effects on AMF (Siddiqui and Mohmood 1996; Fracchia et al. 1998; Godeas et al. 1999; Green et al. 1999; Martínez-Medina et al. 2009). The effects of such interactions should be studied because they could either enhance or inhibit the beneficial effects of the individual species (Barea et al. 1997, 2002; Saldajeno et al. 2008).

In general, AMF and *T. harzianum* inoculation has resulted in an improvement in shoot weight and nutritional status (Martinez-Medina et al. 2009). Haggag and Abd-El-Latif (2001) found that the combined inoculation of *G. mosseae* and *T. harzianum* enhanced the growth of geranium plants; the main plausible explanation being that *T. harzianum* by itself solubilises inorganic nutrients and increases plant nutrient uptake (Altomare et al. 1999; De Silva et al. 2000). Altomare et al. (1999) hypothesised that phosphorus could be solubilised and stored in the *Trichoderma* biomass and then released in a readily available form in close proximity to the roots after lyses of the aged mycelium. Inoculation with *T. harzianum* produces a negative effect on nitrogen uptake in low levels of fertilisation doses, because of competition between plant roots and *T. harzianum*, since, in many cases, microorganisms are superior competitors (Kaye and Hart 1997; Hodge et al. 2000). Martinez-Medina et al. (2009) showed a general increase in the plant nitrogen content when co-inoculated with *T. harzianum* and different AMFs. As mentioned before, it was shown that co-inoculation with *T. harzianum* and *G. mosseae* was more effective than any other combination tested with regard to increases in the uptake of nutrients.

A systemic effect of the AMF on the IAA and ACC balance in the host plant has been shown (Meixner et al. 2005). On the other hand, plants co-inoculated with AMF and *T. harzianum* showed a different hormone profile due to mediation of systemic increases in the IAA, ACC and ABA concentrations. Interaction with auxin signalling has been proposed as a plant growth promotion mechanism for *T. virens*, as a dose-dependent effect of IAA on increasing biomass production has been shown by Contreras-Cornejo et al. (2009). It has been proposed that ethylene may affect shoot and root growth through the stimulation of auxin biosynthesis (Ruzicka et al. 2007). It has also been taken into account that *T. harzianum* itself is able to modify the shoot balance of stress- and defence-related hormones to a great extent. Three major signal molecules are known to be involved in the systemic defence responses of plants: SA, commonly shown to be involved in systemic acquired resistance (SAR), and JA and ethylene, which are involved in induced systemic resistance (ISR) (Abeles et al. 1992; Sticher et al.1997; Pieterse et al. 2003). Shoresh et al. (2005) suggested the involvement of JA and ethylene in the protective effect conferred by *T. asperellum* on cucumber, but did not observe any variation in the SA content. Martinez et al. (2001) instead suggested that the SA and ethylene pathways together coordinate the activation of defence mechanisms via an interaction between the two signalling pathways, since SA causes an inhibition of ethylene production. These findings, together with the increased *F. oxysporum* populations observed in the treatment involving inoculation with *T. harzianum*, indicate that the protective effect against *Fusarium* wilt conferred to melon plants by *T. harzianum* is not only due to direct antagonism but also to a plant-mediated phenomenon (Harman et al. 2004; Shoresh et al. 2005).

The accumulation of SA or pathogenesis-related (PR) proteins, or the expression of marker genes associated with systemic acquired resistance, has not been reported in systemic tissues of AM plants (Pozo et al. 2009). Therefore, it has been suggested that priming is the main mechanism that operates in the induction of resistance by AMF (Cordier et al. 1998; Pozo et al. 2002; Pozo and Azcón-Aguilar 2007; Zhu and Yao 2004; Whipps 2004). It is plausible that AMFs repress SA- as well as JA-dependent systemic pathways stimulated by *T. harzianum* in the host, in order to achieve a compatible interaction. This supports the hypothesis of Pozo and Azcón-Aguilar (2007) that a functional mycorrhiza implies partial suppression of SA-dependent responses in the plant, which is compensated by an enhancement of JA-regulated responses in the roots. Hence, in plants co-inoculated with *G. intraradices* and *T. harzianum*, the additive effect on the suppression of disease caused by *F. oxysporum* seems to be due to a positive interaction between different plant protection mechanisms that are mediated by both agents, which may include a priming response mediated by the AMF as well as a systemic SA-dependent response mediated by *T. harzianum*.

This multiple-player crosstalk normally occurs in nature among plants and rhizosphere microorganisms (Barea et al. 2002; Kohler et al. 2007; Saldajeno et al. 2008; Martinez-Medina et al. 2009). These interactions may result in a different plant status from that associated with the single interaction between the plant and each microorganism (Yedidia et al. 2003; Garmendia et al. 2005; Marra et al. 2006). In nature, in diseased plants, the SA, JA, ACC and ABA concentrations increase in the shoot; *T. harzianum*-inoculated plants show a similar hormonal profile, thus indicating the involvement of similar pathways in both interactions (*T. harzianum* and pathogen plant reaction). There is extensive evidence that *Trichoderma* spp. are able to elicit the defence systems of a number of plant species (Hanson and Howell 2004; Shoresh et al. 2005; Vinale et al. 2008b). ACC and ABA accumulation in plant shoots has been shown to be lower when *T. harzianum* or *G. intraradices* and the pathogen interacted with the plant, compared to the pathogen alone, thus confirming that the presence of antagonists modifies the effect of a pathogen on the hormonal profile of a plant. Moreover, when *T. harzianum* interacts with the pathogen, a decrease in SA is observed in diseased plants. Marra et al. (2006) have shown that, in bean plants, *Rhizoctonia solani* or *Botrytis cinerea* pathogens and *Trichoderma* lead to a change in proteins involved in the response to pathogen attack, when the beneficial fungus is present. These authors observed that when both *Trichoderma* and the pathogen interacted with the plant, several PR proteins were upregulated less than by the pathogen alone. *T. harzianum* plants may have two effects: (a) related to basal resistance, as increased SA and JA concentrations as observed in healthy *T. harzianum*-inoculated plants, and (b) hormone modulating through ABA-, ethylene- and SA-dependent responses to pathogen attack, which resulted in the activation of a more appropriate response. The resistance against *F. oxysporum* mediated by *G. intraradices* instead seems to include mechanisms that are independent of SA and JA signalling, although *G. intraradices* is also able to attenuate the plant defence response to pathogen attack, which results in attenuated ACC and ABA accumulation in the shoot.

Therefore, the synergistic effect observed in plants co-inoculated with *G. intraradices* and *T. harzianum* seems to be due to a positive interaction between different plant protection mechanisms mediated by the two agents, which may include an improvement in plant nutrition, damage compensation, competition for colonisation sites or photosynthates, changes in the root system, changes in rhizosphere microbial populations and activation of plant defence mechanisms (Pozo and Azcon-Aguilar 2007). All the gathered information points out that the mechanisms involved in the control of *Fusarium* wilt by *T. harzianum* in melon plants seem to be related not only to an induction of plant resistance but also to its capacity to attenuate the hormonal disruption caused by the disease (Martínez-Medina et al. 2010, 2011, 2014). However, the resistance mediated by *G. intraradices* against *F. oxysporum* seems to include mechanisms that are independent of SA and JA signalling. A synergistic effect on disease control has been found for the *T. harzianum–G. intraradices–F. oxysporum* tripartite interaction. However, no synergistic effect has been observed on either the attenuation of the hormone disruption induced by the pathogen or on the SA , JA or ACC mediated plant response. Therefore, the synergistic effect pertaining to biocontrol capacity seems to be the result of numerous modes of action exhibited by each microorganism, which are different from the plant-mediated mechanism.

It has often been reported that the plant response to mycorrhizal fungi can vary according to the AMF species (Martinez-Medina et al. 2009, 2013). Moreover, the accumulation pattern of the transcript of defence-related genes varies among the different plant–AMF symbioses (Pozo et al. 2002; Gao et al. 2004; Grunwald et al. 2009). Campos-Soriano et al. (2012) have shown that rice root colonisation by *G. intraradices* promotes the systemic induction of genes that play a regulatory role in the host defence response. In *Medicago* species, transient increases in different flavonoid/isoflavonoid compounds were found to depend on the plant and fungal genotypes (Harrison and Dixon 1993, 1994; Volpin et al. 1994, 1995). These findings reinforce the idea that differences in the colonisation dynamics of AMF species could explain their dissimilar behaviour as plant bioprotectors (Martínez-Medina et al. 2009).

Arbuscular mycorrhizal (AM) fungi as well as fungal biological control agent such as *Trichoderma* spp. have shown beneficial effects on plant growth and health. The study of the interactions between AM fungi, fungal antagonists and plants is of great interest. Before using them it is in fact important to understand whether and how the two different beneficial fungal components can influence each other and/or the host plant. In this chapter, the attention is focused on the plant responses to the inoculation with AM fungi and *Trichoderma* spp., alone or in combination, with particular attention to the impact on the plant hormonal balance.

## 7.4 Conclusion

Specific arbuscular mycorrhizal and *Trichoderma* species can be used for sustainable agriculture by utilising their beneficial effects on plant growth and health; but before using them, it is in fact important to understand whether and how the two different beneficial fungal components can influence each other and/or the host plant. It is demonstrated that the interactions between them are direct, and also it is remarkable that the crosstalk between AMF–*Trichoderma* is also through the plant by phytohormonal signalling, such as by salicylic and jasmonic acid, ethylene, ACC, IAA or ABA.

## References

Abeles FB, Morgan PW, Saltveit ME Jr (1992) Ethylene in plant biology. Academic, San Diego

Akiyama K, Matsuzaki K, Hayashi H (2005) Plant sesquiterpenes induce hyphal branching in arbuscular mycorrhizal fungi. Nature 435:824–827

Altomare C, Norvell WA, Björkman T, Harman GE (1999) Solubilization of phosphates and micronutrients by the plant–growth promoting and biocontrol fungus *Trichoderma harzianum* Rifai 1295–22. Appl Environ Microbiol 65:2926–2933

Avis TJ, Gravel V, Antoun H, Tweddell RJ (2008) Multifaceted beneficial effects of rhizosphere microorganisms on plant health and productivity. Soil Biol Biochem 40:1733–1740

Azcón–Aguilar C, Bago B (1994) Physiological characteristics of the host plant promoting and undisturbed functioning of the mycorrhizal symbiosis. In: Gianinazzi S, Schepp H (eds) Impact of arbuscular mycorrhizas on sustainable agriculture and natural ecosystems. Birkhauser, Basel

Azcón–Aguilar C, Barea JM (1996) Arbuscular mycorrhizas and biological control of soil–borne plant pathogens – an overview of the mechanisms involved. Mycorrhiza 6:457–464

Barea JM, Azcón–Aguilar C, Azcón R (1997) Interactions between mycorrhizal fungi and rhizosphere microorganisms within the context of sustainable soil–plant systems. In: Gange AC, Brown VK (eds) Multitrophic interactions in terrestrial systems. Blackwell Science, Cambridge

Barea JM, Azcón R, Azcón–Aguilar C (2002) Mycorrhizosphere interactions to improve plant fitness and soil quality. Anton Leeuw 81:343–351

Benítez T, Rincón AM, Limón Antonio MC, Codón C (2004) Biocontrol mechanisms of *Trichoderma* strains. Int Microbiol 7:249–260

Bi HH, Song YY, Zeng RS (2007) Biochemical and molecular responses of host plants to mycorrhizal infection and their roles in plant defense. Allelopathy J 29:15–27

Brimner T, Boland GJ (2003) A review of the non–target effects of fungi used to biologically control plant diseases. Agric Ecosyst Environ 100:3–16

Calvet C, Barea JM, Pera J (1992) *In Vitro* interactions between the vesicular–arbuscular mycorrhizal fungus *Glomus mosseae* and some saprophytic fungi isolated from organic substrates. Soil Biol Biochem 8:775–780

Calvet C, Pera J, Barea JM (1993) Growth response of marigold (*Tagetes erecta* L.) to inoculation with *Glomus mosseae*, *Trichoderma aureoviride* and *Pythium ultimum* in a peat–perlite mixture. Plant Soil 148:1–6

Campos–Soriano L, García–Martínez J, San Segundo B (2012) The arbuscular mycorrhizal symbiosis promotes the systemic induction of regulatory defence–related genes in rice leaves and confers resistance to pathogen infection. Mol Plant Pathol 13:579–592

Camprubí A, Calvet C, Estaún V (1995) Growth enhancement of *Citrus reshni* after inoculation with *Glomus intraradices* and *Trichoderma aureoviride* and associated effects on microbial population and enzyme activity in potting mixes. Plant Soil 173:233–238

Caron M (1989) Potential use of mycorrhizae in control of plant borne diseases. Can J Plant Pathol 11:177–179

Chet I (1987) *Trichoderma* – application, mode of action, and potential as a biocontrol agent of soilborne plant pathogenic fungi. In: Chet I (ed) Innovative approaches to plant disease control. Wiley, New York

Chet I, Inbar J, Hadar Y (1997) Fungal antagonists and mycoparasites. In: Wicklow S (ed) Mycota environmental and microbial relationships, vol IV. Springer–Verlag, Berlin, pp 165–184

Contreras–Cornejo HA, Macías–Rodríguez L, CortésPenagos C, López–Bucio J (2009) *Trichoderma virens*, a plant beneficial fungus, enhances biomass production and promotes lateral root growth through an auxin–dependent mechanism in *Arabidopsis*. Plant Physiol 149:1579–1592

Cordier C, Pozo MJ, Barea JM, Gianinazzi S, Gianinazzi–Pearson V (1998) Cell defense responses associated with localized and systemic resistance to *Phytophthora* induced in tomato by an arbuscular mycorrhizal fungus. Mol Plant Microbe Interact 11: 1017–1028

De Meyer G, Bigirimana J, Elad Y, Höfte M (1998) Induced systemic resistance in *Trichoderma harzianum* T39 biocontrol of *Botrytis cinerea*. Eur J Plant Pathol 104:279286

De Silva A, Patterson K, Rothrock C, Moore J (2000) Growth promotion of highbush blueberry by fungal and bacterial inoculants. Hortscience 35:1228–1230

Fernández I, Merlos M, López–Ráez JA, Martínez–Medina A, Ferrol N, Azcón C, Bonfante P, Flors V, Pozo MJ (2014) Defense related phytohormones regulation in arbuscular mycorrhizal symbioses depends on the partner genotypes. J Chem Ecol 40:791–803

Filion M, St–Arnaud M, Fortin JA (1999) Direct interaction between the arbuscular mycorrhiza fungus *Glomus intraradices* and different rhizosphere microorganisms. New Phytol 141:525533

Fiorilli V, Catoni M, Miozzi L, Novero M, Accotto GP, Lanfranco L (2009) Global and cell–type gene expression profiles in tomato plants colonized by an arbuscular mycorrhizal fungus. New Phytol 184: 975–987

Foo E, Ross JJ, Jones WT, Reid JB (2013) Plant hormones in arbuscular mycorrhizal symbioses: an emerging role for gibberellins. Ann Bot 111:769–779

Fracchia S, Mujica MT, García–Romera I, GarcíaGarrido JM, Martin J, Ocampo JA, Godeas A (1998) Interactions between *Glomus mosseae* and arbuscular mycorrhizal sporocarp–associated saprophytic fungi. Plant Soil 200:131–137

Gao LL, Knogge W, Delp G, Smith FA, Smith FS (2004) Expression patterns of defense–related genes in different types of arbuscular mycorrhizal development in wild–type and mycorrhiza–defective mutant tomato. Mol Plant-Microbe Interact 17:1103–1113

Garmendia I, Aguirreolea J, Goicoechea N (2005) Defence–related enzymes in pepper roots during interactions with arbuscular mycorrhizal fungi and or *Verticillium dahliae*. Biocontrol 51:293–310

Godeas A, Fracchia S, Mujica MT, Ocampo JA (1999) Influence of soil impoverishment on the interaction between *Glomus mosseae* and saprobe fungi. Mycorrhiza 9:185–189

Gravel V, Antoun H, Tweddell RJ (2007) Growth stimulation and fruit yield improvement of greenhouse tomato plants by inoculation with *Pseudomonas putida* or *Trichoderma atroviride*: possible role of indole acetic acid (IAA). Soil Biol Biochem 39:1968–1977

Green H, Larsen J, Olsson PA, Jensen DF, Jakobsen I (1999) Suppression of the biocontrol agent *Trichoderma harzianum* by mycelium of the arbuscular mycorrhizal fungus *Glomus intraradices* in root–free soil. Appl Environ Microbiol 65:1428–1434

Grunwald U, Guo W, Fischer K, Isayenkov S, Ludwig–Mueller J, Hause B, Yan X, Küster H, Franken P (2009) Overlapping expression patterns and differential transcript levels of phosphate transporter genes in arbuscular mycorrhizal, Pi–fertilised and phytohormone–treated *Medicago truncatula* roots. Planta 229:1023–1034

Gutjahr C (2014) Phytohormone signaling in arbuscular mycorrhiza development. Curr Opin Plant Biol 20:26–34

Haas JH, Krikun J (1985) Efficacy of endomycorrhizal–fungus isolates and inoculum quantities required for growth response. New Phytol 100:613–621

Haggag WM, Abd–El–Latif FM (2001) Interaction between vesicular arbuscular mycorrhizae and antagonistic biocontrol microorganisms on controlling root rot disease incidence of geranium plants. On Line J Biol Sci 1:1147–1153

Hanson LE, Howell CR (2004) Elicitors of plant defense responses from biological control strains of *Trichoderma virens*. Phytopathology 94:171–176

Harman GE (2000) Myth and dogmas of biocontrol changes in perceptions derived from research on *Trichoderma harzianum* T–22. Plant Dis 84:377–393

Harman GE, Howell CR, Viterbo A, Chet I, Lorito M (2004) *Trichoderma* species opportunistic, avirulent plant symbionts. Nat Rev Microbiol 2:3456

Harrison MJ, Dixon RA (1993) Isoflavonoid accumulation and expression of defense gene transcripts during the establishment of vesicular arbuscular mycorrhizal associations in roots of *Medicago truncatula*. Mol Plant-Microbe Interact 6:643–654

Harrison M, Dixon R (1994) Spatial patterns of expression of flavonoid/isoflavonoid pathway genes during interactions between roots of Medicago truncatula and the mycorrhizal fungus Glomus versiforme. Plant J 6:9–20

Hause B, Maier W, Miersch O, Kramel R, Strack D (2002) Induction of jasmonate biosynthesis in arbuscular mycorrhizal barley roots. Plant Physiol 130:1213–1220

Hodge A, Stewart J, Robinson D, Griffiths BS, Fitter AH (2000) Competition between roots and soil micro–organisms for nutrients from nitrogen–rich patches of varying complexity. J Ecol 88:150–164

Howell CR (2003) Mechanisms employed by *Trichoderma* species in the biological control of plant diseases: the history and evolution of current concepts. Plant Dis 87:4–10

Howell CR, Hanson LE, Stipanovic RD, Puckhaber LS (2000) Induction of terpenoid synthesis in cotton roots and control of *Rhizoctonia solani* by seed treatment with *Trichoderma virens*. Phytopathology 90:248–252

Jeffries, Barea JM (2001) Arbuscular mycorrhiza – a key component of sustainable plant–soil ecosystems. In: Hock B (ed) The mycota Vol. IX: fungal associations. Springer–Verlag, Berlin

Kaye JP, Hart SC (1997) Competition for nitrogen between plants and soil microorganisms. Trends Ecol Evol 12:139–143

Kohler J, Caravaca F, Carrasco L, Roldan A (2007) Interactions between a plant–growth–promoting rhizobacterium, an AM fungus and a phosphate–solubilising fungus in the rhizosphere of Lactuca sativa. Appl Soil Ecol 35:480–487

Korolev N, David DR, Elad Y (2008) The role of phytohormones in basal resistance and *Trichoderma*–induced systemic resistance to *Botrytis cinerea* in *Arabidopsis thaliana*. Biocontrol 53:667–683

Lace B, Genre A, Woo S, Faccio A, Lorito M, Bonfante P (2014) Gate crashing arbuscular mycorrhizas: in viv-imaging shows the extensive colonization of both symbionts by *Trichoderma atroviride*. Environ Microbiol Rep 7:64–67

Linderman RG (1992) Vesicular–arbuscular mycorrhizae and soil microbial interactions. In: Bethlenfalvay GJ, Linderman RG (eds) Mycorrhizae in sustainable agriculture, Special publication no. 54. Agronomy Society of America, Madison

Liu J, Maldonado–Mendoza I, Lopez–Meyer M, Cheung F, Town CD, Harrison MJ (2007) Arbuscular mycorrhizal symbiosis is accompanied by local and systemic alterations in gene expression and an increase in disease resistance in the shoots. Plant J 50:529–544

Ludwig–Müller J, Bennett R, García–Garrido J, Piché Y, Vierheilig H (2002) Reduced arbuscular mycorrhizal root colonization in *Tropaeolum majus* and *Carica papaya* after jasmonic acid application cannot be attributed to increased glucosinolate. J Plant Physiol 159:517–523

Marra R, Ambrosino P, Carbone V (2006) Study of the three–way interaction between *Trichoderma atroviride*, plant and fungal pathogens by using a proteomic approach. Curr Genet 50:307–321

Martinez C, Blanc F, Le Claire E, Besnard O, Nicole M, Baccou JC (2001) Salicylic acid and ethylene pathways are differentially activated in melon cotyledons by active or heat–denatured cellulase from *Trichoderma longibrachiatum*. Plant Physiol 127: 334–344

Martinez A, Obertello M, Pardo A, Ocampo JA, Godeas A (2004) Interactions between *Trichoderma pseudokoningii* strains and the arbuscular mycorrhizal fungi *Glomus mosseae* and *Gigaspora rosea*. Mycorrhiza 14:79–84

Martínez–Medina A, Pascual JA, Lloret E, Roldán A (2009) Interactions between arbuscular mycorrhizal fungi and *Trichoderma harzianum* and their effects on *Fusarium* wilt in melon plants grown in seedlings nurseries. J Sci Food Agric 89:1843–1850

Martínez–Medina, Pascual JA, Pérez–Alfocea F, Albacete A, Roldan A (2010) *Trichoderma harzianum* and *Glomus intraradices* modify the hormone disruption induced by *Fusarium oxysporum* infection in melon plants. Phythopathology 100:682–688

Martínez–Medina A, Roldán A, Pascual JA (2011) Interaction between arbuscular mycorrhizal fungi and *Trichoderma harzianum* under conventional and low input fertilization field condition in melon crops: growth response and *Fusarium* wilt biocontrol. Appl Soil Ecol 47:97–105

Martínez–Medina, Fernández I, Sánchez–Guzmán MJ, Jung SC, Pascual JA, Pozo MJ (2013) Deciphering the hormonal signalling network behind the systemic resistance induced by *Trichoderma harzianum* in tomato. Front Plant Sci 4:206

Martínez–Medina A, Alguacil MDM, Pascual JA, Van Wees SC (2014) Phytohormone profiles induced by *Trichoderma* isolates correspond with their biocontrol and plant growth–promoting activity on melon plants. J Chem Ecol 40:804–815

McAllister CB, García–Romera I, Godeas A, Ocampo JA (1994) Interactions between *Trichoderma koningii*, *Fusarium solani* and *Glomus mosseae*: effects on plant growth, arbuscular mycorrhizas and the saprophyte inoculants. Soil Biol Biochem 26:1363–1367

McAllister CB, Garcia–Romera I, Martín J, Godeas A, Ocampo JA (1995) Interaction between *Aspergillus niger* van Tiegh and *Glomus mosseae* (Nicol and Gerd) Gerd & Trappe. New Phytol 129:309–316

McAllister CB, Garcia–Garrido JM, Garcia–Romera I, Godeas A, Ocampo JA (1996) *In vitro* interactions between *Alternaria alternata*, *Fusarium equiseti* and *Glomus mosseae*. Symbiosis 20:163–174

McAllister CB, Garcia–Garrido JM, Garcia–Romera I, Godeas A, Ocampo JA (1997) Interaction between *Alternaria alternata* or *Fusarium equiseti* and *Glomus mosseae* and its effects on plant growth. Biol Fertil Soils 24:301–305

Meixner C, Ludwig–Müller J, Miersch O, Gresshoff P, Staehelin C, Vierheilig H (2005) Lack of mycorrhizal autoregulation and phytohormonal changes in the supernodulating soybean mutant nts1007. Planta 222: 709–715

Pieterse CMJ, Van Pelt JA, Verhagen BWM, Ton J, van Wees SCM, Léon–Kloosterziel KM, van Loon LC (2003) Induced systemic resistance by plant growth promoting rhizobacteria. Symbiosis 35:3954

Pozo MJ, Azcón–Aguilar C (2007) Unraveling mycorrhiza–induced resistance. Curr Opin Plant Biol 10:393–398

Pozo MJ, Cordier C, Dumas–Gaudot E, Gianinazzi S, Barea JM, Azcón–Aguilar C (2002) Localized versus systemic effect of arbuscular mycorrhizal fungi on defence responses to *Phytophthora* infection in tomato plants. J Exp Bot 53:525–534

Pozo MJ, Verhage A, García–Andrade J, García JM, Azcón–Aguilar C (2009) Priming plant defence against pathogens by arbuscular mycorrhizal fungi. In: Azcón–Aguilar C, Barea JM, Gianinazzi S, Gianinazzi–Pearson V (eds) Mycorrhizas – functional processes and ecological impact. Springer, Berlin

Ruzicka K, Ljung K, Vanneste S, Podhorsk'a R, Beeckman T, Friml J, Benková E (2007) Ethylene regulates root growth through effects on auxin biosynthesis and transport dependent auxin distribution. Plant Cell 19:2197–2212

Saldajeno MGB, Chandanie WA, Kubota M, Hyakumachi M (2008) Effects of interactions of arbuscular mycorrhizal fungi and beneficial saprophytic mycoflora on plant growth and disease protection. In: Siddiqui ZA, Akhtar MS, Futai K (eds) Mycorrhizae: sustainable agriculture and forestry. Springer Sciences, The Netherlands

Shoresh M, Yedidia I, Chet I (2005) Involvement of jasmonic acid/ethylene signaling pathway in the systemic resistance induced in cucumber by *Trichoderma asperellum* T203. Phytopathology 95:76–84

Siddiqui ZA, Mahmood I (1996) Biological control of *Heterodera cajani* and *Fusarium udum* on pigeonpea by *Glomus mosseae*, *Trichoderma harzianum* and *Verticillium chlamydosporium*. Isr J Plant Sci 44: 49–56

Sticher L, Mauch–Mani B, Métraux JP (1997) Systemic acquired resistance. Anni Rev Phytopathol 35:235–270

Tarbell TJ, Koske RE (2007) Evaluation of commercial arbuscular mycorrhizal inocula in a sand/peat medium. Mycorrhiza 18:51–56

Toussaint JP (2007) Investigating physiological changes in the aerial parts of AM plants: what do we know and where should we be heading? Mycorrhiza 17:349–353

Tsavkelova EA, Cherdyntseva TA, Netrusov AI (2005) Auxin production by bacteria associated with orchid roots. Microbiology 74:46–53

Vassilev N, Vassileva M, Nikolaeva I (2006) Simultaneous P–solubilizing and biocontrol activity of microorganisms: potentials and future trends. Appl Microbiol Biotechnol 71:137–144

Vázquez MM, César S, Azcón R, Barea JM (2000) Interactions between arbuscular mycorrhizal fungi and other microbial inoculants (*Azospirillum*, *Pseudomonas*, *Trichoderma*) and their effects on microbial population and enzyme activities in the rhizosphere of maize plants. Appl Soil Ecol 15:261–272

Vierheilig H (2004) Regulatory mechanisms during the plant–arbuscular mycorrhizal fungus interaction. Can J Bot 82:1166–1176

Vinale F, Sivasithamparam K, Ghisalberti EL, Marra R, Barbetti MJ, Li H, Woo SL, Lorito M (2008a) A novel role for *Trichoderma* secondary metabolites in the interactions with plants. Physiol Mol Plant Pathol 72: 80–86

Vinale F, Sivasithamparam K, Ghisalberti EL, Marra R, Woo SL, Lorito M (2008b) *Trichoderma*–plant– pathogen interactions. Soil Biol Biochem 40:1–1

Volpin H, Elkind Y, Okon Y, Kapulnik Y (1994) A vesicular arbuscular mycorrhizal fungus *Glomus intraradices* induces a defence response in alfalfa roots. Plant Physiol 104:683–689

Volpin H, Phillips DA, Oken Y, Kapulnik Y (1995) Suppression of an isoflavonoid phytoalexin defense response in mycorrhizal alfalfa roots. Plant Physiol 108:1449–1454

Whipps JM (2004) Prospects and limitations for mycorrhizas in biocontrol of root pathogens. Can J Bot 82: 1198–1227

Yedidia I, Shoresh M, Kerem Z, Benhamou N, Kapulnik Y, Chet I (2003) Concomitant induction of systemic resistance to *Pseudomonas syringae* pv. *lachrymans* in *Cucumber* by *Trichoderma asperellum* (T–203) and accumulation of phytoalexins. Appl Environ Microbiol 69:7343–7353

Zhu HH, Yao Q (2004) Localized and systemic increase of phenols in tomato roots induced by *Glomus versiforme* inhibits *Ralstonia solanacearum*. J Phytopathol 152:537–542

# Bioformulations of Novel Indigenous Rhizobacterial Strains for Managing Soilborne Pathogens

# 8

Pervaiz A. Abbasi, Salah Eddin Khabbaz, and Liang Zhang

## Contents




**Abstract**

The antagonistic potential of indigenous strains of *Bacillus*, *Paenibacillus*, and *Pseudomonas* isolated from potato rhizosphere or soil and soybean plants to manage soil and root diseases of vegetable crops is presented here as a case study. These bacterial strains were also characterised for production or activities of antibiotics, metabolites, volatiles, phytohormones, and lytic enzymes. In agar plate assays, most of the strains showed broad-spectrum antagonistic activity against various selected fungal and oomycete pathogens. Irradiated peat was found to be a suitable carrier material for preparing formulations of these bacteria and delivering them on seed, root, or substrate. In pot experiments, irradiated peat formulations of these bacteria provided control of *Pythium* damping-off and root rot and *Phytophthora* blight and root rot of cucumber, *Rhizoctonia* damping-off of radish, and *Fusarium* crown and root rot and *Fusarium* wilt of tomato. Bacterial treatments also resulted in higher fresh weights of plants produced in pathogen-infested substrate. In micro-plot



P.A. Abbasi (✉)
Atlantic Food and Horticulture Research Centre, Agriculture and Agri-Food Canada, 32 Main Street, Kentville, Nova Scotia B4N 1J5, Canada
e-mail: Pervaiz.Abbasi@agr.gc.ca

S.E. Khabbaz • L. Zhang
Southern Crop Protection and Food Research Centre, Agriculture and Agri-Food Canada, 1391 Sandford St., London, Ontario N5V 4T3, Canada




N.K. Arora et al. (eds.), *Bioformulations: for Sustainable Agriculture*,
DOI 10.1007/978-81-322-2779-3_8

trials, coating of seed potato tubers with irradiated peat formulation of some of these antagonistic bacteria reduced scab severity on daughter tubers and increased tuber weights when treated tubers were planted in a potato soil with a history of common scab disease. These disease suppression and plant-growth promotion activities of various bacterial strains might be related to the production of different antibiotics, secondary metabolites, lytic enzymes, phytohormones, siderophores, and volatiles. The results of these studies indicate that several of the bacterial strains isolated from local sources showed their potential use as biofungicides for protection of seedling damping-off and root rot and tuber diseases caused by various soilborne pathogens.

## 8.1 Background

The traditional disease control methods based on chemical fungicides have been very effective in managing soilborne plant pathogens, but they also have ecological and environmental concerns, such as contamination of soil, water, and food material, resistance development in pathogens, and nontarget effects on beneficial soil microbes. During the past 30 years, significant research efforts have been dedicated toward the development of biological-based disease management strategies as alternative to chemical pesticides (Gerhardson 2002; Weller et al. 2002; Fravel 2005; Haas and Défago 2005; Lugtenberg and Kamilova 2009; Olubukola 2010). As a result, biological control has become an ecologically sound and environmentally friendly method for managing diseases caused by soilborne plant pathogens. Plant-growth promoting rhizobacteria (PGPR) have been successfully used as potential biocontrol agents of plant pathogens (Weller 1988; Raaijmakers et al. 2009). These beneficial bacteria can colonise plants and inhibit pathogenic microorganisms by producing an array of antibiotics, metabolites, lytic enzymes, and phytohormones and by fixing nitrogen and solubilising phosphate (Knowles 1976; Neilands 1981; Nautiyal et al. 2002; Kim et al. 2008; Athukorala et al. 2009; Santoyo et al. 2012). Beneficial bacteria can also provide suppression of plant diseases by inducing systemic resistance in host plants (Ramamoorthy et al. 2001; Bakker et al. 2003).

Disease in nature is the exception and not the rule, and natural biological control may be playing a part in that exception. This implies that sources of biofungicides are widespread and their population can either be exploited and enriched or potential candidates can be isolated, tested in vivo and in vitro for broad-spectrum anti-pathogen activities, formulated in some suitable carrier material and evaluated again for disease suppression, and finally redeployed. So the soils that are exposed to pathogens and soils that are not exposed to pathogens can be good source of biofungicides. In soils that are exposed to pathogens, biological control may be occurring naturally leading to low disease occurrence on host plants even though the pathogen inoculum in the soil is high. These are also known as suppressive soils (Cook and Baker 1983), such natural soils suppressive to soilborne pathogens have been discovered worldwide (Cook and Baker 1983; Alabouvette 1999; Weller et al. 2002), and indigenous soil microbial communities generally contribute to disease suppression by reducing and competing with pathogen populations (Weller et al. 2002; Bonanomi et al. 2010). It is also possible to exploit natural disease-suppressive conditions by soil incorporation of organic materials (Mathre et al. 1999; Abbasi et al. 2008; Abbasi 2011; Bernard et al. 2012) and other agronomic practices. Organic soil amendments which act primarily as nutrient source for crop production can also act as disease management tools. Disease management by organic amendments may occur by increasing the populations and activities of biocontrol agents and by enhancing their disease control effect (Mathre et al. 1999).

Soilborne plant diseases such as wilts, scab, damping-off, and root rots are major yield-limiting factors in agricultural, ornamental, and horticultural crops. These diseases can cause substantial economic losses to growers every year. Chemical fumigants may provide control of some of these diseases but they can be expensive, undesirable, and not always available or effective. Beneficial microbial communities from rhizosphere and endophytes play a central role in improving plant growth and health. In this chapter, we present recent research findings from our laboratory on the isolation and characterisation of beneficial rhizobacteria from indigenous resources as potential biofungicides for managing soil and root diseases of vegetable crops.

## 8.2 Isolation and Selection of Antagonists

Increasing demand for biological-based disease control strategies has prompted searches for novel strains of biocontrol agents from native resources and to develop them as biofungicides. Biological control is also a key component of organic amendment-mediated disease management, particularly those diseases caused by soil-borne pathogens. Incorporation of organic materials to soils prior to planting provides suppression of plant diseases and in most cases disease suppression occurs through enhanced microbial activity in the amended soil (Abbasi et al. 2008; Abbasi 2011). We found that such amended soils are a good source for isolating biocontrol strains (Abbasi 2013).

Thirteen bacterial strains were chosen after preliminary agar plate screening for further characterisation of antagonistic and biocontrol potential (Table 8.1 and Fig. 8.1). Eight of these bacterial strains (Khabbaz and Abbasi 2014; Zhang et al. 2015) were isolated from soil samples collected from potato field plots treated with an organic soil amendment for establishing disease-suppressive conditions against soilborne potato diseases (Abbasi 2013). In addition, two of the bacterial strains were isolated from leaves of soybean plants grown in pots containing soil from a field in a greenhouse, and three previously uncharacterised strains selected were from culture collection (Zhang et al. 2015). The antagonistic activity of these bacterial strains or their volatiles was determined on agar plates in co- or dual-culture or sealed double plate assays against various fungal and oomycete pathogens. The cell-free culture filtrates (CF) of five selected strains were also tested against these plant pathogens on PDA plates. All 13 bacterial stains displaying broad-spectrum antagonistic activities against the selected pathogens in plate assays were identified based on their sequences of the 16S rRNA gene after polymerase chain reaction

**Table 8.1** Bacterial strains assessed for antagonistic and biocontrol potential

| Biocontrol strain | Reference |
|---|---|
| *Bacillus amyloliquefaciens* 1A-48 | Zhang et al. (2015) |
| *Bacillus amyloliquefaciens* 1B-14 | Zhang et al. (2015) |
| *Bacillus amyloliquefaciens* 1B-23 | Zhang et al. (2015) |
| *Pseudomonas chlororaphis* 1B-26 | Zhang et al. (2015) |
| *Bacillus subtilis* 8B-1 | Khabbaz and Abbasi (2014) |
| *Pseudomonas* sp. 8D-45 | Khabbaz and Abbasi (2014) |
| *Pseudomonas fluorescens* 9A-14 | Khabbaz and Abbasi (2014) |
| *Bacillus amyloliquefaciens* 9A-31 | Zhang et al. (2015) |
| *Pseudomonas chlororaphis* SL5 | Zhang et al. (2015) |
| *Paenibacillus polymyxa* SL23 | Zhang et al. (2015) |
| *Paenibacillus polymyxa* #50 | Zhang et al. (2015) |
| *Paenibacillus polymyxa* #53 | Zhang et al. (2015) |
| *Pseudomonas fluorescens* PEF-5 #18 | Zhang et al. (2015) |

**Fig. 8.1** Graphical description for selection and screening of bacterial antagonists as potential biofungicides

(PCR) amplification (Khabbaz and Abbasi 2014; Zhang et al. 2015).

## 8.3 Characterisation of Bacterial Strains for Antagonistic Activities

An understanding of the mechanisms of pathogen inhibition, disease suppression, and plant-growth promotion by biocontrol strains is necessary for their development as biofungicides. Therefore, all 13 bacterial strains were also characterised for their ability to produce different types of antibiotics, lytic enzymes, secondary metabolites, volatiles, or phytohormones. PCR analysis showed the presence of the genes involved in the biosynthesis of antibiotics such as bacillomycin, bacilysin, iturin A, surfactin, and fengycin in most of the *Bacillus* strains and the genes for biosynthesis of pyoluteorin, pyrrolnitrin, and 2,4-diacetylphloroglucinol (DAPG) in most of the *Pseudomonas* strains. However, the actual production of these antibiotics by bacterial strains was not assayed in this study. Using various agar plate assays, these bacterial strains were also assayed for their ability to produce hydrogen cyanide (HCN), siderophore, β-1,3-glucanase, chitinase, protease, indole acetic acid (IAA), and salicylic acid (SA) and to fix nitrogen and solubilise phosphate (Khabbaz et al. 2015; Zhang et al. 2015).

The results of the agar plate assays indicated that nine of the antagonistic bacterial strains produced protease, eight strains produced HCN, seven strains showed β-1,3-glucanase activity, and four strains showed chitinase activity and produced siderophore (Khabbaz et al. 2015; Zhang et al. 2015). Ten bacterial strains were able to produce phytohormones such as IAA and SA, eight strains showed the ability to fix nitrogen, and three strains showed phosphate solubilisation activity (Khabbaz et al. 2015; Zhang et al. 2015). However, any specific role of these various activities produced by the antagonistic bacterial strains needs to be established for pathogen inhibition, disease suppression, and plant-growth promotion.

## 8.4 Preparation of Bioformulations of Antagonistic Bacteria

Formulation of antagonistic bacteria in a suitable carrier material is necessary for their effective delivery as well as for improving their shelf life (Vidhyasekaran et al. 1997; Bashan et al. 2014). A suitable carrier material is one that is easily available, economical, and easy to use. A carrier material should stabilise the populations of the injected or inoculated bacteria in the formulations during production and storage phases giving them a longer shelf life. Formulations can also protect the inoculated organisms from adverse environmental effects and may enhance their activity and effectiveness for plant-growth promotion and disease suppression (Van Veen et al. 1997; Jones and Burges 1998). As described below, the formulations of antagonistic bacteria were prepared in irradiated peat (in most experiments) or talc powder (in some preliminary studies only). Antagonistic bacteria were applied as preplant or post-plant treatments to the potting mix or as seed or root treatments.

### 8.4.1 Irradiated Peat Formulation

A sterile irradiated peat (Becker Underwood Company, Saskatchewan Canada) was used in most experiments for preparing formulations of antagonistic bacterial strains. Bacterial strains were grown in Luria-Bertani (LB) or nutrient broth-yeast extract (NBY) broth and their cells were harvested by centrifugation and resuspended in sterile distilled water (DW). The concentration of bacteria in the suspension was estimated by spectrophotometer readings and by dilution plating on NBY or LB agar plates. Sterile irradiated peat (120 g) was mixed with autoclaved LB or NBY broth (70 mL) and respective bacterial suspension (5 mL at $10^8$–$10^9$ CFU $mL^{-1}$) under sterile conditions and incubated for 24 h. The formulated irradiated peat was later air-dried in a laminar flow hood to a workable moisture level and kept in polythene bags and used for the treatments immediately or as needed. To prepare formulation containing a mixture of bacterial strains, the antagonistic bacteria were grown separately in LB or NBY broth and equal volume (v/v) of each bacterial suspension was mixed with irradiated peat and broth as described above.

### 8.4.2 Talc Formulation

The formulations of antagonistic bacterial strains were also prepared in talc powder for some initial experiments. The talc formulation was prepared with sterilised talc powder as described previously (Vidhyasekaran and Muthamilan 1995). Talc powder (1 kg) was sterilised by autoclaving and mixed with LB broth (400 mL). Calcium carbonate (15 g) to neutralise the pH and carboxymethyl cellulose (10 g) as adhesive were added to the mixture under sterile conditions. Bacterial strains were grown in flasks containing LB or NBY broth as described above and cells were centrifuged, resuspended in sterile DW, and the concentration estimated as described above. Bacterial suspension ($10^8$–$10^9$ CFU $mL^{-1}$) was added to the talc-broth mixture under sterile conditions and incubated for 24 h. The formulated product was dried in a laminar flow hood as described above. Talc formulations containing a mixture of bacterial strains were also prepared as described above.

### 8.4.3 Shelf Life of Formulated Products

The shelf life of irradiated peat and talc formulations, prepared with three bacterial strains *Pseudomonas fluorescens* 9A-14, *Pseudomonas* sp. 8D-45, and *Bacillus subtilis* 8B-1 and stored at room temperature for 6 months, was determined by following the populations of bacterial strains on NBY agar plates by dilution plating (Khabbaz and Abbasi 2014). One-gram samples were drawn after 10- to 30-day intervals for preparing tenfold dilutions and plating on NBY plates. Colonies on plates were counted after 3 days of incubation and expressed as CFU $g^{-1}$ product.

The results of shelf-life experiments indicated that the initial bacterial populations of the three strains were slightly higher in formulations prepared with irradiated peat (9.89–10.11 log CFU $g^{-1}$) as compared to in the talc powder formulations (9.86–9.98 log CFU $g^{-1}$) although both products were injected with similar bacterial concentrations. The viability of injected bacteria in both irradiated peat and talc formulations gradually declined overtime during the 6-month storage period; however, the population densities of all three bacteria remained above $10^8$ CFU $g^{-1}$ in both products throughout the storage period. The viability of antagonistic bacteria did not significantly change during the first 20 days of storage in irradiated peat formulation and during the first 10 days in talc formulation compared to their respective initial populations at day zero. In irradiated peat formulation, the bacterial populations dropped to 9.0 log CFU $g^{-1}$ after 2 months and to 8.0 log CFU $g^{-1}$ after 3 months of storage at room temperature. In talc formulation, the bacterial populations dropped to 9.0 log CFU $g^{-1}$ after 40–50 days and to 8.0 log CFU $g^{-1}$ after 70 days to 3 months of storage. These results indicated that the formulated products should be applied immediately after preparation (irradiated peat within 20 days and talc within 10 days) for achieving maximum initial population densities of antagonistic bacteria.

## 8.5 Application of Antagonistic Bacteria

The biocontrol potential of antagonistic bacterial strains was evaluated for disease suppression and plant-growth promotion in potting substrates or soil infested with pathogen inoculum in growth-room or greenhouse pot assays (Abbasi et al. 2004), as described below. To achieve this, bacterial strains were grown in NBY broth, and their cell suspensions or bioformulations prepared in irradiated peat and talc were either applied directly to potting substrates as preplant amendment or post-plant drench treatments or as seed or root treatments.

### 8.5.1 Substrate Application

The potting mix used in growth-room pot assays was a commercial peat-based mix, Pro-Mix BX containing 75–85 % sphagnum peat moss, horticultural grade perlite and vermiculite, macro- and micronutrients, dolomitic and calcitic limestone, and a wetting agent (Premier Horticulture Inc., Rivière-du-Loup, Quebec, Canada). Controlled slow-release fertiliser Osmocote 14-14-14 (N-P-K) (Scotts-Sierra Horticultural Products Co., Marysville, Ohio) was added to the potting mix or soil at 4 g $kg^{-1}$ mix. Bacterial strains were applied to the pathogen-infested potting mix either as cell suspensions in sterile water (100 mL aliquots at $5 \times 10^8$ CFU $mL^{-1}$) or as formulations in irradiated peat and talc (10 % m/m mix) prior to planting (preplanting amendment). Bacterial cell suspensions were also applied as drench treatments immediately after planting. The preplanting treatments were mixed with the infested potting mix and incubated overnight at 24 °C before planting. The post-planting drench treatments of bacteria were diluted with water to the desired concentration ($5 \times 10^8$ CFU $mL^{-1}$) and applied as 100 mL aliquots immediately after planting or 24 h after pathogen infestation.

### 8.5.2 Seed and Root Application

Cucumber (*Cucumis sativus* L. 'Straight Eight') or radish (*Raphanus sativus* L. 'Early Scarlet Globe') seeds were used in all growth-room experiments. Seeds were soaked in the bacterial suspensions ($5 \times 10^8$ CFU $mL^{-1}$) for 2 h, air-dried in a laminar flow hood, and immediately planted in the pathogen-infested potting mix. Seeds were coated with talc and irradiated peat formulations of antagonistic bacteria by mixing the pre-wetted seeds with the bacterial bioformulations. The treated seeds were immediately sown in the infested mix. Seeds soaked in sterile water were used as control, whereas seeds treated with metalaxyl 48 % (10 mg $L^{-1}$) or benomyl 50 % WP (0.6 g $kg^{-1}$) served as the fungicide control. Roots of tomato (*Solanum lycopericum* L. 'Bonny Best') were treated with irradiated peat formulations of the selected bacterial strains to control *Fusarium* crown and root rot and *Fusarium* wilt. Tomato seedlings were grown in vermiculite plug trays in a growth room for 4 weeks. They were gently removed from the trays and their roots after washing with running water were treated with the control and inoculated irradiated peat by coating. Treated seedlings were immediately planted in pathogen-infested potting mix or soil in pots and placed in a greenhouse.

## 8.6 Evaluation of Antagonistic Bacteria for Disease Suppression

The biocontrol potential of the selected antagonistic bacterial strains was evaluated for suppression of seedling damping-off and root rot of cucumber [*Pythium ultimum* Trow and *Phytophthora capsici* Leonian], seedling damping-off of radish [*Rhizoctonia solani* Kühn], *Fusarium* crown and root rot of tomato [*Fusarium oxysporum* Schlechtend.:Fr. f. sp. *radicis-lycopersici* WR Jarvis and RA Shoemaker], and *Fusarium* wilt of tomato [*Fusarium oxysporum* f. sp. Schlechtend.:Fr. *lycopersici* (Sacc.) WC Snyder and HN Hans] in growth-room or greenhouse pot assays and common scab of potato [*Streptomyces scabies* (Thaxter) Waksman and Henrici, syn: *S. scabiei* (Trüper and dè Clari 1997)] in micro-plots. Bacterial treatments were applied as cell suspension or as talc and irradiated peat formulation either directly to the potting mix or on seeds or roots. Disease conditions were created artificially by inoculating potting mix or soil with pathogen inoculum (1.25–3 % m/m) grown on pearl millet seed (Borrego-Benjumea et al. 2014), except for the potato scab study in micro-plots where a naturally infested soil from a field with a history of potato scab was used. The pearl millet seed inoculum of pathogens was mixed with potting mix or soil and incubated overnight prior to planting.

### 8.6.1 *Pythium* Damping-Off and Root Rot of Cucumber

The biocontrol potential of the three selected antagonistic bacteria, *P. fluorescens* 9A-14, *Pseudomonas* sp. 8D-45, and *B. subtilis* 8B-1, was evaluated for suppression of seedling damping-off and root rot of cucumber as preplant amendment and seed treatments in a *P. ultimum*-infested potting mix under growth-room conditions (Khabbaz and Abbasi 2014). In initial experiments, only cell suspensions of the three bacteria were used either as preplant amendment or post-plant drench treatments. The irradiated peat and talc formulations of the three bacteria were used as preplant amendment treatment, whereas irradiated peat formulation of the three bacteria was used as seed treatment.

#### 8.6.1.1 Preplant Substrate Treatment

Application of cell suspensions of the three antagonistic bacteria to the *P. ultimum*-infested potting mix, either prior to planting as an amendment or after planting as a drench treatment, provided equal control of *Pythium* damping-off and root rot of cucumber seedlings and showed similar increase in plant fresh weights. In comparison to the infested control, the preplanting bacterial treatments showed a 125–290 % increase in healthy cucumber seedlings and a 27–50 % reduction in disease severity, whereas the post-planting bacterial treatments showed a 173–273 % increase in healthy seedlings and a 27–45 % reduction in disease severity. Both preplant and post-plant bacterial treatments to the pathogen-infested mix also significantly increased plant fresh weights compared to the pathogen-infested and noninfested controls.

Bioformulation of the three antagonistic bacteria prepared in irradiated peat or talc was also evaluated as preplant amendment treatments for suppression of *Pythium* damping-off and root rot of cucumber seedlings (Khabbaz and Abbasi 2014). As seen with cell suspensions, treatment of pathogen-infested potting mix with irradiated peat and talc formulations of all three bacteria also showed a significant increase in healthy cucumber seedlings and plant fresh weights and a significant decrease in damping-off and root rot severity compared to the infested controls. A preplant amendment of untreated control irradiated peat and talc did not provide any protection of cucumber seedlings from damping-off and root rot. The disease protection by bacterial bioformulation and metalaxyl treatments was comparable. The preplant amendment treatment of irradiated peat formulation of Pf 9A-14 showed a significant increase in plant fresh weights of cucumber compared to all other treatments.

#### 8.6.1.2 Seed Treatment

Seed treatment of control irradiated peat and talc did not provide any protection of cucumber seedlings from damping-off and root rot. Under a very high disease condition (97 % diseased plants in the infested control), all three bacterial and metalaxyl seed treatments were equally effective in protecting cucumber seedlings from damping-off and root rot with 66–75 % healthy seedlings with cell suspensions, 78–94 % healthy with talc formulation, 78–97 % healthy with irradiated peat formulation, and 84 % healthy with the metalaxyl seed treatment. Bacterial and metalaxyl seed treatments also showed a 53–69 % reduction in damping-off and root rot severity (Khabbaz and Abbasi 2014) and a four- to sixfold increase in plant fresh weights compared to the infested controls.

Seed treatment of a mixture of two or three strains of antagonistic bacteria in irradiated peat formulation was also investigated for suppression of *Pythium* damping-off and root rot of cucumber. Seed treatments with metalaxyl and irradiated peat formulations of all three antagonistic bacteria as single or mixtures of two or three strains suppressed damping-off and root rot of cucumber seedlings and showed a 590–810 % increase in healthy seedlings, a 43–57 % reduction in damping-off and root rot severity, and a 92–245 % increase in plant fresh weights as compared to the infested controls. The seed treatment with a mixture of three bacterial strains (*P. fluorescens* 9A-14, *Pseudomonas* sp. 8D-45, and *Bacillus subtilis* 8B-1) was the best treatment that showed a 245 % increase in plant fresh weights as compared to the infested control and a 61 % increase as compared to the noninfested control.

The biocontrol potential of another three antagonistic strains, *P. fluorescens* PEF-5 #18, *Pseudomonas chlororaphis* SL5, and *Paenibacillus polymyxa* #50, was also evaluated for suppression of *Pythium* damping-off and root rot of cucumber in growth-room assays. Bacteria were applied on seeds as irradiated peat formulations. All three bacterial treatments showed less disease severity, more healthy plants, and higher plant fresh and dry weights as compared with the non-treated seeds grown in pathogen-infested potting mix (Abbasi and Zhang unpublished data).

### 8.6.2 *Phytophthora* Damping-Off and Root Rot of Cucumber

The irradiated peat and talc formulations of the three antagonistic bacteria, *P. fluorescens* 9A-14, *Pseudomonas* sp. 8D-45, and *B. subtilis* 8B-1, were also evaluated for suppression of *Phytophthora* damping-off and root rot of cucumber seedlings as preplant amendment and seed treatments in a *P. capsici*-infested potting mix in growth-room pot assays (Khabbaz et al. 2015).

#### 8.6.2.1 Preplant Substrate Treatment

In a *P. capsici*-infested potting mix, preplant treatments with cell suspension or irradiated peat and talc formulations of all three antagonistic bacteria also protected cucumber seedlings from *Phytophthora* damping-off and root rot and produced 69–85 % healthy seedlings compared to 6 % healthy in the infested controls. Disease control by antagonistic bacteria was comparable to the metalaxyl treatment. Plants produced in pathogen-infested potting mix pretreated with cell suspension or irradiated peat or talc formulations of the three bacteria showed significantly higher fresh weights compared to the plants produced in infested and noninfested control mixes and in infested mix pretreated with metalaxyl.

#### 8.6.2.2 Seed Treatment

Seed treatments of cell suspension or irradiated peat and talc formulations of all three antagonistic bacteria also protected cucumber seedlings from *Phytophthora* damping-off and root rot and showed 72–85 % healthy seedlings compared to 3–6 % healthy in the infested controls. Seed treatment of irradiated peat and talc formulation of *P. fluorescens* 9A-14 provided better disease protection than the metalaxyl seed treatment. All bacterial seed treatments also showed significantly higher plant weights compared to infested and noninfested controls and the metalaxyl seed treatment.

Seed treatments with mixtures of antagonistic bacteria in irradiated peat formulation were also evaluated for suppression of *Phytophthora* damping-off and root rot of cucumber (Khabbaz et al. 2015). Seed treatments with irradiated peat formulations of a mixture of two or three bacteria showed 63–84 % healthy seedlings as compared to 22 % healthy in infested control and 50 % healthy in metalaxyl seed treatment. All bacterial seed treatments also showed a reduction in damping-off and root rot severity and an increase in plant fresh weights compared to the infested controls. The bacterial seed treatment containing a mixture of three bacteria was the best treatment that showed a significant increase in plant fresh weights as compared to all other treatments.

In order to improve the disease suppression by antagonistic bacteria, acibenzolar-*S*-methyl (ASM) or Actigard (50 WP; Syngenta Crop Protection Inc., Guelph, Canada), a chemical activator of disease resistance in plants (Oostendorp et al. 2001; Anith et al. 2004; Park et al. 2013), was used as seed treatment in combination with *P. fluorescens* 9A-14 (Pf 9A-14). The aqueous suspension of ASM (30 mg a.i. $L^{-1}$) was used as seed treatment alone or in combination with Pf 9A-14 ($5 \times 10^8$ CFU $mL^{-1}$). Both as single and mixture treatments, ASM and Pf 9A-14 protected cucumber seedlings from *Phytophthora* damping-off and showed a 280–461 % increase in healthy seedlings, 40–44 % reduction in damping-off severity, and a 112–217 % increase in plant fresh weights as compared to the untreated infested control (Fig. 8.2).

There were no significant differences between Pf 9A-14 and ASM treatments in suppressing *Phytophthora* damping-off of cucumber seedlings when used alone, but Pf 9A-14 seed treatment resulted in higher plant fresh weights over ASM treatment (Fig. 8.2). The mixture treatment (Pf 9A-14 + ASM) did not show a significant improvement over the individual treatments for increasing healthy cucumber seedlings or reducing disease severity, but showed a significant increase in plant fresh weights over the individual treatments (Fig. 8.2).

The effectiveness of four additional antagonistic bacterial strains, *P. chlororaphis* SL5, *B. amyloliquefaciens* 1A-48 and 1B-14, and *P. fluorescens* PEF-5 #18, was also evaluated as seed treatment for suppression of *Phytophthora* damping-off and root rot of cucumber in pathogen-infested potting mix. After 2 weeks of planting, plants in pathogen-infested potting mix sown with seeds treated with irradiated peat formulation of all four bacteria consistently showed less disease severity, more healthy plants, and higher plant fresh and dry weights compared to the plants in pathogen-infested mix sown with untreated control seeds (Abbasi and Zhang unpublished data).

### 8.6.3 *Rhizoctonia* Seedling Damping-Off of Radish

The biocontrol potential of three antagonistic bacteria *P. fluorescens* 9A-14, *Pseudomonas* sp. 8D-45, and *B. subtilis* 8B-1 was also evaluated for suppression of damping-off of radish seedlings as preplant amendment and seed treatments in a *R. solani*-infested potting mix under growth-room conditions (Khabbaz et al. 2015).

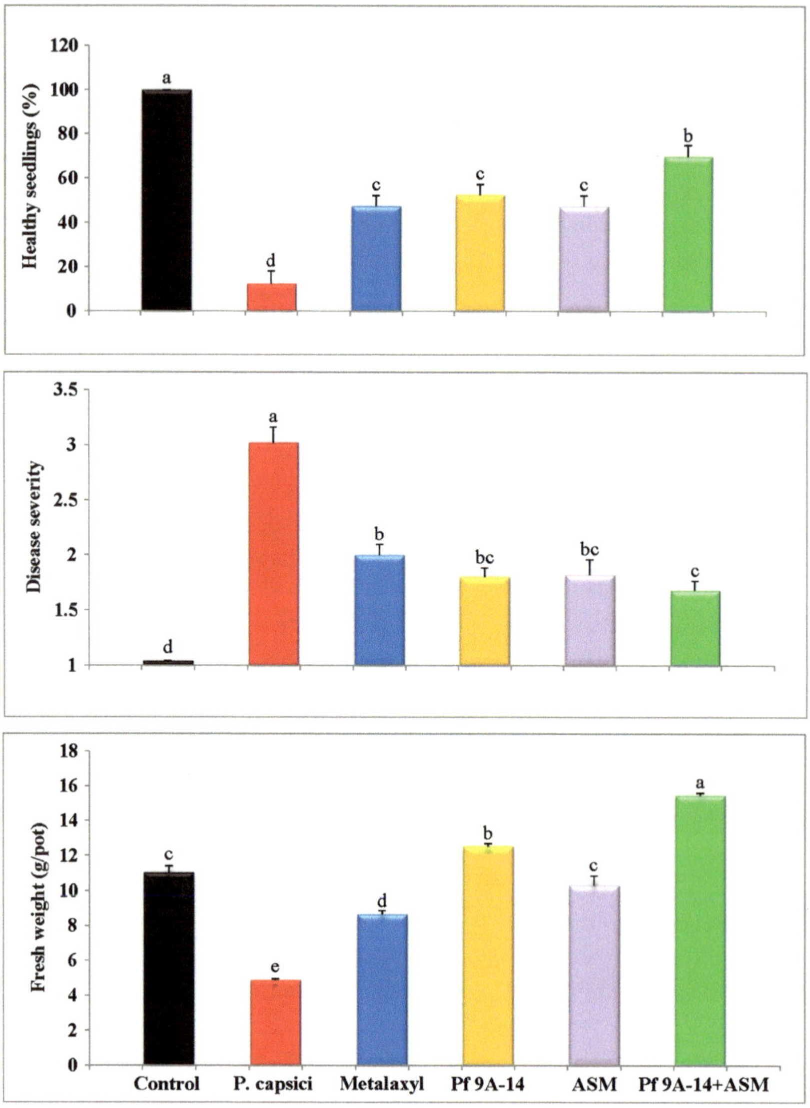


**Fig. 8.2** Effect of seed treatment of acibenzolar-*S*-methyl (*ASM*) and *P. fluorescence* Pf 9A-14 alone or in combination on healthy seedlings, *Phytophthora* damping-off and root rot severity, and plant fresh weights of cucumber in an infested peat mix under growth-room condition. Values (± SE) are the mean of four replicates. Means with a common letter are not significantly different according to Fisher's protected least significant difference test at $P \leq 0.05$

#### 8.6.3.1 Preplant Substrate Treatment

Preplant application of cell suspension or irradiated peat and talc formulations of all three antagonistic bacteria to the *R. solani*-infested potting mix showed a modest protection of radish seedlings from damping-off with 18–27 % healthy seedlings in potting mix treated with bacteria as compared to the 5 % healthy in the infested controls. Comparatively, benomyl preplant treatment was more effective in increasing the percentage of healthy seedlings than the antagonistic bacteria. Preplant treatments of all three antagonistic bacteria and benomyl also increased fresh weights of radish plants compared to the infested controls.

#### 8.6.3.2 Seed Treatment

The irradiated peat formulations of three antagonistic bacteria applied alone or in mixture as seed treatments also showed a modest protection of radish seedlings from damping-off. In pots sown with bacterial-treated seeds, 25–51 % seedlings were healthy as compared to the 5 % healthy in the infested controls. All bacterial seed treatments showed a reduction in *Rhizoctonia* damping-off severity and an increase in plant

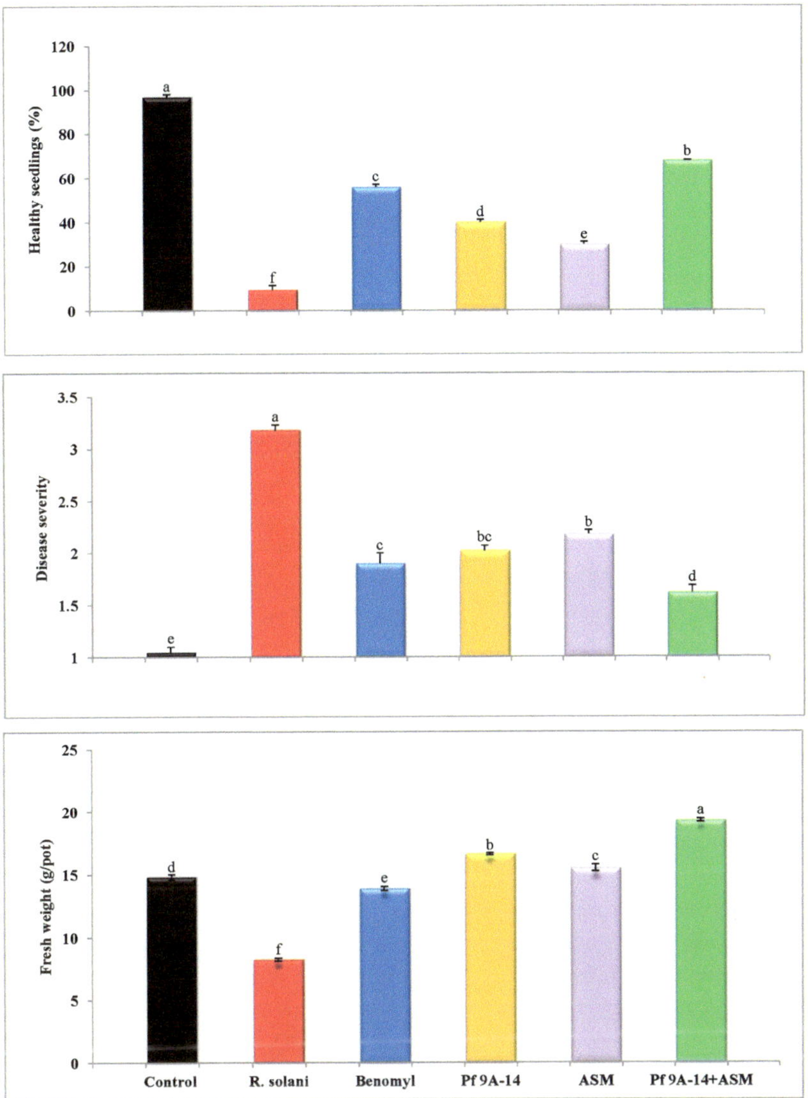


**Fig. 8.3** Effect of seed treatment of acibenzolar-*S*-methyl (*ASM*) and *P. fluorescens* Pf 9A-14 alone or in combination on healthy seedlings, *Rhizoctonia* damping-off and root rot severity, and plant fresh weights of radish in a *R. solani*-infested potting mix under growth-room condition. Values ± (SE) are the mean of five replicates. Means with a common letter are not significantly different according to Fisher's protected least significant difference test at $P \leq 0.05$

fresh weights compared to the untreated infested controls. The bacterial seed treatments containing mixtures of any two antagonistic bacteria showed similar percentage of healthy seedlings. However, the seed treatment containing a mixture of three bacteria significantly increased plant fresh weights as compared to all other treatments.

As disease protection with antagonistic bacteria was only modest, an improvement in suppression of *Rhizoctonia* damping-off of radish by antagonistic bacteria was attempted with combined seed treatment of ASM and Pf 9A-14 in growth-room pot assays. Actigard or ASM was dissolved in water at 30 mg a.i. $L^{-1}$ and used as seed treatment alone or in combination with Pf 9A-14 ($5 \times 10^8$ CFU $mL^{-1}$). Radish seeds treated with ASM and Pf 9A-14 alone or as mixture significantly increased the percentage of healthy seedlings, decreased damping-off severity, and increased plant fresh weights compared to the infested control (Fig. 8.3). The combined ASM and Pf 9A-14 seed treatment showed a significant improvement over the individual treatments and protected radish seedlings from damping-off by increasing the percentage of healthy radish seedlings, reducing disease severity, and increasing plant fresh weights (Fig. 8.3). The disease protection achieved with the mixture treatment (ASM + Pf 9A-14) was also better than benomyl seed treatment (Fig. 8.3).

### 8.6.4 *Fusarium* Crown and Root Rot (FCRR) of Tomato

The biocontrol potential of the six selected antagonistic bacterial strains (*P. fluorescens* 9A-14, *B. subtilis* 8B-1, *Pseudomonas* sp. 8D-45, *P. polymyxa* #53, *P. chlororaphis* SL5, and *P. fluorescens* PEF-5 #18) was investigated for suppression of FCRR in *Forl*-infested potting mix and a field soil in greenhouse pot assays (Zhang et al. 2015). Bacteria were applied as irradiated peat formulation on roots of tomato seedlings which were immediately planted in the infested potting mix or soil.

Tomato seedlings treated with irradiated peat formulation of the six bacteria and grown in *Forl*-infested potting mix for 25 days showed a 40–63 % reduction in FCRR severity, a 169–274 % increase in plant height, and a 101–126 % increase in root length as compared to the non-treated control seedlings grown in *Forl*-infested mix. The bacterial roots treatments also showed a significant increase in plant fresh and dry weights as compared to the untreated infested controls. The effects of most bacterial treatments were better than or comparable to the benomyl treatment for plant height, root length, fresh or dry weights, and FCRR severity. The plants treated with bacterial strains *Pseudomonas* sp. 8D-45 and *P. fluorescens* PEF-5#18 and grown in *Forl*-infested mix showed a 63–66 % increase in fresh and 29 % increase in dry weights as compared to untreated plants grown in noninfested mix.

Similar results were also achieved in *Forl*-infested field soil (Zhang et al. 2015). Tomato plants treated with irradiated peat formulation of bacteria and grown in *Forl*-infested soil for 25 days showed a 40–60 % reduction in FCRR severity, a 17–53 % increase in plant height, and a 265–300 % increase in root length as compared to the untreated plants grown in *Forl*-infested field soil. All bacterial and benomyl treatments also showed similar results for FCRR severity, root length, and plant fresh and dry weights in field soil as well. Plants treated with bacterial and benomyl treatments and produced in *Forl*-infested soil showed higher fresh and dry weights as compared to the untreated control plants produced in *Forl*-infested soil, and such plants also had longer roots as compared to untreated control plants grown in pathogen noninfested soil.

### 8.6.5 *Fusarium* Wilt of Tomato

The biocontrol potential of ten antagonistic bacteria (Table 8.2) was also evaluated for suppression of *Fusarium* wilt in greenhouse pot assays. Tomato seedlings were treated with irradiated peat formulation of the selected bacterial strains and grown in *Fol*-infested potting mix for 25 days before evaluating for disease severity and growth parameters. Plants treated with bacterial formulations and planted in *Fol*-infested potting mix showed a 61–74 % reduction in severity of *Fusarium* wilt (Table 8.2). These treated plants also showed a 607–673 % increase in fresh weights and a 300–400 % increase in dry weights as compared to the untreated plants grown in pathogen-infested mix. Plants treated with bacterial formulations and planted in *Fol*-infested potting mix showed a 17–28 % increase in plant fresh weights and a 25 % increase in plant dry weights (six bacteria) compared to the untreated plants grown in pathogen noninfested control soil (Table 8.2).

### 8.6.6 Common Scab of Potato

Common scab and *Verticillium* wilt are two main soilborne diseases of potatoes (*Solanum tuberosum* L.) causing significant crop losses to growers and both these diseases often occur in the same field. These two diseases and the potato plant are a good model system to evaluate disease control effects of biofungicides and other products. Seed potato tubers also offer good surface area for application of microbial biocontrol agents.

The biocontrol potential of the selected ten antagonistic bacterial strains was also assessed

**Table 8.2** Effect of irradiated peat formulation of selected antagonistic bacterial strains as tomato root application on plant fresh and dry weights and severity of *Fusarium* wilt in a potting mix infested with *Fusarium oxysporum* f. sp. *lycopersici* (*Fol*) in a greenhouse

| Treatment | Fresh weight (g) | Dry weight (g) | Disease severity |
|---|---|---|---|
| Control | 9.1 ± 0.4 b | 0.4 ± 0.0 b | 0 ± 0.0 c |
| *Fol* | 1.5 ± 0.1 c | 0.1 ± 0.0 d | 4.6 ± 0.0 a |
| Benomyl | 10.5 ± 0.5 a | 0.4 ± 0.0 b | 1.4 ± 0.2 b |
| *Paenibacillus polymyxa* #50 | 10.9 ± 0.4 a | 0.5 ± 0.0 a | 1.4 ± 0.2 b |
| *Paenibacillus polymyxa* #53 | 11.2 ± 0.6 a | 0.4 ± 0.0 b | 1.8 ± 0.2 b |
| *Bacillus amyloliquefaciens* 1A-48 | 11.2 ± 0.3 a | 0.5 ± 0.0 a | 1.2 ± 0.4 b |
| *Bacillus amyloliquefaciens* 1B-14 | 11.6 ± 0.4 a | 0.5 ± 0.0 a | 1.6 ± 0.2 b |
| *Bacillus amyloliquefaciens* 1B-23 | 11.4 ± 0.4 a | 0.4 ± 0.0 b | 1.4 ± 0.2 b |
| *Pseudomonas chlororaphis* 1B-26 | 10.7 ± 0.4 a | 0.5 ± 0.0 a | 1.8 ± 0.2 b |
| *Bacillus subtilis* 8B-1 | 11.3 ± 0.6 a | 0.5 ± 0.0 a | 1.6 ± 0.2 b |
| *Pseudomonas* sp. 8D-45 | 11.2 ± 0.8 a | 0.5 ± 0.0 a | 1.8 ± 0.2 b |
| *Pseudomonas fluorescens* 9A-14 | 10.6 ± 0.5 a | 0.4 ± 0.0 b | 1.6 ± 0.2 b |
| *Bacillus amyloliquefaciens* 9A-31 | 10.9 ± 0.6 a | 0.4 ± 0.0 b | 1.8 ± 0.2 b |

Values ± standard errors are the average of five replicates, if followed within a column by a common letter are not significantly different according to Fishers protected least significant difference test at $P \leq 0.05$

for suppression of common scab of potato in the micro-plots. A sandy-loam soil for these studies was brought from a commercial potato field with a history of common scab disease. Seed potato 'Yukon Gold' tubers were coated with irradiated peat formulation of the bacterial strains prior to planting in a potato soil. Untreated tubers or tubers treated with non-inoculated control irradiated peat served as control. Five treated and untreated tubers were planted in early June in each of the four replicate micro-plots per treatment. Daughter tubers were harvested in late September and assessed for disease severity and deep-pitted lesions (Abbasi 2013). Harvested tubers were also weighed to determine total yield.

The disease level on daughter tubers harvested from the control micro-plots was very high and most of the tubers had deep-pitted lesions. Under these high disease conditions, seven bacterial treatments provided a modest disease control and showed a 13–39 % reduction in severity of common scab on daughter tubers as compared to the control treatment. Three bacterial strains showed no reduction of scab severity on daughter tubers. Seed treatment of most bacterial strains increased tuber weights compared to the untreated control treatment. Only one strain showed a reduction in deep-pitted lesions on tubers (Abbasi et al. unpublished data).

## 8.7 Conclusions and Next Steps

Indigenous antagonistic bacterial strains displayed the potential to be developed as biofungicides to control various plant diseases caused by soilborne pathogens. Under controlled-environment conditions, bacterial strains provided good control of *Pythium* or *Phytophthora* damping-off, root rot of cucumber, *Fusarium* crown and root rot and *Fusarium* wilt of tomato. The disease control effect of bacterial strains against *Rhizoctonia* damping-off was only modest. Under micro-plot conditions, some bacterial strains provided only modest control of common scab of potato in a very high disease level natural soil. Irradiated peat was found to be a suitable carrier material for delivery of biocontrol bacteria on seed, root, or directly to potting substrate as an amendment. Antagonistic bacteria also showed a significant effect on plant-growth promotion and increased plant fresh or dry weights. In some cases, the disease protection and plant-growth promotion effect by antagonistic bacteria was comparable or even better than the chemical fungicides. Most of the *Bacillus* and *Paenibacillus* strains showed the capacity to produce antibiotics such as bacillomycin, bacilysin, iturin A, surfactin, and fengycin, whereas most of the *Pseudomonas* strains possessed antibiotic biosynthetic genes for pyoluteorin, pyrrolnitrin, and 2,4-DAPG. However, the actual production of these antibiotics by antagonistic bacterial strains was not confirmed. Most bacterial strains showed production of lytic enzymes, HCN, IAA, and SA. Some strains also produced siderophores and volatile compounds. Antagonistic and plant-growth promotion activities of these bacterial strains might be related to the production of several types of antibiotics, lytic enzymes, phytohormones, secondary metabolites, siderophores, and volatile compounds.

Although the biocontrol potential of antagonistic bacteria for suppression of soilborne diseases was demonstrated under controlled-environment, unpasteurized or unsterilized potting substrate and field soil was used in all pot experiments and should be containing the natural competing organisms. The bioformulations of these antagonistic bacterial strains prepared in irradiated peat can now be evaluated under field conditions. Therefore, future studies should be directed at confirming the disease control potential of these antagonistic bacteria under natural conditions of disease development in the field. Studies should also be aimed at confirming the specific modes of disease suppression and plant-growth promotion by antagonistic bacteria and integrating these biocontrol agents with other disease management options.

**Acknowledgements** Technical assistance was provided by Brian Weselowski, Bruce McPherson, and Igor Lalin. The funding for the research work reported in this book chapter was provided by Agriculture and Agri-Food Canada.

## References

Abbasi PA (2011) Exploiting and understanding disease suppressing effects of fish emulsion for soil-borne and foliar diseases. Am J Plant Sci Biotechnol 5(Special Issue 2):61–68

Abbasi PA (2013) Establishing suppressive conditions against soilborne potato diseases with low rates of fish emulsion applied serially as a pre-plant soil amendment. Can J Plant Pathol 35:10–19

Abbasi PA, Conn KL, Lazarovits G (2004) Suppression of *Rhizoctonia* and *Pythium* damping-off of radish and cucumber seedlings by addition of fish emulsion to peat mix or soil. Can J Plant Pathol 26:177–187

Abbasi PA, Lazarovits G, Conn KL (2008) Enhancing biological control of soilborne plant diseases by organic soil amendments. In: Barka EA, Clément C (eds) Plant-microbe interactions. Research Signpost, Kerala, pp 319–343

Alabouvette C (1999) *Fusarium* wilt suppressive soils: an example of disease-suppressive soils. Aust Plant Pathol 28:57–64

Anith K, Momol M, Kloepper J, Marios J, Olson S, Jones J (2004) Efficacy of plant growth-promoting rhizobacteria, acibenzolar-*S*-methyl, and soil amendment for integrated management of bacterial wilt on tomato. Plant Dis 88:669–673

Athukorala SN, Fernando WG, Rashid KY (2009) Identification of antifungal antibiotics of *Bacillus* species isolated from different microhabitats using polymerase chain reaction and MALDI-TOF mass spectrometry. Can J Microbiol 55:1021–1032

Bakker PAHM, Ran LX, Pieterse CMJ, van Loon LC (2003) Understanding the involvement of rhizobacteria-mediated induction of systemic resistance in biocontrol of plant diseases. Can J Plant Pathol 25:5–9

Bashan Y, de-Bashan LE, Prabhu SR, Hernandez J-P (2014) Advances in plant growth-promoting bacterial inoculant technology: formulations and practical perspectives (1998–2013). Plant Soil 378:1–33

Bernard E, Larkin RP, Tavantzis S, Erich MS, Alyokhin A, Sewell G, Lannan A, Gross SD (2012) Compost, rapeseed rotation, and biocontrol agents significantly impact soil microbial communities in organic and conventional potato production systems. Appl Soil Ecol 52:29–41

Bonanomi G, Antignani V, Capodilupo M, Scala F (2010) Identifying the characteristics of organic soil amendments that suppress soilborne plant diseases. Soil Biol Biochem 42:136–144

Borrego-Benjumea A, Basallote-Ureba MJ, Melero-Vara JM, Abbasi PA (2014) Characterization of Fusarium isolates from asparagus fields in southwestern Ontario and influence of soil organic amendments on Fusarium crown and root rot. Phytopathology 104:403–415

Cook RJ, Baker KF (1983) The nature and practice of biological control of plant pathogens. American Phytopathological Society, St. Paul, 539 pp

Fravel DR (2005) Commercialization and implementation of biocontrol. Annu Rev Phytopathol 43:337–359

Gerhardson B (2002) Biological substitutes for pesticides. Trends Biotechnol 20:338–343

Haas D, Défago G (2005) Biological control of soil-borne pathogens by fluorescent pseudomonads. Nat Rev Microbiol 3:307–319

Jones KA, Burges HD (1998) Technology of formulation and application. In: Burges HD (ed) Formulation of microbial biopesticides. Kluwer Academic Publishers, Dordrecht/Boston, pp 7–30

Khabbaz SE, Abbasi PA (2014) Isolation, characterization, and formulation of antagonistic bacteria for the management of seedlings damping-off and root rot disease of cucumber. Can J Microbiol 60:25–33

Khabbaz S, Zhang L, Cáceres LA, Sumarah M, Wang A, Abbasi PA (2015) Characterization of antagonistic Bacillus and Pseudomonas strains for biocontrol potential and suppression of damping-off and root rot diseases. Ann Appl Biol 166:456–471

Kim YC, Jung H, Kim KY (2008) An effective biocontrol bioformulation against phytophthora blight of pepper using growth mixtures of combined chitinolytic bacteria under different field conditions. Eur J Plant Pathol 120:373–382

Knowles JW (1976) Microorganisms and cyanide. Bacteriol Rev 40:652–680

Lugtenberg BJJ, Kamilova F (2009) Plant growth promoting rhizobacteria. Annu Rev Microbiol 63:541–556

Mathre DE, Cook RJ, Calla NW (1999) From discovery to use: traversing the world of commercializing biocontrol agents for plant disease control. Plant Dis 83:972–983

Nautiyal CS, Johri JK, Singh HB (2002) Survival of the rhizosphere-competent biocontrol strain *Pseudomonas fluorescens* NBRI2650 in the soil and phytosphere. Can J Microbiol 48:588–601

Neilands JB (1981) Microbial iron compounds. Annu Rev Biochem 50:715–731

Olubukola OB (2010) Beneficial bacteria of agricultural importance. Biotechnol Lett 32:1559–1570

Oostendorp M, Kunz W, Dietrich B, Staub T (2001) Induced disease resistance in plants by chemicals. Eur J Plant Pathol 107:19–28

Park K, Park J-W, Lee S-W, Balaraju K (2013) Disease suppression and growth promotion in cucumbers induced by integrating PGPR agent *Bacillus subtilis* strain B4 and chemical elicitor ASM. Crop Prot 54:199–205

Raaijmakers JM, Paulitz TC, Steinberg C, Alabouvette C, Moënne-Loccoz Y (2009) The rhizosphere: a playground and battlefield for soilborne pathogens and beneficial microorganisms. Plant Soil 321:341–361

Ramamoorthy V, Viswanathan R, Raguchander T, Prakasam V, Samiyappan R (2001) Induction of systemic resistance by plant growth promoting rhizobacteria in crop plants against pests and diseases. Crop Prot 20:1–11

Santoyo G, Orozco-Mosqueda MDC, Govindappa M (2012) Mechanisms of biocontrol and plant growth-promoting activity in soil bacterial species of *Bacillus* and *Pseudomonas*: a review. Biocontrol Sci Tech 22:855–972

Trüper HG, dè Clari L (1997) Taxonomic note: necessary correction of specific epithets formed as substantives (nouns) "in apposition". Int J Syst Bacteriol 47:908–909

Van Veen JA, Van Overbeek LS, Van Elsas JD (1997) Fate and activity of microorganisms introduced into soil. Microbiol Mol Biol Rev 61:121–135

Vidhyasekaran P, Muthamilan M (1995) Development of formulations of *Pseudomonas fluorescens* for control of chickpea wilt. Plant Dis 79:782–786

Vidhyasekaran P, Rabindran R, Muthamilan M, Nayar K, Rajappan K, Subramanian N, Vasumathi K (1997) Development of a powder formulation of *Pseudomonas fluorescens* for control of rice blast. Plant Pathol 46:291–297

Weller DM (1988) Biological control of soilborne plant pathogens in the rhizosphere with bacteria. Annu Rev Phytopathol 26:379–407

Weller DM, Raaijmakers JM, Gardener BBM, Thomashow LS (2002) Microbial populations responsible for specific soil suppressiveness to plant pathogens. Annu Rev Phytopathol 40:309–348

Zhang L, Khabbaz SE, Wang A, Li H, Abbasi PA (2015) Detection and characterization of broad-spectrum anti-pathogen activity of novel rhizobacterial isolates and suppression of Fusarium crown and root rot disease of tomato. J Appl Microbiol 118:685–703

# Bio-based and Reduced-Risk Strategies for the Management of *Phytophthora* Blight and Root Rot of Pepper

9

Yan Ma, Qiujun Wang, Yun Cao, Guangfei Wang, Dejie Guo, and Pervaiz A. Abbasi

## Contents



**Abstract**

*Phytophthora* blight and root rot caused by *Phytophthora capsici* affects several solanaceous and cucurbit hosts worldwide. The disease has become a constraint in cultivation of chili pepper in China and is an emerging problem in Canada and the USA affecting pepper, tomato, cucumber, and other cucurbits. Several bio-based approaches such as biofungicides; soil amendments of rapeseed meal, composts, and manures; soil solarization; grafting; and reduced-risk chemicals have been investigated for the management of *Phytophthora* blight and root rot of pepper. The biocontrol potential of fungal antagonists such as *Penicillium striatisporum* Pst10, *Trichoderma harzianum*, and *Trichoderma hamatum* 382 and bacterial antagonists such as *Bacillus amyloliquefaciens* BS211, *Pseudomonas corrugata*, and other *Pseudomonas* and *Bacillus* strains was shown for suppression of pathogen growth and disease. Soil amendment of rapeseed meal, manures, and composts also provided suppression of *Phytophthora* blight and root rot under greenhouse and field conditions. Some combinations of biocontrol agents and soil amendments provided enhanced disease control. Soil solarization can be very effective in reducing pathogen inoculum from the top

Y. Ma • Q. Wang • Y. Cao • G. Wang • D. Guo
Institute of Agricultural Resource and Environment, Jiangsu Academy of Agricultural Sciences, Nanjing 210014, China

P.A. Abbasi (✉)
Atlantic Food and Horticulture Research Centre, Agriculture and Agri-Food Canada, 32 Main Street, Kentville, NS B4N 1J5, Canada
e-mail: Pervaiz.Abbasi@agr.gc.ca





N.K. Arora et al. (eds.), *Bioformulations: for Sustainable Agriculture*,
DOI 10.1007/978-81-322-2779-3_9

10-cm layer of soil, but this strategy may be only effective in countries where summer months are very hot. Anaerobic soil disinfestation using wheat straw as the carbon source followed by flooding for 2–3 weeks reduced pathogen populations in soil and also disease incidence on pepper. Reduced-risk chemicals such as phosphonates and Actigard and grafting of resistant rootstocks also provided disease control and offer additional sustainable methods for the management of *Phytophthora* blight and root rot of pepper. Integration of various management strategies may be more effective for managing disease problems such as *Phytophthora* blight and root rot.

## 9.1 Introduction and Background

*Phytophthora* blight and root rot is caused by a soilborne oomycete pathogen *P. capsici* Leonian. The pathogen has a wide host range infecting a variety of plant species including pepper, tomato, eggplant, cucumber, and other cucurbits worldwide and causing symptoms such as damping-off, blight, and crown, root, or fruit rots depending on the developmental stage of the plant at the time of infection (Ristaino and Johnston 1999; Hausbeck and Lamour 2004). *Phytophthora* blight and root rot is also a common and destructive disease of greenhouse and field-grown chili and sweet peppers (Fig. 9.1) in China and worldwide. In China, the disease has become a serious constraint in cultivation of chili peppers causing serious economic losses to growers in several provinces every year. This disease situation may have been established due to continuous cultivation of peppers in greenhouses for many years. *Phytophthora* blight and root rot is an emerging disease in Canada and the USA affecting several solanaceous and cucurbit crops both under field conditions and in commercial greenhouses. With the phasing out of methyl bromide and due to concerns of negative impacts of chemical pesticides on the environment and human health, the development of sustainable, eco-friendly, and alternative methods of managing soilborne diseases such as *Phytophthora* blight and root rot has become a top priority (Colla et al. 2012).

Phosphorous acid and the phosphorous acid-based fungicides such as phosphonates or phosphites are reduced-risk chemicals as there are no known harmful effects associated with their use (Guest and Grant 1991). Phosphorous acid and phosphonates fall under the category of natural products. Phosphonates and phosphites have been very effective against diseases caused by oomycete plant pathogens. Although their compatibility with microbial biofungicides for disease suppression has not been reported, phosphonates have been shown to induce symbiotic infections by mycorrhizal fungi (Jabaji-Hare and Kendrick 1987; Howard et al. 2000). Application of a phosphonate formulation as a pre-planting amendment, post-planting drench, or seed treatment provided effective control of *Pythium* damping-off of cucumber (Abbasi and Lazarovits 2005, 2006a) and as an application to soil prior to planting or after planting provided suppression of clubroot of bok choy and cabbage under micro-plot and field conditions (Abbasi and Lazarovits 2006b). Acibenzolar-*S*-methyl (ASM) or Actigard is a chemical activator and inducer of systemic resistance in plants (Oostendorp et al. 2001). Both these chemicals can be used as sustainable disease management approaches.

There is also a growing interest in the use of microbial-based biofungicides as bioformulations and agricultural by-products as biofumigants for replacement of chemical pesticides to manage soilborne diseases and pests. There have been considerable efforts devoted toward developing biofungicides for managing soilborne diseases, and several bacterial and fungal antagonists have been reported as potential candidates (Ezziyyani et al. 2007). Similarly, soil incorporation of organic material containing agricultural by-products and manures and grafting also show tremendous potential of managing such soilborne diseases (Gilardi et al. 2013). In the case of biofumigation, the suppression of soilborne pathogens and pests is mainly accomplished by release of inhibitory chemicals during microbial

**Fig. 9.1** *Phytophthora* blight and root rot affecting sweet pepper (**a–d**) and chili pepper (**e** and **f**)

decomposing of organic material (Matthiessen and Kirkegaard 2006). Several plant species including mustards, oilseed radish, rapeseed, Sudan grass, and pearl millet have been used as rotation or cover crops for their biofumigation effect (Hansen and Keinath 2013). As cover crops, these plants can be disked back into the soil as green manure followed by plastic mulching to retain volatiles for maximum effect. Most of the nutrients will also be retained in the soil.

Alternative and sustainable strategies have also been used for the management of *Phytophthora* blight and root rot of pepper, and considerable research efforts and resources were put forward in search for potential biofungicides or suitable soil amendments from locally available resources. Soil solarization and anaerobic soil disinfestation are also used as alternative strategies for the management of *Phytophthora* blight and root rot of pepper. The reduced-risk chemicals such as phosphonates and Actigard, as well as grafting technology, were also investigated as alternative management solutions for this disease. In this chapter, an overview of the work on the management of *Phytophthora* blight and root rot of pepper with bio-based approaches such as biofungicides; soil amendments of rapeseed meal, composts, and manures; solarization; grafting; and reduced-risk chemicals is discussed.

## 9.2 Management of *Phytophthora* Blight and Root Rot of Pepper

Control of *Phytophthora* blight and root rot is very complicated, and several sustainable disease management options have been investigated including cultural practices, soil amendments, resistant cultivars, biofungicides, solarization, and grafting (Ristaino and Johnston 1999; Granke et al. 2012; Gilardi et al. 2013). Some of these sustainable disease management strategies are described in this review. For instance, all the disease management work on chili pepper was carried out in China using locally available resource materials such as bioformulations with indigenous biofungicides, soil biofumigation with locally available agricultural by-products such as rapeseed meal and manures, and soil amendment of wheat straw followed by flooding. The disease management work on bell or sweet pepper was reported from Canada, the USA, Europe, and Korea.

### 9.2.1 Bioformulation of Fungal Antagonists

During the search for potential biocontrol agents, an isolate of *P. striatisporum* Pst10 was recovered from the rhizosphere of chili peppers grown in a greenhouse in Nanjing, China (Ma et al. 2008). On agar plate assays, the isolate Pst10 was shown to inhibit the growth of mycelium of *Phytophthora* spp., *Cladosporium cucumerinum*, and *Sclerotinia sclerotiorum*. In subsequent in vitro pathogen inhibition tests, the culture filtrates (CF) of Pst10 were used. The CF of Pst10 was obtained after growing the fungal culture in potato dextrose broth in a shaker for a week and by removing mycelium and spores after filtration and centrifugation. In in vitro tests, the CF inhibited the mycelium growth and also inhibited formation and/or germination of sporangia and spores of *P. capsici*. The CF completely inhibited the growth of mycelium, and a 20-fold dilution of CF inhibited formation and germination of sporangia and spores. A 100-fold dilution of CF also caused abnormal effects on the pathogen mycelium. The antifungal substances from organic solvent extracts of CF of Pst10 were isolated using thin-layer chromatography (TLC) approach. Three potential compounds isolated from TLC plate also showed antifungal activity against *P. capsici*.

The next challenging work was how to exploit the biocontrol potential of this native isolate obtained from the rhizosphere of chili peppers to manage *Phytophthora* blight and root rot. Both fungal propagules and CF of Pst10 showed the potential to be used as active ingredients for growth suppression of *P. capsici*. The ability of Pst10 for suppression of *Phytophthora* blight and root rot in chili peppers was tested in greenhouse pot assays with soil infested with sporangia of *P. capsici*. Conidia and CF of Pst10 were mixed with the pathogen-infested soil. Pots were planted with chili pepper seedlings cv. Sujiao No. 2 and placed in a greenhouse. Plants were assessed for disease incidence after 7 and 14 days. The treatment of pathogen-infested soil with conidia and CF of Pst10 significantly reduced the incidence of *Phytophthora* blight and root rot of chili. More than 70–80 % plants were still healthy with both Pst10 treatments after 14 days of planting.

A suitable substrate was needed that could serve as a delivery material for Pst10 and enhance its colonization and effectiveness. Composted pig manure was the obvious and likely choice as it was easily available from local sources. The colonization of composted pig manure by Pst10 was assessed in pots containing soil from a local vegetable farm. The soil was inoculated with Pst10 at $4 \times 10^6$ conidia per g dry soil and amended with composted pig manure at 1 % on a weight basis. The pots were also planted with chili pepper seedlings. The colony-forming units (CFUs) of Pst10 were determined in soil samples taken from pot experiments on a selective medium (Ma et al. 2008) after 3, 10, 20, and 40 days of planting. We found that the composted pig manure slowly increased the colonization of the chili rhizosphere by Pst10. The CFUs of Pst10 were 3 times higher after 10 days and 17 times higher after 20 days of planting in composted pig manure-amended soil as compared to Pst10-amended soil. However, after 40 days of planting, the Pst10 CFUs in composted pig manure-amended soil decreased but still 4 times higher than the Pst10-amended soil.

Composted pig manure seems to enhance or at least maintain the original inoculated population density of Pst10 in the amended soil. This may be a key factor for a successful biological control agent to establish and proliferate in soil and rhizosphere. It was interesting to know the impact of combined treatments of Pst10 and composted pig manure on the incidence of *Phytophthora* blight and root rot of chili. In greenhouse pot experiments, the simultaneous or combined application of Pst10 (conidia + CF) and composted pig manure to pathogen-infested soil was proved to be the best treatment that significantly reduced the incidence of *Phytophthora* blight and root rot of chili. More than 92 % plants were still healthy with the combined treatment after 14 days of planting.

An isolate of *T. harzianum* showed some success in suppressing the growth of *P. capsici* and controlling *Phytophthora* blight and root rot of pepper in Spain (Sid Ahmed et al. 1999). This antagonistic isolate also showed synergistic activity with a bacterial antagonist *Streptomyces rochei* as a combined treatment greatly reducing the pathogen population from soil (Ezziyyani et al. 2007). Treatment of pepper seeds and roots with *T. harzianum* reduced stem necrosis caused by *P. capsici* (Sid Ahmed et al. 2000). Several biocontrol mechanisms for the suppression of pathogen growth and disease by *T. harzianum* have been suggested including direct inhibition of pathogen growth by production of inhibitory compounds, competition for nutrients and niche, and induction of resistance in pepper plants against pathogen (Sid Ahmed et al. 2000).

Another fungal antagonist *T. hamatum* 382 was reported for its biocontrol potential against *P. capsici* in a greenhouse study from Ohio (Khan et al. 2004). In this study, *T. hamatum* 382 was inoculated onto a compost-amended substrate. The fungal antagonist reduced the severity of *Phytophthora* damping-off and root rot of cucumber, and induced resistance was suggested as a mechanism of this disease reduction.

### 9.2.2 Bioformulation of Rhizobacteria

An antagonistic strain of *B. amyloliquefaciens* BS211 was isolated from the root of healthy pepper in a field severely affected by *Phytophthora* blight and root rot in Huaian, Jiangsu province of China (Wang et al. 2014a). In dual-culture plate assays, the strain BS211 was antagonistic to mycelium growth of *P. capsici* (Wang et al. 2014a) and other plant pathogenic fungi. The biocontrol potential of BS211 was evaluated in greenhouse pot experiments for suppression of *Phytophthora* blight and root rot disease of chili pepper in a pathogen-infested soil. Soil used in these pot experiments was collected from a pepper field in Huaian and had a sandy loam texture with pH 7.8. The soil was first inoculated with spore suspension of *P. capsici* at $10^3$ CFU $g^{-1}$ soil, and the pathogen-infested soil was later mixed with the biocontrol bacteria BS211 at $10^7$ CFU $g^{-1}$ soil. The seedlings of chili pepper cv. Sujiao No. 5 grown in a greenhouse mix up to four-leaf stage were transplanted into pots. Plants were rated visually for the incidence of *Phytophthora* blight and root rot disease after 7, 15, 25, and 35 days of transplanting and the percentage of infected plants calculated. After 7 days of planting, all plants grown in *P. capsici* + BS211-treated soil were healthy compared to plants grown in *P. capsici*-infested soil only (0 vs. 15 % disease incidence, respectively). The incidence of *Phytophthora* blight and root rot gradually increased in *P. capsici*-infested soil reaching 100 % after 35 days of planting. Disease incidence also steadily increased in the *P. capsici* + BS211 treatment reaching a maximum of 20 % after 35 days of planting (80 % plants were still healthy).

The pathogen *P. capsici* has a broad host range, and it also affects cucumbers, melons, and other cucurbits. The same pathogen also causes seedling damping-off and root rot diseases in cucumbers. The cucumber-*Phytophthora* has been a very good model system to assess biocontrol potential of rhizobacteria, and during that screening process, we found several novel strains of rhizobacteria belonging to genera *Pseudomonas* and *Bacillus* that showed biocontrol potential for suppression of *Phytophthora* damping-off and root rot of cucumber (Khabbaz et al. 2015). The formulations of the bacterial strains were prepared in irradiated peat or talc powder and applied directly to the pathogen-infested potting substrate or on seed. A more detailed description of this work can be found in Chapter No. 8 in this book.

Strains of rhizobacteria were also investigated for suppression of *Phytophthora* blight and root rot on bell pepper in Korea. In this work, bacteria were screened using radicle and seedling assays and tested for suppression of *Phytophthora* blight and root rot of pepper under artificial and natural inoculation conditions (Sang et al. 2008). Bacteria were applied as drench and root treatments. Artificial inoculations were made with *P. capsici* zoospores 5–7 days after bacterial treatments. The results of both controlled environment and field conditions consistently showed less incidence and severity of *Phytophthora* blight and root rot on bell pepper plants treated with the two *P. corrugata* strains CCR04 and CCR80. This reduction in disease was later shown to be related to bacterial colonization of pepper roots (Sang and Kim 2014).

Another study from Korea used a field-effective biocontrol strategy against *Phytophthora* blight and root rot complex of pepper caused by three pathogens (Kim et al. 2008). In that study, a combination of three strains of chitinolytic bacteria was selected based on their activity against three pathogens in the complex to control *Phytophthora* blight complex under greenhouse and field conditions. The combination of three bacterial strains was more effective than any single strain in suppressing *Phytophthora* blight and root rot in pot assays. Bioformulations of the combined strains prepared in a chitin-based growth medium resulted in effective control in field applications. The study also suggested that the effectiveness of the formulated product for the management of control of *Phytophthora* blight and root rot of pepper could be further improved by integrating with other management strategies such as crop rotation and soil solarization and by altering timing of application.

### 9.2.3 Biofumigation of Rapeseed Meal

Incorporation of *Brassica* plant residues or seed meal to soil has been shown to suppress soilborne pathogens and pests by the effect of biofumigation (Matthiessen and Kirkegaard 2006). To determine the biofumigation effect of rapeseed meal on *Phytophthora* blight and root rot of pepper, a field experiment was conducted in a pepper-growing area of Huaian, Jiangsu province of China. The rapeseed meal used in the field experiment was obtained by grinding dried seed containing rapeseed pods in a warring blender. This seed meal had a total glucosinolate content of 85.71 μmol $g^{-1}$ and 5.04 % of total N. The rapeseed meal (0.4 %, w/w) and dazomet (0.03 %, w/w) were incorporated in the field plots to a depth of 20 cm using a rotary tiller, and plots were immediately irrigated to ensure high soil moisture for hydrolysis of glucosinolates. Subsequently, the plots were covered with plastic sheets for about 20 days before transplanting with 5-week-old red pepper seedlings. Plots were irrigated immediately after planting and daily after that as needed. Pepper seedlings were monitored for disease development, and the cumulative number of infected plants was calculated from the day after transplanting until pepper harvest. Plants were designated infected with *P. capsici* if they showed pathogen signs such as white mycelia at the stem base and symptoms commonly associated with the disease, such as yellowing of leaves and wilting. Disease incidence from each plot was calculated and expressed as the percentage of diseased plants.

The results of the field experiment indicated a positive effect of biofumigation with soil incorporation of rapeseed meal on suppression of *Phytophthora* blight and root rot disease of chili pepper. The plots treated with rapeseed meal showed less disease incidence and an increase in pepper yield as compared to the control and dazomet treatments (Wang et al. 2014b). The effects of soil amendment on communities of bacteria and fungi in soil were also assessed by denaturing gradient gel electrophoresis, and the results indicated that the biofumigation of rapeseed meal increased bacterial diversity and decreased fungal diversity (Wang et al. 2014b). The incidence of disease on chili peppers was negatively correlated with bacterial diversity in soil and positively correlated with fungal diversity in soil. The evaluation of chemical properties of the amended soils indicated that rapeseed meal amendment increased total N and nitrate-N contents of soil as well as available P and K contents. The field study also found a significant correlation between microbial community structures in the amended soil and chemical properties of the soil.

Overall, the results of this study indicated that soil biofumigation of rapeseed meal reduced incidence of *Phytophthora* blight and root rot disease of chili pepper. This disease control effect could be due to release of fumigants (Larkin and Griffin 2007) after breakdown of rapeseed meal in soil that may be toxic to pathogen and also by indirect effects of induced changes in the structure and composition of soil microbial communities. On another note, rapeseed meal is a high-nitrogen-containing product, and its amendment to soil could also lead to release of ammonia and nitrous acid which can kill *Phytophthora* propagules (Tsao and Oster 1981). This needs to be further investigated as it depends on the application rate of the nitrogenous amendment and other soil factors (Tenuta and Lazarovits 2002). An accurate mode of action of a soil amendment will be very helpful for its optimal use and for maximizing its disease-controlling effect.

### 9.2.4 Combinations of Biofumigation and Bioformulation

The effects of a combination of rapeseed meal and biofungicide *B. amyloliquefaciens* BS211 for control of *Phytophthora* blight and root rot of pepper on soil bacterial community structure were determined in greenhouse experiments (Wang et al. 2014a). The soil for the pot experiment was collected from a field in Huaian, Jiangsu province, where the field had been under repeated pepper cultivation since 1980. The soil was infested with spore suspension of *P. capsici* at a concentration of $10^3$ CFU $g^{-1}$ soil. At 1 day after infesting soil with *P. capsici*, the rapeseed meal was incorporated into soil at a rate of 4 g $kg^{-1}$ dry soil, and the appropriate quantity of water was added to adjust soil moisture content to 50 % of the water holding capacity. This moisture level was maintained by regularly adding water when required. Amended soil was later transferred to 5-L pots and covered with double-layered plastic film to minimize fumigant emissions. Pots were transferred in a growth room and incubated at 25 ± 2 °C for 20 days. The plastic cover was maintained for 20 days, and after its removal, the soil was gently mixed to ensure the release of any residues of isothiocyanate. Antagonistic bacteria BS211 was grown separately in nutrient broth in 5-L fermentation tanks. The bacterial cells were suspended in 0.01 M phosphate buffer, and the concentration of bacteria was adjusted at $10^9$ CFU $ml^{-1}$ using a spectrophotometer. Immediately after rapeseed incorporation, the suspension of BS211 was added to the soil at a final concentration of $10^7$ CFU $g^{-1}$ soil.

Analysis of the disease incidence showed that the rapeseed meal alone and in combination with BS211 significantly reduced disease incidence of

*Phytophthora* blight and root rot in pepper plants relative to the control treatment (Wang et al. 2014a). The disease incidence in the control treatment was 26.7 %, whereas there was no disease in the rapeseed meal and rapeseed meal + BS211 treatments after 15 days of soil treatment. After 20 days of soil treatment, 66.7 % of the pepper plants in the control treatment were diseased, whereas the disease incidence in the rapeseed meal treatment was 12.5 % and in the rapeseed + BS211 treatment was 6.3 %.

### 9.2.5 Soil Amendments of Composts and Manures

Anaerobic digestion of manures for biogas generation is a promising way to treat large amounts of animal manures. The slurry obtained after digestion should be safe to apply in the fields for crop production. Manures and manure slurries have been shown to promote plant growth and control plant-parasitic nematodes (Jothi et al. 2003). In China, the effects of locally available manures for suppression of *Phytophthora* root rot of chili pepper were evaluated in greenhouse pot assays. Their impacts on mycelium growth and spore germination of *P. capsici* were investigated in agar plate tests.

The results indicated that the anaerobically digested slurry (ADS) can inhibit the mycelium growth and zoospore germination of *P. capsici* and application of slurry to infested soil significantly reduced the incidence of *Phytophthora* blight and root rot in pepper plants (Cao et al. 2013, 2014). The effects on zoospore germination and disease incidence were greater with anaerobically digested pig slurry than anaerobically digested dairy slurry (Cao et al. 2014). While application of raw pig and dairy slurry had no significant suppressive effect on *Phytophthora* blight and root rot, pepper plants grown in soil treated with anaerobically digested pig and dairy slurries showed significantly higher shoot biomass than those grown in soil treated with corresponding raw slurry and negative control. ADS also significantly increased the numbers of soil bacteria, fungi, actinomycete, and antagonistic microorganisms including *Pseudomonas fluorescens* and *Trichoderma* spp. The PCR-DGGE analysis also showed an improvement of the bacterial and fungal diversities in the ADS treatments. Anaerobically digested pig slurry application could enhance the activities of peroxidase and catalase and hydrogen peroxide content while decreasing malondialdehyde content in chili pepper leaves (Cao et al. 2014).

Similarly, composts or compost extracts have been shown to suppress *Phytophthora* blight and root rot of pepper (Kim et al. 1997a; Sang et al. 2010). Composts and various soil amendments showed variable results in disease suppression at Florida (Kim et al. 1997a). In greenhouse pot assays, various composts and soil amendments were incorporated into soil prior to planting pepper seedlings and inoculations with *P. capsici* zoospores to assess their impact on *Phytophthora* blight and root rot. Chitosan, crab shell waste, and citrus pulp with molasses reduced incidence and severity of the disease on bell pepper, but there were no reductions in soil populations of *P. capsici*. There was also high microbial activity in the amended soils. It was speculated that the observed disease reductions may have been due to induced resistance in pepper and increased microbial activity. However, some of these same composts and soil amendments yielded variable results under field conditions (Kim et al. 1997b). Some amendments were not incorporated in the field soil. For instance, roots of pepper seedlings were dipped in chitosan (0.2 % w/v) solution and planted in the field.

Extracts of composts from six different commercial compost facilities in Korea were investigated whether they can be used to control root and foliar infections in pepper plants by *P. capsici* (Sang et al. 2010). In in vitro assays, four out of 47 compost extracts inhibited zoospore germination, germ tube elongation, mycelium growth, and population of *P. capsici*. The selected compost extracts also showed a reduction in incidence and severity of the disease in the seedling and plant assays.

In a greenhouse study from Spain, Nunez-Zofio et al. (2011) tested semi-composted mixture of horse manure and chicken litter and a non-composted mixture of sheep manure and chicken litter followed by plastic mulching for reduction of *Phytophthora* blight and root rot of pepper (Nunez-Zofio et al. 2011). They found a reduction of 86 % and 65 % in disease incidence, and this reduction was linked to an increase in soil microbial activity. Organic soil amendments in conjunction with plastic mulching can be a good strategy for the control of *Phytophthora* crown and root rot of pepper particularly under temperate climate. The study also suggested that the reduction in disease incidence was partly due to a decrease in viability of pathogen oospores possibly by the production of ammonia. An increase in soil microbial activity and functional diversity may also have resulted in pathogen suppressiveness.

### 9.2.6 Soil Solarization

Soil solarization is another sustainable approach applied in several countries where possible with an aim to reduce the pathogen inoculum from soil and reduce *Phytophthora* blight and root rot of pepper. In some cases, soil solarization was also used in combination with soil organic amendments for better results. Soil is actually heated using a clear plastic as a mulch cover by trapping the sun's energy. Solarization may be very effective in killing pathogen inoculum at the top soil layer to a depth of 10 cm or less. The soil disinfestation is only possible during hottest summer months and in abundant sunshine. In the open fields, the use of soil solarization may be limited to small areas only. However, solarization is gaining popularity in greenhouses or plastic houses and high tunnel hoop houses for vegetable production. In Italy, soil solarization has been very effective to control *Phytophthora* blight in plastic houses (Polizzi et al. 1994).

In a 2-year field study from Turkey, soil solarization with clear plastic for 8 weeks was shown to reduce the incidence of *Phytophthora* blight and root rot of pepper comparable to the methyl bromide treatment (Yucel 1995). Soil temperature in the top 5-cm soil was recorded as 47 °C in their study. Similar soil solarization temperatures were also reported to a depth of 10 cm from a Florida study, and at that depth, the solarization treatment reduced *P. capsici* populations similar to the methyl bromide treatment (Coelho et al. 1999). Soil temperature dropped to 41 °C at 25-cm depth, and soil solarization was not effective in reducing pathogen populations at those depths (Coelho et al. 1999). In another study from Florida, soil solarization was examined in combination with soil amendment of urban plant debris (Chellemi 2006). *Phytophthora* blight was significantly less in solarized plots.

### 9.2.7 Anaerobic Soil Disinfestation

Anaerobic soil disinfestation is another bio-based and nonchemical alternative method of soil fumigation. This preplant treatment works by creating anaerobic conditions in the soil. The soil is incorporated with easily decomposable organic material such as fresh plant material, crop residue, wheat or rice bran, rice straw, and molasses and later irrigated or flooded. After treatment of soil with anaerobic soil disinfestation, it can be solarized with plastic mulch. This combination can provide very effective control of soilborne plant pathogens. Anaerobic soil disinfestation can be applied in cloudy periods or in areas where sunlight is low as it does not require high solar radiation. In that way, it has advantage over soil solarization. In a US field study, a combination of anaerobic soil disinfestation using molasses as the carbon source and solarization was suggested as an effective strategy to maintain pepper and eggplant yields in the absence of soil fumigants (Butler et al. 2014).

Soil incorporation of wheat straw followed by flooding for extended periods of time is a common practice for chili pepper production in China. A study was carried out to determine the impact of soil incorporation of wheat straw and flooding on soil properties, population of *P. capsici*, and the growth and yield of chili pepper in high tunnel fields (Gu et al. 2014). In a laboratory experiment, the effects of different rates of wheat straw on soil physical properties and *P. capsici* population were studied under different flooding times. Wheat straw and flooding significantly decreased soil EC but increased organic matter, available phosphorus, available potassium, organic acids, and phenolic acid contents. The number of *P. capsici* was reduced significantly in both flooding only and flooding plus wheat straw in comparison with the moist soil. Application of 0.25 % wheat straw with flooding for 10 or 14 days strongly inhibited the growth of *P. capsici*. However, inhibition of *P. capsici* declined with increasing rates of wheat straw. Field tests showed that soil incorporation of wheat straw at 400 kg per 667 $m^2$ followed by flooding for 20 days not only increased contents of ammonium nitrogen, available phosphorus, and available potassium but also effectively reduced disease incidence of *Phytophthora* blight and root rot on chili pepper, promoting pepper growth and enhancing yield of chili pepper by 12 %.

### 9.2.8 Reduced-Risk Chemicals

Phosphonates are systemic and very stable in plants so they required less application. Although phosphonates are categorized as natural products, they are currently not registered for organic crop production. The disease management work on sweet or bell pepper was carried out in Canada with a new liquid formulation of phosphonate. In our previous work, phosphonate provided effective control of some soilborne diseases of vegetable crops when applied as an amendment prior to planting or as a drench after planting or as seed treatment (Abbasi and Lazarovits 2005, 2006a). It also suppressed clubroot on bok choy and cabbage when applied to the infested soil before or after planting (Abbasi and Lazarovits 2006b). Phosphorous acid can be taken up by plants as phosphonate ions and can provide control of oomycete diseases of agronomical and horticultural crops; however, it is not a substitute for phosphate fertilization (Förster et al. 1998).

The effectiveness of phosphonate to control *Phytophthora* blight and root rot on sweet or bell pepper was shown in a greenhouse potting mix or sandy loam soil artificially infested with soil inoculum of *P. capsici* in growth room pot assays (Abbasi et al. 2011). Phosphonate treatments (0.05 %, 0.1 %, and 0.2 % active ingredient (a.i.)) were applied as post-planting drenches after transplanting pepper seedlings in the infested potting mix or soil. The pathogen inoculum was incorporated into potting mix or soil, and the infested mix or soil was then placed into plug trays. Six-week-old pepper seedlings "Early Calwonder" were transplanted and immediately treated with phosphonate drench solutions. Trays were kept in a growth room, and plants were rated 6 weeks later for incidence and severity of *Phytophthora* blight and root rot using a modified 1–4 scale (Silvar et al. 2005). Disease incidence was expressed as a percentage of plants showing disease symptoms.

The bell pepper plants grown in the infested potting mix or soil that received a drench application of phosphonate showed less incidence and severity of *Phytophthora* blight root rot compared to the control plants receiving no treatment (Abbasi et al. 2011). In both peat-based mix and soil, disease protection increased with the increasing phosphonate concentration. The 0.2 % phosphonate treatment provided the best disease control with more than 90 % of the plants remaining healthy after 6 weeks of treatment compared to none in the control. In the untreated infested control, all the plants were diseased, and most were dead within 2 weeks after transplanting.

Another chemical Actigard is an activator of resistance in plants against diseases, and its potential for the management of *Phytophthora* blight and root rot disease on peppers was investigated as foliar sprays (Matheron and Porchas 2002). In greenhouse pot experiments, the pepper plants were either inoculated with *P. capsici* or grown in soil naturally infested with the pathogen and sprayed with Actigard. The plants sprayed with Actigard and grown in naturally infested field soil showed better survival compared with non-treated plants. These greenhouse trials showed potential of Actigard as chemical management tool for *Phytophthora* blight and root rot on peppers, but the efficacy of this product needs to be confirmed under field conditions. Phytotoxicity to pepper plants may be another issue that needs to be addressed as well.

### 9.2.9 Grafting of Resistant Rootstocks

Grafted vegetable technology is one option which has the potential to increase resistance in the vegetables against plant diseases and, thereby, reducing pesticide usage. The use of grafted plants can dramatically increase the productivity of field-grown vegetables under conventional and organic management. This technology is currently popular in Asia but is gaining interest in Europe and North America. The high cost of grafting may be the main reason of slow adoption of this technology for broader applications. In recent years, the interest in the use of resistant rootstock particularly for grafting has significantly increased for crops such as tomato, pepper, and melon. In countries like Canada and Italy, a big number of grafted plants are used every year but mostly for greenhouse vegetable production.

Grafting is mainly used to increase host resistance against root rots and wilt diseases and to increase yield. In a study from Italy, pepper plants grafted onto partially resistant and resistant rootstocks were investigated for the management of *Phytophthora* blight and root rot disease under artificially and naturally infested field conditions (Gilardi et al. 2013). Grafting of pepper plants on resistant rootstocks provided the best control of *Phytophthora* blight and root rot under both conditions. The study also looked at improving the disease control effect of plants grafted on susceptible and partially resistant rootstocks in combination with compost soil treatment. Compost soil amendment did not show a significant effect under artificially infested conditions, but was a significant factor in reducing disease incidence and improving disease control under naturally infested field conditions. Therefore, there is always possibility to integrate compost or other soil amendments with grafting, particularly when cultivars susceptible to *Phytophthora* blight are grown or when pepper plants grafted onto partially resistant rootstocks were used.

## 9.3 Conclusions and Next Steps

Bio-based approaches including indigenous biofungicides as bioformulations; rapeseed meal, composts, and manures as soil biofumigation; grafting technology; and reduced-risk chemicals such as phosphonates and Actigard can be used as additional disease management strategies to control *Phytophthora* blight and root rot of pepper. Some fungal and bacterial antagonists isolated from pepper rhizosphere showed biocontrol potential for the management of *Phytophthora* blight and root rot disease of pepper. Soil amendments of rapeseed meal, composts, and manures also provided suppression of *Phytophthora* blight and root rot under greenhouse and field conditions possibly by altering the composition and structure of soil microbial communities and by direct toxicity to pathogen. Similarly, some combinations of soil amendments and biofungicides also showed less disease incidence on pepper plants in greenhouse pot assays and in the field. Soil solarization can be very effective in reducing pathogen inoculum from the top 10-cm layer of soil, and its effect can be enhanced with other strategies such as soil amendments. This strategy will be only effective in hot summer months. Anaerobic soil disinfestation using wheat straw as a carbon source followed by extended periods of flooding is another effective way of reducing pathogen inoculum from soil and also reducing disease incidence on plants. Anaerobic soil disinfestation can also be combined with solarization for enhanced effect. Grafted plants on resistant rootstocks also offer sustainable management solution if the associated costs could be lowered. Phosphonate post-planting treatments and Actigard foliar sprays also showed a reduction in the severity and incidence of *Phytophthora* blight and root rot of pepper under growth room conditions. Adoption of any single method for managing diseases such as *Phytophthora* blight can be a challenge. Therefore, future considerations should be focused on developing a holistic approach including several management options built in the production system to reduce overall losses from soilborne diseases such as *Phytophthora* blight and root rot.

**Acknowledgments** Technical assistance in China was provided by Anyi Hu and Yueding Xu, and the research work reported here was completed by funding from the research programs in nonprofit industries (agricultural sector) and Ministry of Agriculture of China (201103004, 201203050). Technical assistance in Canada was provided by Brian Weselowski and several summer students (Antoine Hnain, Bethany Bradford, Halema Khan, Lindsay Knezevich, Nicole Korba, Meghan Braun, and Victoria Hewelett), and the funding for research was provided by Agriculture and Agri-Food Canada.

## References

Abbasi PA, Lazarovits G (2005) Effects of AG3 phosphonate formulations on incidence and severity of *Pythium* damping-off of cucumber seedlings under growth room, microplot, and field conditions. Can J Plant Pathol 27:420–429

Abbasi PA, Lazarovits G (2006a) Seed treatment with phosphonate (AG3) suppresses Pythium damping-off of cucumber seedlings. Plant Dis 90:459–464

Abbasi PA, Lazarovits G (2006b) Effect of soil application of AG3 phosphonate on the severity of clubroot of bok choy and cabbage caused by *Plasmodiophora brassicae*. Plant Dis 90:1517–1522

Abbasi PA, Lazarovits G, Weselowski B (2011) Effectiveness of AG3 phosphonate formulation in suppressing phytophthora blight in cucumber and bell pepper plants under growth room conditions. Can J Plant Pathol 33:150–158

Butler DM, Kokalis-Burelle N, Albano JP, McCollum TG, Muramoto J, Shennan C, Rosskopf EN (2014) Anaerobic soil disinfestation (ASD) combined with soil solarization as a methyl bromide alternative: vegetable crop performance and soil nutrient dynamics. Plant Soil 378:365–381

Cao Y, Chang Z, Wang J, Ma Y, Fu G (2013) The fate of antagonistic microorganisms and antimicrobial substances during anaerobic digestion of pig and dairy manure. Bioresour Technol 136:664–671

Cao Y, Chang Z, Wang J, Ma Y, Yang H, Fu G (2014) Potential use of anaerobically digested manure slurry to suppress Phytophthora root rot of chili pepper. Sci Hortic 168:124–131

Chellemi DO (2006) Effect of urban plant debris and soil management practices on plant parasitic nematodes, Phytophthora blight and Pythium root rot of bell pepper. Crop Prot 25:1109–1116

Coelho L, Chellemi DO, Mitchell DJ (1999) Efficacy of solarization and cabbage amendment for the control of *Phytophthora* spp. in North Florida. Plant Dis 83:293–299

Colla P, Gilardi G, Gullino ML (2012) A review and critical analysis of the European situation of soilborne disease management in the vegetable sector. Phytoparasitica 40:515–523

Ezziyyani M, Requena ME, Egea-Gilabert C, Candela ME (2007) Biological control of Phytophthora root rot of pepper using *Trichoderma harzianum* and *Streptomyces rochei* in combination. J Phytopathol 155:342–349

Förster H, Adaskaveg JE, Kim DH, Stanghellini ME (1998) Effect of phosphite on tomato and pepper plants and on susceptibility of pepper to Phytophthora root and crown rot in hydroponic culture. Plant Dis 82:1165–1170

Gilardi G, Baudino M, Moizio M, Pugliese M, Garibaldi A, Gullino ML (2013) Integrated management of *Phytophthora capsici* on bell pepper by combining grafting and compost treatment. Crop Prot 53:13–19

Granke LL, Quesada-Ocampo LM, Lamour K, Hausbeck MK (2012) Advances in research on *Phytophthora capsici* on vegetable crops in the United States. Plant Dis 95:1588–1600

Gu Z, Ma Y, An X, Wang G, Sun D, Wang Q (2014) Effects of wheat straw with flooding on soil properties and Phytophthora blight control in continuous chili pepper cropping field. J Agro Environ Sci 33:1762–1769

Guest DI, Grant B (1991) The complex action of phosphonates as antifungal agents. Biol Rev 66:159–187

Hansen Z, Keinath A (2013) Increased pepper yields following incorporation of biofumigation cover crops and the effects on soilborne pathogen populations and pepper diseases. Appl Soil Ecol 63:67–77

Hausbeck MK, Lamour KH (2004) *Phytophthora capsici* on vegetable crops: research progress and management challenges. Plant Dis 88:1292–1303

Howard K, Dell B, Hardy GE (2000) Phosphite and mycorrhizal formation in seedlings of three Australian Myrtaceae. Austr J Bot 48:725–729

Jabaji-Hare SH, Kendrick WB (1987) Response of an endomycorrhizal fungus in *Allium porrum* L. to different concentrations of the systemic fungicides, metalaxyl (Ridomil) and fosetyl-Al (Aliette). Soil Biol Biochem 19:95–99

Jothi G, Pugalendhi S, Poornima K, Rajendran G (2003) Management of root-knot nematode in tomato *Lycopersicon esculentum* Mill with biogas slurry. Bioresour Technol 89:169–170

Khabbaz SE, Zhang L, Cáceres LA, Sumarah M, Wang A, Abbasi PA (2015) Characterization of antagonistic *Bacillus* and *Pseudomonas* strains for biocontrol potential and suppression of damping-off and root rot diseases. Ann Appl Biol 166:456–471

Khan J, Ooka JJ, Miller SA, Madden LV, Hoitink HAJ (2004) Systemic resistance induced by *Trichoderma hamatum* 382 in cucumber against Phytophthora crown rot and leaf blight. Plant Dis 88:280–286

Kim KD, Nemec S, Musson G (1997a) Control of Phytophthora root and crown rot of bell pepper with composts and soil amendments in the greenhouse. Appl Soil Ecol 5:169–179

Kim KD, Nemec S, Musson G (1997b) Effects of composts and soil amendments on soil microflora and Phytophthora root and crown rot of bell pepper. Crop Prot 16:165–172

Kim YC, Jung H, Kim KY, Park SK (2008) An effective biocontrol bioformulation against Phytophthora blight of pepper using growth mixtures of combined chitinolytic bacteria under different field conditions. Eur J Plant Pathol 120:373–382

Larkin RP, Griffin TS (2007) Control of soilborne potato diseases using Brassica green manures. Crop Prot 26:1067–1077

Ma Y, Chang Z, Zhao J, Zhou M (2008) Antifungal activity of *Penicillium striatisporum* Pst10 and its biocontrol effect on Phytophthora root rot of chili pepper. Biol Control 44:24–31

Matheron ME, Porchas M (2002) Suppression of Phytophthora root and crown rot on pepper plants treated with acibenzolar-*S*-methyl. Plant Dis 86:292–297

Matthiessen JN, Kirkegaard JA (2006) Biofumigation and enhanced biodegradation: opportunity and challenge in soilborne pest and disease management. Crit Rev Plant Sci 25:235–265

Nunez-Zofio M, Larregla S, Garbisu C (2011) Application of organic amendments followed by soil plastic mulching reduces the incidence of *Phytophthora capsici* in pepper crops under temperate climate. Crop Prot 30:1563–1572

Oostendorp M, Kunz W, Dietrich B, Staub T (2001) Induced disease resistance in plants by chemicals. Eur J Plant Pathol 107:19–28

Polizzi G, Agosteo GE, Cartia G (1994) Soil solarization for the control of *Phytophthora capsici* on pepper. Acta Hortic 366:331–338

Ristaino JB, Johnston SA (1999) Ecologically based approaches to management of Phytophthora blight on bell pepper. Plant Dis 83:1080–1089

Sang MK, Kim KD (2014) Biocontrol activity and root colonization by *Pseudomonas corrugata* strains CCR04 and CCR80 against Phytophthora blight of pepper. BioControl 59:437–448

Sang MK, Chun S, Kim KD (2008) Biological control of Phytophthora blight of pepper by antagonistic rhizobacteria selected from a sequential screening procedure. Biol Control 46:424–433

Sang MK, Kim JG, Kim KD (2010) Biocontrol activity and induction of systemic resistance in pepper by compost water extracts against *Phytophthora capsici*. Phytopathology 100:774–783

Sid Ahmed A, Pe'rez-Sanchez C, Egea C, Candela ME (1999) Evaluation of *Trichoderma harzianum* for controlling root rot caused by *Phytophthora capsici* in pepper plants. Plant Pathol 48:58–65

Sid Ahmed A, Pe'rez-Sanchez C, Candela ME (2000) Evaluation of induction of systemic resistance in pepper plants (*Capsicum annuum*) to *Phytophthora capsici* using *Trichoderma harzianum* and its relation with capsidiol accumulation. Eur J Plant Pathol 106:817–824

Silvar C, Díaz J, Merino F (2005) Real-time polymerase chain reaction quantification of *Phytophthora capsici* in different pepper genotypes. Phytopathology 95:1423–1429

Tenuta M, Lazarovits G (2002) Ammonia and nitrous acid from nitrogenous amendments kill the microsclerotia of *Verticillium dahliae*. Phytopathology 92:255–264

Tsao PH, Oster JJ (1981) Relation of ammonia and nitrous-acid suppression of Phytophthora in soils amended with nitrogenous organic-substances. Phytopathology 71:53–59

Wang Q, Ma Y, Wang G, Gu Z, Sun D, An X, Chang Z (2014a) Integration of biofumigation with antagonistic microorganism can control Phytophthora blight of pepper plants by regulating soil bacterial community structure. Eur J Soil Biol 61:58–67

Wang Q, Ma Y, Yang H, Chang Z (2014b) Effect of biofumigation and chemical fumigation on soil microbial community structure and control of pepper Phytophthora blight. World J Microbiol Biotechnol 30:507–518

Yucel S (1995) A study on soil solarisation and combined with fumigant application to control Phytophthora crown blight (*Phytophthora capsici* Leonian) on peppers in the East Mediterranean region of Turkey. Crop Prot 14:653–655

# Formulation of *Pochonia chlamydosporia* for Plant and Nematode Management 10

Aurelio Ciancio, Mariantonietta Colagiero, Isabella Pentimone, and Laura Cristina Rosso

## Contents




**Abstract**

Soil microorganisms play a key role in plant nutrition and health, interacting with soil pests with beneficial, antagonistic effects. The nematode parasitic and root endophytic fungus *Pochonia chlamydosporia* has been extensively studied in the last years for exploitation, due to its multiple behaviours in soil and the rhizosphere. The fungus has a complex biology and can act as a biological control agent of phytonematodes, as a plant growth promoter or as a soil saprotroph. In this review we consider several aspects concerning its production and application as a nematode and plant management tool, including biodiversity and trophic specialisation. Formulations of *P. chlamydosporia* have already reached the industrial stage. Commercial products are available for biological control of root-knot or cyst nematodes or plant growth promotion in intensive to peri-urban cropping systems. Aspects related to the fungus biology, production substrates, industrial scale-up and conservation methods are examined. Finally, potential in nematode management is discussed.



A. Ciancio (✉) • M. Colagiero • I. Pentimone • L.C. Rosso
Istituto per la Protezione Sostenibile delle Piante (IPSP), Consiglio Nazionale delle Ricerche, Via G. Amendola 122/D, 70126 Bari and URT IPSP, 86035 Larino (CB), Italy
e-mail: aurelio.ciancio@ipsp.cnr.it



N.K. Arora et al. (eds.), *Bioformulations: for Sustainable Agriculture*,
DOI 10.1007/978-81-322-2779-3_10

## 10.1 Introduction

Technical progress in soil management and plant protection relies on knowledge concerning many microorganisms, including the use and application of biological control agents (BCAs) as possible alternatives to pesticides. Biological pest management is a promising technology in crop protection since it can provide sustainable and safer alternatives to many chemicals. This approach also became necessary due to some circumstances, related to consumers' demand for organic food or to the recent withdrawal by the European Union (EU) and other agencies of many pesticides and nematicides, due to environment and human safety issues.

Although promising, an organic approach is often difficult to apply in plant protection. In fact, in spite of the number of BCAs or plant growth-promoting microorganisms (PGPMs) investigated in the last decades, and of the efforts underpinning effective application protocols, many problems/pests remain yet difficult to manage. Furthermore, basic background data on the biology, behaviour and biochemistry of useful soil BCAs are often lacking or limited to a restricted number of species.

Among nematode BCAs, *P. chlamydosporia* (Goddard) Zare and W. Gams (*Hypocreales*: *Clavicipitaceae*) has been investigated for more than three decades. It is a widespread soil fungus, acting as a facultative parasite of phytonematode eggs, often associated to many phytonematodes of economic importance. It has also been found as a parasite of animal parasitic nematodes, of other soil invertebrates like snail eggs or of oospores (Kerry 2000). *P. chlamydosporia* is a rhizosphere inhabitant and also an endophyte, capable to colonise epidermal and cortex root cells (Bordallo et al. 2002; Maciá-Vicente et al. 2009; Lopez-Llorca et al. 2010). Studies carried out in vitro or in field conditions showed that it also acts as a PGPM (Maciá-Vicente et al. 2009; Escudero and Lopez-Llorca 2012).

This species is present in two forms, *P. chlamydosporia* var. *chlamydosporia*, isolated from many regions worldwide, including Northern Europe and Mediterranean areas, and *P. chlamydosporia* var. *catenulata* (Kamyschko ex Barron and Onions) Zare and Gams, found in Latin America (Kerry 2000; Franco-Navarro et al. 2008; Manzanilla-López et al. 2013). Variability for parameters like egg parasitism or rhizosphere colonisation was experimentally shown among isolates, and a selection for parasitism capacity or other properties (i.e. soil or root colonisation or propagules produced) may be needed for practical exploitation and prior introduction in soil (Bourne et al. 1996; Kerry 2000; Franco-Navarro et al. 2008; Siddiqui et al. 2009; Vieira Dos Santos et al. 2013). The fungus has been tested as a BCA of many nematode pests, including the cereal cyst nematode (CCN) *Heterodera avenae* (Kerry et al. 1984; Kerry 2000), root-knot nematodes (RKNs) *Meloidogyne* spp. (de Leij and Kerry 1991; Verdejo-Lucas et al. 2003; Bontempo et al. 2014) and the false root-knot nematode *Nacobbus aberrans* (Flores-Camacho et al. 2007; Franco-Navarro et al. 2008).

The efficacy of *P. chlamydosporia* as a BCA of *H. avenae* was indirectly demonstrated by applying formalin (at a concentration lethal only to fungi) to soil parcels in which the fungus was present. Subsequent measures of nematode densities during the following months showed the eventual increase of the *H. avenae* population, as resulting from the elimination of the *P. chlamydosporia* antagonistic action (Kerry 2000). Applications with this fungus are at an advanced phase and in some cases already allowed an industrial production. First formulations have been made commercially available in EU (Italy, Spain), Latin America (Cuba, Brazil), Australia or Asia (India, China) (Kerry and Hidalgo-Díaz 2004; Mo et al. 2005).

## 10.2 Biology of *P. chlamydosporia*

### 10.2.1 Isolation and Detection

*P. chlamydosporia* was originally described as a member of the genus *Verticillium*, due to the presence of verticillated phialidic conidiophores each bearing at their apex one cluster of subglobose, ovoid to ellipsoidal conidia

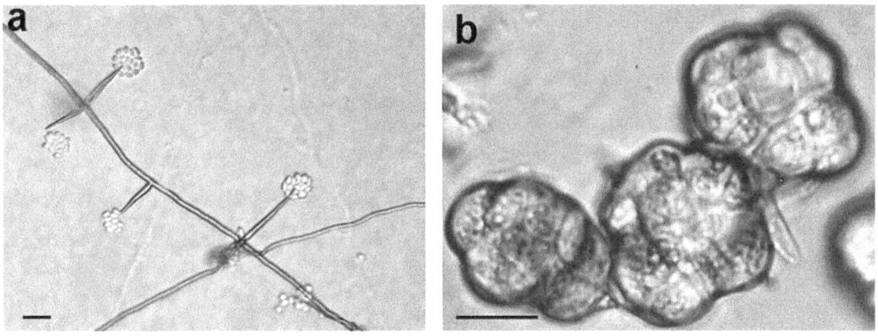


**Fig. 10.1** Hyphae of *P. chlamydosporia* var. *chlamydosporia* with conidiophores bearing apical conidial clusters, visible as adhering to the agar surface. (**a**) Chlamydospores. (**b**) *Scale bars* = 10 μm

measuring 2.5–5.5 × 1.5–3.0 μm (Fig. 10.1a). A specific character of the fungus, helpful for its identification and detection on water agar (WA, 8–10 g agar $L^{-1}$ distilled water) or other culture media, is the production of dictyochlamydospores, durable resting propagules also known as chlamydospores (Fig. 10.1b), which confer a yellowish-creamy colour to the mycelium, after prolonged culture.

The fungus taxonomy and phylogenetic position were clarified through studies based on ribosomal RNA sequences, which allowed the erection of the new genus *Pochonia* in which the species was finally placed (Zare et al. 2001). These studies showed that the new genus was phylogenetically distinct from the true *Verticillium*, which includes only plant-pathogenic lineages. Recent studies based on whole-genome sequence and transcriptome data also revealed the close proximity of *Pochonia* to the clavicipitaceous genera *Metarhizium*, which includes endophytic and insect parasitic species, and *Epichloe*, an endophyte of winter grasses (Larriba et al. 2012, 2014).

*P. chlamydosporia* can be easily isolated from hyphae emerging from nematode eggs or fragmented cysts or egg masses, as well as from root debris and soil particles, once these have been spread on the surface of a nutrient poor medium, i.e. WA, allowing a slow growth of fungi to have a clear view of the developing hyphae. A semi-selective medium was also developed to facilitate the isolation of *P. chlamydosporia* from soil or roots or for counting the number of colony-forming units (CFUs), a procedure needed when studying the fungus density and multiplication rate (de Leij and Kerry 1991).

For production of monoconidial isolates, the clusters of conidia have to be located using a stereoscope. They are visible at the apex of the conidiophores (Fig. 10.1a) formed along the hyphae that emerge from the inoculum, i.e. egg masses, soil particles or root debris. The hyphae often protrude on the WA surface, forming aerial networks or isolated hyphal bridges. After inspecting the plate and once locating the fungus, a conidial cluster can be gently intercepted with a needle holding at its tip a glued eyelash. This simple tool may be sterilised by a quick immersion in a few ml of 95 % ethanol and air dried for a few seconds before use. Alternatively, a tungsten wire (obtained from a broken bulb) can be fused on the glass stirred tip of a Pasteur pipette and then sterilised by heating on a flame before use. It is necessary to rapidly cool the wire by a quick immersion in a few ml of 95 % ethanol or sterile distilled water (SDW) and let it become dry and cool for about 10 s, before aseptically touching the conidial cluster. The intercepted conidia adhering to the sterile eyelash or wire tip can be rapidly smeared and dispersed on a suitable culture medium (usually potato dextrose agar (PDA) or cornmeal agar (CMA)) on which they will germinate, originating new colonies in the following 3–4 days once stored at 25–26 °C. At this moment the small colonies can be observed on the Petri dish using a stereoscope. To produce a monoconidial isolate, a marginal fragment from each of the smallest, isolated colonies appearing on the medium can be removed with a sterile needle tip, to be transferred on a new plate with a suitable growth medium. The colonies can be identified at this moment as a new entry in the collection by assigning them a unique reference code, adding related data and other information to the collection database (collector's name, date, original sample location, substrate type, etc.).

### 10.2.2 Metabolism

Several factors like soil nutrients and root exudates affect *P. chlamydosporia* growth and behaviour. Like other parasitic fungi, the excess of compounds rich in C may affect production of key enzymes involved in nematode infection, whereas a high organic matter content of soil may stimulate growth, without any increase in biological control efficacy (Kerry and Bourne 1996). Carbon (C) sources readily available for metabolism and unfavourable pH in the rhizosphere or nematode egg mass may compromise nematode parasitism (Ward et al. 2012).

Studies on parasitic fungi showed that the chemical composition of the host surface is a key inducer for production of specific hydrolytic enzymes during infection by nematode-trapping fungi (Tunlid and Jansson 1991). In the *P. chlamydosporia*–nematode interaction, the molecular response of the fungus involves many genes, including an alkaline serine protease (VCP1), an enzyme specifically degrading proteins from the outer vitelline layers of the egg. This single-copy gene is expressed early during infection (Segers et al. 1996; Morton et al. 2003) and shows single nucleotidic polymorphisms associated to different host nematode species that reflect a long-term, genetic adaptive process that occurred on evolutionary scales (Morton et al. 2003). VCP1 induction is one of the first steps in a complex cascade of regulated genes leading to egg parasitism (Lopez-Llorca et al. 2010; Ward et al. 2012).

Given the sensitivity of the fungus to external biochemical signals affecting its behaviour, it is important to know the nature and effect of these stimuli to apply the mycelium or propagules in conditions that are most suitable for nematode biological control. Two isolates of *P. chlamydosporia* with different host origins, IMI 380407 from RKN and IMI 331547 from the potato cyst nematode (PCN) *Globodera pallida*, were assayed in their saprotrophic-to-parasitic transition, with varying nutritional conditions and sampling times. Treatments included nutrient-rich or nutrient-poor media or a poor medium with PCN or RKN eggs. Expression studies showed that the transcript-derived fragments (TDFs) obtained were affected by different nutritional conditions (starvation in the presence/absence of eggs or nutrient-rich medium). The changes in gene regulation reflected specific systems or functional links, with differences in TDF levels between pathogenic and saprophytic growth. The genes expressed during parasitism for both isolates were involved in cellular signal regulation and transport, regulation of other genes, DNA repair and other functions. Multivariate analysis showed co-expression of genes associated to egg parasitism, together with constitutive genes, related to general metabolism. The differential expression of parasitism-related genes suggested a network of induced/repressed products, playing a role in fungal signalling and infection (Rosso et al. 2011).

Clusters identified by a multivariate analysis in *P. chlamydosporia* IMI 331547 included a phytase expressed in starvation and early after contact with RKN nematode eggs, not linked to other metabolic pathways, and probably accounting for a distinct activity. Parasitism-related TDFs also included a phospholipase D1, expressed in rich nutritional conditions and early after contact with eggs, and a monooxygenase expressed after contact with eggs. Starvation and presence of nematode eggs also induced a bZIP transcription factor not related to parasitism (Rosso et al. 2011).

Assays on healthy or RKN-parasitised plants also showed the induction, 4 and 8 h after incubation of *P. chlamydosporia* with eggs, of the bZIP transcription factor and the phytase-related gene. Mycelial growth stimulation was observed in the presence of 5–10 ppm of phytic acid, suggesting that the fungus can use this molecule as a P source. Nematode-free tomato plants alone induced, at 4 h incubation, a phytase transcription sixfold higher than control. Changes in transcription were not significant when the fungus was in the presence of nematode-parasitised plants. The expression of a phospholipase D was, after 8 h incubation in the presence of nematode eggs, 900-fold higher than in fungus alone. Further assays performed at different pHs or in the presence of glucose and $NH_4^+$ showed that the bZIP and phytase transcripts reflect early changes in the fungus metabolism (Rosso et al. 2014).

Parasitism-induced bZIP transcription factors are dimers that regulate genes active in early infection events by recognising DNA palindrome sequences (i.e. CREB, ATF2 or CCAAT/enhancer-binding protein (C/EBP)) (Thiel et al. 2005). These proteins transactivate genes containing cAMP response element (CRE) motifs in their regulatory regions, connecting cellular stimulation with transcription of CRE-containing genes (Bakker and Parker 1991). Studies on *Metarhizium anisopliae* showed that cAMP is involved in the formation of the appressorium, together with $Ca^{+2}$ (Clarkson and Charnley 1997). St Leger et al. (1991) proposed that hyphal differentiation begins after a change in the membrane potential is induced by a contact, possibly through a mechanosensitive ion channel resulting in a disruption of the apical $Ca^{+2}$ gradient required for polar growth. Elevated levels of cAMP coincided with the formation of the appressorium and of a protein kinase identified in the plasma membrane of *Metarhizium anisopliae*, whose substrate could be a CREB factor promoting genes involved in the appressorium formation. In *M. anisopliae* this is a PR1 serine protease, whose level increased tenfold within 24 h of contact with the insect cuticle (Clarkson and Charnley 1997). The bZIP transcript might hence be involved in a similar mechanism in *P. chlamydosporia*, including the activation of specific egg-degrading enzymes (Ward et al. 2012). In line with this possibility, the bZIP expression is reduced in the presence of D-(+)-glucose and enhanced by $NH_4^+$ (Rosso et al. 2014).

Nematode parasitism and P mobilisation open intriguing scenarios for practical exploitation of *P. chlamydosporia*, suggesting that multiple potential benefits can be derived by this species. Recently, *P. chlamydosporia* genes expressed when endophytic on barley showed production of several proteases, hydrolases and other secreted proteins involved in transport during root colonisation (Larriba et al. 2014). Observations suggest a long-term, stable evolutive adaptation of *P. chlamydosporia* to the soil and rhizosphere environments, characterised by specific interactions with the host plant and complex relationship involving contacts with roots and/or nematode eggs.

The phylogenetic proximity of *P. chlamydosporia* to *Metarhizium* spp. also suggests possible ecological and metabolic similarities. Studies on *M. robertsii* revealed that these entomopathogens have a complex endophytic relationship with their host plant, involving the return to the colonised roots of the N metabolites acquired by the insects when feeding on them. This task is achieved by means of a translocation carried out through the *Metarhizium* endophytic phase, once the fungus parasitises the insects' larvae present in the rhizosphere (Behie et al. 2012). We hypothesise that a similar behaviour might be present in *P. chlamydosporia* when in the presence of (or in contact with) phytoparasitic nematodes, with a direct role in plant nutrition, a part of its direct biological control capacities. If further data will support this hypothesis, then parasitism must be considered as part of a more complex lifestyle, unifying the tri-trophic root, nematode and fungus interactions in a nutrient-based, cyclic relationship sustaining the plant growth.

Considering the feasibility of massal production of *P. chlamydosporia* and that chlamydospores and other propagules may facilitate conservation and dispersal in soil, the fungus holds potential both as a BCA and PGPM. However, for optimal production in fermenters or other solid substrates, data on metabolic preferences of isolates must be produced, since differences may occur for some basic parameters, i.e. growth rates (Fig. 10.2a).

The capability of isolates to metabolise sugars and C sources commonly used in substrates may

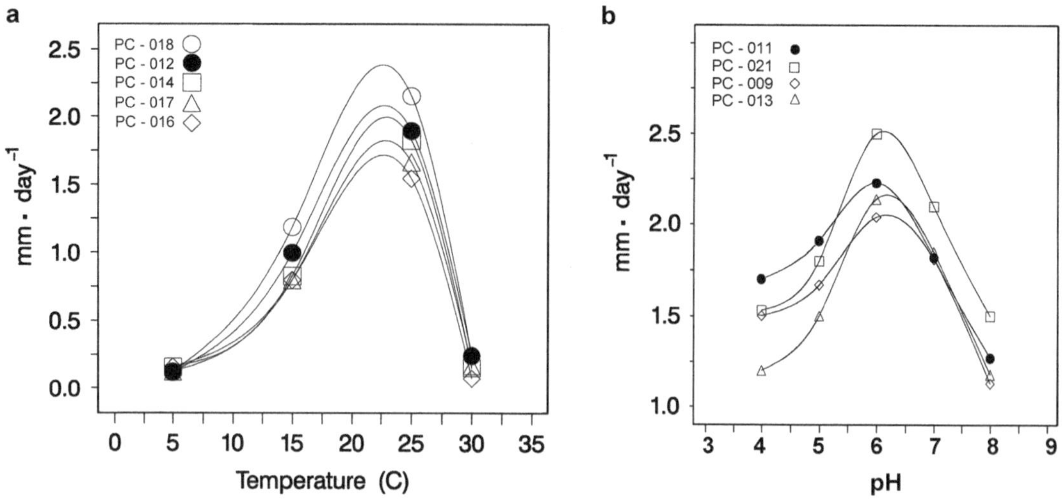


**Fig. 10.2** Effect of (**a**) temperature and (**b**) pH on radial growth on cornmeal agar of some isolates of *P. chlamydosporia* var. *chlamydosporia* collected in Italy

yield variations that can affect production. In *P. chlamydosporia* fermentation assays with isolate collected in China and carried out with different C sources, highest conidial production was observed at a C/N ratio of 10:1 (pH 3.7), whereas maximum biomass production was obtained with a C/N ratio of 40:1 (pH 6.8). When testing the fungus growth, sporulation of *P. chlamydosporia* HSY-12-14 was highest on media with 6 g $l^{-1}$ C (C/N = 40:1), or 8 g·$l^{-1}$ C (C/N = 20:1 or 40:1) for conventional production on a single medium, and with 8 g·$l^{-1}$ C (C/N = 10:1) when using two different sequential media (Gao and Liu 2010). Initial pH also affects mycelial growth and conidial production, and optimal values reported were 3.5–4.5 for sporulation and 5–6 for growth. Maximum conidial production was obtained with initial pH 4.0 whereas the maximum biomass was achieved at pH 6.0 (Mo et al. 2005). The latter was also the preferential pH for optimal radial growth observed for some isolates collected in Italy (Fig. 10.2b).

C and N sources affect *P. chlamydosporia* metabolism, with consequences on its growth rate and conidiation. Among the 18 different N sources tested on an Arkansas isolate, casein showed best growth rates on Blackburn and Hayes (1966) solid medium, whereas L-aspartic acid and DL-glutamic acid showed highest conidiation. In liquid cultures, peptone showed highest growth, whereas ammonium nitrate reduced growth and conidiation in solid medium, with minimal growth of *P. chlamydosporia* also in liquid medium (Liu and Chen 2003). Among C sources, D-(−)-fructose, D-(+)-galactose, melibiose and D-(−)-ribose showed highest growth rates and conidiation of *P. chlamydosporia*, followed by D-(+)-glucose, D-(+)-xylose, D-mannitol, sucrose, D-(+)-trehalose and maltose. Among nine vitamins tested, folic acid showed highest growth rate and myo-inositol the optimal conidiation (Liu and Chen 2003).

Like many other groups of invertebrate parasitic fungi, *Pochonia* spp. also are characterised by the production of biologically active secondary metabolites. These include pochonins, a group of antiviral and antiparasitic compounds also known as monordens or resorcylic acid lactones (also referred to as radicicol). They were exclusively produced in submerged cultures by *P. chlamydosporia* and some other species of the genus (Stadler et al. 2003). Isolate P0297 of *P. chlamydosporia* var. *catenulata* produced monorden in methyl-alpha-D-glucopyranoside (MGP) medium, with yields of 15–20 mg·$L^{-1}$ (Hellwig et al. 2003). Further secondary metabolites of members of the genus include a phomalactone with nematicidal properties (Khambay et al. 2000), aurovertin B and citreoviridin A produced by *Pochonia suchlasporia* var. *catenata* (Gams and Dackmann) Zare and Gams (Stadler et al. 2003) and four aurovertins produced by *P. chlamydosporia* (Niu et al. 2010). Isolate TAMA 87 of *P. suchlasporia* var. *suchlasporia* also produces pochonicine, a pyrrolizidine alkaloid acting as a potent inhibitor of beta-N-acetylglucosaminidase, a chitinolytic enzyme involved in degradation of oligomers produced by chitinase, with potential in defence from insects or other fungi (Usuki et al. 2009).

## 10.3 Biological Control and Management

### 10.3.1 Soil Biodiversity and Crop Production

Thanks to their richness in microbial species, soil habitats sustain a web of complex trophic networks (Sleator et al. 2008). However, the effects of this complexity are often neglected, in spite of the many factors and benefits present in the rhizosphere. The positive interaction between soil microbial species and plants is indeed a fundamental component of soil fertility underpinning crop productivity. In spite of the advancements achieved in the last century through the use of chemical fertilizers and pesticides, food production is still dependent on the soil microbiota. Furthermore, several elements that play a fundamental role in sustaining plant nutrition will face, in this century, a worrying shortage, and first of all will be the phosphate sources. It is expected that solutions to emerging problems in plant nutrition and soil pest management will possibly arise by the broad range of soil microbial species and/or by enhancing the activity of beneficial microorganisms selected and applied for this purpose (Altieri 1999; Avis et al. 2008; Adesemoye et al. 2009; Pareg and McMillan 2015). This assumption further underlines the actual need for producing detailed information, either on the density regulation mechanisms active in the rhizosphere or on the biology, ecology and biochemistry of the species involved in these interactions.

DNA-based studies on soil microbial diversity showed that it is extremely rich in species and that their numbers can even appear underestimated, due to many taxa not yet described or difficult to culture (Fierer at al. 2012; Hirsch and Mauchline 2012). Recent advances in soil metagenomic studies showed a range of operational taxonomic units (OTUs) varying in the order of $10^3$ to $10^4 \cdot g^{-1}$ of soil (Roesch et al. 2007; Lauber et al. 2009; Frisli et al. 2013). A lower level of phylogenetic diversity, however, is found in soil when compared to other environments, i.e. marine or sediment habitats (Lozupone and Knight 2008). In any case, due to its biodiversity, soil is a rich container for a wide range of microbial by-products and metabolites, including enzymes, antibiotics or siderophores (Martínez-Viveros et al. 2010; Kumar et al. 2012). All these components play a role in the rhizosphere and affect the success of practices aiming at the introduction and establishment of fungi like *P. chlamydosporia* or other nematophagous species. However, soils often show a buffering capacity as concerns their microbial constituents. A mycostasis effect has been often invoked to explain failures in establishing a nematophagous species or the reduced efficacy of selected isolates, in comparison to the promising results found in greenhouse or semi-controlled assays (Lopez-Llorca et al. 2008).

There are several convergent observations on plant growth promotion effects exerted by *P. chlamydosporia*. This fungus can survive in soil in absence of nematodes as a saprotroph for several months. It is also an efficient coloniser of the root surface (Fig. 10.3) and a true root endophyte (Lopez-Llorca et al. 2002; Maciá-Vicente et al. 2009; Manzanilla-Lopez et al. 2011, 2013). When endophytic, either direct or indirect effects may underpin growth stimulation, including the early induction of a wide range of plant defence and resistance genes (Rosso et al. 2013).

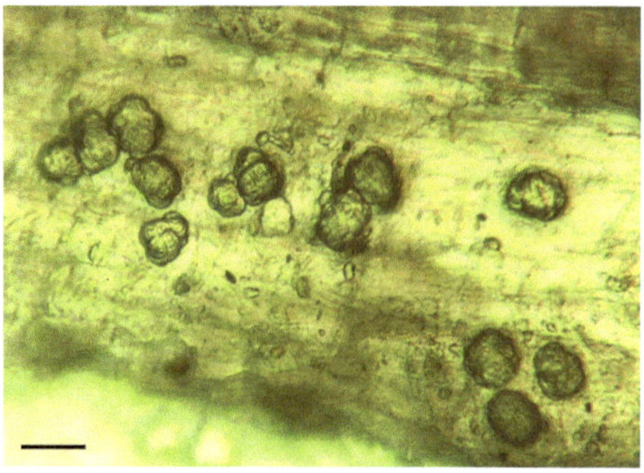

**Fig. 10.3** Chlamydospores of *P. chlamydosporia* var. *chlamydosporia* colonising a tomato root. *Scale bar* = 25 μm

### 10.3.2 Plant Protection and Nematode Management

For decades, since the late 1940s, phytonematodes have been managed by means of chemical treatments based on a variety of nematicides like chlororganics, carbamates, fosforganics, halogenated hydrocarbons, methyl bromide or chloropicrin. These tools, although effective in controlling nematode pests in the short term, often showed a series of negative feedbacks, the most common of which involved the density changes, if not the extinction, of one or more microbial soil components, responsible for root nutrition and/or natural pest regulation. The effects of 2,4-D-(1,3-dichloropropene) in decreasing the numbers of nitrifying bacteria has been documented at this regard (Elliot et al. 1977), as also the mass extinction of any living organism in soil, including beneficial species, due to methyl bromide.

Microbial predators and parasites have been applied since the early 1970s in attempts aiming at safer and effective methods relying on biological control of nematodes. However, soil environment is highly variable and complex, and the reproducibility of results, as usual in agriculture, has been difficult to achieve. This effect is mainly due to the occurrence of several, distinct factors affecting the success (or failure) of the strategies applied. Apart from mycostasis, an important variable involves the microorganism and isolate applied. For *P. chlamydosporia*, a scrutiny of published data shown in the literature for nematode management shows different levels of biological control efficacy, related to the isolate used, the target host species or the application method and dosage (Table 10.1).

*P. chlamydosporia* is often associated to parasitise nematode eggs, from which it can be easily isolated. The fungus also colonises the egg masses of RKNs as well as the cysts and egg sacs of PCN, CCN and other cyst nematodes. The egg parasitic activity starts through hyphal penetration and eventual digestion of the eggshell, occurring early after contact of the hyphae with the egg surface. The contact eventually originates fungal appressoria allowing the penetration of the vitelline layers and the formation of an internal swollen bulb, from which new invasive hyphae are formed. The egg invasion step is biochemically very active and characterised by the induction and production of specific enzymes, i.e. chitinases and VCP1, specifically involved in the process (Lopez-Llorca et al. 2008). The invasion of the embryo then follows that is rapidly lysed and destroyed by the developing mycelium (Fig. 10.4).

*P. chlamydosporia* has also been applied to control animal parasitic nematodes (Ferreira et al. 2011; Silva et al. 2011) or other plant-pathogenic fungi. Together with *Pochonia rubescens* and *Lecanicillium lecanii*, isolate 4624 was capable to reduce root colonisation and damage caused by the fungus *Gaeumannomyces graminis* var. *tritici*, a severe pathogen of wheat, barley and other crops and causal agent of take-all disease (Monfort et al. 2005).

Apart from nematode eggs, there are also reports of parasitic activity on adult females, in particular for CCN and other cyst nematodes whose early infection was found to be correlated to higher levels of biological control (de Leij and Kerry 1991; Kerry 2000). Occasionally, parasitism of juveniles may be observed, in relation to the invasion of eggs that are close to hatching.

Density-dependent mechanisms may explain nematode regulation by the soil microflora or by specific antagonists and provide a basic perspective when evaluating BCAs like

**Table 10.1** Nematode regulation effects observed for various treatments with different isolates of *Pochonia chlamydosporia* and target nematodes

| Treatments | Target nematode | Efficacy[a] | References |
|---|---|---|---|
| *P. chlamydosporia* | *Heterodera avenae* | 24–98 (E) | Kerry et al. (1984) |
| *P. chlamydosporia* CMI 334168 | *Meloidogyne incognita* | 68–72 (J + E) | de Leij et al. (1992a) |
| | | 81 (F) | |
| *P. chlamydosporia* IMI 331547 | *Meloidogyne incognita* | 27–61 (E) | de Leij et al. (1992b) |
| | | 13–45 (J) | |
| *P. chlamydosporia* | *Meloidogyne incognita* | 37–44 (E) | Kerry and Bourne (1996) |
| | | 81 (F) | |
| *P. chlamydosporia* with *Hirsutella rhossiliensis* | *Meloidogyne hapla* | 74 (G) | Chen et al. (1999) |
| *P. chlamydosporia* CMI 334168 | *Meloidogyne hapla* | 29–57 (J2) | Viaene and Abawi 2000 |
| | | 52–71 (E) | |
| *P. chlamydosporia* var. *catenulata* Res 392 | *Meloidogyne incognita* | 17 (J2) | Atkins et al. (2003) |
| *P. chlamydosporia* IMI 331547 with oxamyl | *Meloidogyne javanica* | 20–25 (G) | Verdejo-Lucas et al. (2003) |
| | | 10–16 (E) | |
| *P. chlamydosporia* with Surinam grass | *Meloidogyne javanica* | 28 (G) | Dallemole-Giaretta et al. (2011) |
| *P. chlamydosporia* | *Globodera rostochiensis* | 65 (E) | Muthulakshmi et al. (2012) |
| | | 53 (F) | |

Target stages: *E* eggs, *F* females, *J2* juveniles, *G* gall index
[a]Impact of treatments expressed as percentage of corresponding controls.

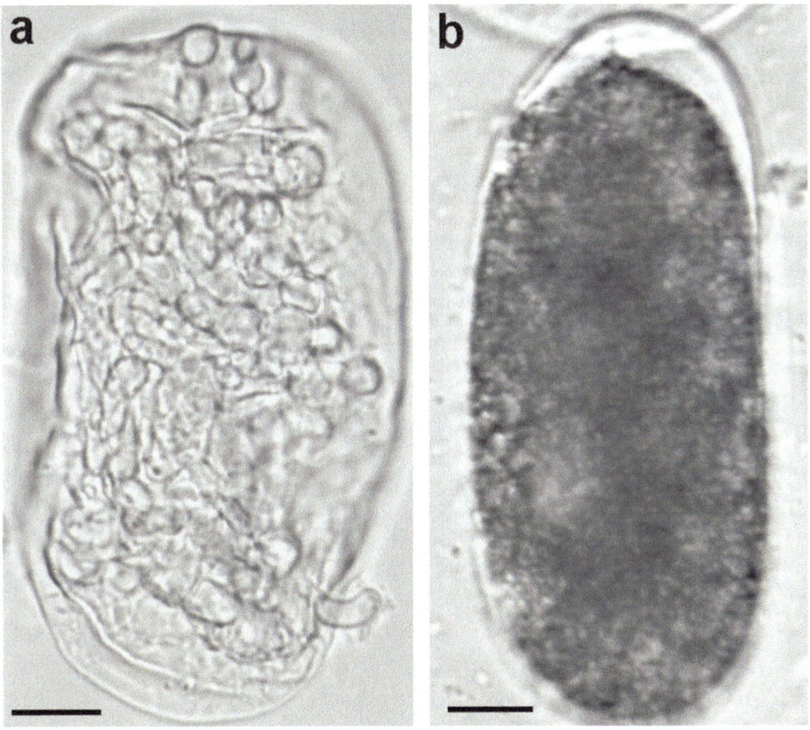


**Fig. 10.4** Hyphae of *P. chlamydosporia* var. *chlamydosporia* within a (**a**) RKN-parasitised egg and (**b**) healthy egg with developing embryo. *Scale bars* = 10 μm

*P. chlamydosporia* (Kerry 2000). There are several reports confirming the beneficial effects of treatments with *P. chlamydosporia* applied in field conditions. Replicated data reinforce the observations carried out in controlled trials and the beneficial effects of the fungus observed in Northern Europe, in CCN-infested cereal fields and in Mediterranean crops mainly for RKNs (Kerry 2000). However, no antagonist may produce a mass extinction of its host in a short term, unless its density is artificially increased through inundative treatments, the host and parasite coexistence being the most common rule. Also, depending on the soil structure and texture, varying saturation thresholds may be imposed to the fungus by self-density dependence and other physical factors that may limit the total number of propagules and active mycelial parts that can be present and active in the rhizosphere. After this threshold is achieved, it is non-useful to increase the density of propagules, since the biomass introduced will be lost, becoming food for other soil organisms or undergoing other processes determining a density decline. The extent of the microbial competition in the rhizosphere is a factor that must be evaluated, as well as the marginal yield increase or benefits achievable by the application, that will start to decline after a given density threshold has been reached.

In general, users must be aware that the nematode population changes need a given time period before any effect of *P. chlamydosporia* can be scored. In any case, it is almost impossible to eradicate a nematode after its spread in a cultivated field, in particular for RKNs that are polyphagous and tend to remain in infestation foci all around the cultivated parcels, on weeds or volunteer plants. Differing from chemicals, BCAs are living forms subject to their own limiting factors and as sensitive to soil conditions as many other soil microorganisms. Their use should then aim at keeping the residual nematode population far below its damaging threshold density, for a time period covering the different plant phenologic stages.

Combined use of *P. chlamydosporia* with chemicals and other microorganisms has also been tested. When applied on potato at 10 kg·ha$^{-1}$ for PCN management with *Pseudomonas fluorescens* and *Trichoderma viride*, with or without carbofuran, an Indian isolate of the fungus showed a 70 % tuber weight increase and a cyst population decrease by around 72 % (Muthulakshmi et al. 2012). In a 2-year study with alternated tomato and lettuce crops infested by *Meloidogyne javanica*, *P. chlamydosporia* IMI 331547 was used alone or in combination with oxamyl. The fungus, applied at a rate of $1.3–2.4 \cdot 10^6$ chlamydospores per plant, was re-isolated from parasitised nematode eggs up to 9 months after application to crops. Treatments with oxamyl showed the highest reduction of galls and egg densities when compared to the fungus alone or untreated controls (Table 10.1). Results showed, however, low rates of egg parasitism and variability between the two experimental sites investigated (Verdejo-Lucas et al. 2003).

Growth promotion effects have been reported either in pot studies or field conditions, with or without nematodes. In a pot study, increased shoot and root weights were observed in barley (Maciá-Vicente et al. 2009). In field conditions in Southern Italy, an increase of around 30 % of yield, achieved through a higher fruit size, was found for zucchini plants treated in a greenhouse with a liquid commercial product based on isolate DSM 26985, given at transplant and at 15- or 30-day intervals (Pietrantonio et al. 2013).

## 10.4 Industrial Production, Quality Control and Storage

Essential components for biological control programmes are the quality and availability of adequate amounts of inoculum. These may be obtained through primary mass production systems that should be compatible with low industrial production costs for subsequent commercial development and field application. The production of experimental formulations may be carried out starting from inoculum mainly produced in flasks for work in field assays (Fig. 10.5). A larger-scale production system has to be used instead to start an industrial

**Fig. 10.5** Small-scale production of mycelium and propagules of *P. chlamydosporia* var. *chlamydosporia* on a sand–maize mix in flasks, for subsequent use in an experimental biological control assay of RKN

production cycle. In the latter case, more intensive and economically viable methods must be followed to achieve market sales. The production processes for anamorphic fungi commonly found to be practical on a pilot scale is the diphasic liquid–solid fermentation technique, in which the fungus is first grown in liquid media and then added to a solid substrate for conidiation (Lomer et al. 1997).

### 10.4.1 Liquid Culture

In mass production cycles, a liquid starting culture is useful to achieve a rapid mycelial growth that will produce the inoculum for the subsequent solid-phase cycle. The advantages of this procedure are (a) a fast-growing and therefore less expensive production method; (b) enhanced competitiveness due to the possibility of controlling the fungus, reducing the risks that contaminating microorganisms reach the solid substrate; (c) an easier quality check of the liquid culture when testing for contaminations originating from the slope culture; (d) a more rapid colonisation of the solid substrate when the actively growing liquid culture is distributed on it; and (e) an even coverage of the substrate, resulting in homogeneous growth and maximum conidiation.

The industrial production of mycelium often starts in liquid culture in large fermentation vessels which have electronic controls and monitoring. However, aerial conidia are not produced in liquid but require an air-substrate interface on which the propagules can be formed. For *P. chlamydosporia* the chlamydospores are difficult to produce in liquid, but can be obtained from a solid-phase substrate inoculated with conidia or chlamydospores (Montes de Oca 2004). However, a patent (US-20080213867-A1) claimed a method for liquid fermentation in a medium including triglycerides and other fats and C sources that allows production of more than $10^6$ chlamydospores $ml^{-1}$.

It is important to check if, in the early growth phases, distinct clumps of mycelium are formed within an otherwise clear broth. This form of growth is less productive than a homogenous broth colonisation obtained by evenly dispersed mycelial and hyphal fragments. To encourage even distribution of mycelium and hyphal fragments in the liquid broth, the temperature should be tightly controlled and a spore suspension should be used to inoculate the flasks/containers in preference to agar blocks. Fermentation systems could be set up when a large quantity of liquid inoculum is required and a filtered aeration unit (air pump or compressor) is important for optimal production. A simple system would consist of a single unit capable of running multiplex fermenter vessels at any one time (Taborsky 1992).

### 10.4.2 Solid Substrates

For mass production a solid phase may provide a physical support for the fungus to produce aerial conidia or chlamydospores (the propagules which are best suited for storage and formulation in oil). Usually, the substrate is a cereal or a mix with a cereal by-product such as rice, millet and maize or wheat bran. Since these are natural products, their nutrient status and microbial components are undefined. The fungus will use a certain proportion of the nutrients supplied by these products during growth and conidiation. In some ways, the structure of the substrates is more important than the nutrients they supply. An ideal substrate will provide a high surface-to-volume ratio, with individual particles kept separated to provide interparticle spaces for aeration and formation of conidia and other propagules. For this reason, broken sterilised cereals are often the preferred substrate, providing a large conidiation surface. If prepared carefully, the grain fragments remain separated from each other after autoclaving and inoculation (Hadapad and Zebitz 2006).

For chlamydospore mass productions, several media can be used, based on barley, rice, wheat or corn. A production medium for *P. chlamydosporia* can be obtained by mixing milled barley with fine sand in 1:1 proportion, followed by autoclaving and drying to the appropriate moisture level (Kerry and Bourne 2002). After inoculation, the fermenter bags/vessels are sealed and turned several times to distribute the inoculum evenly and break up any lumps. Fermenter bags/vessels are inspected for contamination and turned for at least two times during the incubation period to increase aeration and break up any clump eventually formed during the incubation process. The effect on final yields of turning/mixing the substrate during incubation should be evaluated. Good records should be maintained regarding the time at which bags are manipulated during the production process (Jenkins et al. 1998).

Whatever formulation is required, the product must first be dried to ensure that the propagules remain viable during storage. This process can be speeded up by a drier or an air-conditioning system in the incubation room or by accelerating ventilation with a fan. This will depend on the local environmental conditions, being careful to operate the dryer in a cool room to prevent overheating of the conidia during the process. Conidiation typically takes place after approximately 10 days, and for *P. chlamydosporia* on solid substrate, the medium must be kept for 1 month before chlamydospore extraction. Once the conidiation process is complete and spores have been formed over the whole surface of the solid substrate, the next step in the process will depend largely on the intended use of the end product. At this point it is important to ensure that the method of extraction is compatible with the formulation. If, for example, the conidia are to be formulated in oil for ultralow-volume (spraying) application, a very fine powder with uniform particle size is required. Therefore, the extraction process must be either efficient or selective for conidia that have to be separated from the cereal dust. In case the intended use is for direct application to soil, the conidia may be left on the substrate and the whole product of the fermentation may be used for direct application.

For chlamydospore extraction, the colonised medium can be homogenised and washed in the feeding tank of an appropriate separation apparatus, and the spores can be then stored at 4 °C. For conidia-producing fungi, when a minimal moisture content (5 %) is required, the conidia powder can be extracted by filtering or other kinds of devices and packaged in moisture-proof packaging such as aluminium laminate sachets. In this way, conidia have a long shelf life and do not require storage below 0 °C. For *P. chlamydosporia* a formulation of pellets can be produced using a mycelium suspension mixed with Na alginate and diatomite clay at a 10 % (w/v) final concentration (Duan et al. 2008; Walker and Connick 1983).

### 10.4.3 Culture Maintenance

Effective maintenance of stock cultures is essential for quality control, validation of the production methodology applied and research purposes. Maintaining backup sources of inoculum as a master culture is essential for many reasons, since the isolate original culture may be required for (a) research purposes, (b) preparation of inoculated samples and specimens for quality control and training purposes, (c) conservation of reference strains for development and validation of any new methods or (d) production of cultures for microbiological assays. There are a large number of long-term culture maintenance techniques that can be employed for the preservation of valuable fungal material. Freeze-drying lyophilisation is the most satisfactory method since it is a form of suspended metabolism. A simple and reliable method (used for entomopathogenic fungi) consists in drying a large number of infected insect cadavers prior to re-isolation of the fungus of interest. The sporulated cadavers should be dried over silica gel to a constant weight and stored in a freezer in a sealed plastic bag, always with silica gel. This material is likely to remain viable for many years and single insects can be removed when a fresh source of inoculum is required (Grivell and Jackson 1969). However, the equivalent for *P. chlamydosporia* would be the storage of RKN egg masses or cysts colonised by the fungus, a method that may appear unsatisfactory given the small dimensions of nematodes and, considering their habitat, the possibility of external contaminations by other soil microorganisms.

Sterile soil has found wide use for stock culture maintenance of spore-forming microorganisms and appears particularly suitable for preservation of fungi. For storage of *P. chlamydosporia* at room temperature, sterilised soil can be used (Green 2008). The fungus can also be stored by pouring a spore suspension in sterile distilled water (SDW) on silica gel particles, later dried and stored at 80 °C.

A simpler method based on SDW, originally described by A. Castellani in 1939 and applied to many fungal groups (Burdsall and Dorworth 1994; Nakasone et al. 2004; Diogo et al. 2005), is also suitable for *P. chlamydosporia*. It has been successfully applied in our laboratory also for *Hirsutella* spp. and other soil fungi. The method is based on agar discs that are aseptically removed from a culture dish using the large end of a Pasteur pipette, flame sterilised. The plugs are then packed in cryo vials for prolonged storage at room temperature, tightly capped to avoid mite contaminations. The vials allow the fungi to be recovered after one or more years. In alternative, a spore suspension in SDW can be used.

### 10.4.4 Quality Control (QC)

Quality control (QC) should be carried out either during the production process or on the resulting product at the end of the multiplication cycle. Both checks are essential for successful production of high-quality and viable propagules that have to be free of potentially dangerous contaminants. One important objective of the QC procedures carried out during production and related inspection and monitoring is to avoid entrance of contaminants into the system. QC protocols must be incorporated in the routine of the production cycle in order to identify the contaminants and their source, when detected. QC carried out on the final product is designed to ensure that it meets predetermined standards, including parameters like propagule and mycelium viability, virulence and other characters, i.e. colour, moisture content (for long shelf life) as well as those directly related to the cycle efficiency, like the number of propagules $g^{-1}$ of formulation or substrate.

### 10.4.5 In-Process QC

QC procedures should also be used to monitor/exclude the presence of contaminants entering the system during the downstream processing and/or formulation phases and to provide evidence, through the corresponding analysis reports that they did not occur nor multiply during the fermentation process. Standard QC procedures include measures for detecting large-scale problems such as sterilisation failure and the introduction of contaminants due to in-process handling of the solid substrate. It is important to ensure that contamination is detected as early as possible and that its origin is identified so that subsequent problems can be avoided. A formal set of QC checks must be incorporated into the production process. During the incubation and manipulation phases, any contamination present in the fermenter must be rapidly identified, and contaminated batches and containers must be discarded (Jenkins et al. 1998).

Analyses include monitoring of the conidial inoculum purity, screening of the liquid media or autoclaved cereal substrates for sterility prior to inoculation and purity checks on liquid or solid substrates postinoculation. Liquid cultures can be checked by aseptically taking a small sample for examination under the microscope for visible signs of contamination. The sterility of the solid substrate can be verified using un-inoculated bags as controls for each autoclave and batch. Usually sterility screenings take place by plating on generalist media. Agar plate cultures are checked visually for the presence of contaminants before use as inoculum for liquid cultures. For propagule yields, although counting *P. chlamydosporia* CFUs on its selective culture medium may be possible, this approach may show some difficulties, i.e. species-/strain-specific differences that should be known. Furthermore, the procedure is extremely time-consuming, and the results are not immediately available.

Accurate identification and quantification protocols are the starting point to develop standards for biological purity, including assessment of the production process to establish that harmful microorganisms or other *P. chlamydosporia* isolates with distinct characteristics, proceeding from other production cycles, are unlikely to be present in the final product. DNA-based technologies involving the use of the polymerase chain reaction (PCR) allow the development of detection techniques that are independent from the culture properties and/or density of target fungi. PCR-based detection assays have been described and applied to a wide range of fungi, including some *P. chlamydosporia* isolates (Hirsch et al. 2001; Ciancio et al. 2005; Atkins et al. 2005, 2009; Cordier et al. 2007; Rosso et al. 2007). To check for bacterial contaminants, PCR with universal 18S bacterial primers like 27F-1492R may be used.

However, for routine use of PCR methods with the fungus isolate, due to the reaction sensitivity and specificity, appropriate selective target regions and primer sets unequivocally assigning the amplicon to the specific isolate produced are needed. Data also must be known on best reaction conditions. Variability is known in *P. chlamydosporia* as concerns specific genes that can be used for detection. Single nucleotidic polymorphisms (SNPs) were found in VCP1. This enzyme is active during the digestion of the egg vitelline and chitinous layers and its sequence is therefore a useful parameter in the choice of the isolate to use. It can also be used for species confirmation. When used in PCR-based assays, this single-copy gene has important practical implications for the identification of the fungus. Isolates of *P. chlamydosporia* proceeding from eggs of *M. incognita*, *Heterodera schachtii* or *Globodera rostochiensis* have SNPs in their VCP1 gene reflecting variations in their amino acid composition and enzyme molecular structure, which depend on the host they were isolated from (Morton et al. 2003; Rosso and Ciancio 2005). This gene may hence be exploited as a specificity marker in production, as well as in monitoring, after the fungus is released in the soil environment (Rosso and Ciancio 2005).

A further source of genetic variability is represented by the presence and possible activation of transposable elements. Genome-wide reshufflings induced by transposons may be difficult to detect, since their activation and effects may change the functionality of genes without a direct phenotypic evidence. At least two different transposons, a polyprotein (closest GenBank acc. CAB91877, with function as protease, integrase, reverse transcriptase) and a Fot5 transposase (closest GenBank acc. CAE55867), have been initially detected in *P. chlamydosporia* (Rosso et al. 2011). However, further transcriptomic studies on endophytic metabolism of *P. chlamydosporia* DSM 26985 showed 154 - transposon-related transcripts, with 6 terms for "transposon", 142 for "transposase-like protein" and 6 for "retrotransposable element-like" (Ciancio et al. 2013).

### 10.4.6 Record Keeping

Records must be kept for each production batch since the initial step of inoculum production and substrate inoculation. Batches are split according to the autoclave used for sterilisation of the solid substrate, and separated records must be kept for each. Details must be provided on date of inoculation, number of bags and unique tags, results of QC analyses and incidence of contamination during incubation, type of contaminant and final yield of conidia from all bags extracted. It should also be possible to trace each batch produced back to the original stock culture. Details are normally transferred and stored in a digital format, following packaging of each batch, so that production data are archived either as hard or electronic copies. They can be analysed periodically to evaluate, i.e. the cycle efficiency and the effects of minor substrate or process differences, on the final yield.

### 10.4.7 QC on Final Product

Efficacy is the most critical factor to ensure product performance and long-term acceptance by customers. Knowledge on the isolate ecology and on the effects of environmental factors is essential to ensure reliable field performance and targeted application of the highest-quality product. In general, some simple tests for viability may be set up to measure the product efficiency. A relative quick and accurate procedure to test viability is to determine the propagule germination rate. A relatively quick and accurate method to test viability of *P. chlamydosporia* is to determine the percentage of conidia and chlamydospore germination by plating a diluted suspension of propagules on sorbose agar with antibiotics. The medium is based on 12 g agar, 2 g sorbose, distilled water up to 1 L, adding after autoclaving and cooling 50 mg of each antibiotic (streptomycin sulphate, chlortetracycline and chloramphenicol). If testing for presence of contaminating bacteria, the antibiotics should not be used. Viable and nonviable spores can be counted under a microscope following their incubation at 25–26 °C for a week (Kerry and Bourne 2002).

The virulence of *P. chlamydosporia* can be verified in a bioassay using nematode eggs (Kerry and Bourne 2002). Parasitism is determined by spreading 200–300 RKN or PCN eggs on the surface of a WA culture with antibiotics and then incubated at 24–26 °C for a week. The WA surface must be covered with 2–3-day-old fungal colonies previously inoculated by smearing a small culture block with a sterile loop. Parasitism can be assessed by counting the first 50–100 eggs encountered when scanning the dish under a light microscope at 125–250×, scoring the number of eggs with evidence of parasitism (Fig. 10.4). Variability in bioassays depends on a number of factors including the natural diversity of the isolates or of the nematode host, and the original purity of the population used, that may be contaminated by other indigenous fungal populations or bacteria. Care should be taken to avoid use of nematode populations with other associated parasites. The eggs should be preferably obtained by dissolving egg masses of possibly pure nematode populations in NaOCl (Hussey and Barker 1973), followed by several washings with SDW, before aseptical transfer to the WA. Prior inspection of eggs on a water temporary slide in light microscopy at 125–250× will allow evaluation of incidence of other associated fungal or bacterial species. Some dishes without *P. chlamydosporia* inoculation should also be used as control, to exclude contaminations during the subsequent handling.

### 10.4.8 Storage

Shelf life is a critical limiting factor for biological control products. It should last as long as possible and, preferably, should not require refrigeration or special treatments (Jones and Burges 1998). Many factors affect the shelf life of fungal products, including nutrition, culture conditions during fermentation, methods of downstream processing and harvesting, product moisture content, formulation and packaging (Magan 2001). Longevity of fungal products in storage can be determined through equations and experimental tests defining constants for propagules, as per standard methods for longevity of other propagules, i.e. seeds (Roberts and Ellis 1989; Hong et al. 1997). Effects of storage conditions on the survival for *P. chlamydosporia* were measured monthly for 6 months and shelf life was significantly improved at low temperatures and low water activity ($a_w$) (<0.33). Vacuum and N increased the fungus viability, whereas $CO_2$ increased the activity, as compared to ambient air. Optimal parameters for *P. chlamydosporia* in formulation of alginate pellets included a temperature range of 4–20 °C, $a_w = 0.12$, and a N-filled atmosphere (Duan et al. 2008).

## 10.5 Formulation and Application

The number of patents reporting one or more nematode BCAs increased during the last decade, in relation to market demand and potential of microorganisms for industrial and commercial exploitation (Fig. 10.6). Although some of these records refer other uses of the species of interest (i.e. the patents listing *Cordyceps* for medical uses that also report its imperfect stage *Hirsutella*), patent trends reflect changes in the approaches for phytonematode control, and the growth of industrial sectors vacated to promotion and exploitation of alternative management methods. Formulates containing propagules and hyphae of *P. chlamydosporia*, alone or in combination with *Arthrobotrys oligospora* or other fungi, have been produced and tested in Italy either as a dry powder (Fig. 10.7a) or liquid products with vegetable oil (Fig. 10.7b). For field or greenhouse treatments, the fungus may be given to plants by a variety of methods, i.e. liquid sprays, direct addition of chlamydospores or with water suspensions of mycelial extracts to roots. The density of application suggested in many studies is $5 \times 10^3$ chlamydospores $g^{-1}$ of soil (Kerry 2000). Hence, for equivalent field applications, an estimate of the soil volume explored by the roots is required. An aqueous mixture of the inoculum may be applied for direct plant inoculation keeping soil moisture at field capacity and dispersing the product on soil or roots, after their exposure.

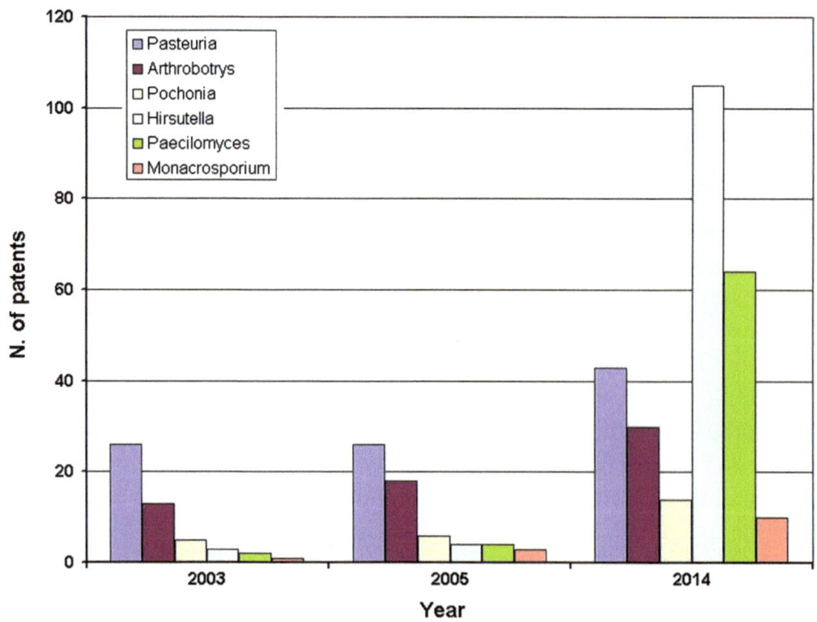


**Fig. 10.6** Number of patents retrieved using genera of nematode BCAs (*inset*) as query term per year (Source: http://www.european-patent-office.org/index.htm)

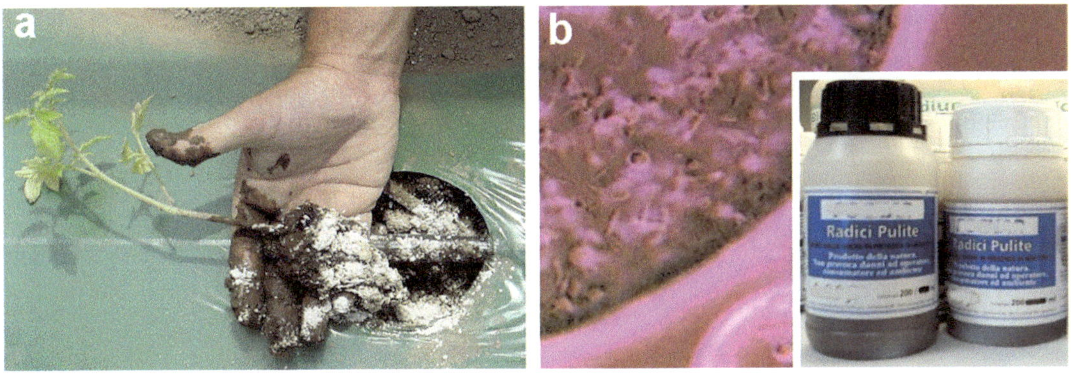


**Fig. 10.7** Treatment of tomato seedlings using a powder formulate with *P. chlamydosporia* propagules. (**a**) Liquid commercial products with propagules and (**b**) hyphal fragments suspended in an oil drop as seen at 250×, sold in sealed plastic containers (*inset*)

*P. chlamydosporia* has no phytotoxicity nor does it exert negative effects on the plant, and treatments may be also performed on seedlings before or during the transplant operations. In a greenhouse, for post-transplant application or for treatments of more advanced phenologic stages, the product may be suspended in water and distributed through the irrigation system with a low operating pressure.

Commercial and market issues for nematode BCAs are related to the market dimension, to the presence of other commercial products and/or microorganisms, to the type of formulations available and to the farmers' perspectives for a successful replacement of synthetic pesticides. Market demands to be met concern technical tools to sustain plant nutrition and manage pests and pathogens (cyst and root-knot nematodes, in particular). To develop, on a microbiological basis, innovative plant growth-promoting products based on fungi like *P. chlamydosporia*, it is necessary to produce a large amount of information and to deploy research efforts on the occurrence, ecology, biology, production and testing of the isolates available, as well as on effective formulations and field efficacy. The type of market that the project aims to reach is represented by the sector of technical tools (biopesticides and/or soil amendment products) for farmers and producers of the agricultural sectors of specialised horticulture, including either organic or conventional crops. In the latter, a wide market space has been opened as a consequence of the ban of methyl bromide and other pesticides. As concerns the organic sector, attention is focussed on producers that, although representing (with a number of farmers around, i.e. 50,000–60,000 only in Italy) a minor fraction of the global market, are more interested in these products and potentially more receptive to their use. Finally, a detailed knowledge is required on the *P. chlamydosporia* isolates to use or produce, since their specificity may become a source of concern if the fungus is applied to control an unsuitable host or it is distributed in suboptimal operational conditions.

## 10.6 Conclusions

Our current knowledge on the biodiversity, role and potentials of rhizosphere microorganisms is still far from being exhaustive, and a significant effort will be required to get deeper and detailed information, useful for application in a wide range of agroecosystems. The exploitation of root functional species like *P. chlamydosporia* aims at restoring soil equilibrium or enhancing nematode control capacity, but still requires many research efforts. Data on behaviour in soil are fundamental for industrial exploitation of the fungus. To complete this task, a great help proceeds from the fast advancements in transcriptome and genome sequencing (Ciancio et al. 2013; Larriba et al. 2014). This knowledge will highlight the many factors underpinning the specific activity of *P. chlamydosporia* isolates, as well as provide a knowledge-based platform to sustain its use in food production in many cropping systems, including low-input agriculture.

## References


Adesemoye A, Torbert H, Kloepper JW (2009) Plant growth-promoting rhizobacteria allow reduced application rates of chemical fertilizers. Microb Ecol 58:921–929

Altieri MA (1999) The ecological role of biodiversity in agroecosystems. Agric Ecosyst Environ 74:19–31

Atkins SD, Hidalgo-Diaz L, Kalisz H, Mauchline TH, Hirsch PR, Kerry BR (2003) Development of a new management strategy for the control of root-knot nematodes (*Meloidogyne* spp.) in organic vegetable production. Pest Manag Syst 59:183–189

Atkins SD, Clark I, Pande S, Hirsch PR, Kerry BR (2005) The use of real-time PCR and species-specific primers for the identification and monitoring of *Paecilomyces lilacinus*. FEMS Microbiol Ecol 51:257–264

Atkins SD, Peteira B, Clark IM, Kerry BR, Hirsch PR (2009) Use of real-time quantitative PCR to investigate root and gall colonisation by co-inoculated isolates of the nematophagous fungus *Pochonia chlamydosporia*. Ann Appl Biol 155:143–152

Avis TJ, Gravel V, Antoun H, Tweddell RJ (2008) Multifaceted beneficial effects of rhizosphere microorganisms on plant health and productivity. Soil Biol Biochem 40:1733–1740

Bakker O, Parker MG (1991) CAAT/enhancer binding protein is able to bind to ATF/CRE elements. Nucleic Acids Res 19:1213–1217

Behie SW, Zelisko PM, Bidochka MJ (2012) Endophytic insect-parasitic fungi translocate nitrogen directly from insects to plants. Science 336:1576–1577

Blackburn F, Hayes WA (1966) Studies on the nutrition of *Arthrobotrys oligospora* Fres. and *A. robusta* Dudd. I. The saprophytic phase. Ann Appl Biol 58:43–50

Bontempo AF, Fernandes RH, Lopes J, Freitas LG, Lopes EA (2014) *Pochonia chlamydosporia* controls *Meloidogyne incognita* on carrot. Australas Plant Pathol 43:421–424

Bordallo JJ, Lopez-Llorca LV, Jansson HB, Salinas J, Persmark L, Asensio L (2002) Colonization of plant roots by egg-parasitic and nematode-trapping fungi. New Phytol 154:491–499

Bourne JM, Kerry BR, de Leij FAAM (1996) The importance of the host plant on the interaction between root-knot nematodes (*Meloidogyne* spp.) and the nematophagous fungus, *Verticillium chlamydosporium* Goddard. Biocontrol Sci Tech 6:539–548

Burdsall HH, Dorworth EB (1994) Preserving cultures of wood-decaying Basidiomycotina using sterile distilled water in cryovials. Mycologia 86:275–280

Chen J, Abawi GS, Zuckerman BM (1999) Suppression of *Meloidogyne hapla* and its damage to lettuce grown in a mineral soil amended with chitin and biocontrol organisms. J Nematol 31(4S):719–725

Ciancio A, Loffredo A, Paradies F, Turturo C, Finetti-Sialer M (2005) Detection of *Meloidogyne incognita* and *Pochonia chlamydosporia* by fluorogenic molecular probes. OEPP/EPPO Bull 35:157–164

Ciancio A, Colagiero M, Ferrara M, Nigro F, Pentimone I, Rosso LC (2013) Transcriptome changes in tomato roots during colonization by the endophytic fungus *Pochonia chlamydosporia*. 5th Congress European Microbiologists (FEMS), 21–25 July, Leipzig, Germany

Clarkson JM, Charnley AK (1997) News insights into the mechanisms of fungal pathogenesis in insects. Trends Microbiol 4:197–203

Cordier C, Edel-Hermann V, Martin-Laurent F, Blal B, Steinberg C, Alabouvette C (2007) SCAR-based real time PCR to identify a biocontrol strain (T1) of *Trichoderma atroviride* and study its population dynamics in soils. J Microbiol Methods 68:60–68

Dallemole-Giaretta R, Freitas LG, Lopes EA, Ferraz S, de Podestá GS, Agnes EL (2011) Cover crops and *Pochonia chlamydosporia* for the control of *Meloidogyne javanica*. Nematology 13:919–926

de Leij FAAM, Kerry BR (1991) The nematophagous fungus *Verticillium chlamydosporium* Goddard, as a potential biological control agent for *Meloidogyne arenaria* (Neal) Chitwood. Rev Nématol 14:157–164

de Leij FAAM, Kerry BR, Dennehy JA (1992a) The effect of fungal application rate and nematode density on the effectiveness of *Verticillium chlamydosporium* as a biological control agent for *Meloidogyne incognita*. Nematologica 38:112–122

de Leij FAAM, Davies KG, Kerry BR (1992b) The use of *Verticillium chlamydosporium* Goddard and *Pasteuria penetrans* (Thorne) Sayre & Starr alone and in combination to control *Meloidogyne incognita* on tomato plants. Fundam Appl Nematol 15:235–342

Diogo HC, Sarpieri A, Pires MC (2005) Fungi preservation in distilled water. An Bras Dermatol 80:591–594

Duan W, Yang E, Xiang M, Liu X (2008) Effect of storage conditions on the survival of two potential biocontrol agents of nematodes, the fungi *Paecilomyces lilacinus* and *Pochonia chlamydosporia*. Biocontrol Sci Tech 18:605–612

Elliot JM, Marks CF, Tu CM (1977) Effects of certain nematicides on soil nitrogen, soil nitrifiers, and populations of *Pratylenchus penetrans* in flue-cured tobacco. Can J Plant Sci 57:143–154

Escudero N, Lopez-Llorca LV (2012) Effects on plant growth and root-knot nematode infection of an endophytic GFP transformant of the nematophagous fungus *Pochonia chlamydosporia*. Symbiosis 57:33–42

Ferreira SR, Araújo JV, Braga FR, Araujo JM, Carvalho RO, Silva AR, Frassy LN, Freitas LG (2011) Ovicidal activity of seven *Pochonia chlamydosporia* fungal isolates on Ascaris suum eggs. Trop Anim Health Prod 43:639–642

Fierer N, Leff JW, Adams BJ, Nielsen UN, Bates ST, Lauber CL, Owens S, Gilbert JA, Wall DH, Caporaso JG (2012) Cross-biome metagenomic analyses of soil microbial communities and their functional attributes. Proc Natl Acad Sci U S A 109:21390–21395

Flores-Camacho R, Manzanilla-López RH, Cid del Prado-Vera I, Martínez-Garza A (2007) Control de *Nacobbus aberrans* (Thorne) Thorne y Allen con *Pochonia chlamydosporia* (Goddard) Gams y Zare. Rev Mex Fitopatol 25:26–34

Franco-Navarro F, Vilchis-Martínez K, Miranda-Damián J (2008) New records of *Pochonia chlamydosporia* from Mexico: isolation, root colonization and parasitism of *Nacobbus aberrans* eggs. Nematropica 39:133–142

Frisli T, Haverkamp THA, Jakobsen KS, Stenseth NC, Rudi K (2013) Estimation of metagenome size and structure in an experimental soil microbiota from low coverage next generation sequence data. J Appl Microbiol 114:141–151

Gao L, Liu X (2010) Effects of carbon concentrations and carbon to nitrogen ratios on sporulation of two biological control fungi as determined by different culture methods. Mycopathologia 169:475–481

Green LH (2008) Culturing and preserving microorganisms. In: Goldman E, Green LH (eds) Practical handbook of microbiology, IIth edn. CRC Press, Boca Raton, pp 31–35

Grivell AR, Jackson JF (1969) Microbial culture preservation with silica gel. J Gen Microbiol 58:423–425

Hadapad AB, Zebitz CP (2006) Mass production, survival and evaluation of solid substrate inocula of *Beauveria brongniartii* (Saccardo) Petch against *Holotrichia serrata* (Coleoptera: Scarabaeidae). Commun Agric Appl Biol Sci 71(2B):433–441

Hellwig V, Mayer-Bartschmid A, Müller H, Greif G, Kleymann G, Zitzmann W, Tichy HV, Stadler M (2003) Pochonins A-F, new antiviral and antiparasitic resorcylic acid lactones from *Pochonia chlamydosporia* var. *catenulata*. J Nat Prod 66:829–837

Hirsch PR, Mauchline TH (2012) Who's who in the plant root microbiome? Nat Biotechnol 30:961–962

Hirsch PR, Atkins SD, Mauchline TH, Morton OC, Davies KG, Kerry B (2001) Methods for studying the nematophagous fungus *Verticillium chlamydosporium* in the root environment. Plant Soil 232:21–30

Hong TD, Ellis RH, Moore D (1997) Development of a model to predict the effect of temperature and moisture on fungal spore longevity. Ann Bot 79:121–128

Hussey RS, Barker KR (1973) A comparison of methods of collecting inocula of *Meloidogyne spp.* including a new technique. Plant Dis Rep 57:1025–1028

Jenkins NE, Heviefo G, Langewald J, Cherry AJ, Lomer CJ (1998) Development of mass production technology for aerial conidia of mitosporic fungi for use as mycopesticides. Biocontrol News Inform Serv 19:21N–31N

Jones KA, Burges HD (1998) Technology of formulation and application. In: Burges HD (ed) Formulation of microbial biopesticides. Kluwer Academic Publishers, Dordrecht, pp 7–30

Kerry BR (2000) Rhizosphere interactions and the exploitation of microbial agents for the biological control of plant-parasitic nematodes. Annu Rev Phytopathol 38:423–424

Kerry BR, Bourne JM (1996) The importance of rhizosphere interactions in the biological control of plant parasitic nematodes – a case study using *Verticillium chlamydosporium*. Pestic Sci 47:69–75

Kerry BR, Bourne JM (eds) (2002) A manual for research on *Verticillium chlamydosporium* a potential biological control agent for root knot nematodes. IOBC/WPRS, Gent, 84 pp

Kerry BR, Hidalgo-DÚaz L (2004) Application of *Pochonia chlamydosporia* in the integrated control of root-knot nematodes on organically grown vegetable crops in Cuba. In: Multitrophic interactions in soil and integrated control. IOBC/WPRS bulletin 27. International Organisation for Biological and Integrated Control, West Palaeartic Regional Section, Dijon, pp 123–126

Kerry BR, Simon A, Rovira AD (1984) Observations on the introduction of *Verticillium chlamydosporium* and other parasitic fungi into soil for control of the cereal cyst-nematode *H. avenae*. Ann Appl Biol 105:509–516

Khambay BPS, Bourne JM, Cameron S, Kerry BR, Zaki M, Javed CS (2000) A nematicidal metabolite from *Verticillium chlamydosporium*. Pest Manage Sci 56:1098–1099

Kumar P, Dubey RC, Maheshwari DK (2012) *Bacillus* strains isolated from rhizosphere showed plant growth promoting and antagonistic activity against phytopathogens. Microbiol Res 167:493–499

Larriba E, Martín Nieto J, Lopez-Llorca LV (2012) Gene cloning, molecular modeling, and phylogenetics of serine protease P32 and serine carboxypeptidase SCP1 from nematophagous fungi *Pochonia rubescens* and *Pochonia chlamydosporia*. Can J Microbiol 58:815–827

Larriba E, Jaime MD, Carbonell-Caballero J, Conesa A, Dopazo J, Nislow C, Martín-Nieto J, Lopez-Llorca LV (2014) Sequencing and functional analysis of the genome of a nematode egg-parasitic fungus, *Pochonia chlamydosporia*. Fungal Genet Biol 65:69–80

Lauber C, Knight R, Hamady M, Fierer N (2009) Soil pH as a predictor of soil bacterial community structure at the continental scale: a pyrosequencing-based assessment. Appl Environ Microbiol 75:5111–5120

Liu XZ, Chen SY (2003) Nutritional requirements of *Pochonia chlamydospora* and ARF18, fungal parasites of nematode eggs. J Invertebr Pathol 83:10–15

Lomer CJ, Prior C, Kooyman C (1997) Development of *Metarhizium* spp. for the control of locusts and grasshoppers. Mem Entomol Soc Can 171:265–286

Lopez-Llorca LV, Bordallo JJ, Salinas J, Monfort E, López-Serna ML (2002) Use of light and scanning electron microscopy to examine colonisation of barley rhizosphere by the nematophagous fungus *Verticillium chlamydosporium*. Micron 33:61–67

Lopez-Llorca LV, Maciá-Vicente JG, Jansson HB (2008) Mode of action and interactions of nematophagous fungi. In: Ciancio A, Mukerji KG (eds) Integrated management and biocontrol of vegetable and grain crops nematodes. Springer, Dordrecht, 356 pp

Lopez-Llorca LV, Gómez-Vidal S, Monfort E, Larriba E, Casado-Vela J, Elortza F, Jansson HB, Salinas J, Martín-Nieto J (2010) Expression of serine proteases in egg-parasitic nematophagous fungi during barley root colonization. Fungal Genet Biol 47:342–351

Lozupone C, Knight AR (2008) Species divergence and the measurement of microbial diversity. FEMS Microbiol Rev 32:557–578

Maciá-Vicente J, Rosso LC, Ciancio A, Jansson HB, Lopez-Llorca LV (2009) Colonisation of barley roots by endophytic *Fusarium equiseti* and *Pochonia chlamydosporia*: effects on plant growth and disease. Ann App Biol 155:391–401

Magan N (2001) Physiological approaches to improving the ecological fitness of fungal biocontrol agents. In: Jackson TMC, Magan N (eds) Fungi as biocontrol agents: progress, problems and potential. CAB International, Wallingford, pp 239–251

Manzanilla-López RH, Esteves I, Powers SJ, Kerry BR (2011) Effects of crop plants on abundance of *Pochonia chlamydosporia* and other fungal parasites of root-knot and potato cyst nematodes. Ann Appl Biol 159:118–129

Manzanilla-López RH, Esteves I, Finetti-Sialer MM, Hirsch PR, Ward E, Devonshire J, Hidalgo-Díaz L (2013) *Pochonia chlamydosporia*: advances and challenges to improve its performance as a biological control agent of sedentary endo-parasitic nematodes. J Nematol 45:1–7

Martínez-Viveros O, Jorquera MA, Crowley DE, Gajardo G, Mora ML (2010) Mechanisms and practical considerations involved in plant growth promotion by rhizobacteria. J Soil Sci Plant Nutr 1:293–319

Mo MH, Xu CK, Zhang KQ (2005) Effects of carbon and nitrogen sources, carbon-to-nitrogen ratio, and initial pH on the growth of nematophagous fungus *Pochonia chlamydosporia* in liquid culture. Mycopathologia 159:381–387

Monfort E, Lopez-Llorca LV, Jansson HB, Salinas J, Park JO, Sivasithamparam K (2005) Colonisation of seminal roots of wheat and barley by egg-parasitic nematophagous fungi and their effects on *Gaeumannomyces graminis* var. *tritici* and development of root-rot. Soil Biol Biochem 37:1229–1235

Montes de Oca N (2004) Buenas prácticas de fabricación para la obtención de un bionematicida a partir de la cepa Vcc 108 de *Pochonia chlamydosporia* var. *catenulata*. Ph.D. thesis, Universidad Agraria de la Habana, Centro Nacional de Sanidad Agropecuaria, Havana, Cuba

Morton CO, Mauchline TH, Kerry BR, Hirsch PR (2003) PCR-based DNA fingerprinting indicates host-related genetic variation in the nematophagous fungus *Pochonia chlamydosporia*. Mycol Res 107:198–205

Muthulakshmi M, Kumar S, Subramanian S, Anita B (2012) Compatibility of *Pochonia chlamydosporia* with other biocontrol agents and carbofuran. J Biopestic 5(Suppl):243–245

Nakasone KK, Peterson SW, Jong SC (2004) Preservation and distribution of fungal cultures. In: Mueller GM, Bills GF, Foster MS (eds) Biodiversity of fungi – inventory and monitoring methods. Elsevier, Amsterdam

Niu XM, Wang YL, Chu YS, Xue HX, Li N, Wei LX, Mo MH, Zhang KQ (2010) Nematode toxic aurovertin-type metabolites from a root-knot nematode parasitic fungus *Pochonia chlamydosporia*. J Agric Food Chem 58:828–834

Pereg L, McMillan M (2015) Scoping the potential uses of beneficial microorganisms for increasing productivity in cotton cropping systems. Soil Biol Biochem 80:349–358

Pietrantonio L, Di Cillo P, Pignoli G, Torzi E, Rosso L, Ciancio A, Colagiero M (2013) Green technologies for sustainable management of field crops: use and potentialities of the fungus *Pochonia chlamydosporia*. Italian Forum on Industrial Biotechnology and Bioeconomy (IFIB), 22–23 October, Naples

Roberts EH, Ellis RH (1989) Water and seed survival. Ann Bot 63:39–52

Roesch LFW, Fulthorpe RR, Riva A, Casella G, Hadwin AKM, Kent AD, Daroub SH, Camargo FAO, Farmerie WG, Triplett EW (2007) Pyrosequencing enumerates and contrasts soil microbial diversity. ISME J 1:283–290

Rosso L, Ciancio A (2005) Indagini sulla specificità parassitaria di *Pochonia chlamydosporia*. Nematol Mediterr 33(Suppl):87–90

Rosso LC, Ciancio A, Finetti-Sialer M (2007) Application of molecular methods for detection of *Pochonia chlamydosporia* from soil. Nematropica 37:1–8

Rosso LC, Finetti-Sialer MM, Hirsch PR, Ciancio A, Kerry BR, Clark IM (2011) Transcriptome analysis shows differential gene expression in the saprotrophic to parasitic transition of *Pochonia chlamydosporia*. Appl Microbiol Biotechnol 90:1981–1994

Rosso LC, Pentimone I, Colagiero M, Ferrara M, Nigro F, Ciancio A (2013) Expression of *Meloidogyne incognita* resistance genes induced by endophytic *Pochonia chlamydosporia* in tomato. Nematropica 43:322

Rosso LC, Colagiero M, Salatino N, Ciancio A (2014) Effect of trophic conditions on gene expression of *Pochonia chlamydosporia*. Ann Appl Biol 164:232–243

Segers R, Butt TM, Kerry BR, Beckett A, Peberdy JF (1996) The role of the proteinase VCP1 produced by the nematophagous *Verticillium chlamydosporium* in the infection process of nematode eggs. Mycol Res 100:421–428

Siddiqui IA, Atkins SD, Kerry BR (2009) Relationship between saprotrophic growth in soil of different biotypes of *Pochonia chlamydosporia* and the infection of nematode eggs. Ann Appl Biol 155:131–141

Silva AR, Araujo JV, Braga FR, Benjamim LA, Souza DL, Carvalho RO (2011) Comparative analysis of destruction of the infective forms of *Trichuris trichiura* and *Haemonchus* contortus by nematophagous fungi *Pochonia chlamydosporia*; *Duddingtonia flagrans* and *Monacrosporium thaumasium* by scanning electron microscopy. Vet Microbiol 147:214–219

Sleator RD, Shortall C, Hill C (2008) Metagenomics. Lett Appl Microbiol 47:361–366

St Leger RJ, Charnley AK, Cooper RM (1991) Kinetics of the digestion of insect cuticles by a protease (Pr1) from *Metarhizium anisopliae*. J Invertebr Pathol 57:146–147

Stadler M, Tichy HV, Katsiou E, Hellwig V (2003) Chemotaxonomy of *Pochonia* and other conidial fungi with *Verticillium*-like anamorphs. Mycol Prog 2:95–122

Taborsky V (1992) Small-scale processing of microbial pesticides. FAO agricultural services bulletin 96, Rome

Thiel G, Al Sarraj J, Vinson C, Stefano L, Bach K (2005) Role of basic region leucine zipper transcription factors cyclic AMP response element binding protein (CREB), CREB2, activating transcription factor 2 and CAAT/enhancer binding protein $\alpha$ in cyclic AMP response element-mediated transcription. Neurochemistry 92:321–336

Tunlid A, Jansson S (1991) Proteases and their involvement in the infection and immobilization of nematodes by the nematophagous fungus *Arthrobotrys oligospora*. Appl Environ Microbiol 57:2868–2872

Usuki H, Toyo-oka M, Kanzaki H, Okuda T, Nitoda T (2009) Pochonicine, a polyhydroxylated pyrrolizidine alkaloid from fungus *Pochonia suchlasporia* var. *suchlasporia* TAMA 87 as a potent b-N-acetylglucosaminidase inhibitor. Bioorg Med Chem 17:7248–7253

Verdejo-Lucas S, Sorribas FJ, Ornat C, Galeano M (2003) Evaluating *Pochonia chlamydosporia* in a double-cropping system of lettuce and tomato in plastic houses infested with *Meloidogyne javanica*. Plant Pathol 52:521–528

Viaene NM, Abawi GS (2000) *Hirsutella rhossiliensis* and *Verticillium chlamydosporium* as biological control agents of *Meloidogyne hapla* on lettuce. J Nematol 32:85–100

Vieira Dos Santos MC, Esteves I, Kerry B, Abrantes I (2013) Biology, growth parameters and enzymatic activity of *Pochonia chlamydosporia* isolated from potato cyst and root-knot nematodes. Nematology 15:493–504

Walker HL, Connick WJ (1983) Sodium alginate for production and formulation of mycoherbicides. Weed Sci 31:333–338

Ward E, Kerry BR, Manzanilla Lopez RH, Mutua G, Devonshire J, Kimenju J, Hirsch PR (2012) The *Pochonia chlamydosporia* serine protease gene *vcp1* is subject to regulation by carbon, nitrogen and ph: implications for nematode biocontrol. PLoS One 7: e35657

Zare R, Gams W, Evans HC (2001) A revision of *Verticillium* section Prostrata V. The genus *Pochonia*, with notes on *Rotiferophthora*. Nova Hedwigia 73:51–86

# 11 Improvement of Crop Protection and Yield in Hostile Agroecological Conditions with PGPR-Based Biofertilizer Formulations

Dilfuza Egamberdieva and Anthony O. Adesemoye

## Contents




**Abstract**

Intensive research attempts are underway to mitigate the impacts of climate change, improve salt tolerance and disease resistance of plants using organic farming practices, including biofertilizer which eventually improve degraded soils. In addition, it will form part of integrated environmentally friendly approach for nutrient management and sustainability in ecosystem functions. The use of such microbial inoculants as biofertilizers or biopesticides portends a great promise for controlling disease, improving plant health and soil productivity under environmentally stressed conditions. Stress-tolerant microorganisms with plant-stimulating properties are being discovered, selected and tested under field conditions and the number of successful applications is increasing. Formulation of microorganisms with various carrier materials enables long-term storage and protects them from various stress factors. This review summarizes the current status of microbial inoculants usage and prospects in crop cultivation and crop stress management, with particular attention to arid stress agro-ecological conditions.



D. Egamberdieva (✉)
Leibniz Centre for Agricultural Landscape Research (ZALF), Institute for Landscape Biogeochemistry, Müncheberg 15374, Germany
e-mail: dilfuza.egamberdieva@zalf.de

A.O. Adesemoye
Plant Pathology Department, West Central Research & Extension Center, University of Nebraska–Lincoln, North Platte, NE 69101, USA
e-mail: tony.adesemoye@unl.edu



N.K. Arora et al. (eds.), *Bioformulations: for Sustainable Agriculture*,
DOI 10.1007/978-81-322-2779-3_11

## 11.1 Introduction

The increase in harsh abiotic stresses such as drought, salinity and abrupt changes in temperature is part of the main consequences of climate change and has led to the loss of soil organic matter and other forms of soil degradation that negatively affect agricultural productivity (Ahmad et al. 2012). Agricultural development is considered as the main sector in reducing hunger and poverty but the decrease in productivity of agricultural lands due to abiotic stresses has a direct impact on the standard of living of the more than seven billion people around the world (Munns and Tester 2008). Drought and salinity continue to have significant impacts in both developed and developing countries and are regarded as major factors in crop losses and diseases (Mantri et al. 2012; Hameed et al. 2014). Salinity alone affects 33 % of the potentially arable land area of the world, whereas 950 million ha of salt-affected lands occur in arid and semi-arid regions (UNEP 2008). Major factors increasing salinity include irrigation of cultivated lands with saline water, poor cultural practices and low precipitation. Climate change will result in a change towards a more arid climate, which is conducive to salt accumulation (Othman et al. 2006).

Another important consequence of climate change (especially temperature) and abiotic stresses is increased infection/infestation from pathogens and pests (Korus et al. 2015). For example, high temperatures may make crops to grow faster with reduced seed maturity time and reduced yield and may help certain diseases. On the other hand, low temperatures promote some other diseases and extremely low temperatures and/or snow may lead to frost bite that could predispose crops to diseases afterwards. Scientific projections have indicated that climate change will continue to have major impacts on crops productivity and yield across the world. Recently, a Global Alliance for Climate-Smart Agriculture was launched by world leaders during the United Nations Climate Summit in New York on September 23, 2014, involving multiple stakeholders, farmers, 20 governments and 30 organizations for incorporation of climate-smart approaches within food and agriculture systems (http://www.un.org/climate change/summit/action–areas). Also, intensive research attempts are underway to mitigate the impacts of climate change, improve salt tolerance and management of plant diseases using farming practices, such as application of farm manure, compost, biofertilizer, recycling of crop residues and/or green manures or a combination of these practices in integrated systems, which will eventually improve soils (Pathma and Sakthivel 2012; Adesemoye and Egamberdieva 2013).

These farm practices have been considered as supplements or possible alternatives to chemical fertilizers. The practices will increase soil health and productivity, crop production in sustainable organic farming and hold a great promise to improve crop yields through environmentally sustainable and better nutrient supplies and use (Wu et al. 2005; Adesemoye et al. 2008; Hashem et al. 2014). Bacterial fertilizers are usually prepared as carrier-based inoculants containing effective microbes which include rhizobia and plant growth-promoting rhizobacteria belonging to genera such as *Bacillus, Pseudomonas, Azotobacter*, *Azospirillum*, *Bradyrhizobium*, *Enterobacter*, *Mezorhizobium*, etc. (Stephens and Rask 2000; Rinu and Pandey 2009; Egamberdiyeva et al. 2004; Egamberdiyeva 2007). Formulation of microorganisms with various carrier materials enables long-term storage and high effectiveness of biofertilizers (Marin 2006; Trivedi et al. 2012).

The microorganisms are able to colonize plant roots, stimulate root growth, make soil nutrients available to the plant root, protect plants from various diseases and enhance plant tolerance to various environmental stresses including drought, salinity, high or low temperatures and heavy metals (Egamberdiyeva and Hoflich 2002, 2003; Adesemoye and Kloepper 2009). In addition, they may help change nutrients that will otherwise be bound up in soil to available forms for plants, through solubilization of minerals and increase the concentrations of P, N, K and Mg in plant root and shoots (Son et al. 2006). Based on these properties, microbial inoculants are considered as environmentally friendly microbial technologies for sustainable agriculture (Fravel 2005; Hashem et al. 2014). However, extreme environmental conditions such as high or low temperature, water and salt stress are deleterious for the survival and effective functioning of the introduced microbial inoculants (Pandey et al. 1998).

The different carrier materials for bacterial inoculants such as peat, charcoal/biochar, methyl ethyl cellulose and vegetable oils that support higher populations of bacteria and display much longer shelf lives are important for plant production in arid and semi-arid soils, which may also suffer from high salinity, water stress, temperature stress, low organic matter and nutrients (Stephens and Rask 2000; Ardakani et al. 2010).

The application of bacterial inoculants for improvement of plant may not always be successful or could produce variable results in different locations and cropping systems. In extremely cold conditions, many gram negative bacteria may not survive to enhance crop performance and gram positive bacteria may become dormant through sporulation. Low population of the inoculated bacterial strain in the rhizosphere of plants grown under hostile environmental conditions may occur. It is reported that salinity and desertification cause a great disturbance to plant–microbe interactions and inhibit microbial activity in the soil (Vardharajula et al. 2011; Egamberdieva 2012).

It is notable that microbes which are adapted to stress conditions will be beneficial to the growth and yield of various agriculturally important plants, including in salinated arid and semi-arid soils (Fu et al. 2010; Turan et al. 2012; Berg et al. 2013). Identifying such microbial strains and/or using such microbial inoculants as biofertilizers or biopesticides portends a great promise for controlling disease and improving plant health and soil productivity under environmentally stressed conditions. In addition it will form part of the needed environmentally friendly approach for nutrient management, climate-smart agriculture and sustainability of ecosystems. This review summarizes the current status of microbial inoculants usage and prospects in crop cultivation and crop stress management, with particular attention to arid and semi-arid agro-ecological conditions.

## 11.2 Beneficial Microbes

The plant root is associated with different microbial communities, and these microbes can be beneficial, neutral or pathogenic and compete for nutrients and niches under strongly varying conditions (Lugtenberg and Kamilova 2009; Berg et al. 2013). Beneficial rhizosphere bacteria are of two general types, those forming a symbiotic relationship with the plant and those that are free-living in the soil and root (Hameed et al. 2014; Egamberdieva et al. 2013, 2014). Plant growth-promoting rhizobacteria (PGPR) are free-living and play important roles in crop production as they: (i) stimulate root growth, (ii) make soil nutrients available to the plant root and/or fix nitrogen from the air to improve soil fertility, (iii) suppress soil-borne pathogens and (iv) alleviate abiotic stresses in plants (Egamberdiyeva and Hoflich 2004; Hashem et al. 2015; Shrivastava et al. 2015).

Some of the plant beneficial bacteria which can enhance plant growth, induce systemic resistance against diseases and enhance tolerance of plants to salt stress include species of *Pseudomonas*, *Bacillus*, *Azotobacter*, *Azospirillum*, *Flavobacterium*, *Stenotrophomonas* and rhizobia (Nabti et al. 2010). The utilization of root-associated bacteria that interact with plants by mitigating the effects of various stresses opens a novel, inexpensive and advanced technology for combating the problems of salinity (Arshad et al. 2008) and other climate related problems; thus, it continues to stimulate research and commercial interests. It has been shown that most PGPR isolates from environmental stress conditions showed certain enhancement of tolerance e.g., salt tolerance due to efficient metabolic processes that help them to adapt under these environments (Egamberdiyeva 2005; Wu

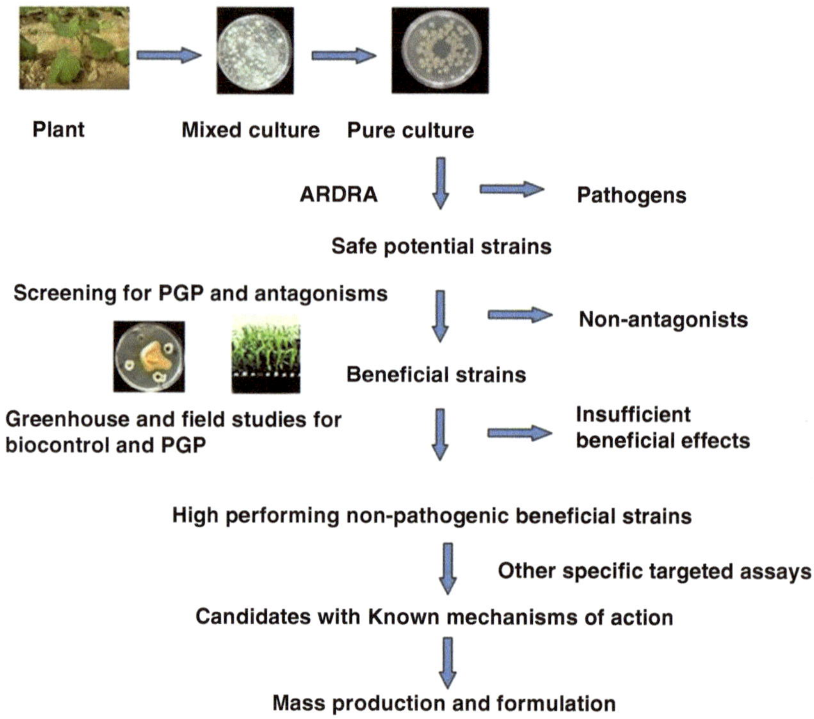


**Fig. 11.1** Screening for plant growth promoting and biocontrol strains for the development of microbial inoculants

et al. 2011; Garg and Baher 2013). However, there is concern with variability in the effectiveness of biologicals when used in different conditions or cropping systems. The success of microbial inoculants applied as plant growth stimulator or biocontrol agents to multiple conditions therefore depends on the collection strategy and the screening process (Fravel 2005; Adesemoye personal communication). The screening flowchart (Fig. 11.1) is an effective procedure that could be used for identifying salt-tolerant bacterial inoculants, which are competitive colonizers, as well as for drought- or cold-tolerant bacteria for the formulation of microbe-based technology and could also be adapted to other environmental conditions.

There are many reports on the improvements of plant growth, development and nutrient uptake of cotton (*Gossypium hirsutum*) (Egamberdiyeva and Hoflich 2004), basil (*Ocimum basilicum*) (Golpayegani and Tilebeni 2011), wheat (*Triticum aestivum*) (Nabti et al. 2010; Turan et al. 2012) and goat's rue (*Galega officinalis* L) (Egamberdieva et al. 2013) under drought, cold and/or salt stress conditions. The plants treated with salt-tolerant PGPR contained lower $Na^+$ and $Cl^-$ contents and higher nitrogen, phosphorus, potassium (NPK) contents compared with un-inoculated plants under salt stress (Yildirim et al. 2011). In another study, Mishra et al. (2011) selected a total of 12 pseudomonads, which were the top PGPR strains from a total of 534 psychro-tolerant bacteria collected at high altitudes in north-western Indian Himalayan region and tested them on wheat. They reported that the strains promoted the growth of wheat and helped in tolerance to cold stress. It has also been observed that mycorrhizal fungi and *Trichoderma* inoculants significantly increased plant height, dry weight, number of nodules, pods and yield of chickpea in comparison to un-inoculated control under stress conditions (Patel et al. 2012; Garg and Baher 2013; Hameed et al. 2014).

Several mechanisms have been proposed to be used by PGPR for plant growth promotion such as nitrogen fixation, synthesis of phytohormones, osmoprotectants, exopolysacharides and 1-aminocyclopropane-1-carboxylate (ACC) deaminase and solubilization of minerals such as phosphorus and potassium (Berg et al. 2013;

Goswami et al. 2013; Sharma et al. 2013). Stimulation of plant growth and yield of various plants by inoculation with PGPR having indole acetic acid (IAA) producing ability has been repeatedly documented, particularly when the plants were subjected to stressful growth conditions (Nadeem et al. 2010). There are also several reports which indicate improved plant growth and development by co-inoculated PGPR (Valverde et al. 2005; Egamberdieva et al. 2010, 2013). Combined inoculation of rhizobacteria strains which produce IAA and have ACC–deaminase activity with *Bradyrhizobium* enhanced the nodulation and plant growth in mung bean (Shaharoona et al. 2006), goats rue (Egamberdieva et al. 2013) and chickpea (Khurana and Sharma 2000) compared with single inoculation.

There are also many reports on the plant growth promoting and biological control activity of *Bacillus* species (Marques et al. 2010; Araujo et al. 2012). The organisms were able to tolerate high temperature, pH and osmotic conditions and were well adapted to the arid- and salt-affected environments (Islam et al. 2013; Ashwini and Srividya 2014). This enabled these bacteria to be effective as plant growth stimulators and biocontrol agents under many environmental conditions (saline, drought, heavy metal contamination, etc.). Other mechanisms of PGPR include exhibition of antagonistic activity against phytopathogenic microorganisms by producing siderophores, cell wall-degrading enzymes, antibiotics and cyanide; induction of systemic resistance; and competition for nutrients and niches (Lugtenberg and Kamilova 2009; Glick 2010; Egamberdieva et al. 2011). Possible mechanisms involved in the establishment of a successful interaction between PGPR and plant roots under hostile environmental conditions have been reviewed and discussed by Adesemoye and Egamberdieva (2013).

## 11.3 Carrier-Based Preparations of Microbial Inoculants

Arid and semi-arid soils are mostly characterized by water deficiencies, prone to salinity, high or low temperatures and poor nutrients which pose additional challenges for the use of bio-inoculants (Bashan 1998; Egamberdiyeva 2007). The survival and colonization of microbial inoculants in the plant root is critical for them to be able to have significant effect on plant growth and health of crops under hostile environments (Rekha et al. 2007; Khavazi et al. 2007). In arid soils, due to poor bacterial survival in the soil, failure of plant growth promotion and biological control of plant diseases by microbial inoculants are often observed (Van Dyke and Prosser 2000). Improved methods of formulation and use of carrier materials need to be considered to assure efficient delivery of bio-inoculants in crop production systems in stressful conditions (Bashan 1998; Egamberdieva et al. 2011).

The carrier materials used for the bacterial inoculants protect them from various stress factors and prolong shelf life (Kumar et al. 2007; Ardakani et al. 2010). Therefore, carrier is important for the viability of the inoculum in an appropriate formulation. Several materials have been evaluated as potential carriers for bacterial inoculants, including alginate beads (Trivedi et al. 2005), peat (Albareda et al. 2008), biochar (Hale et al. 2015), perlite (Daza et al. 2000; Khavazi et al. 2007) and vermiculite (Sangeetha 2012). Arora et al. (2014) studied other materials including bagasse, saw dust and wood ashes as possible microbial inoculant carriers, and they found that coriander husk, saw dust and bagasse materials were able to sustain the highest viable cell number of rhizobial and *Pseudomonas* strains. Calcium alginate (CA) gel, a biodegradable microcapsule, is another material that has been widely utilized as a carrier for bacterial immobilization, which may protect the cells and has long-term effects under hostile environments (Wu et al. 2011).

The survival rates of microbial inocula in the carrier materials differ, e.g., perlite inoculants can maintain a higher population of microorganisms such as *Rhizobium leguminosarum* bv. phaseoli, *Rhizobium tropici*, *Bradyrhizobium japonicum* and *Bacillus megaterium* than peat inoculants at room temperature for 6 months period (Daza et al. 2000). Trivedi et al. (2005) observed maximum numbers of alginate-based PGPR *Bacillus subtilis* and *Pseudomonas corrugata* in the rhizosphere of 6-week grown maize (*Zea mays*), compared to a charcoal-based formulation. Wu et al. (2011) studied the improvement in the survival rate of *Klebsiella oxytoca* Rs-5 under salinity stress conditions through the preparation of microcapsules using sodium alginate and calcium chloride. They observed that the bacteria released from microcapsules reached up to $10^{10}$ cfu/g when immersed in physiological saline for 3 weeks. When biochar (pine wood) was used as a carrier, significantly greater *Enterobacter cloacae* UW5 populations were detected in soil after 4 weeks as compared to direct soil inoculation with liquid inoculums (Hale et al. 2015). Peat moss which contains high availability of labile carbon and high nitrogen content was the best for survival of *E. cloacae* UW5 after 4 weeks in non-sterile soil (Hale et al. 2015). The talc powders containing 1 % carboxy methyl cellulose were also used as carrier for bacterial inoculants with broad-spectrum antifungal activity and suppression of fungal pathogens (Negi et al. 2005).

Albareda et al. (2008) studied the survival of several PGPR strains including *Bradyrhizobium*, *Mesorhizobium* and *Sinorhizobium* strains in different carrier materials and liquid formulations such as bagasse, cork compost, attapulgite, sepiolite, perlite and amorphous silica. They observed that compost and perlite support higher survival rate of PGPR strains compared to other materials, whereas the viable numbers of *Sinorhizobium fredii* SMH12 and *B. japonicum* USDA110 strains were greater than $10^9$ and $5x10^8$ cells $g^{-1}$, respectively.

Ben Rebah et al. (2002) studied wastewater sludge as a carrier for *Sinorhizobium meliloti* and observed that dewatered sludge supported rhizobia survival. Sludge was considered as non-toxic and low cost raw material for microbial inoculants. Ardakani et al. (2010) used bentonite as a mineral carrier for the formulation of *Pseudomonas fluorescens*, which supported more stimulatory effect compared to peat and rice bran formulated inoculants.

## 11.4 The Use of Microbial Inoculants as Biofertilizers in Hostile Environmental Conditions

The positive effects of PGPR have been observed on various agricultural crops and medicinally important plants under different hostile environmental conditions, including arid- and salt-affected soils (Bano and Yasmeen 2010; Berg et al. 2013; Egamberdieva and Lugtenberg 2014). In a study by Wu et al. (2011) mentioned earlier, the effects of inoculation with free and alginate-encapsulated cells of *K. oxytoca* Rs-5 on cotton germination, root, shoot length and dry weight under salinity stress were examined. They observed that microcapsulated Rs-5 inoculant increased plant growth and biomass as compared with those uninoculated plant and free cells of Rs-5.

Bharathi et al. (2004) studied talk-based formulation of antagonistic bacteria for the management of fruit rot infection in chillies. The combined inoculation of *P. fluorescens* (Pf-1) with *B. subtilis* increased germination of chille by 43 % compared to uninoculated seeds. In addition talk-formulated bacterial strains increased shoot length (56.14 %), number of flowers (270.59 %), fruits (133.3 %) and also reduced fruit rot infection in chillies caused by *Colletotrichum capsici* by 51 % over control. The increase in chitinase, β-1, 3 glucanase, peroxidase, polyphenol oxidase, phenylalanine ammonia lyase and phenol accumulation in plants treated with mixed formulation were also reported (Bharthi et al. 2004).

Evidently, the use of combined inoculation for crop production could result in a significant improvement in various growth parameters (Egamberdieva et al. 2013; Yasmin et al. 2013). Mixed applications of inoculants have been shown to be more effective as biocontrol agents and plant growth stimulators under various environmental conditions (Adesemoye et al. 2008; Adesemoye et al. 2009; Malusá et al. 2012). Formulated microbial inoculants (AMF and *Pseudomonas*) increased plant growth and uptake of nitrogen (N), potassium (K), phosphorus (P), calcium (Ca) and magnesium (Mg) in sunflower under saline (EC = 7.6 dS $m^{-1}$) calcareous soil condition. In addition, microbial preparation reduced Cl uptake by the plants which enhanced the nutrition condition of the plant (Shirmardi et al. 2010). Sharma et al. (2013) observed increased germination, root and shoot length, fresh weight and proline content of chickpea seedlings by *Bacillus* sp. and *Pseudomonas* sp. under osmotic potential of up to 0.4 MPa over un-inoculated control. The strains were able to produce IAA and showed P solubilizing activity.

Elkoca et al. (2010) demonstrated increased P, K and micro-nutrient in common bean as a result of *Bacillus* and *Azospirillum* inoculations. In another study, formulation of PGPR which produced ACC deaminase showed positive effects on plant growth of chickpea (Roopa et al. 2012) resulting in increased number of nodules, root, shoot growth and yield of plant under stress conditions. The peat formulated *Mezorhizobium ciceri* significantly induced nodule numbers on the roots of chickpea cultivars, which showed a high correlation with shoot and root weights (Egamberdieva et al. 2014). Inoculation of plants with *M. ciceri* IC53 formulation significantly increased shoot and root dry matter by an average of 20 %, (pod number and yield of chickpea) above the untreated plant under arid saline conditions.

Karlidag et al. (2013) also observed positive effect of microbial inoculants on strawberry growth and yield under salt-affected soil conditions consecutively in 2 years of field experiments. The highest efficiency in alleviating salinity stress on the yield and nutrient (N, P, K, Ca, Mg, sulfur (S) and manganese (Mn)) uptake of strawberry plants was obtained from *Kocuria erythromyxa* EY43 treatment. The combined application of compost and microbial inoculant (*Pseudomonas*) increased the plant growth and uptake of minerals like P, Mn, K, Zn and Fe in chickpea plants (Sahni et al. 2008). Peat-based inoculants of *P. fluorescens* increased plant height (58.1 %), number of cobs per plant (19.5 %) and number of grain rows per cob (28.2 %) of corn as compared to the control plant grown under drought conditions (Zafar-ul-Hye et al. 2014). N'cho et al. (2013) studied the effects of microbial inoculants including *Bradyrhizobium* spp. (RACA 6), arbuscular mycorrhizal fungi (Rhizatech) and *Trichoderma harzianum* (Eco-T) on soybean growth and yield in northern Guinea savannah. Microbial inoculants increased plant growth, yield and nodulation of soybean up to 45 %.

Rajput et al. (2013) evaluated microbial inoculant *Planococcus rifietoensis* for the improvement of plant growth and yield of wheat (*T. aestivum*) improvement under salinity stress conditions. According to their observations, salinity reduced growth and yield parameters of wheat (up to 60 %) in field experiments, whereas microbial inoculants enhanced plant height (29 % over the control), biomass (36 % over the control), growth (37 % over the control), straw weight (50 % over the control) and yield (38 % over the control) by alleviating the toxic effects of salinity. Under salt stress arid conditions, corn–molasses based microbial inoculants including *Pseudomonas extremorientalis* TSAU20, *P. fluorescens* PCL1751 and *Stenotrophomonas rhizophila* e–p10 caused statistically significant increases in plant height and yield of cucumber fruit up to 32 % compared to the un-inoculated control plants (Egamberdieva et al. 2011). In wheat and barley, Turan et al. (2012) tested five boron treatments and four PGPR strains – *B. megaterium* M3, *B. subtilis* OSU142, *Raoultella terrigena* and *Azospirillum brasilense* sp245 – at 5/2 °C 8-h (day/night) temperature and showed that the PGPR treatments alleviated the deleterious effect of low temperature on both crops.

In a study to identify novel PGPR that can retain activity at low and warm temperatures and help winter wheat in cold-hardiness, Adesemoye (personal communication) carried out a year round collection and screening of authochtonous bacteria from wheat roots in the western part of Nebraska, United States. The region produces high amount of wheat yearly and it is a semi-arid climate where frost bite/damages sometimes occur. Organisms of interest collected so far and being tested in the ongoing study include strains of *Bacillus safensis*, *B. megaterium* and *Bacillus pumilus*. The use of microbial inoculants in arid climatic conditions continues to expand. For example in Iran, Arvin et al. (2012) reported that *Pseudomonas* inoculants increased yield and seed oil content of *Brassica* oilseed rape in field trials under arid soil conditions. Also, microbial preparations based on *Azotobacter* strain Azo-8 increased plant growth and yield of wheat grown under arid conditions (Singh et al. 2013). In another study, *Azospirillum* inoculants combined with P-solubilizing bacteria improved plant growth, fiber yield and fiber quality of cotton under field conditions (Dhale et al. 2011).

In previous field experiments, we have observed plant growth stimulating efficiency of phosphate solubilizing bacterial (PSB) inoculant *Rhizobium meliloti* combined with phosphorite on cotton (Poberejskaya et al. 2003). The PSB combined with phosphorite showed a significant effect on dry matter accumulation in leaves, shoot and root of cotton (Table 11.1).

Compared to the control and fertilizer treatments, the PSB combined with phosphorite was superior over the other treatments. Higher effects were found at the maturity stage. In field experiments, all treatments increased yield of cotton in comparison to control plants (Fig. 11.2). Higher yield obtained after treatment with PSB *R. meliloti*. The yield of cotton

**Table 11.1** The effect of phosphate-solubilizing bacteria (PSB) *Rizobium meliloti* URM1 combined with phosphorite on dry matter of cotton (field experiments[a], g/plant)

| | Tillering | | Flowering | | | Maturity | | |
|---|---|---|---|---|---|---|---|---|
| Treatments | Leaves | Stem | Leaves | Stem | Bud | Leaves | Stem | Bud case |
| Control | 8.1 | 7.0 | 46.5 | 10.5 | 15.5 | 54.6 | 39.5 | 35.3 |
| $NP_{superp}K$ | 9.5 | 6.3 | 47.0 | 18.7* | 17.1 | 53.0 | 42.1 | 36.0 |
| $NP_{phosphorite}K$ | 8.9 | 6.0 | 46.9 | 18.0* | 18.4* | 57.4 | 51.0* | 38.3* |
| $NP_{PSB}K$ | 14.6* | 8.7* | 47.0 | 18.9* | 23.4* | 89.1* | 64.8* | 49.5* |

[a]Recommended rates of phosphorus (140 kg P $h^{-1}$, as phosphorite and superphosphate), nitrogen (200 kg N $ha^{-1}$, as ammonium sulphate) and potassium (60 kg K $h^{-1}$, as potassium sulphate) were applied. Treatments were: Control – plants without treatments, $NP_{superphosphate}$ K, $NP_{phosphorite}$ K, $NP_{phosphorite+PSB}$ K, (PSB – phosphate-solubilizing bacteria, *Rizobium meliloti* URM1), *significantly different from control plants at $P < 0.05$

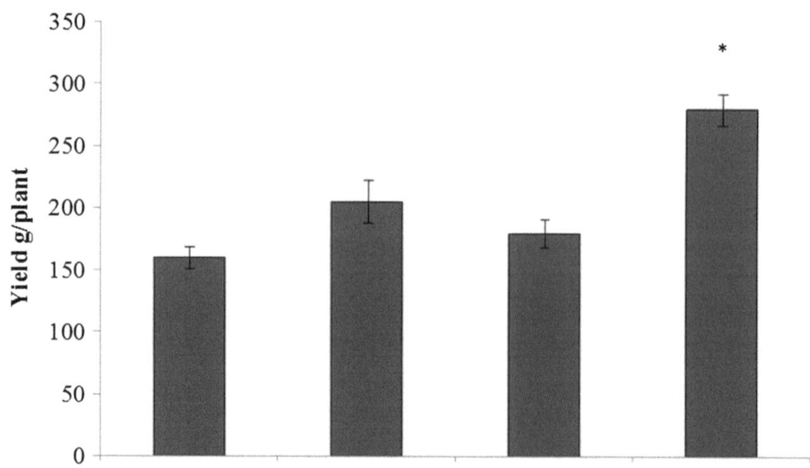


**Fig. 11.2** The effect of phosphate solubilizing bacteria (PSB) *Rizobium meliloti* URM1 combined with phosphorite on cotton fiber yield under semi-arid salinated soil condition. Treatments were: Control – plants without treatments, $NP_{superphosphate}$ K, $NP_{phosphorite}$ K, $NP_{phosphorite+PSB}$ K, (PSB – phosphate-solubilizing bacteria, *Rizobium meliloti* URM1), *significantly different from control plants at $P < 0.05$

**Table 11.2** The effect phosphate-solubilizing bacteria (PSB) *Rizobium meliloti* URM1 combined with phosphorite on N and P uptake of cotton (field experiments[a], N and P content in %)

| | Leaves | | Stem | | Bud case | | Cotton fibers | |
|---|---|---|---|---|---|---|---|---|
| Treatments | N | P | N | P | N | P | N | P |
| Control | 1.45 | 0.51 | 0.68 | 0.21 | 0.78 | 0.19 | 1.78 | 0.81 |
| $NP_{superp}K$ | 1.55 | 0.75* | 0.75 | 0.24 | 0.83* | 0.22 | 1.87 | 0.84 |
| $NP_{phosphorite}K$ | 1.20 | 0.20 | 0.30 | 0.10 | 0.50 | 0.10 | 1.60 | 0.40 |
| $NP_{PSB}K$ | 1.62* | 0.8* | 0.75 | 0.24 | 0.83* | 0.25* | 1.90* | 0.89 |

[a]Recommended rates of phosphorus (140 kg P $h^{-1}$, as phosphorite and superphosphate), nitrogen (200 kg N $ha^{-1}$, as ammonium sulphate) and potassium (60 kg K $h^{-1}$, as potassium sulphate) were applied. Treatments were: Control – plants without treatments, $NP_{superphosphate}$ K, $NP_{phosphorite}$ K, $NP_{phosphorite+PSB}$ K, (PSB – phosphate-solubilizing bacteria, *Rhizobium meliloti* URM1), *significantly different from control plants at $P < 0.05$

increased up to 68 % (275.0 $g^{-1}$ plant) (Poberejskaya et al. 2003).

According to the results obtained, PSB was able to mobilize phosphorus use efficiency in cotton. The phosphorus content was significantly increased in cotton plants when treated with PSB combined with phosphorite (Table 11.2). The standard treatment with fertilizer did not affect P uptake in plants. In another similar study, Shah et al. (2001) reported phosphorus uptake efficiency and yield increases with phosphorus application and with inoculation.

The positive influence of treatments on soil P content is marked (Table 11.3). Soil P content in the variant with PSB reaches 6.0 mg $P_2O_5$/100 g soil. It has been found that the application of phosphorite combined with PSB leads to the increase of P content in soil (tillering, flowering and maturity stages of plants). This result suggests that PSB combined with phosphorite were able to mobilize more P to the plants and improve plant growth. Yasmin et al. (2013) also observed improved yield of cotton under reduced fertilizer conditions by formulations based on *Pseudomonas aeruginosa* Z5 and *Bacillus fusiformis* S10. Similar results were observed for wheat when biofertilizer based on *Azotobacter* in combination with manure improved plant growth, stress tolerance and yield under water stress conditions (Singh et al. 2013).

The immobilized microbial inoculants *A. brasilense* and *Pantoea dispersa* in clay pellets combined with organic olive residue stimulated root dry weight (by 133 % with respect to control plants), shoot dry weight (by 106 % with respect to control plants), P and K content (by 100–70 % with respect to control plants). In addition microbial inoculants increased total C, total organic C and microbial biomass C content and enzyme activities (dehydrogenase, urease and protease) of the rhizosphere of *Cistus albidus* (Schoebitz et al. 2014). In another study, peat-based inoculants of *Rhizobium phaseoli* and *P. fluorescens* reduced adverse effects of salinity on relative water contents, photosynthetic rate, water use efficiency and chlorophyll content of mung beans over the un-inoculated control. In addition, combined inoculation improved the ionic balance and also increased the phosphorus and protein concentrations in grain of mung bean under salt-affected conditions (Ahmad et al. 2013). Peat-based inocula of *R. phaseoli* mitigated salt

**Table 11.3** Phosphorus content in soil as affected by posphate solubilizing bacteria (PSB) *Rizobium meliloti* URM1 combined with phosphorite (field experiments[a], before sowing 1.8 mg $P_2O_5$/100 g soil)

| Treatments | Tillering | Flowering | Maturity |
|---|---|---|---|
| Control | 2.4 | 1.5 | 2.2 |
| $NP_{superp}K$ | 2.8 | 6.0* | 4.7* |
| $NP_{phosphorite}K$ | 2.0 | 1.8 | 1.9 |
| $NP_{PSB}K$ | 5.4* | 6.0* | 4.0* |

[a]Recommended rates of phosphorus (140 kg P $h^{-g}$, as phosphorite and superphosphate), nitrogen (200 kg N $ha^{-1}$, as ammonium sulphate) and potassium (60 kg K $h^{-1,}$ as potassium sulphate) were applied. Treatments were: Control – plants without treatments, $NP_{superphosphate}$ K, $NP_{phosphorite}$ K, $NP_{phosphorite+PSB}$ K, (PSB – posphate solubilizing bacteria, *Rizobium meliloti* URM1), *significantly different from control plants at $P < 0.05$

stress in mung bean and improved growth and yield under saline conditions (Zahir et al. 2010). The plant height was improved by 28.2 %, number of nodules by 71.4 %, plant biomass by 61.2 % and grain yield by 65.3 % compared with untreated control.

## 11.5 Conclusion and Future Perspectives

In arid and semi-arid regions, soil fertility is diminishing gradually due to water stress, salinity, poor nutrients, erosion, high/abrupt changes in temperature and other climatic conditions. There is good evidence that fertilizers based on microbial inoculants could be very useful to improve crop yield and soil productivity. Several materials have been evaluated as carriers for bacterial inoculants, successfully protecting them from various stress factors and prolonging shelf life. Stress-tolerant microorganisms with plant-stimulating properties are being discovered, selected and tested under field conditions and the number of successful applications is increasing. The combination of inoculants has the potential to alleviate abiotic stress, promote plant growth, enhances the root system and nutrient availability that can assist the plant to absorb more nutrients. Those results imply that formulated microbial inoculants could be a useful approach for improving growth and yield of crop plants under hostile environmental conditions. Further research should be directed at optimizing methods for microbial formulations considering agro-ecological conditions and to determine the feasibility of using microbial fertilizers under severe environmental conditions. In addition, further research should be directed at studying the effectiveness of inoculants in integrated systems that examine cropping practices such as fertilizer inputs, tillage practices and crop rotation as they affect introduced microbial inoculants, their survival and plant beneficial properties.

**Acknowledgment** The research activity of Dilfuza Egamberdieva was supported by a Georg Forster Research Fellowship for experienced Researchers (HERMES), Alexander Von Humboldt Foundation.

## References

Adesemoye AO, Egamberdieva D (2013) Beneficial effects of plant growth promoting rhizobacteria on improved crop production: the prospects for developing economies. In: Maheshwari DK, Saraf M, Aeron A (eds) Bacteria in agrobiology: crop productivity. Springer, Berlin/Heidelberg, pp 45–63

Adesemoye AO, Kloepper JW (2009) Plant–microbes interactions in enhanced fertilizer use efficiency. Appl Microbiol Biotech 85:1–12

Adesemoye AO, Torbert HA, Kloepper JW (2008) Enhanced plant nutrient use efficiency with PGPR and AMF in an integrated nutrient management system. Can J Microbiol 54:876–886

Adesemoye AO, Torbert HA, Kloepper JW (2009) Plant growth–promoting rhizobacteria allow reduced application rates of chemical fertilizers. Microb Ecol 58:921–929

Adesemoye AO, Wei H–H, Yuen G. Prospecting for cold–hardy autochthonous novel bacteria in crop root microbiome (personal communication)

Ahmad P, Hakeem KR, Kumar A, Ashraf M, Akram NA (2012) Salt induced changes in photosynthetic activity and oxidative defense system of three cultivars of mustard (Brassica juncea L.). Afr J Biotech 11:2694–2703

Ahmad M, Zahir ZA, Nadeem SM, Nazli F, Jamil M, Khalid M (2013) Field evaluation of *Rhizobium* and *Pseudomonas* strains to improve growth, nodulation and yield of mung bean under salt–affected conditions. Soil Environ 32:158–166

Albareda M, Rodríguez–Navarro DN, Camacho M, Temprano FJ (2008) Alternatives to peat as a carrier for rhizobia inoculants: solid and liquid formulations. Soil Biol Bioch 40:2771–2779

Araujo FF, Souza EC, Guerreiro RT, Guaberto LM, Araujo ASF (2012) Diversity and growth–promoting activities of *Bacillus* sp. in maize. Revista Caatinga Mossoro 25:1–7

Ardakani SS, Hedari A, Tayebi L, Mohammadi M (2010) Promotion of cotton seedlings growth characteristics by development and use of new bioformulations. Int J Bot 6:95–100

Arora NK, Tiwari S, Singh R (2014) Comparative study of different carriers inoculated with nodule forming and free living plant growth promoting bacteria suitable for sustainable agriculture. J Plant Path Microb 5:229

Arshad M, Shaharoona B, Mahmood T (2008) Inoculation with *Pseudomonas* sp. containing ACC–deaminase partially eliminates the effects of drought stress on growth, yield, and ripening of pea (*Pisum sativum* L.). Pedosphere 18:611–620

Arvin P, Vafabakhsh J, Mazaheri D, Noormohamadi G, Azizi M (2012) Study of drought stress and plant growth promoting rhizobacteria (PGPR) on yield, yield components and seed oil content of different cultivars and species of *Brassica* oilseed rape. Ann Biol Res 3:4444–4451

Ashwini N, Srividya S (2014) Potentiality of *Bacillus subtilis* as biocontrol agent for management of anthracnose disease of chilli caused by *Colletotrichum gloeosporioides* OGC1. 3 Biotech 4:127–136

Bano A, Yasmeen S (2010) Role of phytohormones under induced drought stress in wheat. Pak J Bot 42 (4):2579–2587

Bashan I (1998) Inoculants of plant growth–promoting bacteria for use in agriculture. Biotechnol Adv 16:729–770

Ben Rebah F, Prevost D, Tyagi RD (2002) Growth of alfalfa in sludge–amended soils and inoculated with rhizobia produced in sludge. J Environ Qual 31:1339–1348

Berg G, Alavi M, Schmidt CS, Zachow C, Egamberdieva D, Kamilova F, Lugtenberg B (2013) Biocontrol and osmoprotection for plants under saline conditions. In: de Bruijn FJ (ed) Molecular microbial ecology of the rhizosphere. Wiley –Blackwell, Hoboken

Bharathi R, Vivekananthan R, Harish S, Ramanathan S, Samiyappan R (2004) Rhizobacteria–based bio–formulations for the management of fruit rot infection in chillies. Crop Prot 23:835–843

Daza A, Santamaria C, Rodriguez Navarro DN, Camacho M, Orive R, Temprano F (2000) Perlite as a carrier for bacterial inoculants. Soil Biol Bioch 32:567–572

Dhale DA, Chatte SN, Jadhav VT (2011) Response of bioinoculants on growth, yield and fiber quality of cotton under irrigation. Agric Biol J North America. Science Huß, http://www.scihub.org/ABJNA

Egamberdieva D (2012) *Pseudomonas chlororaphis*: a salt tolerant bacterial inoculant for plant growth stimulation under saline soil conditions. Acta Physiol Plant 34:751–756

Egamberdieva D, Lugtenberg B (2014) PGPR to alleviate salinity stress on plant growth. In: Miransari M (ed) Use of microbes for the alleviation of soil stresses, vol 1. Springer, New York, pp 73–96

Egamberdieva D, Berg G, Lindstrom K, Rasanen L (2010) Root colonising Pseudomonas sp. improve growth and symbiosis performance of fodder galega (*Galega orientalis* LAM) grown in potting soil. Eur J Soil Biol 46:269–272

Egamberdieva D, Kucharova Z, Davranov K, Berg G, Makarova N, Azarova T, Chebotar V, Tikhonovich I, Kamilova F, Validov S, Lugtenberg B (2011) Bacteria able to control foot and root rot and to promote growth of cucumber in salinated soils. Biol Fertil Soils 47:197–205

Egamberdieva D, Berg G, Lindström K, Räsänen LA (2013) Alleviation of salt stress of symbiotic *Galega officinalis* L. (goat's rue) by co–inoculation of rhizobium with root colonising *Pseudomonas*. Plant and Soil 369:453–465

Egamberdieva D, Shurigin V, Gopalakrishnan S, Sharma R (2014) Growth and symbiotic performance of chickpea (*Cicer arietinum*) cultivars under saline soil conditions. J Biol Chem Res 31:333–341

Egamberdiyeva D (2005) Characterisation of *Pseudomonas* species isolated from the rhizosphere of plants grown in serozem soil, semi arid region of Uzbekistan. Sci World J 5:501–509

Egamberdiyeva D (2007) The growth and nutrient uptake of maize inoculated with plant growth promoting bacteria affected by different soil types. Appl Soil Ecol 36:184–189

Egamberdiyeva D, Hoflich G (2002) Root colonization and growth promotion of winter wheat and pea by *Cellulomonas* spp. at different temperatures. J Plant Growth Reg 38:219–224

Egamberdiyeva D, Hoflich G (2003) Influence of growth promoting bacteria on the growth of wheat at different soils and temperatures. Soil Biol Biochem 35:973–978

Egamberdiyeva D, Hoflich G (2004) Importance of plant growth promoting bacteria on growth and nutrient uptake of cotton and pea in semi–arid region Uzbekistan. J Arid Envir 56:293–301

Egamberdiyeva D, Qarshieva D, Davranov K (2004) The use of *Bradyrhizobium japonicum* to enhance growth and yield of soybean varieties in Uzbekistan conditions. Plant Growth Reg 23:54–57

Elkoca E, Turan M, Donmez MF (2010) Effects of single, dual and triple inoculations with *Bacillus subtilis, Bacillus megaterium* and *Rhizobium leguminosarum* bv. *phaseoli* on nodulation, nutrient uptake, yield and yield parameters of common bean (*Phaseolus vulgaris* L.). J Plant Nutr 33:2104–2119

Fravel DR (2005) Commercialization and implementation of biocontrol. Ann Rev Phytopath 43:337–359

Fu Q, Liu C, Ding N, Lin Y, Guo B (2010) Ameliorative effects of inoculation with the plant growth–promoting rhizobacterium *Pseudomonas* sp. DW1 on growth of eggplant (*Solanum melongena* L.) seedlings under salt stress. Agric Water Manage 97:1994–2000

Garg N, Baher N (2013) Role of arbuscular mycorrhizal symbiosis in proline biosynthesis and metabolism of *Cicer arietinum* L. (chickpea) genotypes under salt stress. J Plant Growth Regul 32:767–778

Glick BR (2010) Using soil bacteria to facilitate phytoremediation. Biotechnol Adv 28:367–374

Golpayegani A, Tilebeni HG (2011) Effect of biological fertilizers on biochemical and physiological parameters of Basil (*Ociumum basilicm* L.) medicine plant. Amer–Euras J Agric Environ Sci 11:411–416

Goswami D, Vaghela H, Parmar S, Dhandhukia P, Thakker JN (2013) Plant growth promoting potentials of *Pseudomonas* spp. strain OG isolated from marine water. J Plant Interact 8:281–290

Hale L, Luth M, Crowley D (2015) Biochar characteristics relate to its utility as an alternative soil inoculum carrier to peat and vermiculite. Soil Biol Bioch 81:228–235

Hameed A, Egamberdieva D, Abd–Allah EF, Hashem A, Kumar A, Ahmad P (2014) Salinity stress and arbuscular mycorrhizal symbiosis in plants. In: Miransari M (ed) Use of microbes for the alleviation of soil stresses, vol 1. Springer, New York, pp 139–159

Hashem A, Abd_Allah EF, El–Didamony G, Alwhibi Mona S, Egamberdieva D, Ahmad P (2014) Alleviation of adverse impact of salinity on faba bean (*Vicia Faba* L.) by arbuscular mycorrhizal fungi. Pak J Bot 46:2003–2013

Hashem A, Abd_Allah EF, Alqarawi AA, Alwhibi Mona S, Alenazi MM, Egamberdieva D, Ahmad P (2015) Arbuscular mycorrhizal fungi mitigates NaCl induced adverse effects on *Solanum lycopersicum* L. Pak J Bot 47:327–340

Islam MZ, Sattar MA, Ashrafuzzaman M, Berahim Z, Shamsuddoha ATM (2013) Evaluating some salinity tolerant rhizobacterial strains to lentil production under salinity stress. Int J Agric Biol 15:499–504

Karlidag H, Yildirim E, Turan M, Pehluvan T, Donmez F (2013) Plant growth–promoting rhizobacteria mitigate deleterious effects of salt stress on strawberry plants (*Fragaria ananassa*). Hort Sci 48:563–567

Khavazi K, Rejali F, Seguin P, Miransari M (2007) Effects of carrier, sterilization method, and incubation on survival of *Bradyrhizobium japonicum* in soybean (*Glycine max* L.) inoculants. Enzyme Microb Tech 41:780–784

Khurana AS, Sharma P (2000) Effect of dual inoculation of phosphate solubilizing bacteria, *Bradyrhizobium* sp. and phosphorus on nitrogen fixation and yield of chickpea. Indian J Pulses Res 13:66–67

Korus K, Adesemoye AO, Giesler L, Harveson RM, Jackson–Ziems TA, Wegulo SN (2015) Weather variability and disease management strategies. In: Proceedings of the 2015 crop production clinics, University of Nebraska Lincoln Extension Publication, pp 144–146

Kumar B, Trivedi P, Pandey A (2007) *Pseudomonas corrugata*: a suitable bioinoculant for maize grown under rainfed conditions of Himalayan region. Soil Biol Biochem 39:3093–3100

Lugtenberg B, Kamilova F (2009) Plant–growth–promoting–rhizobacteria. Ann Rev Microbiol 63:541–56

Malusá E, Sas–Paszt L, Ciesielska J (2012) Technologies for beneficial microorganisms inocula used as biofertilizers. Sci World J. doi:10.1100/2012/491206

Mantri N, Patade V, Penna S, Ford R, Pang E (2012) Abiotic stress responses in plants: present and future. In: Ahmad P, Prasad MNV (eds) Abiotic stress responses in plants: metabolism, productivity and sustainability. Springer, New York, pp 1–19

Marin M (2006) Arbuscular mycorrhizal inoculation in nursery practice. In: Rai MK (ed) Handbook of microbial biofertilisers. International Book Distributing Co, Lucknow, pp 289–324

Marques APGC, Piresa C, Moreira H, Rangel AOSS, Castro PML (2010) Assessment of the plant growth promotion abilities of six bacterial isolates using *Zea mays* as indicator plant. Soil Biol Biochem 42:1229–1235

Mishra PK, Bisht SC, Ruwari P, Selvakumar G, Joshi GK, Bisht JK, Bhatt JC, Gupta HS (2011) Alleviation of cold stress in inoculated wheat (*Triticum aestivum* L.) seedlings with psychrotolerant *Pseudomonads* from NW Himalayas. Arch Microbiol 193:497–513

Munns R, Tester M (2008) Mechanisms of salinity tolerance. Ann Rev Plant Biol 59:651–681

N'cho CO, Yusuf AA, Ama–Abina JT, Jemo A, Abaidoo RC, Savane I (2013) Effects of commercial microbial inoculants and foliar fertilizers on soybean nodulation and yield in northern Guinea savannah of Nigeria. Inter J Adv Agric Res 1:66–73

Nabti E, Sahnoune M, Ghoul M, Fischer D, Hofmann A, Rothballer M, Schmid M, Hartmann A (2010) Restoration of growth of durum wheat (*Triticum durum* var. waha) under saline conditions due to inoculation with the rhizosphere bacterium *Azospirillum brasilense* NH and extracts of the marine alga *Ulva lactuca*. J Plant Growth Regul 29:6–22

Nadeem SM, Zahir ZA, Naveed M (2010) Microbial ACC deaminase: prospects and applications for inducing salt tolerance in plants. Crit Rev Plant Sci 29 (6):360–393

Negi YK, Kumar J, Garg SK (2005) Cold–tolerant fluorescent *Pseudomonas* isolates from Garhwal Himalayas as potential plant growth promoting and biocontrol agents in pea. Curr Sci 89:2151–2156

Othman Y, Al–Karaki G, Al–Tawaha AR, Al– Horani A (2006) Vaviation in germination and jon uptake in barley genotypes under salinity conditions. World J Agric Sci 2(1):11–15

Pandey A, Sharma E, Palni L (1998) Influence of bacterial inoculation on maize in upland farming systems of the sikkim Himalaya. Soil Biol Bioch 3:379–384

Patel DP, Singh HB, Shroff S, Sahu J (2012) Studies on growth promotion activities of *Trichoderma harzianum* on chickpea. Adv Plant Sci 25:193–195

Pathma J, Sakthivel N (2012) Microbial diversity of vermicompost bacteria that exhibit useful agricultural traits and waste management potential. Springer Plus 1:26

Poberejskaya S, Egamberdiyeva D, Myachina O, Teryuhova P, Seydalieva L, Aliev A, Kim P (2003) Improvement of the productivity of cotton by phosphate solubilizing bacterial inoculants. In: Proceedings of the 6th international symposium and exhibition on environmental contamination in Central and Eastern Europe and the Commonwealth of Independent States, 1–6 September, Prague, Czech Republic

Rajput L, Imran A, Mubeen F, Hafeez FY (2013) Salt–tolerant PGPR strain *Planococcus rifietoensis* promotes the growth and yield of wheat (*Triticum aestivum* L.) cultivated in saline soil. Pak J Bot 45:1955–1962

Rekha PD, Lai WA, Arun AB, Young CC (2007) Effect of free and encapsulated *Pseudomonas putida* CC–FR2–4 and *Bacillus subtilis* CC–pg104 on plant growth under gnotobiotic condition. Bio Res Tech 98:447–451

Rinu K, Pandey A (2009) *Bacillus subtilis* NRRL B–30408 inoculation enhances the symbiotic efficiency of Lens esculenta Moench at a Himalayan location. J Plant Nutr Soil Sci 172:134–139

Roopa B, Maya C, Makari HK (2012) Effect of different PGPR strain along with rhizobium on nodulation and chick pea productivity. Asian J Exp Biol Sci 3:424–426

Sahni S, Sarma BK, Singh DP, Singh HP, Singh KP (2008) Vermicompost enhances performance of plant growth–promoting rhizobacteria in *Cicer arietinum* rhizosphere against *Sclerotium rolfsii*. Crop Prot 27:369–376

Sangeetha D (2012) Survival of plant growth promoting bacterial inoculants in different carrier materials. Int J Pharm Biol Arch 3:170–178

Schoebitz M, Mengual C, Roldán A (2014) Combined effects of clay immobilized *Azospirillum brasilense* and *Pantoea dispersa* and organic olive residue on plant performance and soil properties in the re–vegetation of a semiarid area. Sci Total Environ 466–467:67–73

Shah P, Kakar KM, Zada K (2001) Phosphorus use efficiency of soybean as affected by phosphorus application and inoculation. In: Horst WJ (eds) Plant nutrition– food security and sustainability of agroecosystems. Springer, Dordrecht, pp 670–671

Shaharoona B, Arshad M, Zahir ZA (2006) Effect of plant growth promoting rhizobacteria containing ACC–deaminase on maize (*Zea mays* L.) growth under axenic conditions and on nodulation in mung bean (*Vigna radiata* L). Lett Appl Microbiol 42:155–159

Sharma P, Khanna V, Kumari P (2013) Efficacy of aminocyclopropane–1–carboxylic acid (ACC)–deaminase–producing rhizobacteria in ameliorating water stress in chickpea under axenic conditions. Afr J Microbiol Res 7:5749–5757

Shirmardi M, Savaghebi GR, Khavazi K, Akbarzadeh A, Farahbakhsh M, Rejali F, Sadat A (2010) Effect of microbial inoculants on uptake of nutrient elements in two cultivars of sunflower (*Helianthus annuus* L.) in saline soils. Notulae Scientia Biologicae 2:57–66

Shrivastava S, Egamberdieva D, Varma A (2015) PGPRs and medicinal plants– the state of arts. In: Egamberdieva D, Shrivastava S, Varma A (eds) PGPR and medicinal plants. Springer International Publishing Switzerland. Soil Biology 42

Singh A, Sarma BK, Upadhyay RS, Singh HB (2013) Compatible rhizosphere microbes mediated alleviation of biotic stress in chickpea through enhanced antioxidant and phenylpropanoid activities. Microbiol Res 168:33–40

Son TTN, Diep CN, Giang TTM (2006) Effect of Bradyrhizobia and phosphate solubilizing bacteria application on soybean in rotational system in the Mekong delta. Omonrice 14:48–57

Stephens JHG, Rask HM (2000) Inoculant production and formulation. Field Crop Res 65:249–258

Trivedi P, Pandey A, Palni LM (2005) Carrier–based preparations of plant growth–promoting bacterial inoculants suitable for use in cooler regions. World J Microbiol Biotechnol 21:941–945

Trivedi P, Pandey A, Palni LS (2012) Bacterial inoculants for field applications under mountain ecosystem: present initiatives and future prospects. In: Maheshwari DE (ed) Bacteria in agrobiology: plant probiotics. Springer Berlin Heidelberg, pp 15–44

Turan M, Gulluce M, Şahin F (2012) Effects of plant–growth–promoting rhizobacteria on yield, growth, and some physiological characteristics of wheat and barley plants. Commun Soil Sci Plant Anal 43:1658–1673

UNEP (2008) In dead water. Merging of climate change with pollution, over–harvest, and infestations in the world's fishing grounds. UNEP/GRID–Arendal, Arendal, Norway. Available online at: http://www.grida.no/_res/site/file/publications/InDeadWater_LR.pdf

Valverde A, Velazquez E, Santos FF, Vizcaino N, Rivas R, Mateos PF, Molina EM, Igual JM, Willems A (2005) *Phyllobacterium trifolii* sp. nov., nodulating *Trifolium* and *Lupinus* in Spanish soils. Int J Syst Evol Microbiol 55:1985–1989

Van Dyke MI, Prosser JI (2000) Enhanced survival of *Pseudomonas fluorescens* in soil following establishment of inoculum in a sterile soil carrier. Soil Biol Biochem 32:1377–1382

Vardharajula S, Ali SZ, Grover M, Reddy G, Bandi V (2011) Drought–tolerant plant growth promoting *Bacillus* spp.: effect on growth, osmolytes, and anti–-oxidant status of maize under drought stress. J Plant Interact 6:1–14

Wu SC, Caob ZH, Lib ZG, Cheung KC, Wong MH (2005) Effects of biofertilizer containing N–fixer, P and K solubilizers and AM fungi on maize growth: a greenhouse trial. Geoderma 125:155–166

Wu Z, Zhao Y, Kaleem I, Li C (2011) Preparation of calcium alginate microcapsuled microbial fertilizer coating *Klebsiella oxytoca* Rs–5 and its performance under salinity stress. Eur J Soil Biol 47:152–159

Yasmin S, Hafeez F, Schmid M, Hartmann A (2013) Plant–beneficial rhizobacteria for sustainable increased yield of cotton with reduced level of chemical fertilizer. Pak J Bot 45:655–662

Yildirim E, Turan M, Ekinci M, Dursun A, Cakmakci R (2011) Plant growth promoting rhizobacteria ameliorate deleterious effect of salt stress on lettuce. Sci Res Essays 6(20):4389–4396

Zafar–ul–Hye M, Farooq HM, Zahir ZA, Hussain M, Hussain A (2014) Application of ACC–deaminase containing rhizobacteria with fertilizer improves maize production under drought and salinity stress. Int J Agric Biol 16:591–596

Zahir ZA, Shah MK, Naveed M, Akhtar MJ (2010) Substrate dependent auxin production by *Rhizobium phaseoli* improve the growth and yield of *Vigna radiate* L under salt stress conditions. J Microbiol Biotechnol 20:1288–1294

# 12 Use of Indigenous Cyanobacteria for Sustainable Improvement of Biogeochemical and Physical Fertility of Marginal Soils in Semiarid Tropics

Luigi Paolo D'Acqui

## Contents




**Abstract**

This chapter describes a realistic and innovative approach for promoting a sustainable increase of overall soil fertility of marginal lands in representative semiarid environments in East and Southern Africa (South Africa, Tanzania and Zimbabwe). This realistic and innovative approach was based on the application of indigenous selected strains of nitrogen-fixing cyanobacteria and releasing organic compounds such as exopolysaccharides (EPS) to improve soil fertility and structural stability. To achieve these goals, 17 cyanobacteria strains (over 200 identified in soils of the African countries), able to fix atmospheric nitrogen and to release EPS in considerable amounts, were isolated, purified and grown. Using adequate techniques and procedures, large amounts of cyanobacterial biomasses were produced to be applied to poor soils to ameliorate their quality and to improve their productivity. Laboratory and greenhouse experiments showed that the application of these biomasses significantly improved overall soil fertility and crop yield. In spite of the need to confirm the relevance and persistence of the beneficial effects by further medium and long-term field experiments, the application of selected cyanobacteria strains to marginal lands appeared to be a very promising tool for a



L.P. D'Acqui (✉)
Istituto per lo Studio degli Ecosistemi, CNR, via Madonna del Piano n. 10, 50019 Sesto Fiorentino, Italy
e-mail: dacqui@ise.cnr.it





N.K. Arora et al. (eds.), *Bioformulations: for Sustainable Agriculture*,
DOI 10.1007/978-81-322-2779-3_12

sustainable improvement of fertility and productivity of degraded soils in semiarid tropics.

## 12.1 Introduction

In many tropical countries, the scarcity of productive land and water are the major constraint for the improvement of the standard of living for inhabitants due to the increase of population and, consequently, of food demand. This leads to a progressive use of marginal lands with inherently poor, overall fertility. In these areas clearing and cultivation frequently lead to a rapid decline of natural biochemical fertility as a consequence of loss of nutrients, in particular nitrogen (N) and organic matter (OM), exacerbated by the removal of crop residues from soil used for feeding livestock or for other purposes. Cultivation and the associated loss of OM also promote degradation of physical properties and in particular crusting of the soil surface. This seriously affects soil erosion, effective use of water (whether under irrigation or rain-fed conditions), tillage, seedling emergence and general productivity of the land, especially in low-input agricultural systems of Southern African Tropics (SAT).

An appropriate management of soil and water resources is therefore needed, particularly in arid and semiarid regions. Hence, it is aimed to report how to establish a possible realistic, innovative approach for promoting a sustainable increase of overall soil fertility of marginal lands in representative semiarid environments in East and Southern Africa (South Africa, Tanzania and Zimbabwe) by application of indigenous selected strains of soil cyanobacteria fixing atmospheric $N_2$ and releasing organic compounds (EPS), able to improve soil structural stability.

The presented results mainly come from a study carried out within a project (CYANOSOILS) funded by the European Union. The idea was based on the application of local strains of cyanobacteria, widespread autotrophic soil and water microorganisms, as a low-input tool easily accessible to small-scale farmers in Southern African countries, to improve resilience and overall soil fertility. Many strains of cyanobacteria are able to fix significant amounts of atmospheric $N_2$ and to release EPS, well-known biological amendments to enhance soil physical properties.

Cyanobacteria (oxygenic, photosynthetic bacteria) are known to improve chemical and biological characteristics of the soils, increasing the content of N (by $N_2$-fixing activity) and C (by photosynthesis) and stimulating microbial community activities. The ability to fix $N_2$ (one of the macronutrients often limiting productivity in soils of tropical regions) is shared by many species belonging to the genera *Nostoc*, *Anabaena*, *Fischerella* and *Scytonema*.

The practice of soil inoculation with living $N_2$-fixing cyanobacteria (algalisation or more appropriately cyanobacterisation), has been largely used in Asian countries for many years for increasing soil organic N content mainly in tropical rice cultivation (Venkataraman 1979; Zimmerman 1993; Ghosh and Saha 1997). Data obtained in several experiments suggest an improvement of total N to around 10–30 % in alkaline soils and up to 120 % in neutral soils.

The photosynthetic activity of these organisms involves other beneficial effects such as the increase of soil OM, under favourable conditions of light, moisture and pH, due to the accumulation, as cyanobacterial biomass, up to 450 kg $ha^{-1}$ of organic carbon (OC) and 60 kg $ha^{-1}$ of N. As a consequence, in the long term, a large amount of OC arising from cyanobacterial biomass contributes to a stable C pool in soils. Moreover, the improvement of biological soil conditions by cyanobacterial inoculation results in: (1) a very large increase of total bacterial soil population and specifically of the genus *Azotobacter* with a consequent enhancement of the nitrogenase activity, (2) an increase in the number of actinomycetes and limited increase of fungi, (3) an increase in indigenous cyanobacterial population, (4) an improvement of seedling emergence, (5) biological control of plant pathogenic microbes and (6) a general increase of soil enzymatic activities (Hu et al. 2012).

The positive effects of cyanobacteria can be extended to soil physical properties. In particular, it has been suggested that the role of these microorganisms in the formation of aggregates could be of considerable importance in arid and semiarid tropical soils (Barclay and Lewin 1985; Cogle et al. 1995). In well-aggregated soils, OM and iron oxides bind clay and other particles into water stable structures that resist to the destructive energy of the raindrop impact. Therefore, under conditions of arable agriculture, and particularly in those prevailing in SAT farming systems, soil degradation takes place over time as a result of the progressive loss of OM leading to a substantial weakening of aggregates that become more and more vulnerable to the droplet impact.

Cyanobacteria have been recognised to contribute to soil stabilisation through the production of EPS that have been demonstrated to be among the most effective organic substances for enhancing soil aggregate stability (Rao and Burns 1990; Rogers et al. 1991; Martens and Frankenberger 1992; Rogers and Burns 1994; Falchini et al. 1996; Malam Issa et al. 2007; Maqubela et al. 2012). However, experiments using these microorganisms as soil conditioners are rare, in spite of their potential effectiveness in controlling soil structure. Although growth and activity of cyanobacteria are related to the availability of a sufficient moisture, the importance of soil desiccation and soil water potential on the establishment and activity of inocula and on polysaccharides production have not been specifically investigated.

It can be expected that, if sound studies on the most appropriate conditions for field application in SAT are performed, the inoculation with living cyanobacteria may lead to the establishment and growth of this group of microorganisms in the soil with durable positive effects on overall fertility.

## 12.2 Site Selection, Sampling and Soil Characterisation

The study was conducted in three Southern African countries with semiarid environments: South Africa, Tanzania and Zimbabwe. In South Africa and Zimbabwe, a large number of small-scale farming enterprises were created, and many of these settled in marginal lands because of a massive land-reform programme. In the case of Tanzania, differently, the country had to face with a limited food supply, due to very little commercial agriculture activity and, as a consequence, the need to develop a small-scale agricultural sector to produce efficiently more food, consistently with a good environmental protection ethos. The study sites were selected in representative marginal land areas typical of each country. Seven relatively nutrient poor soils located in important agricultural areas were selected in the three African countries. In Tanzania the sites were Mkindo in Mvomero district and Mafiga located at Sokoine University Farm in Morogoro district. The soils of Mkindo, classified as Gleyic Cambisol, according to World Reference Base for Soil Resource (WRB) classification (IUSS 2014), had been cultivated for 15 years with rice being the crop usually grown. Mafiga site had been cultivated with maize for more than 35 years, with periodic 2-year fallows. This soil was classified as Rhodi-Profondic Lixisol (WRB). In South Africa two soils from the Eastern Cape Province were selected. They were Guquka, located below the escarpment of the Amatola Mountains and Hertzog in the Northern part of the Kat river basin. Guquka soil (Ferric Luvisol, WRB) had been cropped with maize under rain-fed conditions but was on fallow during the time of sampling. Hertzog site was a small-scale irrigation scheme that had also been cropped with maize and was on its second fallow year at the time of the sampling. This soil was defined as Haplic Luvisol in WRB classification. In Zimbabwe, the sampling sites were located in the Central region near Domboshawa village, Chikwaka village and Henderson research station in Mazowe district. The three Zimbabwean sites were under rain-fed maize and represented low-fertility sandy cropping environments occupied by small-scale poor households. These soils were classified in WRB classification as Haplic Lixisol, Chromic Luvisol and Haplic Lixisol, respectively.

**Table 12.1** Main characteristics of the studied soils (average of three replicates) (Source: Pardo et al. 2010)

| Characteristics | Units | Mkindo | Mafiga | Domboshawa | Chikwaka | Hertzog | Guquka |
|---|---|---|---|---|---|---|---|
| Clay | % | 9 | 23 | 17 | 10 | 7 | 13 |
| Silt | % | 30 | 44 | 5 | 2 | 18 | 41 |
| Sand | % | 61 | 33 | 78 | 88 | 75 | 46 |
| pH ($H_2O$) | 1:2.5 | 6.8 | 6.4 | 6.1 | 6.1 | 7.3 | 5.4 |
| Total N | g $kg^{-1}$ | 0.9 | 1.1 | 0.8 | 0.3 | 1.4 | 0.4 |
| Organic C | g $kg^{-1}$ | 16.3 | 12.3 | 10.2 | 6.7 | 16.1 | 6.1 |
| CEC | $mmol_c$ $kg^{-1}$ | 81 | 101 | 21 | 11 | 133 | 18 |
| ESP | % | 1.4 | 1.3 | 1.9 | 3.1 | 0.2 | 4.8 |

*CEC* Cation exchange capacity, *ESP* Exchangeable sodium percentage

Topsoil samples were collected at each site in agricultural areas where soils with chemical or physical constraints occurred. Twelve individual samples, in an area of around 1 ha, collected from each sites were used to prepare three replicate composite samples (D'Acqui et al. 2006; Pardo et al. 2007, 2010).

Soil particle-size distribution and pH measurements were carried out according to Società Italiana della Scienza del Suolo (SISS) methods (1985). Total OC and N content were determined by a Carlo Erba (Milan, Italy) NA 1500 CHNS Analyzer. Cation exchange capacity (CEC) was determined by using 1 M $NH_4OAc$ (Juo et al. 1976). Main soil properties and analyses of the selected soils are reported in Table 12.1.

## 12.3 Isolation, Identification and Characterisation of Indigenous Cyanobacterial Strains Suitable for Soil Inoculation

### 12.3.1 Isolation and Identification of Soil Cyanobacteria

Traditional microbiological techniques were used for the isolation from soil samples of cyanobacteria suitable for soil inoculation with particular reference to $N_2$-fixing activity, EPS production (i.e. cyanobacteria strains with biofertilisation and bioconditioning potential, respectively).

The cyanobacteria strains were isolated from fresh soil samples using BG11 growth medium and strains, able to grow diazotrophically, and were isolated with medium $BG11_0$ (BG-11 without $NaNO_3$) as described by Rippka et al. (1979). Cultures were incubated at 28 °C in a growth chamber with a light of approximately 10–30 μmol photons $m^{-2}$ $s^{-1}$. The enumeration of soil cyanobacteria was assessed after 2 weeks, through both the most probable number (MPN), by using McGrady's tables (McGrady 1915) and the colony-forming units (CFU) counting in Petri dishes. For both enumerations, the BG-11 and $BG\text{-}11_0$ media were used. A Harrison disc was used to randomly pick different colonies that had developed on the Petri dishes based on their diversity in morphology, colour of the colonies and micromorphology of the organisms as observed under a light microscope. The selected colonies were purified by repeatedly streaking on new growth media supplemented with cycloheximide (eukaryote inhibitor). The isolation of cyanobacterial strains in monoculture and identification was carried out according to Castenholz (2001).

More than 200 strains of cyanobacteria and green microalgae (from all countries included in the study) were isolated and conserved in agar slants. After a careful observation under light microscope, around 150 strains were identified and maintained in culture for further studies.

### 12.3.2 Selection of Suitable Cyanobacteria

Selection of cyanobacterial strains with biofertilisation and bioconditioning potential was done by determining the nitrogenase activity of diazotrophic strains and their ability to produce EPS. Nitrogenase activity was determined using acetylene-reduction assay. Liquid cultures were set in Erlenmeyer flasks containing 25 mL of the $BG11_0$ broth inoculated with the cyanobacterial strains and incubated at 30 °C under light (75 μmol photons $m^{-2}$ $s^{-1}$) with continuous shaking at 80 r.p.m. After 10 days of incubation, cells were harvested by centrifugation, resuspended in 2 mL of fresh BG11 medium and incubated in air composed of 13.3 % acetylene in 17 mL Erlenmeyer flasks tightly sealed with rubber stoppers. The amount of ethylene produced after incubation was determined by gas chromatography and expressed as nmol of ethylene produced per g chlorophyll per hour. The strain *Anabaena sp.* PCC 7120, a filamentous, heterocyst-forming cyanobacterium with nitrogenase activity of 17.04 nmol $C_2H_4$ $\mu g^{-1}$ chl $h^{-1}$, was used as a reference (Maqubela et al. 2010). EPS production was qualitatively assessed using the Indian ink staining technique. Cyanobacterial strains with large white zones around their cells were considered to have high EPS content compared with strains with smaller white zones around their cells (Maqubela et al. 2010). On the basis of $N_2$-fixing capability and EPS production, 17 strains were chosen, and only 2–3 strains from each country were used for the experiments (see Table 12.2).

The amount of $N_2$-fixation capacity and EPS production by cyanobacterial strains to be used as biofertilisers and bioconditioners is fundamental but not sufficient for selecting a suitable strain. It is also important to consider the capacity of soil colonisation by the strains in terms of biomass production. Selected strains exhibiting good performances of $N_2$ fixation and EPS production (when grown in the medium) could have problems when inoculated in soil. Therefore, an assessment of the interactions between soil and the strains (selected on the basis of high ability of $N_2$ fixation and EPS production) was carried out. The evaluation of the interactions between native soils and indigenous, inoculated, selected strains was carried out in the lab by measuring: microbial competition, water-stress resistance and growing attitude on soil. On the whole, 32 replicates for each soil were prepared, and half of them were previously sterilised before strains inoculation. Observations of the inoculated cyanobacterial growth and behaviour under different experimental conditions were carried out by visual, light microscope observation and image analyses. EPS production was also tested by light microscope observation after negative staining with Indian ink.

**Table 12.2** Selected strains employed as inoculants for the experiments

| Country | Soil | Strain |
|---|---|---|
| Tanzania | Mkindo | *Nostoc 9v* |
| | Mkindo | *Nostoc 9z* |
| | Mafiga | *Nostoc 11ii* |
| South Africa | Hertzog | *Nostoc 3g* |
| | Guquka | *Nostoc 7e* |
| Zimbabwe | Domboshawa | *Calothrix 16m* |
| | Chikwaka | *Scytonema 19c* |
| | Henderson | *Nostoc 3v* |

The capacity of the selected strains to grow on soil was evaluated as percentage of soil surface coverage (image analysis) in inoculated Petri dishes on four replicates. Two different types of data were obtained: (*a*) qualitative data by visual analysis at each step of the experiment (15, 45 days) and (*b*) quantitative data by measuring the strain-biomass production by a combination of C/N elemental and image analysis for the same time intervals.

The determination of the optimal biomass inoculation rate (3 g $m^{-2}$ of dry weight) was established in pilot experiments.

The experimental design was as follows:

Soil Treatments

- Inoculation with indigenous strains, 1 level (3 g $m^{-2}$ of dw)
- Sterilisation plus inoculum
- Control

Tested on Two Water Conditions

- Continuously moist at 80 % of water-holding capacity (WHC)
- Wet and dry (WD) cycles

**Fig. 12.1** Mkindo soil inoculated with *Nostoc* 9v after 45 days

Replications
– Four

Duration of Experiment
– 45 days

The soil samples were 2 mm sieved and distributed in Petri dishes in amounts of 40 g of dry soil per plate, and half of these Petri dishes were sterilised by twice autoclaving in order to assay the competitiveness of the inoculated strains. Non-inoculated plates were included as control. The strains were inoculated at the rate of 3 g of dry biomass per $m^2$ on the respective original soils. The incubation was performed under artificial lighting of 150 μmol photons $m^{-2}$ $s^{-1}$ with dark cycles (12:12) at about 28 °C for 45 days. The soils in the Petri dishes were maintained continuously moisted at 80 % of the water-holding capacity (WHC) by daily additions of water and monitoring of soil humidity or submitted to wetting and drying (WD) cycles (about every 2–3 days) for the whole period of the experiment, in order to evaluate the water-stress resistance (Fig. 12.1).

The indigenous selected strains *Nostoc* 9v and 9z isolated from Mkindo (Tanzania) and *Nostoc* 11ii isolated from Mafiga (Tanzania) were inoculated on their original soil. Guquka and Hertzog soils (South Africa) were inoculated with their indigenous *Nostoc* strains 7e and 3g, respectively. Domboshawa, Chikwaka and Henderson soils (Zimbabwe) were inoculated with their indigenous strains *Calothrix* 16m, *Scytonema* 19c and *Nostoc* 3v, respectively. Strain *Nostoc* 9v, found to be the best strain in the pilot experiment, was considered as a reference and inoculated not only on the soil of origin but also on the soils from South Africa (Hertzog) and Zimbabwe (Chikwaka) (Table 12.2). This strain possessed a good nitrogenase activity (about 14 nmol μg $chl^{-1}$ $h^{-1}$), fairly comparable with the *Nostoc* PCC 7120 (17 nmol μg $chl^{-1}$ $h^{-1}$) used as a reference.

Biomass production was measured by sampling a known surface area of cyanobacterial biomass developed on a Petri dish using a small metallic cylinder with a cross section of 1 $cm^2$. Samples were dried at 105 °C and analysed by elemental C/N analyser. The obtained data were expressed on biomass basis by considering the relative C and N contents of the different freeze-dried strain biomasses separately measured.

Both isolated and selected strains were maintained in culture on agar slants and grown in liquid cultures under standard conditions in order to obtain sufficient biomass quantities to perform the planned laboratory experiments. The biomasses were grown in big flasks containing liquid $Bg11_0$ medium at 27 °C and cycle of light and dark, besides shaking and bubbling. About 50 % of the isolated cyanobacterial strains were re-plated in the presence of cycloheximide in order to obtain fungi-free cultures. Several techniques were applied to obtain many of the strains in axenic conditions.

## 12.4 In Situ Mass Production of Selected Indigenous Strains for Their Application in Greenhouse and Field Experiments

The massive biomass production of selected cyanobacterial strains was performed using a batch culturing technique in the laboratory and, successively, outdoors. Small samples of selected strains were first cultured in small flasks and then transferred to increasingly larger containers (50, 100, 250, 1000 and 2000 mL conical flasks, 25 L glass bottles and finally 200 L in outdoor ponds (motorised-paddled or masonry)) whenever the cyanobacteria dry matter (DM) concentration reached 1 g $L^{-1}$. Different solutions were adopted in different countries (Fig. 12.2a–d).

**Fig. 12.2** Different solutions adopted in different countries: (**a**) South Africa, motor paddle; (**b**) Tanzania, masonry; (**c, d**) Zimbabwe, masonry, small and large pond

In South Africa it was possible to build a motorised-paddled pond to air the culture for speeding up and promoting the cyanobacteria growth.

In Tanzania and Zimbabwe, the outdoor ponds were built in masonry, and air to the culture was provided by a compressor where electricity was available; otherwise, air providing was performed manually. In Zimbabwe an intermediate small pond was used (Fig. 12.2c) to control and favour the conditions of culture transfer from the lab to outdoor.

## 12.5 Assessment of the Interactions Between Soils and Selected Strains

### 12.5.1 Lab Experiments

#### 12.5.1.1 $N_2$ Fixation and EPS Production

A large morphotype diversity among the *Nostoc* population was observed and confirmed by other features like $N_2$-fixing activity and EPS secretion (Fig. 12.3a, b). *Nostoc* was found to be the best genus compared to others because it fixed a good amount of $N_2$ and produced large amount of EPSs. The nitrogenase activity reached values extraordinarily high in the *Nostoc* strains 7e and 9v (more than 16 and 14 nmol μg $chl^{-1}$ $h^{-1}$, respectively) (Table 12.3). The EPS secreted by *Nostoc* diffused into the medium embedding the entire colony (Fig. 12.3a, b), whilst from *Scytonema* and *Calothrix* surrounded the trichome as a sleeve (Fig. 12.3c).

#### 12.5.1.2 Light Microscope Observation of Cultures in Liquid Media

Qualitative composition of microbial population, determined by observing under light microscope (both the investigated soil samples and the soil enrichment cultures in liquid media) revealed an absolute prevalence of filamentous cyanobacteria, dominated, in order of abundance, by heterocystous (*Nostoc*, *Scytonema*, *Calothrix*, *Cylindrospermum*) and non-heterocystous (*Leptolyngbya*, *Microcoleus*, *Lyngbya*) genera. The enrichment of cultures illustrates the high occurrence of cyanobacteria and green microalgae; instead diatoms were scarcely represented. Some pictures by light microscopy of cyanobacteria in liquid culture enrichments are given in Fig. 12.4.

#### 12.5.1.3 Enumeration of Cyanobacterial Soil Population

The enumeration of the cyanobacterial population present in the samples showed a remarkable abundance of these organisms. This aspect is important as it means that cyanobacteria are not only distributed forming association with biological crust (Belnap and Lange 2001; Garcia-Pichel et al. 2003), but they also can spread as free-living in top soils. Enumeration of cyanobacterial soil population estimated as MPN in liquid media, and those obtained from CFU counting in agarised media in Petri dishes, ranged from $10^2$ to $10^6$ and $10^2$ to $10^5$ per g of soil, respectively. The highest values were observed in the Tanzanian soil from Mkindo cultivated soil, where numerous heterocystous genera, mainly *Nostoc*, were present. Indeed, from Mkindo samples some strains of *Nostoc* were isolated. Strain 9v was characterised by a very good diazotrophic growth, nitrogenase activity (about 14 nmol μg $chl^{-1}$ $h^{-1}$ of ethylene produced) and by a large EPS secretion.

#### 12.5.1.4 Light Microscope Observations of Inoculated Cyanobacteria

Observations of the inoculated cyanobacteria growth and behaviour under different experimental conditions were carried out in order to assess the characteristics of the strains and their interactions with the various soils. The remarks included visual, light microscopy observations of the growth on soil of the indigenous microflora and of the inoculated strains, microbial competition, behaviour under the two different water regimens and EPS production at each step of the experiment (after 15 and 45 days). The capacity of the selected strains to grow on soil was evaluated as percentage of soil surface covering on four replicates of inoculated Petri dishes.

The observation under light microscope of the Petri dishes allowed a deeper comprehension of

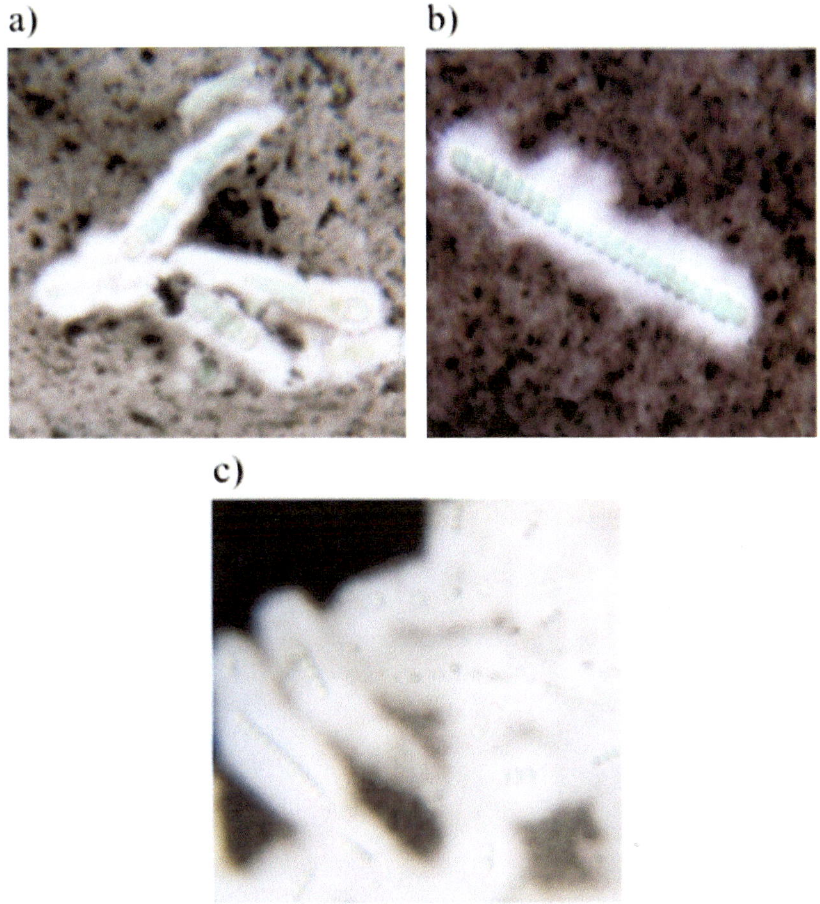

**Fig. 12.3** Indian ink staining. White area represents EPS production; (**a**) *Nostoc* 3v from Guquka (South Africa); (**b**) *Nostoc* 9v from Mkindo (Tanzania); (**c**) *Calothrix* 16m from Domboshawa (Zimbabwe)

**Table 12.3** Nitrogenase activity measured by the acetylene-reduction technique in $BG11_0$-grown cells (nmol μg $chl^{-1}$ $h^{-1}$)

| Country | Strains | Fixed $N_2$ nmol μg $chl^{-1}$ $h^{-1}$ |
|---|---|---|
| Tanzania | *Nostoc* 9x | 7.28 |
| | *Nostoc* 9v | 14.12 |
| | *Nostoc* 9z | 7.51 |
| | *Scytonema* 11ii | 4.95 |
| South Africa | *Nostoc* 3v | 1.33 |
| | *Nostoc* 3g | 4.71 |
| | *Nostoc* 7e | 16.08 |
| Zimbabwe | *Calothrix* 16m | 2.54 |
| | *Scytonema* 19c | 13.65 |

the interactions between strains and soil. The results are reported in succession for each country and inoculated strain (Table 12.2).

***Nostoc* 9v on Mkindo Soil (Tanzania)** Strain *Nostoc* 9v inoculated on the original Tanzanian soil (Mkindo) showed a massive and rapid development in both water regimen conditions (Fig. 12.5, Plate 5a). This would demonstrate a high resistance of 9v against water stress. The water regimen determined the aspect of the development on soil surface: formation of a thick, raised layer that appeared crusted under water-stress conditions (WD cycles) and mucous and shiny under continuously wet conditions. In WHC conditions *Nostoc* filaments appeared embedded in a large amount of EPS; cyanobacterial layer was continuous and entrapped bubbles of photosynthetically evolved oxygen. Under water-stress conditions, the EPS glued soil particles without discontinuity. *Nostoc* 9v in Mkindo showed a very good competitiveness against the indigenous

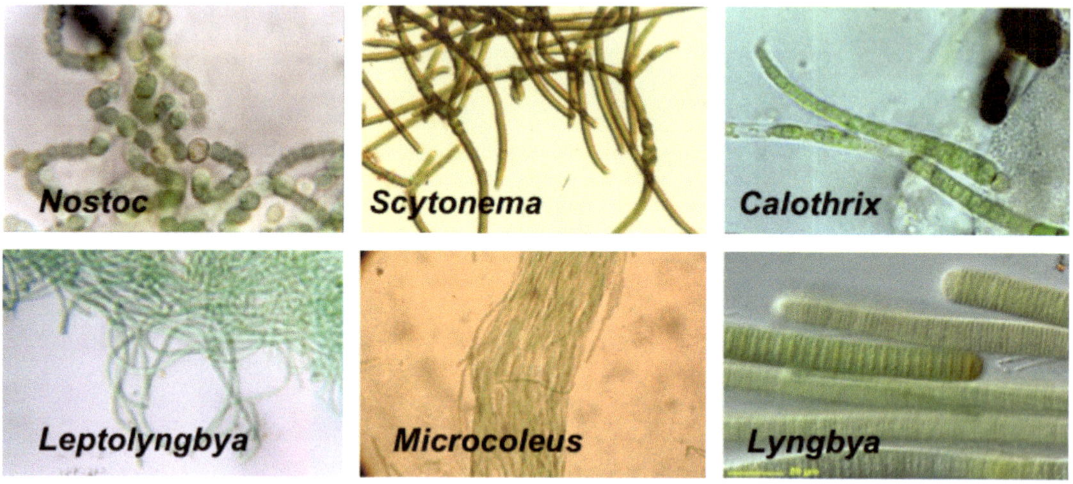


**Fig. 12.4** Main strains occurring in the investigated soil samples (observed under light microscope)

microflora, which was relatively abundant and mainly represented by *Microcoleus*, *Leptolyngbya* and some green algae (*Chlamydomonas*, *Chlorococcum*) (Fig. 12.5, Plate 5b).

***Nostoc* 9z on Mkindo Soil (Tanzania)** *Nostoc* 9z displayed a very good and rapid development under continuously wet conditions, but a not equally good behaviour against water stress, and evidenced a good competitiveness against the indigenous microflora. The most common indigenous organisms were represented by green algae *Chlorococcum*, *Chlorosarcinopsis* and *Trebouxia* and by cyanobacteria (*Lyngbya* and *Microcoleus*). *Nostoc* 9z produced an abundant amount of EPS under WHC conditions forming a thick, raised layer. Under water-stress conditions, *Nostoc* 9z gave rise to a matted development, and the cyanobacterial layer was not uniform indicating a lesser degree of soil particle aggregation mainly due to filaments than to the limited EPS production.

***Nostoc* 11ii on Mafiga Soil (Tanzania)** *Nostoc* 11ii strain inoculated on Mafiga soil was characterised by small-sized cells and showed a limited development and a reduced presence of indigenous microflora. This strain formed a thin layer covering the soil surface better under WD than under WHC conditions and producing a large amount of EPS cementing soil particles. The mucous layer, formed under continuously wet conditions, often entrapped bubbles of photosynthetically evolved oxygen. Cyanobacteria (*Lyngbya*, *Leptolyngbya*, *Microcoleus* and *Nostoc*) and a few green algae (*Chlorococcum*, *Chlorosarcinopsis*) represented the most common indigenous organisms.

***Nostoc* 7e on Guquka Soil (South Africa)** Soil from Guquka inoculated by *Nostoc* 7e presented a scarce indigenous microflora under water-stress conditions and particularly relevant under WHC conditions leading to a reduced development of the inoculated 7e strain (Fig. 12.5, Plate 5c). Hence, 7e strain resulted to have a similar development under WD conditions especially on the sterilised soil. The biomass developed a thin layer that produced small amount of EPS uncapable to form aggregates with soil particles, and its distribution on Petri dishes was discontinuous. Among the indigenous microflora, a branched green alga (*Ulothrix*), *Chlamydomonas* and some cyanobacterial forms were particularly abundant. The reduction, after 45 days of incubation, of the development of the strain inoculated on sterilised soil, could have been due to a successive and massive growth of actinomycetes and fungi.

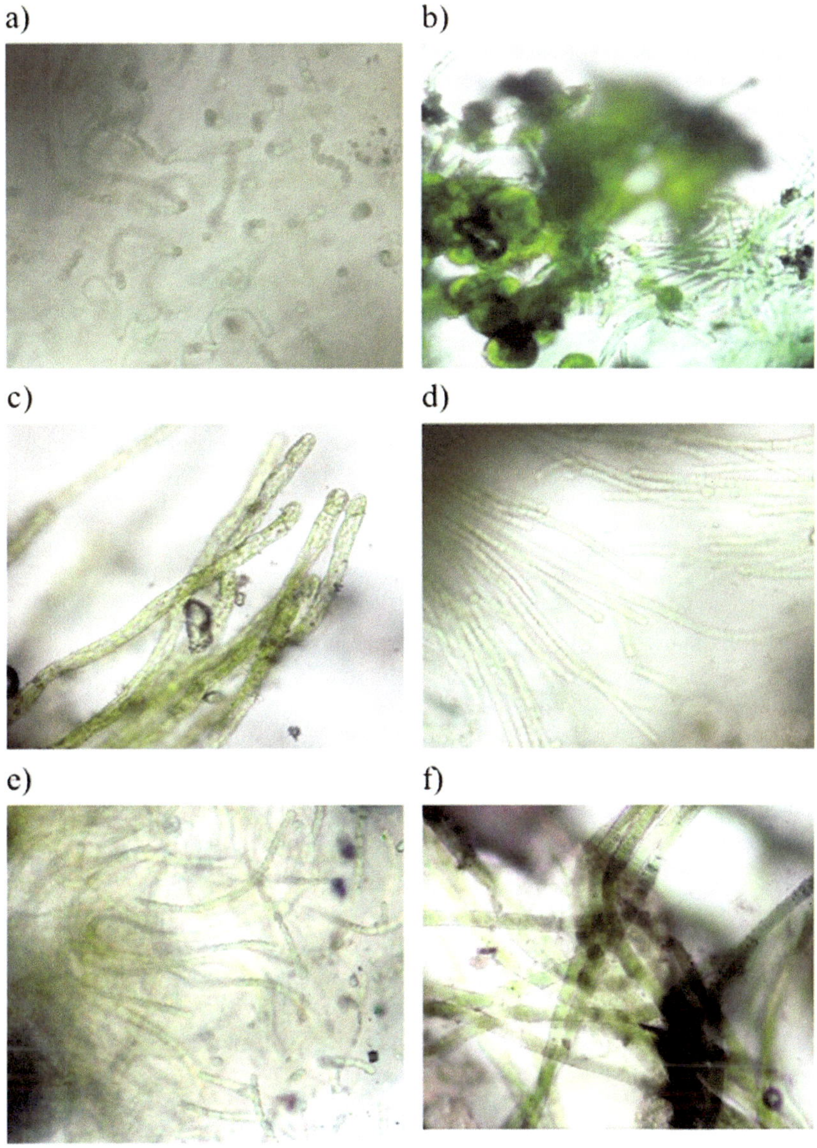


**Fig. 12.5** *Plate* (**a**) *Nostoc* 9v inoculated on Mkindo unsterilised soil from Tanzania under WHC conditions. Note the presence of indigenous cyanobacteria (*Leptolyngbya*); *Plate* (**b**) indigenous microflora, green algae (*Chlorococcum*, *Trebouxia*) and cyanobacteria (*Leptolyngbya*, *Microcoleus*), developed on Mkindo soil from Tanzania under WD conditions; *Plate* (**c**) indigenous microflora represented by green algae (*Ulothrix*) developed on Guquka soil from South Africa under WHC conditions; *Plate* (**d**) *Nostoc* 9v inoculated on Hertzog unsterilised soil from South Africa under WHC conditions; *Plate* (**e**) *Nostoc* 9v inoculated on Chikwaka sterilised soil from Zimbabwe under WD conditions; *Plate* (**f**) indigenous microflora mainly represented by cyanobacteria (*Lyngbya*), developed on Chikwaka soil from Zimbabwe under WD conditions

***Nostoc* 9v on Hertzog Soil (South Africa)** *Nostoc* 9v strain inoculated on the Hertzog soil showed a behaviour similar to the one noticed in the soil of origin, Mkindo (Tanzania), notwithstanding the stark abundance of the indigenous microflora (Fig. 12.5, Plate 5d). *Nostoc* 9v grew very well and showed a good competitiveness and water-stress resistance, confirming the data obtained on the soil of origin. The strain also produced a big amount of EPS, mainly under WHC conditions, forming a thick, crusted, risen up and wrinkled layer sealing soil surface.

***Nostoc* 3g on Hertzog Soil (South Africa)** *Nostoc* 3g strain inoculated on its soil of origin (Hertzog) evidenced a good competitiveness against the indigenous microflora, displaying a massive and rapid development notwithstanding the high presence of the indigenous microflora, mainly represented by a *Nostoc* sp., morphologically different from the inoculated strain. Moreover, *Nostoc* 3g showed a rapid, dense and thick development of biomass in both conditions of water regimen, with a very massive production of EPS. It formed a continuous, mucous layer under WHC and a crusted and discontinuous layer under WD conditions.

***Calothrix* 16m on Domboshawa Soil (Zimbabwe)** The growth of the *Calothrix* 16m strain on its soil of origin (Domboshawa) was scarce and markedly influenced by water conditions. Under WHC conditions the development of biomass was scarce probably due to a rapid development of green algae, being the majority of the indigenous microflora. The growing of the green algae was favoured by the high humidity and prevailed over the inoculated 16m strain. *Calothrix* 16m strain did not produce EPS.

***Nostoc* 9v on Chikwaka Soil (Zimbabwe)** The behaviour of *Nostoc* 9v strain isolated from Mkindo (Tanzania) was similar to that obtained on the original soil (Fig. 12.5, Plate 5e). The strain 9v showed a good development, competitiveness and high EPS production.

***Scytonema* 19c on Chikwaka Soil (Zimbabwe)** *Scytonema* 19c strain, isolated from Chikwaka soil, showed a high degree of soil colonisation under WD condition and efficiency and competitiveness against indigenous microflora mostly represented by filamentous cyanobacteria (*Lyngbya*, *Phormidium*, *Plectonema*) (Fig. 12.5, Plate 5f). *Scytonema* 19c formed (on Petri dishes) a thick mat with densely entangled, continuous and raised filaments. Moreover, 19c strain was characterised by high resistance to water-stress conditions.

***Nostoc* 3v on Henderson Soil (Zimbabwe)** *Nostoc* 3v strain, isolated from Henderson soil, evidenced a massive and rapid development notwithstanding the high presence of the indigenous microflora. The last was represented by a *Nostoc* sp., morphologically different from the inoculated 3v strain, and by *Microcoleus* and other filamentous cyanobacteria. *Nostoc* 3v showed, in both conditions of water regimen, a rapid, dense and thick development of biomass, with a very massive production of EPS with high capacity to bind soil particles. *Nostoc* 3v developed a continuous, mucous layer under WHC conditions and a crusted and discontinuous layer under WD conditions.

In regard to the adaptation capacity of the selected strains to growth on soil, *Nostoc* 9v strain confirmed to be the best among all tested soils, especially when inoculated on the soil of origin (Mkindo). A very high capacity of covering was also shown by the *Nostoc* 3v isolated from Henderson (Zimbabwe); on the contrary, strains *Nostoc* 3g from Hertzog (South Africa) and *Scytonema* 19c from Chikwaka (Zimbabwe) showed only a good colonising capacity. All these strains were also characterised by a very fast growth, being able to cover almost the whole soil surface in about 2 weeks.

The other two *Nostoc* strains, 9z from Mkindo (Tanzania) and 7e from Guquka (South Africa), showed a satisfactory colonising capacity, higher than that of *Calothrix* 16m from Domboshawa (Zimbabwe) and *Nostoc* 11ii from Mafiga (Tanzania).

The response of the strains to the experimental water regimens, i.e. WHC and WD, indicated, as expected, a better growth in high humidity conditions for all strains. However, *Nostoc* 3g from Hertzog (South Africa) showed a good water-stress resistance growing well both under WD and WHC conditions.

The water conditions markedly influenced the appearance and consistency of the cyanobacterial biomass development. When soil samples were subjected to WHC conditions, the development of cyanobacteria formed mucilaginous, thick, sometime raised, mucous and shiny layer, because of the abundance of secreted EPS with bubbles due to the entrapment of the produced oxygen. When soil samples were subjected to WD, cyanobacterial development assumed a crust-like aspect.

In regard to competitiveness degree of the selected strains, *Nostoc* 3g, 3v and 9v were the most effective.

The most frequent groups and genera of the indigenous microflora were represented by cyanobacteria as *Nostoc*, *Microcoleus*, *Lyngbya*, *Leptolyngbya*, and chlorophyta as

*Chlamydomonas*, *Chlorella*, *Chlorococcum*, *Chlorosarcinopsis* and *Trebouxia*. *Nostoc* and *Leptolyngbya* resulted to be more prevalent.

The highest amounts of EPS were secreted under WHC conditions, especially by the *Nostoc* strains 9v, 3v and 3g. The produced EPS promoted the soil particles aggregation. In the case of *Scytonema* 19c, the aggregation was essentially due to the dense entanglement of filaments; instead, for the other strains, the aggregation was due to the gluing effect of EPS.

#### 12.5.1.5 Image Analysis: Quantitative Data

The quantitative analysis confirmed substantially the findings previously described. Data of biomass production, expressed in g m$^{-2}$, were obtained by image analysis combined with analytical C/N data. In Fig. 12.6a, b, the response of the inoculated strains to a water regimen of WHC and to WD cycles, simulating the stress conditions of the natural environment after 15 and 45 days, is shown. As expected, the larger production of biomass was measured in wet conditions for all strains. Only *Nostoc* 3g strain showed a comparable production also in WD conditions, especially in the first 15 days, in spite of the stress constraints. Domboshawa 16 m strain exhibited a very good resistance because the biomass yield did not improve by increased humidity and with time. However, its biomass production was low. In some cases the production of biomass after 45 days was less than after 15 days; this could be explained by the presence of heterotrophic microorganisms and mineralisation processes. In Fig. 12.6c, d the productions of microbial biomass of the different strains inoculated after sterilisation for different soils and different water regimens at the end of the experiment are shown. The biomass growth in sterilised conditions was considered as maximal production without any competition linked to the indigenous microbial population.

The interactions of indigenous microbial activity on the strain could also favour both the

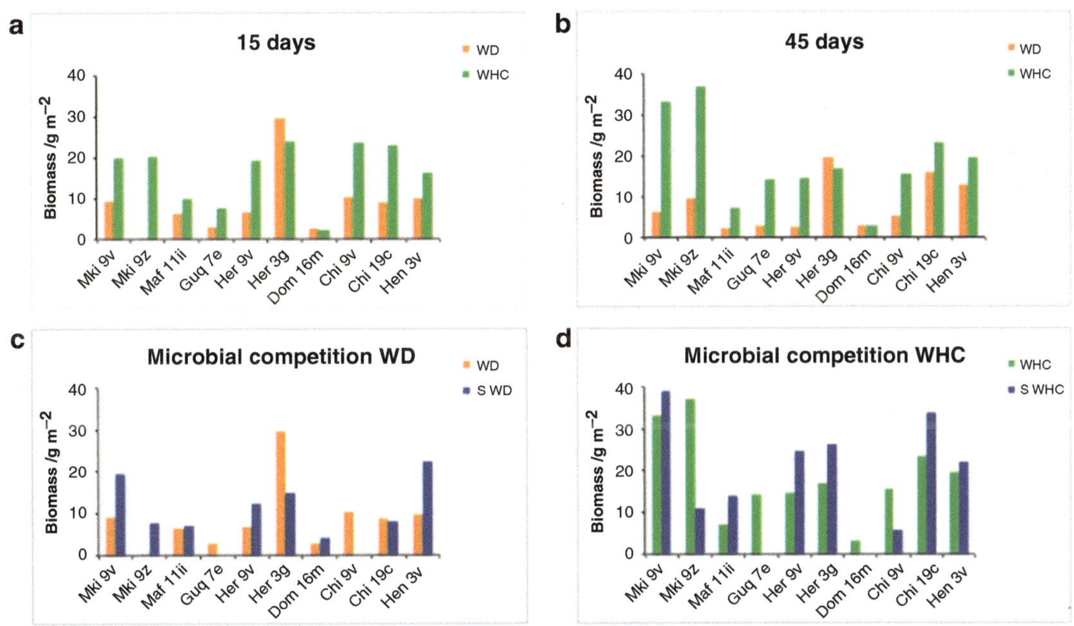


**Fig. 12.6** Interactions between inoculated strains, at different soil humidity regimen, and soil local microbial populations measured in terms of microbial biomass production: (**a**, **b**) biomass production in inoculated soil samples after 15 and 45 days at WD and WHC regimen, respectively; (**c**, **d**) effect of microbial competition on biomass production at the end of experiment (after 45 days) in inoculated soil samples (WD and WHC) at WD and WHC regimen, respectively, and inoculated after sterilisation (S) soil samples (S WD and S WHC) at WD and WHC regimen, respectively

growth of the inoculated cyanobacteria and the local population promoting a synergic action. However, as confirmed by the optical microscope analysis, the presence of inoculated strain was predominant in all samples, also when the contamination of indigenous population was substantial (i.e. Guquka 7e in WHC conditions (Fig. 12.5, Plate c) etc.).

From Fig. 12.6c, it can be observed that Hertzog 3g's indigenous microbial population, in natural conditions (WD), had significant synergic effect in promoting the biomass growth. A similar effect, but to a lesser extent, was found for Chikwaka 9v and 19c. As regards the WHC conditions (Fig. 12.6d), Mkindo 9z's indigenous microbial population had a significant synergic effect in promoting the biomass growth. Also Chikwaka 9v exhibited this effect but to a lesser extent. Mkindo 9v strain displayed a high competiveness (similar biomass production of the respective sterilised sample) and a large amount of biomass production both at different humidity and time conditions, as well as Henderson 3v and Chikwaka 19c but to a lesser extent. When strain 9v was inoculated on other soils, as in the case of Hertzog 9v, a good competiveness was observed as for Mkindo 9v; differently, in the case of Chikwaka 9v, a synergic effect in promoting biomass growth for both humidity regimens was evidenced. However, their biomasses production was relatively lower.

The reference strain *Nostoc* 9v isolated from Mkindo soil (Tanzania) confirmed to possess a very good capacity to colonise all tested soil types. It also exhibited a high competitiveness against the indigenous microflora. The suitability of the genus *Nostoc* for soil inoculation was also evidenced by the performances of the other tested *Nostoc* strains. The property of development in soil, coupled to EPS production and $N_2$-fixing capacities, should ensure a successful introduction in soils of the cyanobacterial biomass both as soil fertiliser and conditioner.

#### 12.5.1.6 Effects on Soil Physical Properties: Microstructure, Soil Structural Stability

The impact of inoculation of cyanobacteria *Nostoc* 9v strain on physical characteristics of poorly aggregated soil from Guquka was investigated in the lab by Malam Issa et al. (2007). In this study the improvement of the soil physical quality was demonstrated to occur few weeks after cyanobacteria inoculation. A dense superficial network of cyanobacterial filaments and extracellular polymer secretions (EPS) covered soil aggregate surface after 4–6 weeks of incubation, and organo-mineral aggregates comprising cyanobacterial filaments and EPS were observed (Fig. 12.7a, b).

The results of aggregate breakdown tests showed no significant difference between non-inoculated samples after 6 weeks, whilst they revealed improvement of aggregate stability for inoculated samples. An increase in aggregate stability by two to four times compared to that of non-inoculated samples (6 weeks following inoculation) was reported (Malam et al. 2007). The improvement of aggregate stability appeared after 2 weeks following inoculation and increased gradually with time and cyanobacterial

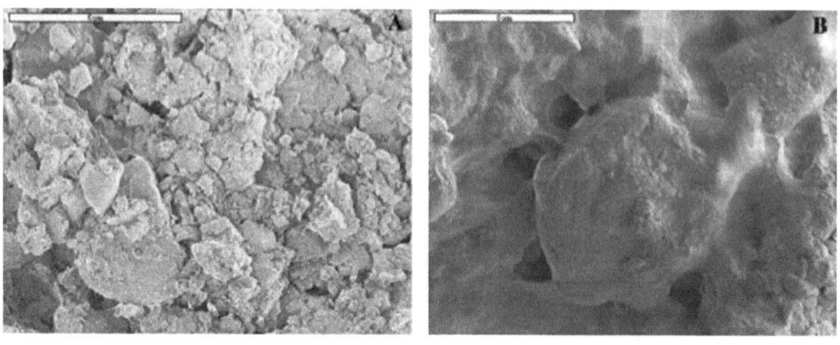


**Fig. 12.7** SEM images. Microstructure and soil aggregate stability of Guquka soil aggregates inoculated by *Nostoc* 9v; (**a**) surface of non-inoculated sample; (**b**) surface of an inoculated sample. *Scale bar*, 5 μm (Source: Malam Issa et al. 2007)

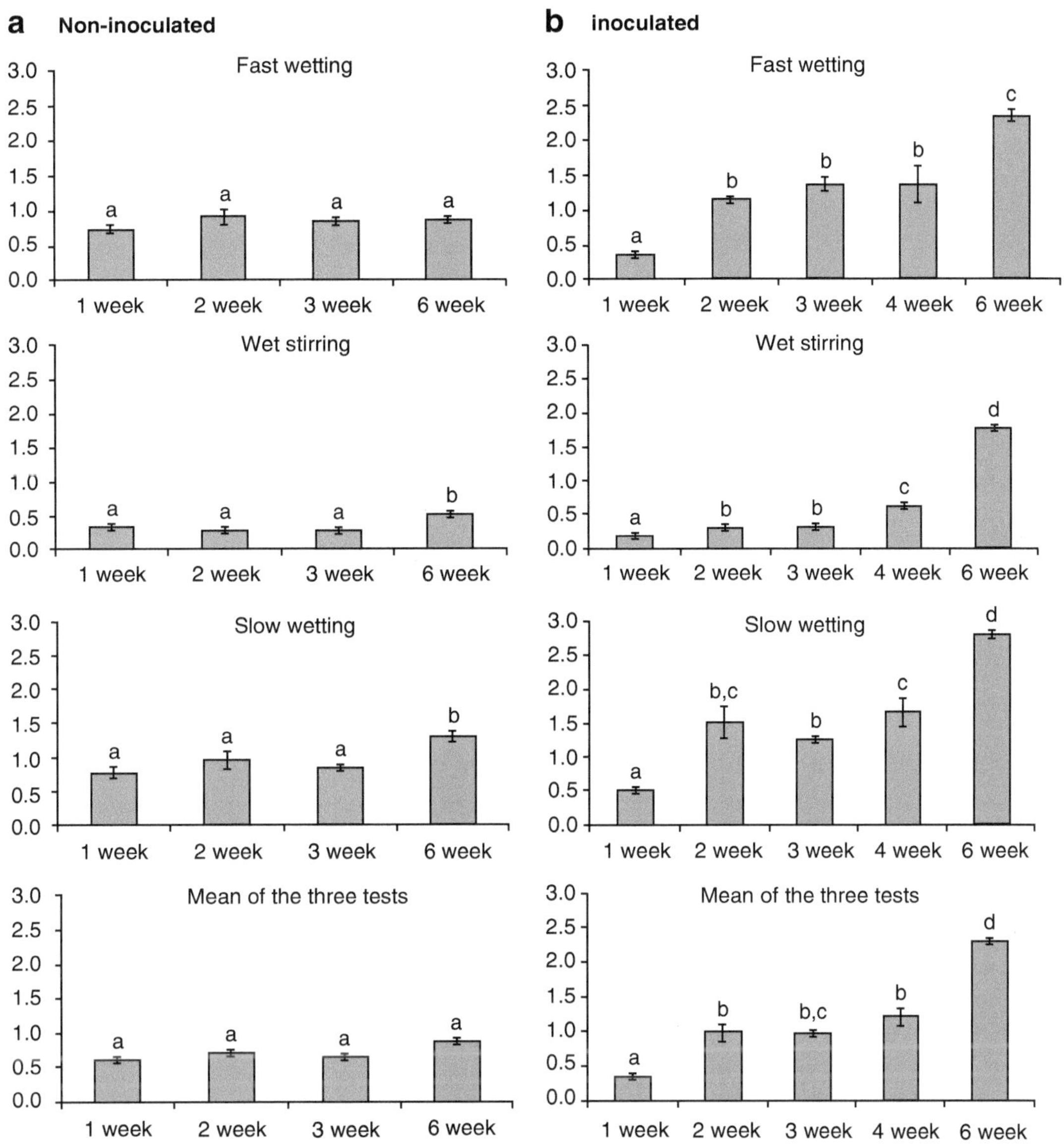


**Fig. 12.8** Soil aggregate stability test. Mean weight diameter of (**a**) non-inoculated samples incubated for 1, 2, 3 and 6 weeks and (**b**) inoculated samples incubated for 1, 2, 3 and 6 weeks (Source: Malam Issa et al. 2007)

growth (Fig. 12.8). The increase of aggregate stability was related to the changes induced in micromorphological characteristics by cyanobacterial filaments and EPS that promoted the coating, enmeshment, binding and gluing of aggregates and isolated mineral particles.

#### 12.5.1.7 Effect on Chemical Properties: Soil Organic Matter Content and Quality

The effect of cyanobacteria *Nostoc* 9v strain on the concentration and composition of soil OM of the degraded soils was studied by Pardo

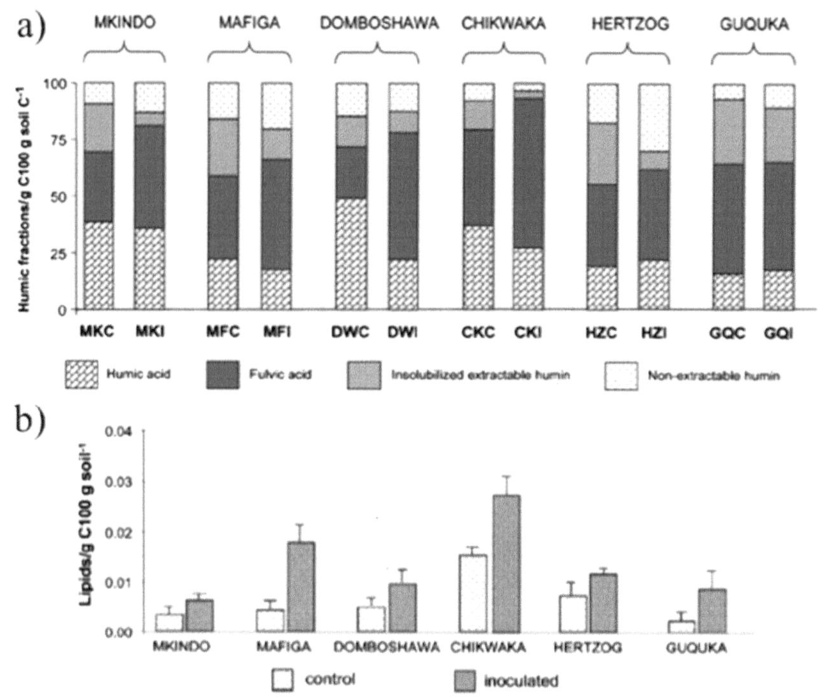


**Fig. 12.9** Effects of cyanobacteria inoculation on soil OM: (**a**) changes in the humic fractions of the control (C) and cyanobacteria inoculated (I) soils; (**b**) lipid content in the control (C) and cyanobacteria inoculated soils (I) (Source Pardo et al. 2010)

et al. (2010). They studied, under laboratory conditions, the effects of *Nostoc* 9v on the soil OM content and quality of different soils from Tanzania, Zimbabwe and South Africa. Soils were inoculated with *Nostoc* 9v and incubated for 3 months, and moisture content was maintained at 60 % of field capacity. *Nostoc* 9v proliferated and colonised the surfaces of all soils very quickly. The results obtained showed significant changes in the quantitative and qualitative characteristics of the soil OM due to cyanobacterial growth (Fig. 12.9a). Cyanobacterial growth resulted in an increase of OC from 0.4 g C $kg^{-1}$ soil to 9.0 $kg^{-1}$ and also affected soil OM characteristics through the incorporation of free, extractable or particulate biomass with a predominantly aliphatic characteristics. Important descriptors of the extent to which cyanobacterial metabolism modified the characteristics of the native soil OM were the changes in the observed amounts of the two humin types (extractable and non-extractable), the increase in lipid concentration (Fig. 12.9b) and the changes in the optical density of the HAs. These results proved the feasibility of improving soil OM content and quality by cyanobacteria-based remediation practices. This fact would be important especially in developing countries where inorganic fertilisers are not commonly used or available.

## 12.5.2 Greenhouse Experiments

In this and the next subchapter, some examples of application of biomasses to soil at different scales (greenhouse and field plots) are reported in order to describe the effects of cyanobacteria inoculation on overall soil fertility.

A study to assess the potential use of cyanobacteria for the enhancement of the structure and fertility of the two degraded arable Hertzog and Guquka soils was carried out by Maqubela et al. (2009) under tunnel house conditions. The effect of inoculation of *Nostoc* 9v strain, on soil C, soil N, mineral N and EPS content and on soil structural stability, and the interactive effect of inoculation of these strains and maize cropping on the above-mentioned soil characteristics were investigated.

The *Nostoc* inoculum, uniformly applied over potted Guquka and Hertzog soils soon after maize germination, established well and improved significantly C, N and EPS contents of the two soils. The highest contents of soil C, soil N and mineral N, however, were found in non-cropped *Nostoc* inoculated soils. This suggests that the cyanobacteria-fixed N in soil was uptaken by maize to improve growth. The *Nostoc* inoculation increased maize dry matter yields by 49 % in Hertzog and 40 % in Guquka soils. Corresponding increases in maize tissue N were 23 % and 14 %, respectively. The aggregate stability of the two soils was also improved as a result of increased production of EPS and soil C in the *Nostoc* inoculated soils. Inoculated soils cropped with maize displayed a lower proportion of stable aggregates presumably due to their low soil C and EPS contents compared to non-cropped soils. Scanning electron microscopy revealed that soil particles and fragments of non-cropped inoculated soils exhibited coatings of extracellular polymeric substances (EPS) with other particles enmeshed in networks of filaments, whilst, by contrast, little or no EPS and/or filaments were observed on cropped and/or non-inoculated soils. Another greenhouse study by Maqubela et al. (2010) on biofertilisation and/or bioconditioning effects in potted crop (maize) soils, as a result of cyanobacteria ability to fix $N_2$ or produce EPS, was carried out on samples from South African soils of Guquka, Hertzog and Qunu villages and Fort Cox College. The results showed the ability of selected strains (3g, 3v and 7e) to influence maize DM and soil C and N contents. Inoculation with cyanobacteria strains increased maize DM and N uptake significantly, on par with the reference strain. These increases were consistent with increases in nitrate-N observed at harvest time in inoculated cropped and non-cropped soils. Strain 7e exhibited larger increases in soil nitrate-N, tissue N and uptake than strain 3g, possibly because of its higher ability to fix $N_2$. Cropping with maize reduced soil total C and N, possibly owing to its negative effects on cyanobacteria establishment.

A trial pot study in a greenhouse in Tanzania (D'Acqui et al. 2004, 2006) was carried out at Sokoine University of Agriculture (SUA), Morogoro, to assess the effect of cyanobacteria inoculation on maize and rice crop biomass production. The Mkindo 9v strain was inoculated on soil samples (packed into 4 L pots) from Mkindo and Mafiga soils. The results showed little effect of the cyanobacteria inoculation on biomass production of first maize and rice crops. Instead, in a second maize crop (planted after the first has been harvested without any additional inoculation), the shoot biomass production was significantly higher ($p < 0.005$) for inoculated soil samples with respect to non-inoculated ones. Similarly the rice resprout crop, i.e. the regrowth of first rice crop (rootstocks, remaining after the harvesting of the shoot, were allowed to sprout and grow for 6 weeks) exhibited a significant increase of the biomass (Fig. 12.10a, b). These findings indicated a residual effect in soil of the biofertilisation, as evidenced by an increase in crop biomass production.

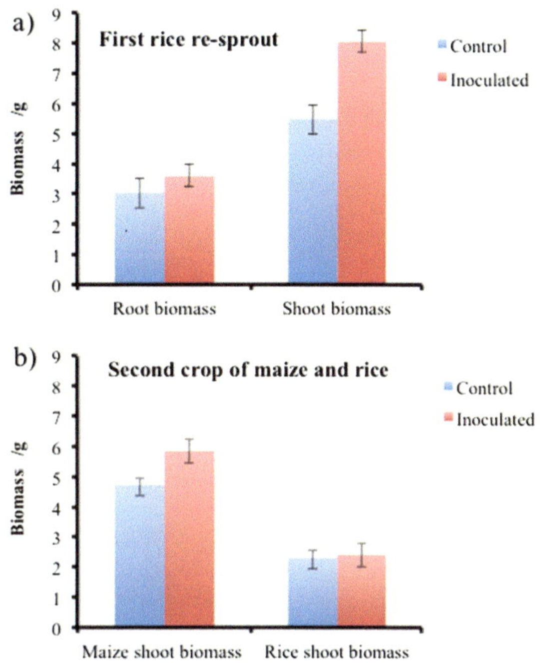


**Fig. 12.10** Effects of soil cyanobacteria inoculation on crop growth: (**a**) effect of inoculation on first rice resprout and (**b**) residual effect of inoculation on shoot biomass of second crop of maize and rice

This effect disappeared for the third maize crop and second rice resprout (D'Acqui et al. 2004, 2006). Greenhouse experiments with rice and maize indicated that inoculation

**Fig. 12.11** Field experiment carried out on *Amaranthus* spp.: (**a**) cropped without inoculation, (**b**) cropped plus inoculation 3 $g^{-2}$ cyanobacteria (**c**) cropped plus inoculation 6 g $m^{-2}$ cyanobacteria (Photo S. Maliondo)

with cyanobacteria had limited effect on the first crop, but the yield of the subsequent crop was significantly increased. The effect of cyanobacteria on crop growth varied with soil types, and this was related to different extents of cyanobacteria establishment. In fact, soils differed in the ability to support biomass production in the pots with the Mafiga soil producing significantly more biomass (9.96 ± 1.30) than Mkindo soil (5.04 ± 1.97).

### 12.5.3 Field Experiments

A field experiment was performed in Mkindo area to evaluate the residual effect of cyanobacteria inoculation on crop (*Amaranthus* spp.) in the field at small-scale (5 × 5 $m^2$) plots in complete randomised block design with four replications (D'Acqui et al. 2004, 2006).

The small plots were initially inoculated with two different rate of indigenous *Nostoc* 9v (3 g $m^{-2}$ and 6 g $m^{-2}$) and cropped with rice. After harvesting, plots were cleared of vegetation and *Amaranthus* spp. crop replanted. The production of biomass observed in the different plots was higher in the plots earlier inoculated than in the not-inoculated ones, and the highest biomass production was related to the highest level (6 g $m^{-2}$) of inoculation (Fig. 12.11). Unfortunately, results of this study were only indicative because of different problems encountered during the experiment (i.e. flooding, illegal harvesting, etc.), some data were missing and as a consequence the statistic treatment of available data was not possible.

Field experiments, in spite of the serious problems encountered, as previously explained, confirmed the positive results of greenhouse tests. In fact, this field experiment carried out for *Amaranthus* spp., a vegetable commonly eaten in these countries, showed the beneficial effect of soil cyanobacteria inoculation on improving crop yield. Moreover, field experiments confirmed that main favourable effects on the crops are transferred to a successive crop as found for trials in greenhouse as well.

## 12.6 Conclusions

Study revealed that a huge number of different indigenous cyanobacteria strains already adapted to their environment could be identified in soils of semiarid environments in East and Southern Africa. This meant that cyanobacteria were not only distributed forming association with biological crust, but they also could spread as free-living cyanobacteria in top soils. A large number of these strains exhibited high capacity of $N_2$ fixation and produced considerable amounts of EPS, thus being able to improve soil fertility.

A number of the identified strains were found to compete favourably against the other microorganisms in soil, to maintain their high capability to fix $N_2$ and to produce EPS under drought conditions.

In the scientific community, there was only a very general awareness of cyanobacteria presence in soils of these African countries, but no attempts, up to now, were made to isolate and purify, by appropriate procedures and techniques, the most productive strains (able to fix $N_2$ and produce EPS) for reducing soil degradation and for a sustainable improvement of overall soil fertility.

Results of the laboratory, greenhouse and field experiments demonstrated that cyanobacteria were capable to grow well even under limiting conditions (soil with low nutrients availability and low pH). In spite of the cyanobacteria activity which decreased the mineral N concentration in soils (as they used these forms of N to grow), the overall soil C and N content increased. Cyanobacteria inoculation modified the soil OM characteristics and increased the more stable humic fractions. Moreover, an important increase in lipid contents was observed. Cyanobacteria inoculation also increased the aggregate's stability of poorly structured soils. The improvement of aggregate's stability appeared shortly after inoculation and increased gradually with time and cyanobacteria growth. The improvement of stability was accompanied by changes of soils surface microstructure and formation of organo-mineral aggregates. The reference strain *Nostoc* 9v isolated from Mkindo soil (Tanzania) confirmed to possess a very good capacity to colonise all tested soil types. It also showed a high competitiveness against the indigenous microflora and capacity of soil particles aggregation owing to the copious EPS secretion. Moreover, in conditions of humidity, Mkindo 9v resulted to have the highest competitiveness. In natural conditions, simulated by wetting and dry cycles, Hertzog 3g, Chikwaka 19c and Henderson 3v strains showed good yield, demonstrating resistance to stress conditions.

The suitability of the genus *Nostoc* for soil inoculation was also evidenced by the performances of the other tested *Nostoc* strains. The property of development in soil, coupled to EPS production and $N_2$-fixing capacities, ensures a successful introduction in soils of the cyanobacterial biomass, both as soil fertiliser and conditioner.

In spite of the need of further medium and long-term experiments for finally confirming the relevance and persistence of the beneficial effects, on the basis of the achieved results, the application of selected cyanobacterial strains seemed to be a very promising tool for a sustainable improvement of fertility and productivity of degraded soils in semiarid tropics. Moreover, taking into account the possible synergic effects of indigenous microbial population on the inoculated strain, the effects of an inoculum constituted not only by one but by more strains suitable to grow together for obtaining a possibly more efficient biofertiliser should be investigated.

These results suggest that by using cyanobacterial soil conditioners, the yield and nutritional value of food crops could be enhanced without using costly fertilisers. Therefore, development of low-cost techniques for screening and culturing of suitable cyanobacteria to be used in countries where crop yields are low in marginal lands and fertilisers are cost-prohibitive is necessary.

**Acknowledgements** This work is based on results from CYANOSOILS project funded by European Union, contract no. ICA4-CT-2001-10058. The author gratefully acknowledges Dr L. Tomaselli for optical microscopy observations, Prof. L. Calamai and Dr B. Pushparaj for the revision of the manuscript and all Partners who took part in the CYANOSOILS project.

## References

Barclay WR, Lewin RA (1985) Microalgal polysaccharide production for the conditioning of agricultural soils. Plant Soil 88:159–169

Belnap J, Lange OL (2001) Biological soil crusts: structure, function, and management. Springer, Berlin

Castenholz RW (2001) General characteristics of cyanobacteria. In: Garrity GM (ed) Bergey's manual of systematic bacteriology, 2nd edn. Springer, New York

Cogle AL, Redy MVR, Rao KPC, Smith GD, McGarry D, Yule DF (1995) The role of biological practices and the soil biota in management of sealing, crusting and hardsetting soils. In: So HB, Smith GD, Raine SR, Schafer BM, Loch RJ (eds) Sealing, crusting and hardsetting soils: productivity and conservation. Australian Society Soil Science, Inc: Queensland Branch, Brisbane

D'Acqui LP, Maliondo SM, Malam Issa O, Le Bissonnais Y, Ristori GG (2004) Influence of indigenous strains of cyanobacteria on physical and biochemical properties of tropical soils. The 4th International Symposium of the Working group MO – ISMOM 2004. Whuan, Cina, Abstracts p 110

D'Acqui LP et al (2006) Use of indigenous $N_2$-fixing cyanobacteria for sustainable improvement of soil biogeochemical performance and physical fertility in semiarid tropics. Final report. EU, Brussels, EU – ICA4-CT-2001-10058, p 300

Falchini L, Sparvoli E, Tomaselli L (1996) Effect of *Nostoc* (cyanobacteria) inoculation on the structure and stability of clay soils. Biol Fertil Soils 23:246–252

Garcia-Pichel F, Belnap J, Neuer S, Schanz F (2003) Estimates of global cyanobacterial biomass and its distribution. Algol Stud 109:213–227

Ghosh TK, Saha KC (1997) Effects of inoculation of cyanobacteria on nitrogen status and nutrition of rice (*Oryza sativa* L.) in an Entisol amended with chemical and organic sources of nitrogen. Biol Fertil Soils 24:123–128

Hu C, Gao K, Whitton BA (2012) Semiarid regions and deserts. In: Whitton A (ed) Ecology of cyanobacteria II: their diversity in space and time. Springer, Dordrecht/Heidelberg/New York/London

IUSS Working Group WRB (2014) World reference base for soil resources 2014. International soil classification system for naming soils and creating legends for soil maps, World soil resources reports N. 106. FAO, Rome

Juo ASR, Ayanlaja SA, Ogunwale JA (1976) An evaluation of cation exchange capacity measurements for soils in the tropics. Commun Soil Sci Plant Anal 7:751–761

Malam IO, Défarge C, Le Bissonais Y, Marin B, Duval O, Bruand A, D'Acqui LP, Nordenberg S, Annerman M (2007) Effects of the inoculation of cyanobacteria on the microstructure and the structural stability of a tropical soil. Plant Soil 290:209–219

Maqubela MP, Mnkeni PNS, Malam Issa O, Pardo MT, D'Acqui LP (2009) *Nostoc* cyanobacterial inoculation in South African agricultural soils enhances soil structure, fertility, and maize growth. Plant Soil 315:79–92

Maqubela MP, Mnkeni PNS, Muchaonyerwa P, D'Acqui LP, Pardo MT (2010) Effects of cyanobacteria strains selected for their bioconditioning and biofertilization potential on maize dry matter and soil nitrogen status in a South African soil. Soil Sci Plant Nutr 56:552–559

Maqubela MP, Muchaonyerwa P, Mnkeni PNS (2012) Inoculation effects of two South African cyanobacteria strains on aggregate stability of a silt loam soil. Afr J Biotechnol 11:10726–10735

Martens DA, Frankenberger WT Jr (1992) Decomposition of bacterial polymers in soil and their influence on soil structure. Biol Fertil Soils 13:65–73

McGrady MH (1915) The numerical interpretation of fermentation-tube results. J Infect Dis 17:183–212

Pardo MT, Almendros G, Kileo E, Maliondo S, Msanya B (2007) Cultivation-induced effects on the soil organomineral matrix and their bearing on crust development on two soil formations from Tanzania. Commun Soil Sci Plan 39:65–81

Pardo MT, Almendros G, Zancada MC, López-Fando C (2010) Biofertilization of degraded Southern African soils with cyanobacteria affects organic matter content and quality. Arid L Res Manag 24:328–343

Rao DLN, Burns RG (1990) The effect of surface growth of blue-green algae and bryophytes on some microbiological, biochemical and physical soil properties. Biol Fertil Soils 9:239–244

Rippka R, Deruelles J, Waterbury JB, Herdman M, Stanier RJ (1979) Generic assignments, strain histories and properties of pure cultures of cyanobacteria. J Gen Microbiol 111:1–61

Rogers SL, Burns RG (1994) Changes in aggregate stability nutrient status, indigenous microbial populations, and seedling emergence following inoculation with *Nostoc muscorum*. Biol Fertil Soils 18:219–215

Rogers SL, Cook KA, Burns RG (1991) Microalgal and cyanobacterial soil inoculants and their effect on soil aggregate stability. In: Wilson WS (ed) Advances in soil organic matter research: the impact of agriculture and the environment. Royal Society of Chemistry, Cambridge

SISS (1985) SISS (Società Italiana della Scienza del Suolo, Italian Society for Soil Science), Metodi normalizzati di analisi del suolo. Edagricole, Bologna

Venkataraman GS (1979) Algal inoculation of rice fields. In: Nitrogen and Rice. IRRI, Los Banos, The Philippines, pp 311–321

Zimmerman WJ (1993) Microalgal biotechnology and applications in agriculture. In: Metting B (ed) Soil microbial ecology. M Dekker, Inc., New York, pp 457–479

# Part III

# Present Scenario and Future

# The Contribution of Secondary Metabolites in the Success of Bioformulations

13

María A. Morel, Célica Cagide, and Susana Castro-Sowinski

## Contents



**Abstract**

Bioformulations are biologically active products containing one or more beneficial microbial strains, commonly used to increase plant growth, soil fertility and suppression of phytopathogens. Many reports support that the use of a combination of beneficial microorganisms has additive or synergistic effects on plant growth and yield. However, the addition of microbial- or plant-produced secondary metabolites to bioformulations may increase agricultural productivity, improving the performance of the inoculation. This chapter focuses on the use of some secondary metabolites (flavonoids, lipochito-oligosaccharides, phytohormones, etc.) in bioformulations, focusing mainly on formulations that improve leguminous crop yields. The information supports that the addition of these molecules may contribute to the sustainable development of new agronomic products.

## 13.1 Introduction: Use of Beneficial Microbes as Bioformulations

Bioformulations are biologically active products containing one or more beneficial microbial strains, embedded in an economical carrier material, that are used to increase plant growth, soil

M.A. Morel • C. Cagide
Unidad Microbiología Molecular, Instituto de Investigaciones Biológicas Clemente Estable, Av. Italia 3318, Montevideo, Uruguay

S. Castro-Sowinski (✉)
Unidad Microbiología Molecular, Instituto de Investigaciones Biológicas Clemente Estable, Av. Italia 3318, Montevideo, Uruguay

Sección Bioquímica y Biología Molecular, Facultad de Ciencias, Universidad de la República, Igua 4225, Montevideo, Uruguay
e-mail: s.castro.sow@gmail.com; scs@fcien.edu.uy


N.K. Arora et al. (eds.), *Bioformulations: for Sustainable Agriculture*,
DOI 10.1007/978-81-322-2779-3_13

fertility, and suppression of phytopathogens (Arora et al. 2010).

The carrier is an abiotic substrate that should deliver viable cells in good physiological conditions and must provide them a protective niche. Different materials can be used as carriers, for example, soils (peat, coal, clays, among others), waste plant materials, inert materials (polymers, vermiculite, perlite), or liquids (Bashan et al. 2014). Most of the investigation done in the field of improving bioformulations has focused on: 1) the improvement of the carrier properties; 2) the search of microbes with better performance on plant yield and; 3) the improvement of the metabolic state of these cells and their capability to use intercellular storage material for survival within the carrier itself (Kadouri et al. 2005).

Due to their conditions of production and storage, bioformulations are stressful environments where bacterial cells must survive many situations, such as desiccation and possibly hot conditions (Kadouri et al. 2005). Bacteria should maintain high survival rates and the ability to improve plant growth during a long period. In order to survive, bacteria use different strategies, such as the production and accumulation of osmolytes or polyhydroxyl alkanoates (PHA). Osmo-adapted microbes that accumulate osmolytes, like trehalose or glycine–betaine, show a higher tolerance to desiccation than non-osmo-adapted cells and significantly improve their plant growth performance (Bonaterra et al. 2005). Cells with higher PHA content can survive longer than those with lower amounts, because PHA provide the cells with the ability to endure a variety of harmful physical and chemical stresses (Kadouri et al. 2005; Morel et al. 2012).

The best-known microbe–plant mutualistic interaction is the diazotrophic microbial association with plants. Diazotrophs are free-living or symbiotic bacteria that fix and reduce atmospheric nitrogen to ammonia (a process known as biological nitrogen fixation; BNF). Examples of diazotrophs are *Azotobacter* (free-living diazotroph), *Azospirillum* (associative symbiont), *Azoarcus* and *Gluconacetobacter diazotrophicus* (endophytic non-nodular symbionts), and rhizobia (nodular symbionts). Some diazotrophs, and other plant growth-promoting bacteria (PGPB), also produce phytohormones, iron-sequestering siderophores, phosphate-solubilizing molecules, and/or 1-aminocyclopropane-1-carboxylate (ACC) deaminase, among other compounds. Examples of non-diazotrophic PGPB are *Pseudomonas* and *Bacillus* (Morel and Castro-Sowinski 2013).

The legume–rhizobia association is the most studied symbiotic microbe–plant interaction and, together with plant–mycorrhizal fungi interactions, is recognized for its importance in sustaining agricultural ecosystems and productivity. During the plant–rhizobia interaction, a complex network of molecular events leads to the nodule formation and finally BNF. The morphological and physiological alterations that conclude in the formation of root nodules involve many molecules such as plant flavonoids and bacterial nodulation factors, chemically identified as lipochitooligosaccharides (LCOs), among others (Morel and Castro-Sowinski 2013).

Much evidence supports that the use of a combination of beneficial microorganisms, showing different mechanisms of plant growth promotion, has additive or synergistic effects on plant growth and yield (Morel et al. 2012). However, new evidence also support that the addition of microbial- or plant-produced secondary metabolites to bioformulations may increase agricultural productivity, improving the performance of the inoculation (Morel et al. 2015). This chapter collects information about the use of some secondary metabolites in both bioformulations and formulations (that do not contain microorganisms), focusing our work mainly on bioformulations that improve leguminous crop yield.

## 13.2 Developing Improved Bioformulations: Future Design and Considerations

The application of PGPB in agricultural systems enhances plant growth, development, and yield of many crops. Their effect has been studied in several field crops (cereals and legumes) in both

**Fig. 13.1** How to improve the inoculation efficiency? Industrial and academic collaboration

extensive and intensive farming conditions for many years. For example, bacteria of the genus *Azospirillum* and rhizobia have been studied for their ability to promote plant growth via the production of phytohormones (as indole-3-acetic acid (IAA) and cytokinins) and/or to fix atmospheric nitrogen (Morel et al. 2012).

During the establishment of the legume–rhizobia association, root colonization by rhizobia is accompanied by important changes in root architecture and plant gene expression, which lead to the nitrogen-fixing phenotype. Even if the compatible combination of rhizobia and plants (each rhizobial strain establishes a symbiosis with only a limited set of host plants and *vice versa*) is used in the field, different reasons may lead to nodulation failure. Thereby, it is extremely important to improve the performance of the inoculation efficiency in different ways, other than the use of some agricultural practices that have economic and environmental negative effects on soils (Bajsa et al. 2013) (Fig. 13.1). This section summarizes the current knowledge about the addition of secondary metabolites in the performance of formulations and bioformulations.

### 13.2.1 Effect of Flavonoids

Flavonoids are important plant-to-bacteria signal molecules involved in the legume–rhizobia symbiotic behavior. They have been proposed as the first molecular signals exchanged between both host plant and symbiont. These root-released molecules interact with the rhizobial NodD protein, triggering the coordinated expression of a series of bacterial nodulation (*nod*) genes. The *nod* genes are responsible for the biosynthesis of LCOs known as Nod factors, which elicit multiple responses required for the nodulation of the appropriate host plants (Morel and Castro-Sowinski 2013). In addition to flavonoids, other root-exuded compounds (such as betaine and aldonic acids) may act as signal molecules that influence the ability of rhizobia to colonize the roots, thus affecting the symbiotic interactions with the plant (Janczarek et al. 2015). Thereby, the potential use of these *nod* gene-inducing compounds (flavonoids and non-flavonoids) could be exploited when designing new bioformulations.

Several studies have documented the addition of flavonoids for enhancing nodulation in legumes. For example, the application of luteolin and naringenin into the soil enhances the nodulation of alfalfa plants (*Medicago sativa*) by *Sinorhizobium meliloti* (Kapulnik et al. 1987; Jain et al. 1990). The direct application of flavonoids into the rhizosphere of inoculated soybean (*Glycine max*) plants increases the grain yield (Zhang and Smith 1996), the nitrogen content of shoots, and the size, number, and weight of nodules (Pan et al. 1998) and also increases the root colonization by mycorrhizas (Xie et al. 1995; Vierheilig et al. 1998). In pea (*Pisum sativum*), the addition of naringenin to the rhizosphere of plants inoculated with *Rhizobium leguminosarum*, grown in vitro, increases the root–shoot ratio, plant biomass, and nodule number (Bandyopadhyay et al. 1996; Novák et al. 2002). It has also been shown that the inoculation of legumes with flavonoid-induced rhizobial cells (inoculation with rhizobia grown in the presence of flavonoids) enhances nodulation, nitrogen fixation, and plant growth, compared with rhizobial cells obtained by fermentation without the addition of flavonoids (Table 13.1).

The addition of flavonoids to inoculated crops enhances the nitrogen fixation (Dashti et al. 2000), improves the rhizobial competitiveness (Pan and

**Table 13.1** The effect of the addition of flavonoids to rhizobial inoculants

| Flavonoid | Legume and rhizobia | Results | References |
|---|---|---|---|
| Genistein | Soybean (*Glycine max*) – *Bradyrhizobium japonicum* | Increase in the number of nodules and nitrogen fixation upon co-inoculation with *Serratia liquefaciens* or *S. proteamaculans* | Dashti et al. (2000) |
| Genistein | Soybean – *B. japonicum* | Increase in the number of nodules and dry matter, under saline/osmotic stress conditions | Muñoz et al. (2014) |
| Genistein | Soybean – *B. japonicum* | Increase in the number and size of nodules, nitrogen fixation, and plant yield, under suboptimal growth temperature | Zhang and Smith (1995) |
| Genistein | Soybean – *B. japonicum* | Increase of protein and grain yield | Zhang and Smith (1996) |
| Genistein | Soybean – *B. japonicum* | Increase in the nodule size, number and weight, shoot weight, and yield, at three levels of nitrogen fertilization | Pan and Smith (2000) |
| Genistein | Soybean – *B. japonicum* | Increase in nodulation and plant yield, under stressful soil salinity and acidity | Miransari and Smith (2007, 2009a) |
| Genistein | Soybean – *B. japonicum* | Increase in nodulation, nitrogen fixation, root and shoot growth, photochemical efficiency, photosynthetic rate, stomatal conductance, and transpiration under salinity stress | Dolatabadian et al. (2012, 2013) |
| Genistein | Soybean – *B. japonicum* | Improvement of nodulation, under water deficit | Nápoles et al. (2009) |
| Genistein | Common bean (*Phaseolus vulgaris* L.) – *R. leguminosarum* | Increase of nodule number, nitrogen plant content, and dry matter | Poustini et al. (2010) |
| Hesperitin and naringenin | Pea (*Pisum sativum*) and lentil (*Lens culinaris*) – *R. leguminosarum* | Increase of nodule number and plant dry matter, under two growth temperature | Begum et al. (2001) |
| Hesperetin and apigenin | Fenugreek (*Trigonella foenum-graecum* L.) – *R. tibeticum* | Increase in total number and fresh weight of nodules, nitrogenase activity, and plant biomass, under salinity stress or in cobalt-contaminated soil | Abd-Alla et al. (2014a or 2014b), respectively |

Smith 2000), and nodulation (Paau et al. 1990; Zhang and Smith 1995; Pan and Smith 2000). The plant inoculation with rhizobial cells that were previously induced with flavonoids during growth significantly alleviates the adverse effects of salinity (Abd-Alla et al. 2014a; Muñoz et al. 2014), acidity (Miransari and Smith 2007, 2009a), low temperature (Zhang and Smith 1995; Begum et al. 2001; Broughton et al. 2003), water stress (Nápoles et al. 2009), and heavy metal toxicity (Abd-Alla et al. 2014b) of soils. In addition, the single application of flavonoids (Bandyopadhyay et al. 1996) into the soil or the inoculation of plants with flavonoid-induced rhizobial cells reduces the inhibitory effect of mineral nitrogen on nodulation (Pan and Smith 2000).

Although flavonoids are expensive, they act at very low concentrations, and their addition during the industrial fermentation of inoculant strains would be economically justified if the performance of the final bioformulation is much improved. Among the molecules involved in the communication between legumes and rhizobia, flavonoids were the first ones produced on an industrial scale for agronomic purposes (Broughton et al. 2003). The first methods for enhancing soybean grown in the field, using genistein/daidzein-induced rhizobial cells were patented by Smith and Zhang (1999) (US5922316A) and commercialized as SoyaSignal™ (Leibovitch et al. 2001).

### 13.2.2 Effect of LCOs

An increasing number of studies demonstrated the influence of the exogenous application of LCOs in the legume–rhizobia symbiosis. These molecules elicit the development of nodules at specific positions in the plant root, and their exogenous addition to rhizobial-inoculated plants promotes nodulation, photosynthesis, plant growth, and seed germination, compared with single rhizobial inoculation (Table 13.2).

The exogenous application of LCOs also affects the growth of non-leguminous plants, mainly accelerating germination and seedling growth (Zhang and Smith 2002). They also enhance corn growth (Souleimanov et al. 2002; Khan et al. 2008) and the yield of tomato fruit (Chen et al. 2007), conifer (*Picea abies*) (Dyachok et al. 2000), and carrot (De Jong et al. 1993). Other examples are the improvement of root growth of *Arabidopsis thaliana* (Khan et al. 2011) and the germination of a variety of economically important plants belonging to diverse families (Prithiviraj et al. 2003).

The data support that the addition of low concentrations of LCOs to bioformulations may increase their performance. However, LCOs have low stability in the soil, probably due to their degradation by plant or bacterial chitinases (Broughton et al. 2003), questioning, in our opinion, the effectiveness of their use in the field. Even though, some companies declare the production of successful bioformulations containing rhizobia and LCOs. For example, the Novozymes patented "LCO promoter technology" product known as a seed treatment (for more information see Sect. 13.3).

### 13.2.3 Effect of Phytohormones

Phytohormones are chemical messengers produced by plants that participate in multiple functions at low concentrations. In general, they are classified into five major classes: auxins, cytokinins (CKs), gibberellins (GAs), abscisic acid (ABA), and ethylene (ET). These phytohormones are involved in plant growth and development mainly due to their ability of inducing cell proliferation and expansion (Perrot-Rechenmann 2010; Davière and Achard 2013). Other phytohormones with relevant functions in plant growth are brassinosteroids, salicylic acid (SA), jasmonates (JAs), polyamines, nitric oxide, and strigolactones, among others (Morel and Castro-Sowinski 2013). In addition to plants, diverse microbial species produce phytohormones (Spaepen et al. 2007; Cohen et al. 2009; Spaepen and Vanderleyden 2011; Morel et al. 2011; Spaepen 2015). The inoculation of plants with these microorganisms has positive effects on plant development and yield, possibly due to their phytohormone production.

The positive effect of phytohormones (from bacterial or plant origin) in plant growth has been described earlier (Ketring and Schubert 1981; Hedin and McCarty 1991, 1994; among others). Then, the following question arises: will the addition of phytohormones increase the performance of bioformulations? The answer is "yes." Many works support the idea that the addition of phytohormones to bioformulations increases plant development and yield when compared with single bioformulations (containing only bacteria). In this section we provide an overview of the effects that phytohormones have on plant growth, briefly exploring reports of their single application in plants (formulations) or in combination with PGPB (bioformulations) (Table 13.3).

#### 13.2.3.1 Effect of Auxins

Bacterial auxins have been deeply studied (Spaepen et al. 2007), IAA being the best characterized one. Examples of auxin-producing microorganisms are *Azospirillum* (Khalid et al. 2011), *Pseudomonas* (Khakipour et al. 2008), *Rhizobium* (Etesami et al. 2009), *Bacillus* (Lim and Kim 2009), and *Delftia* (Morel et al. 2011; Ubalde et al. 2012). IAA promotes plant growth, mainly because this phytohormone increases the root volume, thereby increasing the active area for the absorption of mineral nutrients and water (Morel et al. 2012). IAA also triggers the expression of bacterial genes involved in adhesion, adaptation, and virulence, thus leading to a better plant colonization (Spaepen et al. 2007). Auxins are also involved

**Table 13.2** The effects of the addition of LCOs on legume plant growth and nodulation. *non-host plant or non-nodulate rhizobia; Nod$^-$, mutant cells of rhizobia that failed to form nodules

| Source of LCOs | Plant-rhizobia system | Place of LCOs application | Results | References |
|---|---|---|---|---|
| *R. leguminosarum* bv *viciae* | Vetch (*Vicia sativa*) (uninoculated plants) | Rhizosphere | Production of preinfection structures | van Brussel et al. (1992) |
| *R. leguminosarum* bv *viciae* | Vetch – *R. leguminosarum* | Rhizosphere | Reduction of auxin transport that preceded the root nodule primordium formation | Boot et al. (1999) |
| *Rhizobium* sp. | *Macroptilium atropurpureum* – wild type or mutant (Nod$^-$) – *Rhizobium* sp. | Rhizosphere | Deformation and curling of the root hairs; colonization of roots by mutants cells | Relić et al. (1993) |
| *Rhizobium* sp. | Soybean – *B. japonicum* (Nod$^-$)<br>*Calopogonium caeruleum* – *R. freedi**<br>*Vigna unguiculata* wild type or mutant (Nod$^-$) – *Rhizobium* sp. | Rhizosphere | Formation of pseudonodules in the absence of rhizobia; colonization, nodulation,, and nitrogen fixation by the combinations of LCO with mutant cells | Relić et al. (1994) |
| *S. meliloti* | *M. truncatula* – *S. meliloti* | Rhizosphere or seeds (by soaking) | Increase in number of nodules per plant | Macchiavelli and Brelles-Mariño (2004) |
| *R. leguminosarum* bv. *viciae* | Pea (*Pisum sativum*) (uninoculated plants) | Soaked seeds | Enhanced nitrogenase activity and total plant nitrogen content | Siczek et al. (2013, 2014) |
| *R. leguminosarum* bv. *viciae* | Pea and vetch (*Vicia villosa*) (uninoculated plants) | Soaked seeds | Increase in shoot and root fresh and dry weights and nodule number | Kidaj et al. (2012) |
| *R. leguminosarum* bv. *trifolii* | Clover (*Trifolium pratense*) – *R. leguminosarum* bv. *trifolii* | Soaked seeds | Enhanced clover nodulation and growth of plants | Maj et al. (2009) |
| *R. leguminosarum* bv. *viciae* | Pea (uninoculated plants) | Soaked seeds or foliar application | Increase in number and weight of nodules, chlorophyll content in leaves, and pea yield | Podleśny et al. (2014) |
| *B. japonicum* | Soybean – *B. japonicum* or corn (*Zea mays*) (uninoculated plants) | Rhizosphere or stem injection | Increase in total length and surface area of the roots | Souleimanov et al. (2002) |
| *B. japonicum* | Soybean – *B. japonicum* | Foliar application | Increase in photosynthesis and plant dry weight | Almaraz et al. (2007) and Khan et al. (2008) |
| *B. japonicum* | Soybean (uninoculated plants) | Foliar application | Increase in flower and pod numbers, under water stress | Atti et al. (2005) |
| *B. japonicum* | Soybean (uninoculated plants) | Foliar application | Changes in the global plant gene expression, under suboptimal temperature | Wang et al. (2012) |
| *B. japonicum* | Soybean – Genistein-induced *B. japonicum* cells | Rhizosphere | Increase in calcium uptake into leaves | Supanjani et al. (2006) |

(continued)

**Table 13.2** (continued)

| Source of LCOs | Plant-rhizobia system | Place of LCOs application | Results | References |
|---|---|---|---|---|
| LCO analogue produced by *Serratia proteamaculans* | Soybean – *B. japonicum* | Root or foliar application | Enhancement in nodulation, growth, and seed germination | Bai et al. (2002) |
| *B. japonicum* | Soybean (uninoculated plants) | Rhizosphere | Increase dry matter of aerial and radical parts | Muñoz et al. (2014) |
| *B. japonicum* | Soybean (uninoculated plants) | Rhizosphere | Changes in frequency of root-hair deformations | Duzan et al. (2004) |
| *B. japonicum* | Soybean (uninoculated plants) | Rhizosphere | Induce plant resistance to powdery mildew caused by *Microsphaera diffusa* | Duzan et al. (2005) |
| *Azorhizobium caulinodans* | *Sesbania rostrate* – *A. caulinodans* ( cell defectives in LCOs production) | Rhizosphere | Formation of preinfection structures | D'Haeze et al. (1998) |
| *Rhizobium* sp. | Soybean – *Glomus mosseae* | Watering | Enhance mycorrhizal colonization | Xie et al. (1995) |

in the establishment of the symbiosis between microorganisms and non-leguminous plants, increasing the nitrogen fixation capability (Benson and Silvester 1993; Péret et al. 2008).

Many reports show that the exogenous application of phytohormones (alone or in combination with their biosynthetic precursors) or a mixture of phytohormones increases nodulation rate, number of nodules, dry shoot weight, and/or grain yield of many crops (Table 13.3).

Other reports show that seed priming with phytohormones (IAA, GAs, ABA, and ET) increased the speed and synchronicity of germination and increased crop yields. Possibly, the increase in the rate of amino acids and amide biosynthesis may explain the improvement in germination performance, seedling growth, and grain yield, as shown in maize (*Zea mays* L.) (Tian et al. 2014). The overall results strongly suggest that the application of these phytohormones to seeds might improve germination and finally the crop yield (Table 13.3).

#### 13.2.3.2 Effect of Gibberellins

Gibberellic acid (GA3) is commonly used either alone or with fertilizer to stimulate pasture production (Matthew et al. 2009) and legume crops (Table 13.3). Among commercial products, Nufarm Ltd has developed a water-soluble formulation with GA3 as active constituent (www.nufarm.com.au). However, this phytohormone is sensitive to heat and light and can be quickly degraded in soil, resulting in its low utilization. Thus, farmers repeatedly apply conventional formulations containing GA3, increasing the cost of crop production. A solution to this problem may be the use of controlled release formulations. Liu et al. (2013) successfully prepared a novel gibberellic acid–chitosan (GA3–CS) conjugate and evaluated the controlled release of GA3 in vitro, showing the potential of this strategy for the good use of GA3.

#### 13.2.3.3 Effect of Cytokinins

CKs affect many aspects of growth and plant development, including cell division, stem initiation and growth, leaf senescence, apical dominance, nutrient absorption, gametophyte and embryo development, and response to biotic and abiotic factors. They also play a critical role in nodule formation, triggering the cell division that initiates the nodule development (Tirichine et al. 2007; Argüeso et al. 2012). The efficient use of formulations containing CKs has been shown in some plants as cotton (*Gossypium hirsutum*) (Hedin and McCarty 1991, 1994),

**Table 13.3** The effect of the addition of phytohormones (*IAA* indole-3-acetic acid, *CKs* cytokinins, *GAs* gibberellins, *ABA* abscisic acid, *SA* salicylic acid, *JAs* jasmonates, *ET* ethylene) to crops

| Crop | Phytohormone | Plant effect | References |
|---|---|---|---|
| Wheat (*Triticum aestivum* L.) | IAA+GA+ABA | Increase the speed and synchronicity of germination and grain yield by restructuring tillers and weight in wheat grain | Cai et al. (2014) |
| | ABA – SA | Decrease osmotic stress effects | Marcińska et al. (2013) |
| | CKs | Stimulate amino acid root exudation | Kudoyarova et al. (2014) |
| | JAs | Improve tolerance to salt stress | Qiu et al. (2014) |
| Soybean (*Glycine max.*) | IAA | Increase nodule number and dry weight of roots and higher yield | Sudadi and Suryono (2015) |
| | GAs | Ameliorate the adverse effects of salt stress and restores normal growth and development | Hamayun et al. (2010) |
| | JAs | Alleviate the harmful effects of saline stress | Yoon et al. (2009) |
| | IAA+GAs+ ABA+ET | Increase the speed and synchronicity of germination and crop yield | Tian et al. (2014) |
| Barley (*Hordeum vulgare* L.) | GAs | Increase the rate of germination and suppress inhibitors acting during dormancy | Miransari and Smith (2009b) |
| Rice (*Oryza sativa*) | JAs | Improve tolerance to saline stress; increase ABA endogenous production on plant | Seo et al. (2001) |
| Cotton (*Gossypium hirsutum*) | IAA+GAs | Increase the number of fibers per ovule | Seagull and Giavalis (2004) |
| Grape (*Vitis vinifera*) | ABA | Inhibit the anthocyanin accumulation; improve color | Roberto et al. (2012) |
| Pearl millet (*Pennisetum americanum* L.) | GAs | Increase production of lateral roots | Tien et al. (1979) |
| | CKs | Increase production of lateral roots | |
| | IAA+GAs+CKs | Changes in root morphology; stimulation of root-hair formation, reduction of lateral root production, and main root elongation | |
| Bedding plants | ABA | Reduce the loss of water and extend the shelf life of several plants | Waterland et al. (2010) |
| Muskmelon (*Cucumis melo* L.) | ABA | Excess levels delay post-stress growth and produce chlorosis, during water stress | Agehara and Leskovar (2012) |
| Potato (*Solanum tuberosum* L.) and tomato (*Lycopersicon esculentum* Mill.) | JAs | Protect against the infection with *Phytophthora infestans* | Cohen et al. (1993) |
| Ornamental plants | JAs | Protect against the attack of pathogenic fungi | Meir et al. 1998 |
| *Arabidopsis* | JAs | Reduce development of the disease caused by fungi (*Alternaria brassicicola*, *Botrytis cinerea*, and *Plectosphaerella cucumerina* | Tomma et al. (2000) |

peanut (*Arachis hypogaea*) (Ketring and Schubert 1981), and *Sophora tonkinensis* (Jana et al. 2013), among others (Table 13.3). In wheat, the application of synthetic trans-zeatin or CKs-producing bacteria increased rhizodeposition of amino acids (Kudoyarova et al. 2014).

#### 13.2.3.4 Effect of Ethylene

ET, jasmonic acid (JA), and ABA, usually known as immunity hormones, have a regulatory function during the plant infection with pathogens (Pieterse et al. 2009). ET is also produced in response to biotic and abiotic stresses. High levels of ET inhibit the normal plant growth

(Saleem et al. 2007) and the nodulation process in numerous plant species, as it was shown by Shaharoona et al. (2011). In addition, ET precursors (as ACC) also inhibited the rhizobial infection and decreased the number of nodules (Peters and Crist-Estes 1989; Nukui et al. 2000). Thus, the reduction of endogenous ET level (by the ACC deaminase activities, in roots) suppresses the negative effects of ET (Shaharoona et al. 2011). Some rhizobial strains produce inhibitors of the ET biosynthesis, thereby improving nodulation in legumes, even in presence of ET (Glick 2005; Glick et al. 2007). Up to know, we did not find reports about the exogenous application of molecules that inhibit ET synthesis in inoculated plant.

#### 13.2.3.5 Effect of Abscisic Acid

ABA is involved in seed dormancy and senescence and in the control of stomatal opening in response to environmental changes such as water deficiency. They have negative effects in nodule formation (Liang et al. 2007) and can promote disease or increase plant resistance depending on the plant–pathogen system (Ulferts et al. 2015). Among ABA-producing plant growth-promoting rhizobacteria (PGPR), it has been reported that *A. brasilense* ameliorated the response of *A. thaliana* to drought and stimulated growth and seed yield, among other, mainly by the increase of plants' ABA content (Cohen et al. 2008).

Among other effects (Table 13.3), the exogenous application of ABA reduced the water loss and extended shelf life of several plants, including bedding plants before shipping, thus maintaining their marketability even under severe drought stress conditions (Waterland et al. 2010). Other interesting agronomic use of ABA is its exogenous application to grape berries with patchy skins that improved the fruit color by increasing anthocyanin accumulation (Roberto et al. 2012). In 2013, a patent for the methodology involved in the modification of the sensory characteristics of white grapes and white wine (US8349768 B2) by the application of ABA was registered (Venburg et al. 2013).

The exogenous application of ABA, or SA, also minimizes the negative effect of osmotic stress on wheat seedlings, probably due to an increase in proline and carbohydrate content as well as in antioxidant compounds of plants (Marcińska et al. 2013). In addition to the positive effects of adding ABA, several side effects have been reported on growth and physiology of muskmelon (*Cucumis melo* L.) seedlings. Agehara and Leskovar (2012) also showed that the excess levels of ABA delay post-stress growth and produce chlorosis, during water stress. The authors suggested that the concentration of ABA should be minimized in the formulation.

#### 13.2.3.6 Effect of Jasmonates

The chemical group of JAs includes: JA; JA biosynthetic precursor, the 12-oxophytodenoic acid; JA derivative molecules, such as the methyl jasmonate (MeJA); and the conjugate jasmonate-isoleucine (Ile-Ja). They are molecular signals involved in the plant response to abiotic and biotic stresses, as well as in plant growth and development (Moons et al. 1997; Pedranzani et al. 2003; Wasternack 2007).

The evidence supports that the exogenous application of JAs protects plants from the attack of some fungi (Table 13.3). Briefly, the exogenous application of JAs has important effects in the control of pathogenic fungi and alleviates the detrimental effects caused by some environmental stresses. Among many examples, the foliar application of JAs to potato and tomato plants protects them against the infection with *Phytophthora infestans* (Cohen et al. 1993). Also the postharvest application of JAs to ornamental plants protects against the attack of pathogenic fungi (Meir et al. 1998). Tomma et al. (2000) showed that by spraying JAs to *Arabidopsis* plants, the development of the disease caused by fungi (*Alternaria brassicicola*, *Botrytis cinerea*, and *Plectosphaerella cucumerina*) is reduced. Also the exogenous application of JAs is useful for alleviating the harmful effects of salinity stress in soybean (Yoon et al. 2009) and wheat (Qiu et al. 2014). JAs increase the endogenous level of ABA in rice crops (Seo et al. 2001) and act as inducers of *nod* genes of *B. japonicum* (Mabood and Smith 2005; Mabood et al. 2006).

Many patents documented several methodologies for suppressing diseases and inducing plant defense by the use of JAs in single formulations or in combination with other agri-products (herbicides, pesticides) (such as US5436226A or US20120077674A1, Lulai et al. 1995 and Cargeeg and Seevers 2011, respectively).

### 13.2.4 Addition of Pooled Metabolites

Certainly, the application of phytohormones, flavonoids, or LCOs, increases plant growth and yield (Tables 13.1, 13.2, and 13.3). So, what effect would have the application of a mixture of molecules? This idea was recently developed by Marks et al. (2013) and Morel et al. (2015). Marks et al. (2013) reported that the addition of rhizobial metabolites (produced during the growth of rhizobial cells) enhances grain yields of soybean and maize when inoculated with *Bradyrhizobium* and *Azospirillum* spp., respectively. These enhancements might be explained by the combined effect of the diverse rhizobial secondary compounds (exopolysaccharides, phytohormones, and LCOs). Molla et al. (2001) also found that the addition of a cell-free supernatant of *A. brasilense* Sp7, or IAA, stimulates root growth of soybean, in greenhouse conditions. It has also been shown that the watering of alfalfa plants with the hydroponic cell-free solution obtained during the co-inoculation of alfalfa with *Sinorhizobium* and *Delftia* strains increased plant growth (Morel et al. 2015). The hydroponic solution contained the bacterial- and root-secreted molecules, including flavonoids, phytohormones, and LCOs.

In summary, the inoculation of some plants in combination with the addition of a set of secondary metabolites (plant and bacterial metabolites) enhances plant growth and yield. But the use of commercial bioformulations containing a pool of plant and microbial metabolites is still incipient.

## 13.3 Conclusion

The gathered information shows the huge potential of including secondary metabolites in bioformulations (Fig. 13.2). Although it is

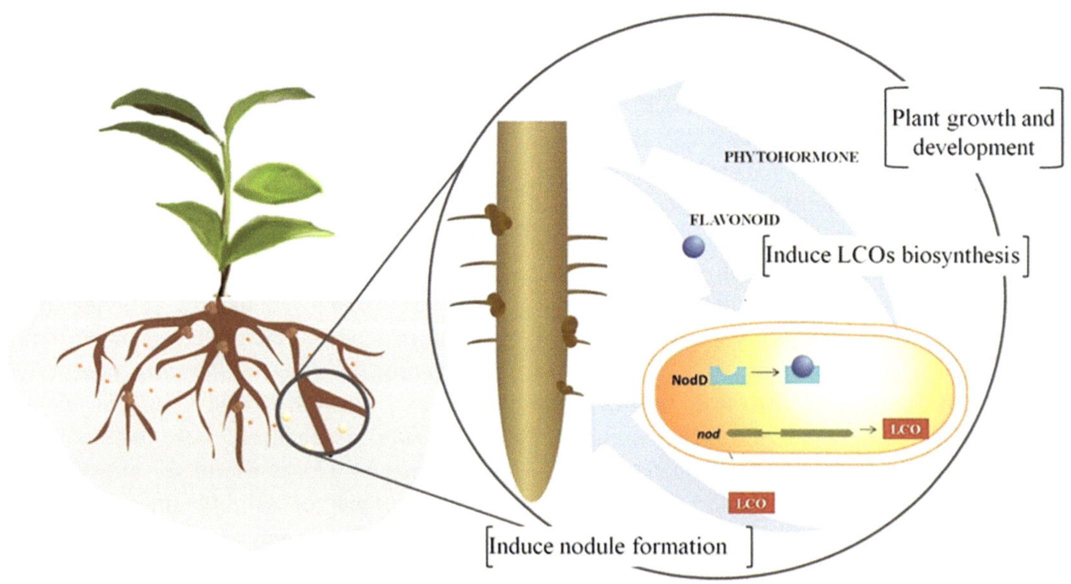


**Fig. 13.2** The effect of secondary metabolites on plant fitness: how to improve bioformulations? The information supports that the addition of LCO, flavonoids, and/or phytohormones improves the performance of bioformulations, rendering better commercial products

**Table 13.4** Commercial formulations containing secondary metabolites. *LCO* lipo-chitooligosaccharides, *GAs* gibberellins, *IAA* indol-3-acetic acid

| Metabolite | Other active components | Crop | Product name | Company | Web page | Country | References |
|---|---|---|---|---|---|---|---|
| LCO | *B. japonicum* | Soybean | Optimize II | Nitragin | www.nitragin.com.ar | Argentina | Torres (2015) |
| *Bioinductors* | *B. japonicum* | Soybean | Signum | Rizobacter | http://www.rizobacter.com | Argentina | Rizobacter.com (2015) |
| LCO | *B. japonicum* + *Penicillium bilaiae* | Soybean | Nitragin Triple | Nitragin | www.nitragin.com.ar | Argentina | Torres (2015) |
| LCO | Aqueous carrier | Corn, Soybean, alfalfa | Ratchet | Novozymes | www.bioag.novozymes.co | USA | Bioag.novozymes.com (2015) |
| LCO | Aqueous carrier | Corn | Torque | Novozymes | www.bioag.novozymes.co | USA | Bioag.novozymes.com (2015) |
| LCO | *Bradyrhizobium* sp. | Peanuts | Optimize | Novozymes | www.bioag.novozymes.co | USA | Bioag.novozymes.com (2015) |
| LCO | *Penicillium bilaiae* + *Rhizobium leguminosarum* | Pea lentil | TagTeamLCO | Novozymes | www.bioag.novozymes.co | USA | Bioag.novozymes.com (2015) |
| LCO | *P. bilaiae* | Wheat, Corn | JumpStart LCO | Novozymes | www.bioag.novozymes.com | USA | Bioag.novozymes.com (2015) |
| LCO | *B. japonicum* | Soybean, Peanuts | Dyna-StartMax | Loveland | www.lovelandproducts.com | USA | Lovelandproducts.com (2015) |
| Isoflavonoids | Aqueous carrier | Cotton | Revv | Novozymes | www.bioag.novozymes.com | USA | Bioag.novozymes.com (2015) |
| GAs | N/D | Blueberries, Cherries, Citrus, Cranberries, Grapes | Falgro 20SP | Fine Americas INC | www.fine-americas.com | USA | Fine Americas (2014) |
| IAA | Mineral nutrients | many crops | Bayfolan | Bayer CropScience AG | www.bayercropscience.com.mx | Mexico | Bayercropscience.com.mx (2015) |
| *Thidiazuron* | N/D | Grapes | Splendor 5 % SC | Bayer CropScience AG | http://www.cropscience.bayer.cl/ | Chile | Cropscience.bayer.cl (2015) |
| *Ethephon* | N/D | Grapes, tomato | Ethrel 48 SL | Bayer CropScience AG | http://www.cropscience.bayer.cl/ | Chile | Cropscience.bayer.cl (2015) |

*N/D* no data

important to conduct more tests, the addition of flavonoids, phytohormones, and/or LCOs increases plant growth and development and finally crop yield, when they were applied alone or in combination with PGPB. This information has led to the generation and use of patents, demonstrating the pivotal importance of the industry–academic collaboration. However, only few manufacturers produce bioformulations containing these secondary metabolites (Table 13.4). Some of these novel products are commercially available, and most of them declare to contain LCOs in single formulations, such as Ratchet® and Torque®, or in bioformulations (contain PGPB also), such as Optimize II®, Signum®, and DynaStartMax®, among others.

Three main groups of molecules involved in the plant–microbe communication have been analyzed in this chapter, but still remains to gather the information about the effect of many other molecules, as extracellular polysaccharides, osmo-protecting molecules, etc.

**Acknowledgments** We thank PEDECIBA-Uruguay (Programa de Desarrollo de las Ciencias Básicas), LAGE y Cia SA (www.lageycia.com), and ANII (Agencia Nacional de Investigación e Innovación; FMV-3-2011-6089) for partial financial support. The work of M.A. Morel and C. Cagide was also supported by ANII. We are especially grateful for Cecilia Hermann's assistance during the search of commercial products. The cartoon or Fig. 13.1 was drawn by Juan José Marizcurrena.

## References

Abd-Alla MH, El-enany AWE, Bagy MK, Bashandy SR (2014a) Alleviating the inhibitory effect of salinity stress on nod gene expression in Rhizobium tibeticum—fenugreek (*Trigonella foenum graecum*) symbiosis by isoflavonoids treatment. J Plant Int 9:275–284

Abd-Alla MH, Bagy MK, El-enany AWES, Bashandy SR (2014b) Activation of *Rhizobium tibeticum* with flavonoids enhances nodulation, nitrogen fixation, and growth of fenugreek (*Trigonella foenum-graecum* L.) grown in cobalt-polluted soil. Arch Environ Cont Tox 66:303–315

Agehara S, Leskovar DI (2012) Characterizing concentration effects of exogenous abscisic acid on gas exchange, water relations, and growth of muskmelon seedlings during water stress and rehydration. J Am Soc Hortic Sci 137:400–410

Almaraz JJ, Zhou X, Souleimanov A, Smith D (2007) Gas exchange characteristics and dry matter accumulation of soybean treated with Nod factors. J Plant Physiol 164:1391–1393

Arca@arca.cl A (2015) Bayer CropScience Chile: Líder en protección de cultivos, productos agricolas y fitosanitarios. [online] Bayercropscience.cl. Available at: http://www.bayercropscience.cl

Argueso CT, Ferreira FJ, Epple P, To JPC, Hutchison CE, Schaller GE, Kieber JJ (2012) Two-component elements mediate interactions between cytokinin and salicylic acid in plant immunity. PLoS Genet 8: e1002448

Arora NK, Khare E, Maheshwari DK (2010) Plant growth promoting rhizobacteria: constrains in bioformulation, commercialization, and future strategies. In: Maheshwari DK (ed) Plant growth and health promoting bacteria microbiology monographs 18. Springer, Berlin, pp 97–116

Atti S, Bonnell R, Prasher S, Smith DL (2005) Response of soybean {*Glycine max* (L.) Merr.} under chronic water deficit to LCO application during flowering and pod filling. Irrig Drain 54:15–30

Bai Y, Souleimanov A, Smith DL (2002) An inducible activator produced by a *Serratia proteamaculans* strain and its soybean growth promoting activity under greenhouse conditions. J Exp Bot 53:1495–1502

Bajsa N, Morel MA, Braña V, Castro-Sowinski S (2013) The effect of agricultural practices on resident soil microbial communities: focus on biocontrol and biofertilization. In: de Bruijn FJ (ed) Molecular microbial ecology of the rhizosphere, vol 2. Wiley-Blackwell, Hoboken, pp 687–700

Bandyopadhyay AK, Jain V, Nainawatee HS (1996) Nitrate alters the flavonoid profile and nodulation in pea (*Pisum sativum* L.). Biol Fert Soils 21:189–192

Bashan Y, de-Bashan LE, Prabhu SR, Hernandez JP (2014) Advances in plant growth-promoting bacterial inoculant technology: formulations and practical perspectives (1998–2013). Plant Soil 378:1–33

Bayercropscience.com.mx (2015) Bayer de México, S.A. de C.V. [online]. Available at: http://www.bayercropscience.com.mx/

Begum AA, Leibovitch S, Migner P, Zhang F (2001) Specific flavonoids induced *nod* gene expression and pre-activated *nod* genes of *Rhizobium leguminosarum* increased pea (*Pisum sativum* L.) and lentil (*Lens culinaris* L.) nodulation in controlled growth chamber environments. J Exp Bot 52:1537–1543

Benson DR, Silvester WB (1993) Biology of *Frankia* strains, actinomycete symbionts of actinorhizal plants. Microbiol Rev 57:293–319

Bioag.novozymes.com (2015) Novozymes BioAg Group. [online] Available at: http://www.bioag.novozymes.com

Bonaterra A, Camps J, Montesinos E (2005) Osmotically induced trehalose and glycine betaine accumulation improves tolerance to desiccation, survival and efficacy of the postharvest biocontrol agent *Pantoea agglomerans* EPS125. FEMS Microbiol Lett 250:1–8

Boot KJ, van Brussel AA, Tak T, Spaink HP, Kijne JW (1999) Lipochitin oligosaccharides from *Rhizobium leguminosarum* bv. *viciae* reduce auxin transport capacity in *Vicia sativa* subsp. *nigra* roots. Mol Plant Microbe Interact 12:839–844

Broughton WJ, Zhang F, Perret X, Staehelin C (2003) Signals exchanged between legumes and *Rhizobium*: agricultural uses and perspectives. Plant Soil 252:129–137

Cai T, Xu H, Peng D, Yin Y, Yang W, Ni Y, Chen X, Xu C, Yang D, Cui Z, Wang Z (2014) Exogenous hormonal application improves grain yield of wheat by optimizing tiller productivity. Field Crops Res 155:172–183

Cargeeg RD, Seevers K (2011) Methods and compositions containing jasmonates or related compounds for promoting biodefense activity in plants. US Patent Application US20120077674 A1

Chen C, McIver J, Yang Y, Bai Y, Schultz B, McIver A (2007) Foliar application of lipo-chitooligosaccharides (Nod factors) to tomato (*Lycopersicon esculentum*) enhances flowering and fruit production. Can J Plant Sci 87:365–372

Cohen Y, Gisi U, Niderman T (1993) Local and systemic protection against *Phytophthora infestans* induced in potato and tomato plants by jasmonic acid and jasmonic metil ester. Phytopathology 83:1054–1062

Cohen AC, Bottini R, Piccoli PN (2008) *Azospirillum brasilense* Sp 245 produces ABA in chemically-defined culture medium and increases ABA content in arabidopsis plants. Plant Growth Regul 54:97–103

Cohen AC, Travaglia CN, Bottini R, Piccoli PN (2009) Participation of abscisic acid and gibberellins produced by endophytic *Azospirillum* in the alleviation of drought effects in maize. Botany 87:455–462

Cropscience.bayer.com (2015) Bayer S.A. [online]. Available at: http://www.cropscience.bayer.cl

D'Haeze W, Gao M, De Rycke R, Van Montagu M, Engler G, Holsters M (1998) Roles for azorhizobial Nod factors and surface polysaccharides in intercellular invasion and nodule penetration, respectively. Mol Plant Microbe Interact 11:999–1008

Dashti N, Prithiviraj B, Zhou X, Hynes RK, Smith DL (2000) Combined effects of plant growth promoting rhizobacteria and genistein on nitrogen fixation in soybean at suboptimal root zone temperatures. J Plant Nutr 23:593–604

Davière JM, Achard P (2013) Gibberellin signaling in plants. Development 140:1147–1151

De Jong AJ, Heidstra R, Spaink HP, Hartog MV, Meijer EA, Hendriks T, Lo Schiavo F, Ab Van Kammen M, De Vries SC (1993) Rhizobium lipooligosaccharides rescue a carrot somatic embryo mutant. Plant Cell Online 5:615–620

Dolatabadian A, Sanavy SAMM, Ghanati F, Gresshoff PM (2012) Morphological and physiological response of soybean treated with the microsymbiont *Bradyrhizobium japonicum* pre-incubated with genistein. S Afr J Bot 79:9–18

Dolatabadian A, Sanavy SAMM, Ghanati F, Gresshoff PM (2013) *Agrobacterium rhizogenes* transformed soybean roots differ in their nodulation and nitrogen fixation response to genistein and salt stress. World J Microbiol Biotech 29:1327–1339

Duzan HM, Zhou X, Souleimanov A, Smith DL (2004) Perception of *Bradyrhizobium japonicum* Nod factor by soybean [*Glycine max* (L.) Merr.] root hairs under abiotic stress conditions. J Exp Bot 55:2641–2646

Duzan HM, Mabood F, Zhou X, Souleimanov A, Smith DL (2005) Nod factor induces soybean resistance to powdery mildew. Plant Physiol Biochem 43:1022–1030

Dyachok JV, Tobin AE, Price NPJ, Von Arnold S (2000) Rhizobial Nod factors stimulate somatic embryo development in *Picea abies*. Plant Cell Rep 19:290–297

Etesami H, Alikhani HA, Akbari AA (2009) Evaluation of plant growth hormones production (IAA) ability by Iranian soils rhizobial strains and effects of superior strains application on wheat growth indexes. World Appl Sci J 6:1576–1584

Fine Americas (2014) Fine Americas Inc. [online]. Available at: http://www.fine-americas.com

Glick BR (2005) Modulation of plant ethylene levels by the bacterial enzyme ACC deaminase. FEMS Microbiol Lett 251:1–7

Glick BR, Cheng Z, Czarny J, Duan J (2007) Promotion of plant growth by ACC deaminase-producing soil bacteria. Eur J Plant Pathol 119:329–339

Hamayun M, Khan SA, Khan AL, Rehman G, Kim YH, Iqbal I, Hussain J, Sohn EY, Lee IJ (2010) Gibberellin production and plant growth promotion from pure cultures of *Cladosporium* sp. MH-6 isolated from cucumber (*Cucumis sativus* L.). Mycologia 102 (5):989–995

Hedin PA, McCarty JC (1991) Effects of kinetin formulations on allelochemicals and agronomic traits of cotton. Agric Food Chem 39:549–553

Hedin PA, McCarty JC (1994) Effects of several commercial plant growth regulator formulations on yield and allelochemicals of cotton (*Gossypium hirsutum* L.). Agric Food Chem 42:1355–1357

Jain V, Garg N, Nainawatee HS (1990) Naringenin enhanced efficiency of *Rhizobium meliloti*-alfalfa symbiosis. World J Microbiol Biotechnol 6:434–436

Jana S, Sivanesan I, Jeong BR (2013) Effect of cytokinins on in vitro multiplication of *Sophora tonkinensis*. Asian Pac J Tropic Biomed 3:549–553

Janczarek M, Rachwał K, Marzec A, Grządziel J, Palusińska-Szysz M (2015) Signal molecules and cell-surface components involved in early stages of the legume–rhizobium interactions. Appl Soil Ecol 85:94–113

Kadouri D, Jurkevitch E, Okon Y, Castro-Sowinski S (2005) Ecological and agricultural significance of

bacterial polyhydroxyalkanoates. Crit Rev Microbiol 31:55–67
Kapulnik Y, Joseph CM, Phillips DA (1987) Flavone limitations to root nodulation and symbiotic nitrogen fixation in Alfalfa. Plant Physiol 84:1193–1196
Ketring DL, Schubert AM (1981) Reproduction of peanuts treated with a cytokinin-containing preparation. Agric J 73:350–352
Khakipour NK, Khavazi Mojallali H, Pazira E, Asadirahmani H (2008) Production of auxin hormone by fluorescent pseudomonads. Am Eur J Agric Environ Sci 4:687–692
Khalid A, Sultana S, Arshad M, Mahmood S, Mahmood T, Siddique MT (2011) Performance of auxin producing rhizobacteria for improving growth and yield of wheat and rice grown in rotation under field conditions. Int J Agric Appl Sci 3:44–50
Khan W, Prithiviraj B, Smith DL (2008) Nod factor [Nod Bj V (C 18: 1, MeFuc)] and lumichrome enhance photosynthesis and growth of corn and soybean. J Plant Physiol 165:1342–1351
Khan W, Costa C, Souleimanov A, Prithiviraj B, Smith DL (2011) Response of *Arabidopsis thaliana* roots to lipo-chitooligosaccharide from *Bradyrhizobium japonicum* and other chitin-like compounds. Plant Growth Regul 63:243–249
Kidaj D, Wielbo J, Skorupska A (2012) Nod factors stimulate seed germination and promote growth and nodulation of pea and vetch under competitive conditions. Microbiol Res 167:144–150
Kudoyarova GR, Melentiev AI, Martynenko EV, Timergalina LN, Arkhipova TN, Shendel GV, Kuz'mina LY, Dodd IC, Veselov SY (2014) Cytokinin producing bacteria stimulate amino acid deposition by wheat roots. Plant Physiol Biochem 83:285–291
Leibovitch S, Migner P, Zhang F, Smith DL (2001) Evaluation of the effect of SoyaSignal technology on soybean yield [*Glycine max* (L.) Merr.] under field conditions over 6 years in eastern Canada and the northern United States. J Agron Crop Sci 187:281–292
Liang YJ, Mitchell DM, Harris JM (2007) Abscisic acid rescues the root meristem defects of the *Medicago truncatula* latd mutant. Dev Biol 304:297–307
Lim JH, Kim SD (2009) Synergistic plant growth promotion by the indigenous auxins-producing PGPR *Bacillus subtilis* AH18 and *Bacillus licheniformis* K11. J Korean Soc Appl Biol Chem 52:531–538
Liu Y, Sun Y, He S, Zhu Y, Ao M, Li J, Cao Y (2013) Synthesis and characterization of gibberellin-chitosan conjugate for controlled-release application. J Biol Macromol 57:213–217
Lovelandproducts.com (2015) Loveland Products Inc. | get growing [online]. Available at: http://www.lovelandproducts.com
Lulai EC, Orr PH, Glynn MT (1995) Natural suppression of sprouting in stored potatoes using jasmonates. U.S. Patent No. 5,436,226
Mabood F, Smith DL (2005) Pre-inoculation of *Bradyrhizobium japonicum* with jasmonates accelerates nodulation and nitrogen fixation in soybean (*Glycine max*) at optimal and suboptimal root zone temperatures. Physiol Plant 125:311–323
Mabood F, Souleimanov A, Khan W, Smith DL (2006) Jasmonates induce Nod factor production by *Bradyrhizobium japonicum*. Plant Physiol Biochem 44:759–765
Macchiavelli RE, Brelles-Mariño G (2004) Nod factor-treated *Medicago truncatula* roots and seeds show an increased number of nodules when inoculated with a limiting population of *Sinorhizobium meliloti*. J Exp Bot 55:2635–2640
Maj D, Wielbo J, Marek-Kozaczuk M, Martyniuk S, Skorupska A (2009) Pretreatment of clover seeds with Nod factors improves growth and nodulation of *Trifolium pratense*. J Chem Ecol 35:479–487
Marcińska I, Czyczyło-Mysza I, Skrzypek E, Grzesiak MT, Janowiak F, Filek M, Dziurka M, Dziurka K, Waligórski P, Juzoń K, Cyganek K, Grzesiak S (2013) Alleviation of osmotic stress effects by exogenous application of salicylic or abscisic acid on wheat seedlings. Int J Mol Sci 14:13171–13193
Marks BB, Megías M, Nogueira MA, Hungria M (2013) Biotechnological potential of rhizobial metabolites to enhance the performance of *Bradyrhizobium* spp. and *Azospirillum brasilense* inoculants with soybean and maize. AMB Express 3:21
Matthew C, Hofmann WA, Osborne MA (2009) Pasture response to gibberellins: a review and recommendations. N Z J Agric Res 52:213–225
Meir S, Droby S, Davidson H, Alsevia S, Cohen L, Horev B, Philosph-Hadas S (1998) Suppression of *Botrytis* rot in cut rose flowers by postharvest application of methyl jasmonate. Postharv Biol Technol 13:235–243
Miransari M, Smith DL (2007) Overcoming the stressful effects of salinity and acidity on soybean nodulation and yields using signal molecule genistein under field conditions. J Plant Nutr 30:1967–1992
Miransari M, Smith DL (2009a) Alleviating salt stress on soybean (*Glycine max* L.) Merr.) –*Bradyrhizobium japonicum* symbiosis, using signal molecule genistein. Eur J Soil Biol 45:146–152
Miransari M, Smith DL (2009b) Rhizobial lipo-chitooligosaccharides and gibberellins enhance barley (*Hordeum vulgare* L.) seed germination. Biotechnology 8:270–275
Molla AH, Shamsuddin ZH, Saud HM (2001) Mechanism of root growth and promotion of nodulation in vegetable soybean by *Azospirillum brasilense*. Commun Soil Sci Plant Anal 32:2177–2187
Moons A, Prinsen E, Bauw G, Montagu MV (1997) Antagonistic effects of abscisic acid and jasmonates on salt stress-inducible transgenic transcripts in rice roots. Plant Cell 9:2243–2259
Morel MA, Castro-Sowinski S (2013) The complex molecular signaling network in microbe-plant interaction. In: Arora NK (ed) Plant microbe symbiosis: fundamentals and advances. Springer, New Delhi

Morel MA, Ubalde MC, Braña V, Castro-Sowinski S (2011) *Delftia* sp. JD2: a potential Cr (VI) reducing agent with plant growth promoting activity. Arch Microbiol 193:63–68

Morel MA, Braña V, Castro-Sowinski S (2012) Legume crops, importance and use of bacterial inoculation to increase the production. In: Goyal A (ed) Crop plant. InTech, Rijeka, pp 217–240

Morel MA, Cagide C, Minteguiaga MA, Dardanelli MS, Castro-Sowinski S (2015) The pattern of secreted molecules during the co-inoculation of alfalfa plants with *Sinorhizobium meliloti* and *Delftia* sp. strain JD2: an interaction that improves plant yield. Mol Plant-Microbe Interact 28:134–142

Muñoz N, Soria-Díaz ME, Manyani H, Sánchez-Matamoros RC, Serrano AG, Megías, Lascano R (2014) Structure and biological activities of lipochitooligosaccharide nodulation signals produced by *Bradyrhizobium japonicum* USDA 138 under saline and osmotic stress. Biol Fert Soils 50:207–215

Nápoles MC, Guevara E, Montero F, Rossi A, Ferreira A (2009) Role of *Bradyrhizobium japonicum* induced by genistein on soybean stressed by water deficit. Span J Agri Res 7:665–671

Novak K, Chovanec P, Škrdleta V, Kropáčová M, Lisá L, Němcová M (2002) Effect of exogenous flavonoids on nodulation of pea (*Pisum sativum* L.). J Exp Bot 53:1735–1745

Nukui N, Ezura H, Yuhashi KI, Yasuta T, Minamisawa K (2000) Effects of ethylene precursor and inhibitors for ethylene biosynthesis and perception on nodulation in *Lotus japonicus* and *Macroptiliumatropurpureum*. Plant Cell Physiol 41:893–897

Paau AS, Bennett ML, Kurtenbach CJ, Graham LL (1990) Improvement of inoculant efficiency by strain improvement and formulation manipulations. In: Gresshoff, Roth, Stacey, Newton (eds) Nitrogen fixation: achievements and objectives. Springer, New York, pp 617–624

Pan B, Smith DL (2000) Preincubation of *Bradyrhizobium japonicum* cells with genistein reduces the inhibitory effects of mineral nitrogen on soybean nodulation and nitrogen fixation under field conditions. Plant Soil 223:235–242

Pan B, Zhang F, Smith DL (1998) Genistein addition to the root medium of soybean at the onset of nitrogen fixation increases nodulation. J Plant Nutr 21:1631–1639

Pedranzani H, Racagni G, Alemano S, Miersch O, Ramirez I, Pena-Cortes H, Taleisnik E, Machado Domenech E, Abdala G (2003) Salt tolerant tomato plants show increased levels of jasmonic acid. Plant Growth Reg 41:149–158

Péret B, Svistoonoff S, Lahouze B, Auguy F, Santi C, Doumas P, Laplaze L (2008) A role for auxin during actinorhizal symbioses formation. Plant Signal Behav 3:34–35

Perrot-Rechenmann C (2010) Cellular responses to auxin: division versus expansion. In: Mark E, Dolf W, Ljung K, Ottoline L (eds) Additional perspectives on auxin signaling. Cold Spring Harbor Laboratory Press 2(5):a001446

Peters NK, Crist-Estes D (1989) Nodule formation is stimulated by the ethylene inhibitor aminoethoxyvinylglycine. Plant Physiol 91:690–693

Pieterse C, Leon-Reyes A, Van der Ent S, Van Wees S (2009) Networking by small-molecule hormones in plant immunity. Nat Chem Boil 5:308–317

Podleśny J, Wielbo J, Podleśna A, Kidaj D (2014) The pleiotropic effects of extract containing rhizobial Nod factors on pea growth and yield. Central Eur J Biol 9:396–409

Poustini K, Mabood F, Smith DL (2010) Preincubation of *Rhizobium leguminosarum* bv. *phaseoli* with jasmonate and genistein signal molecules increases bean (*Phaseolus vulgaris* L.) nodulation, nitrogen fixation and biomass production. J Agr Sci Tech 9:107–117

Prithiviraj B, Zhou X, Souleimanov A, Kahn W, Smith D (2003) A host-specific bacteria-to-plant signal molecule (Nod factor) enhances germination and early growth of diverse crop plants. Planta 216:437–445

Qiu ZB, Guo JL, Zhu AJ, Zhang L, Zhang MM (2014) Exogenous jasmonic acid can enhance tolerance of wheat seedlings to salt stress. Ecotoxicol Environ Saf 104:202–208

Relić B, Talmont F, Kopcinska J, Golinowski W, Promé JC, Broughton WJ (1993) Biological activity of *Rhizobium* sp. NGR234 Nod-factors on *Macroptilium atropurpureum*. Mol Plant Microbe Interact 6:764–774

Relić B, Perret X, Estrada-García MT, Kopcinska J, Golinowski W, Krishnan HB, Pueppke SG, Broughton WJ (1994) Nod factors of *Rhizobium* are a key to the legume door. Mol Microbiol 13:171–179

Rizobacter.com (2015) Rizobacter| Rizobacter global [online]. Available at: http://www.rizobacter.com

Roberto SR, Marinho de Assis A, Yamamoto LY, Miotto LCV, Satto AJ, Koyama R, Genta W (2012) Application timing and concentration of abscisic acid improve color of "Benitaka" table grape. Sci Hortic 142:44–48

Saleem M, Arshad M, Hussain S, Bhatti A (2007) Perspective of plant growth promoting rhizobacteria (PGPR) containing ACC-deaminase in stress agriculture. J Ind Microbiol Biotechnol 34:635–648

Seagull RW, Giavalis S (2004) Pre- and post-anthesis application of exogenous hormones alters fiber production in *Gossypium hirsutum* L. cultivar Maxxa GTO. J Cotton Sci 8:105–111

Seo HS, Song JT, Cheong JJ, Lee YH, Hwang I, Lee JS, Choi YD (2001) Jasmonic acid carboxyl methyltransferase: a key enzyme for jasmonate-regulated plant responses. Proc Natl Acad Sci U S A 98:4788–4793

Shaharoona B, Imran M, Arshad M, Khalid A (2011) Manipulation of ethylene synthesis in roots through bacterial ACC deaminase for improving nodulation in legumes. Crit Rev Plant Sc 30:279–291

Siczek A, Lipiec J, Wielbo J, Szarlip P, Kidaj D (2013) Pea growth and symbiotic activity response to Nod factors (lipo-chitooligosaccharides) and soil compaction. Appl Soil Ecol 72:181–186

Siczek A, Lipiec J, Wielbo J, Kidaj D, Szarlip P (2014) Symbiotic activity of pea (*Pisum sativum*) after application of Nod factors under field conditions. Int J Mol Sci 15:7344–7351

Smith D, Zhang F (1999) Composition for enhancing grain yield and protein yield of legumes grown under environmental conditions that inhibit or delay nodulation thereof. US Patent No. 5.922.316. US Patent and Trademark Office, Washington, DC

Souleimanov A, Prithiviraj B, Smith DL (2002) The major Nod factor of *Bradyrhizobium japonicum* promotes early growth of soybean and corn. J Exp Bot 53:1929–1934

Spaepen S (2015) Plant hormones produced by microbes. In: Lugtenberg B (ed) Principle of plant-microbe interactions. Springer, New York, pp 247–256

Spaepen S, Vanderleyden J (2011) Auxin and plant-microbe interactions. Cold Spring Harb Perspect Biol 3:a001438

Spaepen S, Vanderleyden J, Remans R (2007) Indole-3-acetic acid in microbial and microorganism-plant signaling. FEMS Microbiol Rev 31:425–448

Sudadi, Suryono (2015) Exogenous application of tryptophan and indole acetic acid (IAA) to induce root nodule formation and increase soybean yield in acid, neutral and alkaline soil. Agrivita J Agric Sc 37:1

Supanjani S, Habib A, Mabood F, Lee KD, Donnelly D, Smith DL (2006) Nod factor enhances calcium uptake by soybean. Plant Physiol Biochem 44:866–872

Tian Y, Guan B, Zhou D, Yu J, Li G, Lou Y (2014) Responses of seed germination, seedling growth, and seed yield traits to seed pretreatment in maize (*Zea mays* L.). Sci World J 2014:1–8

Tien TM, Gaskins MH, Hubbell DH (1979) Plant growth substances produced by *Azospirillum brasilense* and their effect on the growth of pearl millet (*Pennisetum americanum* L.). Appl Environ Microbiol 37:1016–1024

Tirichine L, Sandal N, Madsen LH, Radutoiu S, Albrektsen AS, Sato S, Asamizu E, Tabata S, Stougaard J (2007) A gain-of-function mutation in a cytokinin receptor triggers spontaneous root nodule organogenesis. Science 315:104–107

Tomma BPHJ, Eggermont K, Broekaert W, Cammue BPA (2000) Disease development of several fungi on Arabidopsis can be reduced by treatment with methyl jasmonate. Plant Physiol Biochem 38:421–427

Torres S (2015) Nitragin [online] Nitragin.com.ar. Available at: http://nitragin.com.ar

Ubalde MC, Braña V, Sueiro F, Morel MA, Martínez-Rosales C, Marquez C, Castro-Sowinski S (2012) The versatility of *Delftia* sp. isolates as tools for bioremediation and biofertilization technologies. Curr Microbiol 64:597–603

Ulferts S, Delventhal S, Splivallo R, Karlovsky P, Schaffrath U (2015) Abscisic acid negatively interferes with basal defence of barley against *Magnaporthe oryzae*. BMC Plant Biol 15:7

van Brussel AAN, Bakhuizen R, van Spronsen PC, Spaink HP, Tak T, Lugtenberg BJJ, Kijne JW (1992) Induction of pre-infection thread structures in the leguminous host plant by mitogenic lipo-oligosaccharides of *Rhizobium*. Science 257:70–72

Venburg GD, Rath A, Petracek PD (2013) Use of abscisic acid to alter sensory characteristics of white grapes and wine. US Patent No. 5.922.316. US Patent and Trademark Office, Washington, DC

Vierheilig H, Bago B, Albrecht C, Poulin MJ, Piché Y (1998) Flavonoids and arbuscular-mycorrhizal fungi. In: Manthey JA, Buslig BS (eds) Flavonoids in the living system. Springer, New York, pp 9–33

Wang N, Khan W, Smith DL (2012) Changes in soybean global gene expression after application of lipo-chitooligosaccharide from *Bradyrhizobium japonicum* under sub-optimal temperature. PLoS One 7:e31571

Wasternack C (2007) Jasmonates: an update on biosynthesis, signal transduction and action in plant stress response, growth and development. Ann Bot 100:681–697

Waterland NL, Campbell CA, Finer JJ, Jones ML (2010) Abscisic acid application enhances drought stress tolerance in bedding plants. Hort Sci 45:409–413

Xie ZP, Staehelin C, Vierheilig H, Wiemken A, Jabbouri S, Broughton WJ, Vogeli-Lange R, Boller T (1995) Rhizobial nodulation factors stimulate mycorrhizal colonization of nodulating and non nodulating soybeans. Plant Physiol 108:1519–1525

Yoon JI, Hamayun M, Lee SK, Lee IJ (2009) Methyl jasmonate alleviated salinity stress soybean. J Crop Sc Biotechnol 12:63–68

Zhang F, Smith D (1995) Preincubation of *Bradyrhizobium japonicum* with genistein accelerates nodule development of soybean at suboptimal root zone temperatures. Plant Physiol 108:961–968

Zhang F, Smith D (1996) Inoculation of soybean (*Glycine max.* (L.) Merr.) with genistein-preincubated *Bradyrhizobium japonicum* or genistein directly applied into soil increases soybean protein and dry matter yield under short season conditions. Plant Soil 179:233–241

Zhang F, Smith D (2002) Interorganismal signaling in suboptimum environments: the legume-rhizobia symbiosis. Adv Agron y 76:125–161

# Encapsulation Techniques for Plant Growth-Promoting Rhizobacteria

# 14

Mauricio Schoebitz and María Dolores López Belchí

## Contents




**Abstract**

Application of microbial inoculants into the soil can enhance soil properties and plant nutrient acquisition and increase the efficiency of mineral fertilizers and manures. The potential use of microbial inoculants in sustainable agriculture and soil restoration has been gaining increasing interest. Encapsulation of beneficial soil microorganisms has been applied and used in the agricultural industry, particularly in processes such as spray drying, interfacial polymerization, or cross-linking. This chapter presents different techniques for microbial inoculants and their benefits for agricultural and environmental purposes. Techniques include fluidized bed, coacervation, and ionic or inverse gelation. The major topics discussed are conventional inoculants, formulation of microbial inoculants, encapsulation techniques, and application trends. In addition, the use of biochar as inoculant carrier is proposed in order to develop new formulations. This innovative microbial inoculant has many advantages, such as increased water-holding capacity, high internal porosity, and large surface area, while it also provides a suitable habitat for microorganisms to enhance colonization and bacteria protection in the soil.



M. Schoebitz (✉)
Departamento de Suelos y Recursos Naturales, Facultad de Agronomía, Universidad de Concepción, P.O. Box 160-CConcepción, Chile
e-mail: mschoebitz@udec.cl

M.D. López Belchí
Departamento de Producción Vegetal, Facultad de Agronomía, Universidad de Concepción, Av. Vicente Méndez 595, Chillán, Chile





N.K. Arora et al. (eds.), *Biofthe formulations: for Sustainable Agriculture*,
DOI 10.1007/978-81-322-2779-3_14

## 14.1 Introduction

The development of microbial inoculants has been increasing worldwide due to the recognition of the deleterious effects on the environment generated by the excessive and improper application of chemical fertilizers, as well as the improved knowledge about plant-microorganism interactions in the soil (Malusá and Vassilev 2014). In this context, the reduced use of chemical fertilizers with increased applications of plant growth-promoting rhizobacteria (PGPR) is considered as essential to decrease the pressure on the environment derived from current agronomical practices (Malusá et al. 2012).

Sustainable agriculture production and reclamation of degraded soils can be achieved by emphasizing the use of PGPR as microbial inoculants (Schoebitz et al. 2013a; Mengual et al. 2014b). In general, microorganisms promote plant growth in three different ways: synthesizing growth-promoting hormones for the plants, facilitating the uptake of nutrients from the soil, and lessening or preventing plant diseases (Martínez-Viveros et al. 2010). The mechanisms involved in plant growth by PGPR are (1) solubilization of mineral phosphate; (2) biological nitrogen fixation; (3) ability to produce hormones like auxin, i.e., indoleacetic acid (IAA), abscisic acid (ABA), gibberellic acid, and cytokinins; (4) ability to produce 1-aminocyclopropane-1-carboxylate (ACC) deaminase to reduce the level of ethylene in roots of developing plants and thereby increasing root length and growth; (5) antagonism against phytopathogenic bacteria by producing siderophores and antibiotics; and (6) mediated resistance to drought and oxidative stress (Hayat et al. 2012).

Encapsulation of PGPR has been widely used in agriculture to obtain a surrounding structure that allows for protection, release, and functionalization of microorganisms (John et al. 2011). In fact, encapsulation tends to stabilize cells, providing less exposure to abiotic and biotic stresses and potentially enhancing their viability and stability in the production, storage, and handling, while it also confers additional protection during rehydration (Kim et al. 1996).

Numerous studies have focused on developing different encapsulation techniques to improve the survival of microbial inoculants (Herrmann and Lesueur 2013; Schoebitz et al. 2013a; Bashan et al. 2014; Campos et al. 2014). Encapsulation of microorganisms is one of the newest and most efficient techniques. For instance, encapsulation via dripping technique has been employed for the protection of cells to allow for better survival in soil after inoculation (Schoebitz et al. 2012). Encapsulated cells could be released into the target soil in a slow and controlled manner, providing greater long-term effectiveness (John et al. 2011). Nevertheless, a literature survey by Xavier et al. (2004) showed that since the 1980s, most of the studies on beneficial soil microorganisms focused on bacterial genetics and physiology and that research on inoculant formulations represented less than 1 % of the scientific articles on microorganisms. The formulation, storage, and application methods are critical to the success of the product (Jha and Prasad 2006). Some constraints in their widespread use are short shelf life, lack of suitable carrier material, susceptibility to high temperature, problems in transportation, and storage (Ghomarde et al. 2011). However, there is a real need for improving formulations, also developing and commercializing new biofertilizers that could be more effective and stable over time (Herrmann and Lesueur 2013).

Beneficial microorganisms isolated from agricultural lands and crop plants (Bashan and de-Bashan 2005; Park et al. 2005; Lugtenberg and Kamilova 2009; Schoebitz et al. 2009) can decompose organic residues, suppress plant disease and soilborne pathogens, and enhance nutrient cycling and bioactive compounds, such as vitamins, hormones, and enzymes (Singh et al. 2011). Nevertheless, colonization of plant roots by direct inoculation of free microorganisms into the soil is not easy because it is susceptible to environmental variations, such as soil conditions, fluctuation of pH and temperature, humidity, protozoa predation, and salt stress (Wu et al. 2012). This unpredictability in the success of microbial inoculation is mainly

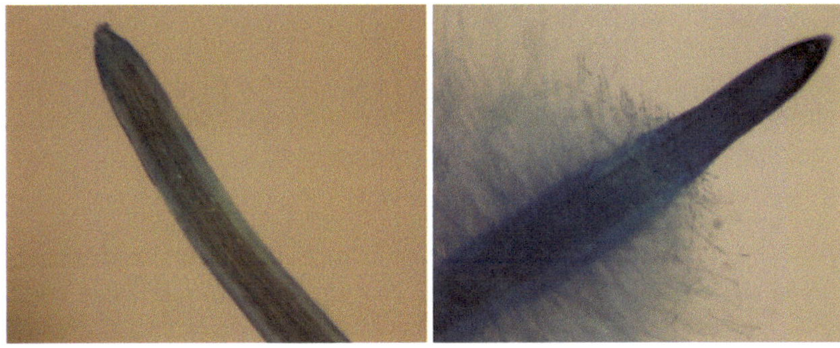

**Fig. 14.1** Effects of rhizobacteria on root growth of tomatoes. Representative pictures of seedlings growing in control plants (*left*) and plant inoculated by *Enterobacter ludwigii* (*right*)

related to the quality of formulations containing an effective bacterial strain and determines the success or failure of a biological agent. Cell immobilization through biodegradable carriers can enhance cell survival so that formulation turns to be the industrial "art" of converting a promising microorganism strain into a commercial inoculant (Bashan 1998). The use of microbial inoculants in agriculture has greatly increased during the past years (Bashan et al. 2014). Microbial inoculants are biological products containing living microorganisms that, when applied to seeds, plant surfaces, or soil, promote growth by several mechanisms, such as nitrogen fixation, phosphate solubilization, and production of various phytohormones, which improve root growth (Fig. 14.1), water absorption, and nutrient uptake (Calvo et al. 2014). In this regard, microbial inoculants have a potential role in developing sustainable agriculture for crop production (Bashan et al. 2004; Rivera-Cruz et al. 2008) and reclamation of degraded soils (Mengual et al. 2014a). Inoculant carriers have been used to improve effectiveness by supplying nutrients, protecting from desiccation, and slowing cell release. Information about the beneficial effects of rhizobacteria on growth promotion is well documented under laboratory and greenhouse conditions. However, reports of rhizobacteria-immobilized inoculants evaluated under field conditions are scarce (Schoebitz et al. 2014). The success of using microbial inoculants introduced into soil requires the survival of an adequate number of cells that can reach suitable habitats where they can thrive (Heijnen and Van Veen 1991). Based on this, formulation is a key factor in the success of microbial inoculants.

## 14.2 Traditional Microbial Inoculants

Peats are the most commonly used carriers to inoculate seed legumes with rhizobia (Denton et al. 2009). Peat is a complex organic material with a high variability. As microbiological contamination decreases the shelf life of the inoculants, this particular situation affects the quality of the final product, its stability during storage, and the survival of microorganisms in the final product (Bashan 1998). For example, Fallik and Okon (1996) found that cell concentration decreased from $10^{10}$ to $10^{5}$ CFU $g^{-1}$ after 6 months of storage with peat inoculated with *Azospirillum brasilense*.

Liquid formulations are inoculants that use broth cultures mainly in water, but also in mineral or organic oils. The seeds are dipped into the inoculant before sowing, or an applicator evenly sprays the liquid inoculant on the seeds (Bashan et al. 2014). Liquid formulations are easier to produce, can be more easily applied by farmers, and present some advantages based on the fact that they use low-cost materials and are easily attainable by small producers (Singleton et al. 2002; Albareda et al. 2008). Liquid inoculants allow direct contact of seeds and microorganisms and, consequently, increase the survival of bacteria on plant roots. However, bacterial survival rates on liquid formulations decrease because this technique does not provide a protective environment for microorganisms and the number of bacteria distributed in each seed is quite heterogeneous. In addition, microorganisms are not sufficiently protected against environmental conditions and contamination during storage, transport, and application into the soil (Bashan et al. 2002). The use of liquid inoculants mainly requires cold conditions for long-term storage to maintain their efficiency and cell viability. Liquid inoculants containing concentrations of $2 \times 10^9$ CFU $mL^{-1}$ and a stable population of microorganisms can be stored for 3 months (Albareda et al. 2008). *Bacillus subtilis* and *Pseudomonas corrugata* in liquid formulation showed a decrease from $10^{10}$ CFU $mL^{-1}$ to $10^7$ CFU $mL^{-1}$ after 6 months of storage at room temperature (Trivedi et al. 2005).

Clays are widely used as support for microbial inoculants and have a long history of usage in various agricultural formulations applied as granules, suspensions, and powder. Clays can act as a desiccant, providing excellent storage for dried inoculants due to large surface area, pore size distribution, and total porosity. Besides, water can be controlled to provide moisture for biologically active formulations (Goss et al. 2003). In addition, clays absorb or distribute dispersing and suspending agents. Some authors have found that clay-based inoculant carriers increased the survival of rhizobia in the soil for 60 days, reporting that these beneficial effects result from the creation of protective microhabitats, which are accessible to bacteria, but inaccessible to predators (Heijnen and Van Veen 1991; Heijnen et al. 1992).

## 14.3 Principle of Microbial Inoculant Formulation

The principle of rhizobacterial inoculants is to protect the microorganisms introduced into the soil and to ensure a gradual and prolonged release (Kim et al. 2012). The degradation rate of the encapsulation matrix is directly related to the biological activity of the microorganisms. In fact, the quicker the degradation of the matrix, the lower the protection and biological activity of the microorganisms, which allows for encapsulation and storage at room temperature for a long period, resulting in a favorable environment for bacteria and reducing the risk of decreased survival. These inoculants can be improved by incorporating essential nutrients for bacterial growth, transforming the capsules in bioreactors, which are capable of increasing the number of encapsulated bacteria and inoculated into the soil. These bacterial inoculants have solved many problems associated with traditional peat inoculants, which originate great variability in peat quality (Deaker et al. 2004). In fact, the encapsulation of rhizobacteria presents numerous advantages, such as the controlled release of bacteria into the soil and the protection of soil microorganisms against biotic and abiotic stresses.

Sodium alginate is the most commonly used material for the encapsulation of cells, enzymes, and biological control agents. Alginate is a natural polymer found in different marine macroalgae. The preparation of beads containing microorganisms involves a multistep procedure conducted at room temperature and with the use of minimal amounts of additional chemicals or equipment (Bashan et al. 2014).

Sodium alginate has been extensively used for the encapsulation of microbial inoculants (Fig. 14.2) due to its simplicity of handling, viscosity, and gel-enhancing properties. Although this is a simple and quick way to obtain beads or capsules, the method presents a major disadvantage referred

**Fig. 14.2** Immobilization of microbial inoculants in calcium alginate beads by ionic gelation technique (beads 2–3 mm in diameter)

to the cell loss that occurs during drying process (Schoebitz et al. 2012). In fact, the main difficulty with encapsulation techniques until now is cell survival during the dehydration process and storage since these techniques exhibit more than 90 % mortality of the initial population of microorganisms. Dehydration is one of the most critical and stressful phases for the bacteria during the encapsulation process. This is a severe problem for non-spore-forming, Gram-negative bacteria, which correspond to most species among the PGPR (Bashan et al. 2014). Besides, the presence of pores in the alginate matrix facilitates the diffusion of hydrophilic molecules (Wu et al. 2014). The incorporation of a filler material into an alginate matrix is a good strategy to solve the above mentioned limitation. Common additives are starch or clays which improve survival, mostly of phosphate solubilizing bacteria and other PGPR (Wu et al. 2012). Besides, adding trehalose and fructose to the culture medium preserves the viability of bacterial cells in a liquid medium and protects cells in the drying process (Schoebitz et al. 2012).

## 14.4 Innovative Techniques for Encapsulation of PGPR

Before describing the encapsulation techniques, it is important to note that three steps are needed to carry out encapsulation of PGPR. Firstly, the nature of the active ingredient that will be incorporated into a matrix (liquid or solid) needs to be identified. Secondly, a mechanical operation needs to be conducted by dispersing or spraying a solution onto solid particles under mechanical stirring. Thirdly, stabilization should be conducted either by a chemical process of polymerization, a physical-chemical process (gelation, coacervation), or a physical process (evaporation, solidification) on a droplet or pellet formed during the second step.

It is well known that encapsulation of living microorganisms is important for the production of special chemicals for industrial and agricultural use. The methods used to immobilize bacterial cells may be classified as physical, chemical, and chemical processes. Physical processes include spray drying, spray chilling/cooling, and fluidized bed; chemical processes include co-crystallization or interfacial polymerization; and chemical processes include coacervation and gelation/inverse (Table 14.1) (Madene et al. 2006). New encapsulation techniques continue to emerge for developing formulations and processes to allow for the improvement of capsule, material properties, and characteristics.

### 14.4.1 Spray Drying

Spray drying is a well-known method of producing a dry powder from a liquid or slurry, which is

**Table 14.1** Comparison of different encapsulation processes

| | Spray drying | Fluidized bed | Interfacial polymerization | Coacervation | Ionic gelation | Inverse gelation |
|---|---|---|---|---|---|---|
| Size <500 μ | Yes | No | Yes | No | No | No |
| Water soluble | Yes | Yes | No | Without cross-linking | No | No |
| Time required | 10–30 s | 10 min–1 h | 10 min–2 h | 12–16 h | 1–2 h | 1–2 h |
| Relative cost | Very low | Low | Low-medium | High | Medium | Medium |
| Advantage | Fast, inexpensive | Flexible, inexpensive | Well-developed process, large batches | Insoluble wall, wall impermeable to hydrophobic molecules | Biocompatible | Biocompatible |
| Disadvantages | Possible dust | Coating fine particles | Core wettability, few wall materials | Aggregation, core wettability, few wall materials, process control | Few wall materials, process control | Few wall materials, process control |

dried by a hot gas. It is a process that involves the dispersion of the bacterial cells in a carrier material forming an emulsion or dispersion, with homogenization of the liquid followed by atomization and spraying of the mixture into a hot chamber (Watanabe et al. 2002). Spray drying, which is a dehydration process, is one of the immobilization techniques used for the production of microbial inoculants that can be incorporated into agricultural and nonagricultural systems. The use of spray drying as a standard production technique is challenging due to its detrimental impact on the cellular integrity of microorganisms during drying (if not correctly optimized) and resultant reduction in the survival of immobilized bacteria (Campos et al. 2014). It has been demonstrated that heat-induced cellular damage is primarily associated with changes in the physical state of the membranes (Behboudi-Jobbehdar et al. 2013). In addition to the carrier-microorganism interaction, the impact of the spray-drying process is associated with process parameters (inlet and outlet air temperature, feed flow rate, residence time at the drying chamber, design parameters of the drying chamber, and temperature of the drying medium) and the biology of the bacteria to be immobilized (species and strain type, adaptation of the bacteria to heat or osmotic stress conditions, growth state of the culture media) (Golowczyc et al. 2010; Schutyser et al. 2012).

One important aspect to consider for this technique is optimal wall material, selecting substances with high solubility in water, low viscosity at high concentration, effective emulsification film-forming characteristics, and efficient drying properties (Reineccius 1988). Some of the advantages of spray-drying techniques are that the process works on a continuous basis, low operating costs, high quality of capsules in good yield, rapid solubility of the capsules, small size, and high stability capsules. However, spray drying also presents some drawbacks, including low uniformity of microcapsule size, limitations in the selection of wall material, and the high temperature required in the process, which may not be suitable for encapsulating bacterial cultures. In addition, this process produces very fine powder, which needs further processing and is inadequate for heat-sensitive material (Risch 1995). Nevertheless, appropriate modification and control of the processing conditions (inlet and the outlet temperatures) achieve viable encapsulated cultures of the particle size distribution required.

As mentioned before, temperature is a key parameter. Some reports have indicated that the survival of bacteria during spray drying decreased with increasing inlet temperature (Mauriello et al. 1999). Studies that dealt with outlet temperatures above 60 °C resulted in poor drying and the humid product often accumulated in the cyclone. Other studies have confirmed these observations, also reporting that the lowest air temperature was associated with the highest survival rate for the microorganisms during the drying process (Gardiner et al. 2000; Golowczyc et al. 2010). In fact, Campos et al. (2014) reached a bacterial survival of 91 % for *Enterobacter* sp. immobilized using sodium alginate and maltodextrin as wall materials (2:13 w/w), feed flow rate of 73 $m^3\ h^{-1}$, inlet temperature of 100 °C, and outlet temperature of 65 °C. In contrast, Schoebitz et al. (2013b) found lower cell survival after the encapsulation of microorganisms such as *Serratia* sp., using only maltodextrin as wall material, inlet temperature of 145 °C, and outlet temperature of 90 °C. In this sense, the type of strain, temperature of the drying process, and formulations of the polymer mixture used as vehicles are important parameters for the encapsulation of beneficial microorganism in order to obtain a successful microbial inoculant. However, despite the potential of this encapsulation technology in the microbial inoculant industry, the use and application of beneficial soil microorganisms to improve soil properties and plant performance is still scarce in agriculture and environmental restoration.

### 14.4.2 Fluidized Bed

The fluid bed technique involves drying, cooling, or coating of particulate materials for a wide range of heat-sensitive products. In fact, fluid bed spray coating is a process used to avoid some problems of the spray-drying technique. In fluidized bed drying, particles to be coated are fluidized by hot air. After that, the coating material is sprayed through a nozzle onto the particles; film formation is started, followed by a sequence of wetting and drying stages. The small droplets of the sprayed liquid reach onto the particle surface and join together, and the solvent evaporates by hot air, while the coating material remains on the particles (Jacquot and Pernetti 2003).

This technology offers a number of advantages. It allows specificity in particle size distribution and yields low porosities of the granules (Uhlemann and Mörl 2000), while it also presents high drying rates, smaller flow area, high thermal efficiency, and simple operation, with lower capital and maintenance costs.

This process of coating operates on particle size less than 100–150 μ. Therefore, uncommon conditions should be considered in order to avoid aggregation of the lower quantity of coated particles (Sparks and Jacobs 1999).

An open cylinder, which is hanging above the fluidization plate and where the fluidization air fed goes to the center region of the fluidization plate, carrying the particles upward through the cylinder was a change introduced by Wurster (1966). The particles are not fluidized in this region, but are simply being sent out while they desiccate. This variation gives superior control of the recirculation action of the particles (Fig. 14.3).

Fluidized bed drying performs better than spray drying because it is considered less stressful for drying microbial cells than spray-drying technology. In addition, it also involves less extreme water loss and temperature gradients (Larena et al. 2003; Morgan et al. 2006).

Herridge and Roughley (1974) compared seed pelleting using a fluidized bed with the conventional rotating drum technique. They observed that fluidized bed produced a firm pellet, but the survival of the inoculum was low probably because the air temperature occasionally reached 35 °C. For this reason, the conditions required to obtain capsules of high integrity may compromise viability of the microbial inoculant (Deaker et al. 2004). This explains why, in spite of being more advantageous, the application of fluidized bed technology to coating some inoculants is less common.

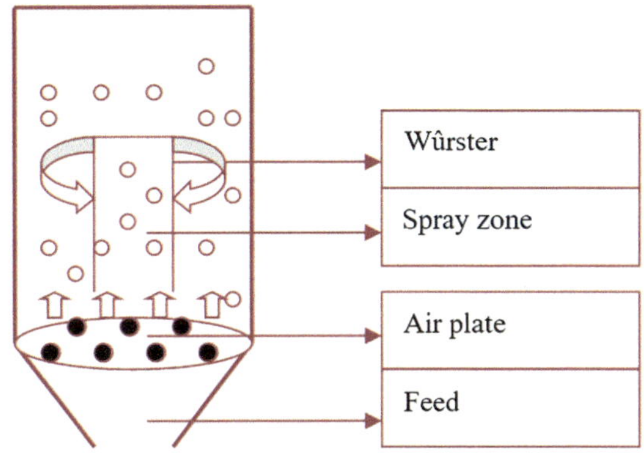


**Fig. 14.3** Fluid bed drying process

### 14.4.3 Interfacial Polymerization

Interfacial polymerization is typically known as a condensation reaction between a diamine or diol and a diacid, where the intermediates are dissolved in a pair of immiscible liquids, one of which is preferably water. To sum up, this technique involves the formation of an emulsion with an aqueous suspension of the cells as the discontinuous phase and organic solvent as the continuous phase. The droplet containing the cells and the reaction is set off when a biocompatible reagent, soluble in the continuous organic phase, is added to the emulsion. This process allows obtaining a high active loading (up to 90 %), but it involves high pH and toxic chemicals, such as sebacoyl chloride (Yeo et al. 2001). Another example of this technique would be microcapsules produced by dripping an alginate suspension (polyanion) in a chitosan solution (polycation), in which the blend of alginate and cells is dripped in a solution of chitosan (acetic acid 1 % at pH 4) with continuous stirring. Chitosan, a water-soluble polymer ($pH < 6$), has been used to microencapsulate some bacteria (Groboillot et al. 1994): a cross-linked chitosan membrane is formed by emulsification/interfacial polymerization using biocompatible reagents with oil-soluble cross-linking agents at low concentrations to minimize cell contact. Nevertheless, the antibacterial property of chitosan could limit its use as a coating material in encapsulation (Sudarshan et al. 1992).

For live cell encapsulation, some synthetic polymers such as nylon or cross-linked polyethylenimine membranes are incompatible due to toxicity of the reagents and hard conditions of encapsulation (Larisch et al. 1994).

Reports using monomers that are nontoxic for microorganisms have been carried out in order to increase their productivity in fermentations with relative success (Groboillot et al. 1993; Hyndman et al. 1993).

### 14.4.4 Coacervation

Coacervation is a widely used method for reversible gelification and microencapsulation of biological material. Coacervation is a phenomenon occurring in colloidal solutions. It consists of the separation from solution of colloid particles, which then agglomerate into a separate, liquid phase, called coacervate (Korus 2001). In general, the core material used in the coacervation must be compatible with the polymer and insoluble in the coacervation medium. In this technique, a core material, such as an oil phase, is dispersed in an aqueous solution of one or more polymers, while pH, ionic strength, temperature, and other parameters can change in the aqueous phase to induce the formation of a second polymer that becomes the wall material. Coacervation can be simple or complex. Simple coacervation involves only one type of polymer with the addition of strongly hydrophilic agents to the colloidal solution, and it is obtained by adding water-miscible non-solvent for the dissolved polymer (e.g., ethanol) or an electrolyte. On the other hand, complex coacervation involves two or more types of polymers and is achieved by lowering the pH of a solution containing two polymers, one of which increases positive charges as a result of the change (high-isoelectric-point gelatin), while the other has only negative charges (e.g., gum arabic) (Bungenberg de Jong 1949). Amiet-Charpentier et al. (1998) showed that due to the complex coacervation, it is possible to obtain a polymer microparticle containing rhizobacteria for inoculation of plants.

As concentration requires the formation of a fine emulsion, which may be different from that needed to increase the yield of microcapsules, optimization of wall material concentration in the emulsification and coacervation process is usually complicated (Nakagawa et al. 2004). Other limitations are evaporation, dissolution of active compound into the processing solvent, and oxidation of product (Flores et al. 1992). Hence, the coacervation method is effective but expensive, and it has some limitations.

### 14.4.5 Ionic Gelation

Ionic gelation produces microparticles through a drop of aqueous solution or suspension that contains the active material and sodium alginate, which is dropped into a solution of calcium chloride to form the capsules (Lim and Sun 1980). When the drop reaches the calcium chloride solution, a membrane of calcium alginate forms instantaneously, maintaining the drop shape in this aqueous/aqueous system. Calcium diffuses in, gelling the entire drop. The drop is then placed in a solution of a polycation that displaces the calcium from the outer surface, forming a permanent membrane. This capsule is then placed in sodium citrate, which slowly solubilizes the calcium through formation of the soluble citrate complex, ungelling the internal portion of the drop. By controlling the molecular weights of the reactants and the times of the reaction, the thickness and size selectivity of the permanent wall can be controlled over a wide range.

The established method produces calcium alginate beads through ionic gelation by dropping an alginate solution into a calcium chloride solution. The main advantage of gel encapsulation is the biocompatibility, although scaling up is difficult and the beads are often porous to cells (Lacroix et al. 1990). A variation of this method is to add another material, such as starch, to improve the process to encapsulate rhizobacteria, as described by Schoebitz et al. (2012). In this method, matrix solution is prepared by mixing alginate and starch (or clay) to improve the survival of rhizobacteria species by encapsulating in alginate beads, which allows the stable production of dried beads that contain a high cellular concentration.

Drying is the most critical step for cell survival in encapsulation methods. The results obtained by Schoebitz et al. (2012) confirm this statement as these authors demonstrated that a large proportion of cells are destroyed during capsule dehydration. In fact, cell mortality during the drying of encapsulated cells has been recognized as a critical point of the encapsulation process by various authors (Bashan et al. 2002; Campos et al. 2014). The survival yield after drying did not exceed 1 % of the initial cell count for various *Azospirillum* and *Pseudomonas* species. Drying involves great physical stress for living cells, including an increase in osmotic pressure that causes a water outflow leading to plasmolysis. Cell death is mainly attributed to the disruption of the plasma membrane (Mille et al. 2002). Nevertheless, the survival of dehydrated cells depends on various factors such as drying conditions, microorganism species, adjuvant used, bacterial strain, and culture conditions (Morgan et al. 2006). It has been determined that drying kinetics is of particular importance for cell survival, showing a detrimental effect on fast drying (Poirier et al. 1997). Beads made of pure alginate contain about 97–98 % water, and such a matrix failed to protect cells during drying as there was a two-log decrease in *A. brasilense* cell number. Using starch in the encapsulation matrix allowed the reduction of water content to 65 % and resulted in a significant increase in cell survival (Schoebitz et al. 2012). This could be due to a decrease in bead drying speed resulting from the lower water content. There are wide range of materials for encapsulation. Clay, skim milk powder, humic acid, and starch have all been used in living cell carrier formulations (Schoebitz et al. 2013a). Starch is an inexpensive material for capsule formation as it is one of the most abundant natural biopolymers (Hickman 1999). It can be used as a nutrient by several soil bacteria, which could help cell release and development in the soil. Previous research carried out on probiotic carriers has found a protective effect of starch due to cell adhesion to granules (Wang et al. 1999; Crittenden et al. 2001). Apart from slowing the drying rate, the ability of granular starch to protect the rhizobacteria from drying stress may be due to cell adhesion to starch. Cell adhesion to starch may depend on the strain encapsulated since there is a relation between adhesion to starch and its use as a carbon substrate by cells (Crittenden et al. 2001). As the best survival rates were obtained with the alginate-starch matrix, various strategies were tested to improve cell survival during drying by using this formulation base. It is generally admitted that stationary-phase cells are more resistant to physical stresses, such as dehydration, than cells harvested in the exponential growth phase (Vriezen et al. 2006). In fact, it has been determined that the optimal growth phase for desiccation survival is largely dependent on microrganisms (Schoebitz et al. 2012). Nevertheless, stationary-phase cells are generally more resistant to physical stresses, such as drying (Vriezen et al. 2006). This is due to the stress response triggered by carbon starvation and exhaustion of available food sources (Morgan et al. 2006).

Protective agents can be added either during the growth of microorganisms or prior to drying. Some molecules are known to work as osmoprotectants for many living cell species. These are mainly reducing sugars (fructose, glucose) and nonreducing sugars (trehalose) (Morgan et al. 2006). The addition of trehalose to the growth medium increased the survival of *Raoultella terrigena* during the drying process. Moreover, adding trehalose to the growth medium presented a much more effective protection against desiccation than adding it to the matrix solution just prior to drying. This may indicate that trehalose must be accumulated by cells to exercise its protective effect (Schoebitz et al. 2012). This result is consistent with the work conducted by Streeter (2003) on the survival of *Bradyrhizobium japonicum* during desiccation. In this study, the accumulation of trehalose within the cell cytoplasm contributed to membrane stabilization during desiccation. However, the accumulation of intracellular trehalose is only possible if the microorganisms cannot use it as a carbon source. Supplementation with fructose did not protect *Azospirillum brasilense* against drying, when added either to the growth medium or to the matrix prior to bead formation. This could be explained by the fact that *A. brasilense* can degrade fructose, using it as a carbon source.

Improving cell survival during encapsulation is not a simple procedure. It depends on various factors: growth media composition, strain and physiological state of cells, and process parameters. Each factor should be optimized to ensure the best inoculant activity after drying and during inoculant storage. By combining these factors, a biodegradable and dried inoculant with a high living cell concentration can be obtained (Schoebitz et al. 2012).

### 14.4.6 Inverse Gelation Microcapsules by Using Alginate

Inverse gelation occurs in a different form. It deals with dropping a calcium suspension in an alginate solution. The conventional method to produce calcium alginate beads through ionic gelation is by dropping an alginate solution into a calcium chloride solution. If the procedure is inversed, that is to say, calcium chloride solution dropped into an alginate solution, aqueous-core calcium alginate capsules are produced (Koyama and Seki 2004; Sasaki et al. 2008). By diffusion in the alginate solution, calcium will be gelifying the alginate and forming a membrane around the droplets. The calcium suspension consists of calcium chloride solution dispersing bacterial cells, and this emulsion finally drips into alginate solution (Abang et al. 2012; López et al. 2012). Same as recommended in ionic gelation, a variation of inverse gelation is performed by using modified starch, which is added in calcium chloride solution or even starch with alginate solution to increase solid content in the membrane.

Abang et al. (2012) studied the effects of process variables on the physical properties of capsules produced by inverse gelation. In this study, alginate was used to form the capsule membrane, and three different methods of incorporating the calcium source in oil were tested. The process variables examined were sodium alginate concentration, calcium chloride concentration, and curing time, while membrane thickness and elastic modulus were the physical properties of the capsules studied.

## 14.5 Reclamation of Degraded Soil by Immobilized PGPR

In recent years, research has shown that microbial inoculants can also play an indirect role on soil degradation and soil fertility. Bioremediation is recognized as an important tool to restore degraded ecosystems. In this sense, plants are essential in the recuperation of degraded soils, and the application of microbial inoculants in reforestation or afforestation of degraded lands is a promising research area (Calvo et al. 2014).

In degraded areas, the establishment of shrubs and trees is difficult due to low soil fertility and climatic conditions characterized by low precipitation and frequent drought periods. Therefore, it is necessary to apply methods to improve soil quality and resistance of planted species to degraded environmental conditions (Mengual et al. 2014b). The establishment of native plant species is widely used for reclaiming degraded lands and constitutes the most effective strategy in degraded areas (Mengual et al. 2014b). Recent studies on reclamation of degraded soils have determined the beneficial effects of the application of organic amendments, for example, alperujo (Schoebitz et al. 2014), sugar beet residue (Mengual et al. 2014a), sheep manure compost, and sewage sludge (Hueso-Gonzalez et al. 2014). Organic amendments have reported beneficial results in the proliferation and development of natural populations of soil microbiota, since the organic residues can be used by soil microorganisms, as substrates and as carbon and energy sources (Medina and Azcón 2010). This effect could be extended to the enhancement of soil enzyme activities, which are key factors that contribute to soilborne microorganism activity and soil fertility (Caravaca et al. 2005). The use of organic waste materials not only increases organic matter and fertility of soils, but it also contributes to the palliation of environmental and economic situations related to waste disposal (Rincón et al. 2006).

Organic amendments could be useful for improving soil quality and when developing afforestation programs in degraded areas. The beneficial effects of the addition of organic amendments in horticultural and revegetation practices have been reported (Alburquerque et al. 2006; Medina et al. 2010). Additionally, the application of organic amendments interacts positively with soil microorganisms. Microbial inoculation increases microbial biomass, soil respiration, and enzyme activities (dehydrogenase, urease, and protease). Increases in enzymatic activities are properties, which are sufficiently sensitive to indicate changes caused by microbial inoculations (Schoebitz et al. 2014). Among the components of the soil microbiota, PGPR are free-living bacteria, which can colonize the rhizosphere and improve root system establishment (Antoun and Kloepper 2001). In this regard, PGPR have a potential role in the establishment of plant cover in degraded environmental conditions (Puente et al. 2004), where they can promote plant growth and improve both water and nutrient uptake (Bashan et al. 2004).

The beneficial effects of PGPR on the growth of plants under laboratory conditions have been well documented. Nevertheless, there is little information regarding the introduction of immobilized PGPR into the soil under field conditions, especially in nonagricultural systems. It has been demonstrated that the use of PGPR improves plant health and growth performance in degraded soils, while it also enhances their tolerance to drought and salinity (de-Bashan et al. 2012).

However, recent studies have indicated that inoculation with immobilized PGPR in field conditions is a useful strategy for the establishment of native shrub and tree species in degraded soils (Mengual et al. 2014b; Schoebitz et al. 2014). Microbial inoculants helped plants to compensate for deficiencies of immobile nutrients, such as phosphate. Therefore, the inoculation with immobilized PGPR can be considered as an effective biotechnological tool for the development of biofertilizers that could partially substitute chemical fertilization. In this sense, the introduction of beneficial microorganisms can improve nutrient availability and thereby increase the efficiency of applied organic amendments (Adesemoye and Kloepper 2009).

## 14.6 Conclusions and Application Trends

Advances in microbial inoculant formulations have been presented, featuring encapsulation materials and techniques used for introducing beneficial microorganisms into the soil. Nevertheless, conventional microbial inoculants are not able to ensure high cell viability during the formulation process. Conventional inoculants need to be stored at room temperatures, avoiding extreme temperature oscillation because shelf life of liquid inoculants in storage conditions is very short and their viability decreases by one or two logs. The use of liquid inoculants does not offer protection to PGPR against soil stresses (nutrient and water deficiency, salinity, pH, predation). Instead, encapsulation provides a niche where PGPR are protected from soil stresses. Furthermore, the liquid inoculum has an instantaneous and very fast release after being introduced into the soil, and the rhizobacteria are delivered only in the initial phase of plant growth.

Alternatively, immobilized inoculants confer a gradual microorganism release that achieves long-term fertilizing effects.

The development of new formulation procedures and carriers will be determined by the demand of the industry, so that the focus of study will probably switch from laboratory scale to industrial production. In recent years, biochar has positioned itself as a beneficial soil material. The mechanisms are multiple and still a subject of intensive research, such as carbon sequestration, soil fertility, nutrient retention, and alteration of soil acidity (Novak et al. 2009; Joseph et al. 2010; Major et al. 2012). For instance, its adsorption and desorption properties are used to support an exchange platform between inorganic and organic compounds into the soil (De Luca et al. 2009). In addition, water-holding capacity and biological properties of the soil are also important (Downie et al. 2009). A favorable structure and morphology of biochar provide a suitable habitat for microorganisms to enhance colonization, growth and multiplication of bacteria, actinomycetes, and arbuscular mycorrhizal fungi (Thies and Rillig 2009). At the same time, the presence of hydrophilic sites on the biochar surface and macropores is relevant to improve soil physical properties, such as drainage and aeration and retention of water (Thies and Rillig 2009). This will produce consequent beneficial interactions with microorganisms of different sizes and adsorption affinities (Muñoz et al. 2014). Indeed, when biochar is sterilized during the pyrolysis process, it offers a great potential as an inoculum carrier. The incorporation of biochar into soils offers the opportunity for simultaneous application of plant beneficial microorganisms (Hale et al. 2014). Biochar properties vary depending on feedstock and production methods, but many biochars have characteristics that are also conducive for use as inoculum carriers, including high internal porosity, large specific surface area, and the ability to adsorb organic compounds and microorganisms (Abit et al. 2012).

The use of PGPR in agriculture is often limited by a low percentage of survival after microorganisms are introduced into the soil. Besides, direct microorganism inoculation of liquid inoculant decreases cell survival, which greatly reduces their vertical transport and the ability to colonize plant roots. For these reasons, the use of carrier materials to improve cell survival and protection could be one of the key factors for successful inoculation of PGPR into the soil. However, even though there are different encapsulation techniques, no formulation of PGPR has been developed by the microbial inoculant industry by this method (Bashan et al. 2014).

Encapsulation represents a wide area of research for the food, pharmaceutical, aquaculture, and cosmetics industries. In fact, different and efficient encapsulation methods have been developed to serve different purposes. Nevertheless, almost no method has been evaluated for the production of microbial inoculants. Many of these promising technologies, which are used in other fields, are worth to be evaluated in greenhouse and field conditions in order to improve the development of microbial inoculant formulations as well as quality of the bioformulations for successful inoculation.

## References


Abang S, Chan ES, Poncelet D (2012) Effects of process variables on the encapsulation of oil in ca-alginate capsules using an inverse gelation technique. J Microencapsul 29:417–428

Abit SM, Bolster CH, Cai P, Walker SL (2012) Influence of feedstock and pyrolysis temperature of biochar amendments on transport of Escherichia coli in saturated and unsaturated soil. Environ Sci Technol 46:8097–8105

Adesemoye AO, Kloepper JW (2009) Plant-microbes interactions in enhanced fertilizer-use efficiency. Appl Microbiol Biotechnol 85:1–12

Albareda M, Rodriguez-Navarro DN, Camacho M, Temprano FJ (2008) Alternative to peat as a carrier for rhizobia inoculants: solid and liquid formulations. Soil Biol Biochem 40:2771–2779

Alburquerque JA, Gonzálvez J, García D, Cegarra J (2006) Composting of a solid olive-mill by-product ("alperujo") and the potential of the resulting compost for cultivating pepper under commercial conditions. Waste Manage 26:620–626

Amiet-Charpentier C, Gadille P, Digat B, Benoit JP (1998) Microencapsulation of rhizobacteria by spray-drying: formulation and survival studies. J Microencapsul 15:639–659

Antoun H, Kloepper JW (2001) Plant growth-promoting rhizobacteria (PGPR). In: Brenner S, Miller JH (eds) Encyclopaedia of genetics. Academic, New York, pp 1477–1480

Bashan Y (1998) Inoculants of plant growth-promoting bacteria for use in agriculture. Biotechnol Adv 16:729–770

Bashan Y, de-Bashan LE (2005) Bacteria/plant growth-promotion. In: Hillel D (ed) Encyclopedia of soils in the environment. Elsevier, Oxford, pp 103–115

Bashan Y, Hernandez JP, Leyva LA, Bacilio M (2002) Alginate microbeads as inoculant carriers for plant growth-promoting bacteria. Biol Fertil Soils 35:359–368

Bashan Y, Holguin G, de-Bashan LE (2004) Azospirillum-plant relationships: physiological, molecular, agricultural, and environmental advances (1997–2003). Can J Microbiol 50:521–577

Bashan Y, de-Bashan L, Prahbu SR, Hernandez JP (2014) Advances in plant growth-promoting bacterial inoculant technology: formulations and practical perspectives (1998–2013). Plant Soil 378:1–33

Behboudi-Jobbehdar S, Soukoulisa C, Yonekuraa L, Fiska I (2013) Optimization of spray-drying process conditions for the production of maximally viable microencapsulated L. acidophilus NCIMB 701748. Dry Technol 31(11):1274–1283

Bungenberg de Jong HG (1949) Morphology of coacervates. In: Kruyt HR (ed) Colloid science. Elsevier, Amsterdam, pp 335–432

Calvo P, Nelson L, Kloepper J (2014) Agricultural uses of plant biostimulants. Plant Soil 383:3–41

Campos DC, Acevedo F, Morales E, Aravena J, Amiard V, Jorquera M, Inostroza N, Rubilar M (2014) Microencapsulation by spray drying of nitrogen-fixing bacteria associated with lupin nodules. World J Microbiol Biotechnol 30(9):2371–2378

Caravaca F, Alguacil MM, Azcón R, Parlade J, Torres P, Roldán A (2005) Establishment of two ectomycorrhizal shrub species in a semiarid site after in situ amendment with sugar beet, rock phosphate, and *Aspergillus niger*. Microb Ecol 49:73–82

Crittenden R, Laitila A, Forssell P, Mättö J, Saarela M, Mattila-Sandholm T, Myllärinen P (2001) Adhesion of Bifidobacteria to granular starch and its implications in probiotic technologies. Appl Environ Microbiol 67:3469–3475

De Luca TH, MacKenzie MD, Gundale MJ (2009) Biochar effects on soil nutrient transformations. In: Lehmann J, Joseph S (eds) Biochar for environmental management science and technology. Earthscan Publisher, London, pp 251–270

Deaker R, Roughley RJ, Kennedy IR (2004) Legume seed inoculation technology – a review. Soil Biol Biochem 36:1275–1288

De-Bashan LE, Hernández JP, Bashan Y (2012) The potential contribution of plant growth-promoting bacteria to reduce environmental degradation – a comprehensive evaluation. Appl Soil Ecol 61:171–189

Denton MD, Pearce DJ, Ballard RA, Hannah MC, Mutch LA, Norng S, Slattery J (2009) A multi-site field evaluation of granular inoculants for legume nodulation. Soil Biol Biochem 41:2508–2516

Downie A, Crosky A, Munroe P (2009) Physical properties of biochar. In: Lehmann J, Joseph S (eds) Biochar for environmental management science and technology. Earthscan Publisher, London, pp 13–32

Fallik E, Okon Y (1996) Inoculants of Azospirillum brasilense: biomass production, survival and growth promotion of Setaria italica and Zea mays. Soil Biol Biochem 28:123–126

Flores RJ, Wall MD, Carnahan DW, Orofino TA (1992) An investigation of internal phase losses during the microencapsulation of fragances. J Microencapsul 3:287–307

Gardiner GE, O'Sullivan E, Kelly J, Auty MA, Fitzgerald GF, Collins JK, Ross RP, Stanton C (2000) Comparative survival rates of human-derived probiotic Lactobacillus paracasei and L. salivarius strains during heat treatment and spray drying. Appl Environ Microbiol 66:2605–2612

Ghomarde V, Deshpande MV, Paknikar KM (2011) Perspectives for nano-biotechnology enabled protection and nutrition of plants. Biotech Adv 29:792–803

Golowczyc MA, Silva J, Abraham GL, De Antoni GL, Teixeira P (2010) Preservation of probiotics strains isolated from kéfir by spray drying. Lett Appl Microbiol 50:7–12

Goss GR, Baldwin HM, Riepl RG (2003) Clays as biological carriers. In: Downer RA, Mueninghoff JC, Volgas GC (eds) Pesticide formulations and delivery systems: meeting the challenges of the current crop protection industry. American Society for Testing and Materials, Dallas, pp 24–34

Groboillot AF, Champagne CP, Darling GD, Poncelet D, Neufeld RJ (1993) Membrane formation by interfacial cross-linking of chitosan for microencapsulation of Lactococcus lactis. Biotechnol Bioeng 42:1157–1163

Groboillot A, Boadi DK, Poncelet D, Neufeld RJ (1994) Immobilization of cells for application in the food industry. Crit Rev Biotechnol 14:75–107

Hale L, Luth M, Kenney R, Crowley D (2014) Evaluation of pinewood biochar as a carrier of bacterial strain Enterobacter cloacae UW5 for soil inoculation. Appl Soil Ecol 84:192–199

Hayat R, Ahmed I, Sheidil RA (2012) An overview of plant growth promoting rhizobacteria (PGPR) for sustainable agriculture. In: Aksoy A, Ashraf M, Ozturk M, Sajid A, Ahmad M (eds) Crop production for agricultural Improvement. Springer, Dordrecht, pp 557–579

Heijnen CE, Van Veen JA (1991) A determination of protective microhabitats for bacteria introduced into soil. FEMS Microbiol Ecol 85:73–80

Heijnen CE, Hok-A-Hin CH, Van Veen JA (1992) Improvements to the use of bentonite clay as a protective agent, increasing survival levels of bacteria introduced into soil. Soil Biol Biochem 24(6):533–538

Herridge DF, Roughley RJ (1974) Survival of some slow-growing Rhizobium on inoculated legume seed. Plant Soil 40:441–444

Herrmann L, Lesueur D (2013) Challenges of formulation and quality of biofertilizers for successful inoculation. Appl Microbiol Biotechnol 97:8859–8873

Hickman MV (1999) Controlled-release pesticide formulations from cornstarch. In: Scher HB (ed) - Controlled-release delivery systems for pesticides. Marcel Dekker, New York, pp 153–171

Hueso-Gonzalez P, Martinez-Murillo JF, Ruiz-Sinoga JD (2014) The impact of organic amendment on forest soil properties under mediterranean climatic conditions. Land Degrad Dev 25:604–612

Hyndman CL, Groboillot A, Poncelet D, Champagne C, Neufeld RJ (1993) Microencapsulation of Lactococcus lactis with cross-link gelatin membranes. J Chem Technol Biotechnol 56:259–263

Jacquot M, Pernetti M (2003) Spray coating and drying processes. In: Nedovic V, Willaert R (eds) Cell immobilization biotechnology. Kluwer Academic Publishers, Dordrecht, pp 343–356

Jha MN, Prasad AN (2006) Efficacy of new inexpensive cyanobacterial biofertilizers including its shelf life. World J Microbiol Biotechnol 22:73–79

John RP, Tyagi RD, Brar SK, Surampalli RY, Prevost D (2011) Bio-encapsulation of microbial cells for targeted agricultural delivery. Crit Rev Biotechnol 31 (3):211–226

Joseph SD, Camps-Arbestain M, Lin Y, Munroe P, Chia CH, Hook J, van Zwieten L, Kimber S, Cowie A, Singh BP, Lehmann J, Foidl N, Smernik RJ, Amonette JE (2010) An investigation into the reactions of biochar in soil. Aust J Soil Res 48:501–515

Kim KI, Baek YJ, Yoon YH (1996) Effects of rehydration media and immobilisation in calcium-alginate on the survival of Lactobacillus casei and Bifidobacterium bifidum. Korean J Dairy Sci 18:193–198

Kim IY, Pusey PL, Zhao Y, Korban SS, Choi H, Kim KK (2012) Controlled release of Pantoea agglomerans E325 for biocontrol of fire blight disease of apple. J Control Release 161:09–15

Korus J (2001) Microencapsulation of flavours in starch matrix by coacervation method. Pol J Food Nutr Sci 51:17–23

Koyama K, Seki M (2004) Evaluation of mass-transfer characteristics in alginate–membrane liquid-core capsules prepared using polyethylene glycol. J Biosci Bioeng 98:114–121

Lacroix C, Paquin C, Arnaud JP (1990) Batch fermentation with entrapped growing cells of Lactobacillus casei. Optimization of the rheological properties of the entrapment gel matrix. Appl Microbiol Biotechnol 32:403–408

Larena I, Melgarejo P, Cal AD (2003) Drying of Conidia of Penicillum oxalicum, a biological control agent against Fusarium wilt of tomato. J Phytopath 151: 600–606

Larisch BC, Poncelet D, Champagne CP, Neufeld RJ (1994) Microencapsulation of *Lactococcus lactis* subsp. cremoris. J Microencapsul 11:189–195

Lim F, Sun AM (1980) Microencapsulated islets as bioartificial endocrine pancreas. Science 210:908–910

López MD, Maudhuit A, Pascual-Villalobos MJ, Poncelet D (2012) Development of formulations to improve the controlled-release of linalool to be applied as an insecticide. J Agric Food Chem 60:1187–1192

Lugtenberg BJJ, Kamilova F (2009) Plant-growth-promoting rhizobacteria. Annu Rev Microbiol 63:541–556

Madene A, Jacquot M, Scher J, Desobry S (2006) Flavour encapsulation and controlled release – a review. I J Food Sci Tech 41:1–21

Major J, Rondon M, Molina D, Riha SJ, Lehman J (2012) Nutrient leaching in a Colombian savana oxisol amended with biochar. J Environ Qual 41:1076–1086

Malusá E, Vassilev N (2014) A contribution to set a legal framework for biofertilisers. Appl Microbiol Biotechnol 98:6599–6607

Malusá E, Sas-Paszt L, Ciesielska J (2012) Technologies for beneficial microorganisms inocula used as biofertilizers. Sci World J 2012:491206. doi:10.1100/2012/491206

Martínez-Viveros O, Jorquera MA, Crowley DE, Gajardo G, Mora ML (2010) Mechanisms and practical considerations involved in plant growth promotion by rhizobacterial. J Soil Sci Plant Nutr 10:293–319

Mauriello G, Aponte M, Andolfi R, Moschetti G, Villani F (1999) Spray-drying of bacteriocin-producing lactic acid bacteria. J Food Prot 62:773–777

Medina A, Azcón R (2010) Effectiveness of the application of arbuscular mycorrhizal fungi and organic amendments to improve soil quality and plant performance under stress conditions. J Soil Sci Plant Nutr 10 (3):354–372

Medina A, Roldán A, Azcón R (2010) The effectiveness of arbuscular-mycorrhizal fungi and Aspergillus niger or Phanerochaete chrysosporium treated organic amendments from olive residues upon plant growth in a semi-arid degraded soil. J Environ Manage 91:2547–2553

Mengual C, Schoebitz M, Azcón R, Roldán A (2014a) Microbial inoculants and organic amendment improves plant establishment and soil rehabilitation under semiarid conditions. J Environ Manage 134:1–7

Mengual C, Roldán A, Caravaca F, Schoebitz M (2014b) Advantages of inoculation with immobilized rhizobacteria versus amendment with olive-mill waste in the afforestation of a semiarid area with Pinus halepensis. Ecol Eng 73:1–8

Mille Y, Beney L, Gervais P (2002) Viability of E. coli after combined osmotic and thermal treatment: a plasma membrane implication. BBA Biomembr 1567:41–48

Morgan CA, Herman N, White PA, Vesey G (2006) Preservation of microorganisms by drying. A review. J Microbiol Methods 66:183–193

Muñoz C, Quilodrán C, Navia R (2014) Evaluation of biochar-plant extracts complexes on soil nitrogen dynamics. Bioenergy 8:377–385

Nakagawa K, Iwamoto S, Nakajima M, Shono A, Satoh K (2004) Microchannel emulsification using gelatin and surfactant-free coacervate microencapsulation. J Colloid Interf Sci 278:198–205

Novak JM, Lima I, Xing B, Gaskin JW, Steiner C, Das KC, Ahmedna M, Rehrah D, Watts DW, Busscher WJ (2009) Characterization of designer biochar produced at different temperatures and their effects on a loamy sand. Ann Environ Sci 3:195–206

Park M, Kim C, Yang J, Lee H, Shin W, Kim S, Sa T (2005) Isolation and characterization of diazotrophic growth promoting bacteria from rhizosphere of agricultural crops of Korea. Microbiol Res 160:127–133

Poirier I, Marechal PA, Gervais P (1997) Effects of the kinetics of water potential variation on bacteria viability. J Appl Microbiol 82:101–106

Puente ME, Bashan Y, Li CY, Lebsky VK (2004) Microbial populations and activities in the rhizoplane of rock-weathering desert plants. I. Root colonization and weathering of igneous rocks. Plant Biol (Stuttg) 6:629–642

Reineccius GA (1988) Spray-drying of food flavours. In: Risch SJ, Reineccius GA (eds) Flavor encapsulation. American Chemical Society, Washington, DC, pp 55–66

Rincón A, Ruíz-Díez B, Fernández-Pascual M, Probanza A, Pozuelo JM (2006) Afforestation of degraded soils with Pinus halepensis Mill: effects of inoculation with selected microorganisms and soil amendment on plant growth, rhizospheric microbial activity and ectomycorrhizal formation. Appl Soil Ecol 34:42–51

Risch SJ (1995) Encapsulation: overview of uses and techiniques. In: Rish SJ, Reineccius GA (eds) Encapsulation and controlled release of food ingredient. American Chemical Society, Washington, DC, pp 2–7

Rivera-Cruz MC, Trujillo-Narcía A, Córdova-Ballona G, Kohler J, Caravaca F, Roldán A (2008) Poultry manure and banana waste are effective biofertilizer carriers for promoting plant growth and soil sustainability in banana crops. Soil Biol Biochem 40:3092–3095

Sasaki E, Kuruyama F, Ida J, Matsuyama T, Yamamoto H (2008) Preparation of microcapsules by electrostática atomization. J Electrost 66:312–318

Schoebitz M, Ribaudo C, Pardo M, Cantore M, Ciampi L, Cura JA (2009) Plant growth promoting properties of a strain of Enterobacter ludwigii isolated from *Lolium perenne* rhizosphere. Soil Biol Biochem 41:1768–1774

Schoebitz M, Simonin H, Poncelet D (2012) Starch filler and osmoprotectants improve the survival of rhizobacteria in dried alginate beads. J Microencapsul 29:532–538

Schoebitz M, López MD, Roldán A (2013a) Bioencapsulation of microbial inoculants for better soil-plant fertilization. A review. Agron Sustain Dev 33:751–765

Schoebitz M, Osman J, Ciampi L (2013b) Effect of immobilized Serratia sp. by spray-drying technology on plant growth and phosphate uptake. Chilean J Agric Anim Sci 29(2):111–119

Schoebitz M, Mengual C, Roldán A (2014) Combined effects of clay immobilized *Azospirillum brasilense* and *Pantoea dispersa* and organic olive residue on plant performance and soil properties in the revegetation of a semiarid area. Sci Total Environ 466–467:67–73

Schutyser AIM, Perdana J, Boom RM (2012) Single droplet drying for optimal spray drying of enzymes and probiotics. Trends Food Sci Tech 27(2):73–82

Singh JS, Pandey VC, Singh DP (2011) Efficient soil microorganisms: a new dimension for sustainable agriculture and environmental development. Agric Ecosyst Environ 140:339–353

Singleton P, Keyser H, Sande E (2002) Development and evaluation of liquid inoculants. Inoculants and nitrogen fixation of legumes in Vietnam. In: Herridge D (ed) ACIAR proceedings. PK Editorial Services, Brisbane, pp 52–66

Sparks RE, Jacobs IC (1999) Selection of coating and microencapsulation processes. In: Scher HB (ed) - Controlled-release delivery system for pesticides. Marcel Dekker, New York, pp 3–29

Streeter JG (2003) Effect of trehalose on survival of *Bradyrhizobium japonicum* during desiccation. J Appl Microbiol 95:484–491

Sudarshan NR, Hoover DG, Knorr D (1992) Antibacterial action of chitosan. Food Biotechnol 6:257–272

Thies JE, Rillig MC (2009) Characteristics of biochar-biological properties. In: Lehmann J, Joseph S (eds) Biochar for environmental management science and technology. Earthscan Publisher, London, pp 85–106

Trivedi P, Pandey A, Palni LMS (2005) Carrier-based preparations of plant growth-promoting bacterial inoculants suitable for use in cooler regions. World J Microb Biot 21:941–945

Uhlemann J, Mörl L (2000) Wirbelschicht-Sprühgranulation. Springer, Berlin

Vriezen JAC, de Bruijn FJ, Nüsslein K (2006) Desiccation responses and survival of Sinorhizobium meliloti USDA 1021 in relation to growth phase, temperature, chloride and sulfate availability. Lett Appl Microbiol 42:172–178

Wang X, Brown IL, Evans AJ, Conway PL (1999) The protective effects of high amylose maize (amylomaize) starch granules on the survival of Bifidobacterium spp. in the mouse intestinal tract. J Appl Microbiol 87:631–639

Watanabe Y, Fang X, Minemoto Y, Adachi S, Matsuno R (2002) Suppressive effect of saturated acyl L-ascorbate on the oxidation of linoleic acid encapsulated with maltodextrin or gum arabic by spray-drying. J Agric Food Chem 50:3984–3987

Wu Z, Guo L, Qin S, Li C (2012) Encapsulation of R. planticola Rs-2 from alginate-starch-bentonite and its controlled release and swelling behavior under simulated soil conditions. J Ind Microbiol Biotechnol 39(2):317–327

Wu Z, He Y, Chen L, Han Y, Li C (2014) Characterization of Raoultella planticola Rs-2 microcapsule prepared with a blend of alginate and starch and its release behavior. Carbohyd Polym 110:259–267

Wurster DE (1966) Particle coating process. US Patent No. 3,253,944, Wisconsin Alumni Research Foundation

Xavier IJ, Holloway G, Leggett M (2004) Development of rhizobial inoculant formulations. Online Crop Manag. https://dl.sciencesocieties.org/publications/cm/abstracts/3/1/2004-0301-06-RV. Accessed 16 Mar 2015

Yeo Y, Back N, Park K (2001) Microencapsulation methods for delivery of protein drugs. Biotechnol Bioproc Eng 6:213–230

# An Overview of Globally Available Bioformulations

# 15

Samina Mehnaz

## Contents




**Abstract**

As the awareness about the environment is increasing among the public, the use of chemicals for the food production is scaring them. They are becoming more interested in organic food. Chemical fertilizers and pesticides are in regular use throughout the world, to increase the yield and to keep the plants and their products safe from diseases.



S. Mehnaz (✉)
Department of Biological Sciences, Forman Christian College (A Chartered University), Ferozepur Road, Lahore 54600, Pakistan
e-mail: saminamehnaz@fccollege.edu.pk



N.K. Arora et al. (eds.), *Bioformulations: for Sustainable Agriculture*,
DOI 10.1007/978-81-322-2779-3_15

These chemicals have their advantages and disadvantages. However, nowadays, environmentalists, researchers, and the public are looking more critically at disadvantages. Fortunately, now we have options for replacing or at least minimizing the use of these chemicals. These alternatives are the bioformulations containing living microbes such as bacteria and fungi. These are commercially available all over the world as biofertilizers and biopesticides/biofungicides. These can be used alone or in combination of chemical fertilizers. This article is a review of globally available bioformulations which are "in use" in different parts of the world as well as those which are under the process of commercialization.

## 15.1 Introduction

Since green revolution, the use of chemical fertilizers has increased tremendously. The use of these chemicals is very important to get higher yield from smaller area, but on the other hand, excessive use of these chemicals is contaminating ground water and soil, decreasing soil fertility and ruining the atmosphere as well. It is becoming a major problem for the public, environmentalist, and industry. As these chemicals are becoming a risk to human health, advanced countries are implementing strong rules and regulations for their use. For Third World countries, cost, preparation, and availability of fertilizers are big issues. Biofertilizer is the most feasible solution to these issues as it is an eco-friendly option and maintains the soil and crop health with increased efficiency. The term "biofertilizers" has been used for several kinds of formulations. In general, anything which is not chemically synthesized, biodegradable, and can be used as a fertilizer, is known as biofertilizer, i.e., compost, humus, animal and human waste, organic matter, etc. However, scientifically, biofertilizer means a fertilizer which contains living organisms and can be broadly classified as nitrogen-fixing and phosphate-solubilizing biofertilizers which contain bacteria or fungi. Recently, zinc and sulfur solubilizers and potash mobilizers have also been identified as biofertilizers.

When hazardous effects of chemical fertilizers are discussed, the contribution of chemical pesticides cannot be ignored as these synthetic formulations are also equally contributing in destroying the soil, water, environment, and human health. When the use of chemical fertilizers increased the vigor and yield of crops, it also increased the diseases and pest attack on the plants. These factors became responsible for losses of billions of dollars, worldwide. As a next step, chemical pesticides, fungicides, herbicides, bacteriocides, etc. were introduced in the market to protect the crop. These chemicals are mostly applied as foliar sprays and as seed treatment and sometimes on fruits as well to protect them against postharvest diseases. Now, biopesticides have been introduced in the market to reduce the use of chemicals. These bioformulations also contain bacteria and fungi, individually or in combination. These can be used in the same way as the chemicals such as foliar spray, as seed dressing, or as soil treatment.

This chapter is focused on all types of bioformulations including biofertilizers and biopesticides, containing bacteria and fungi as main components. Those which do not contain any living organism have not been included. The brief introduction and information about biofertilizers and biopesticides available in different regions of the world, with their mode of action, wherever disclosed, are provided in the following sections.

## 15.2 Novozymes (www.novozymes.com)

It is a very well-known European company, based in Denmark which sells several bio-products to increase fertility and yield of crops as well as control the diseases. Their products are available with and without lipo-chitooligosaccharide (LCO) Promoter

Technology. LCO Promoter Technology is a unique molecule that, when present "at the time of planting," enhances growth processes such as root and shoot development, immediately and independently of variety, soil, and environmental conditions. It provides a healthier start for plants, translating into higher yields and better returns at the end of the season.

The products of Novozymes are available in three categories, i.e., biofertilizers, bio-yield enhancers, and biocontrol products. First two products are commercially available and most of the biocontrol products are under registration process. Only those products which contain microbes are listed below, and detailed information is available on the company's website.

### 15.2.1 Biofertilizers

**JumpStart** contains the fungus *Penicillium bilaii* that makes phosphate available to the plant. Fungus colonizes the plant roots, releases organic compounds into the soil that break the bonds between phosphate and other elements. The plant gets access to more phosphate, and the fungus gets its nutrition from plant, forming a symbiotic relationship. It is available for canola, wheat, and legume crops.

**JumpStart LCO** is a combination of *Penicillium bilaii* and LCO Promoter Technology. *Penicillium bilaii* unlocks bound phosphate, and LCO Promoter Technology increases root and shoot development in the early growth stages.

**TagTeam** is a multi-action inoculant, specifically for legumes. It makes better use of phosphate and provides more fixed nitrogen. It is a combination of rhizobia strains with *Penicillium bilaii*. This product is available in granular, peat, and liquid formulations for use on pea, lentil, chickpea, soybean, and dry bean.

**TagTeam LCO** is a multi-action inoculant that combines the LCO Promoter Technology® and TagTeam, available in liquid and granular formulation, for use on pea, lentil, and soybean.

**Cell-Tech** (N-Prove) contains "nitrogen-fixing bacteria" (name of organism is not disclosed), making nitrogen available for the plant to use.

**Optimize® with LCO Promoter Technology®** combines a nitrogen inoculant with LCO Promoter Technology, available in a liquid formulation for soybean.

**Nitragin Gold®** contains rhizobia strains with patented slow-drying system, assures a high number of bacteria on the seed, and results in high levels of nitrogen fixation for maximum yield, available for alfalfa and sweet clover.

**Apron® XL/Allegiance® FL compatible** Nitragin Gold compatibility with Apron XL and Allegiance FL ensures the combined benefits of an inoculant and a fungicide where required. Both products have high bacterial count for excellent longevity (18 months on alfalfa and sweet clover) and treat alfalfa and sweet clovers (white, yellow, hubam, madrid, bitter, and sour clover).

**RhizoMyco** contains 18 species of endo- and ectomycorrhizae and growth-promoting substances. It is available in a soluble/injectable form to provide broad-spectrum application for increased nutrient update and enhances root systems.

**RhizoPlex** is uniquely formulated with Novozymes proprietary blend of patented bacterial cultures and stress reducing ingredients, plus 18 species of endo- and ectomycorrhiza.

**RhizoMyx** is an endomycorrhiza inoculant designed to improve the plant performance by increasing root development, making nutrients more available.

Legume inoculants **Glycimax®**, **Rhizomax®**, and **Legumax®** effectively supply nitrogen to a wide range of legume crops.

### 15.2.2 Bio-yield Enhancers

**RhizoBio®** is composed of 18 species of endo- and ectomycorrhiza and a complex biostimulant for increased nutrient and water uptake and tolerance to stress.

**RhizoMyx®** is uniquely formulated with nine species of endomycorrhiza and a biostimulant to provide broad-spectrum application for increased nutrient and water uptake and tolerance to stress.

### 15.2.3 Biocontrol Products

**Met52** is a bioinsecticide, containing spores of the soil fungus *Metarhizium anisopliae*. Suspended spores of *Metarhizium* attach to the surface of the target insect, germinate, penetrate the exoskeleton, and grow inside the pest, and death of target pest takes place in few days.

**Taegro** is a bacterial-based biofungicide/bactericide used for suppressing selected soilborne and foliar diseases.

#### 15.2.3.1 Under Registration Process

**TrichoderMax®** contains the fungus *Trichoderma asperellum* for the effective control of a number of significant soilborne crop diseases.

**BoveMax®** is a bioinsecticide containing *Beauveria bassiana* for the control of Broca (*Hedypates betulinus*) infestations in Erva-mate plantations.

**MethaMax®** is a bioinsecticide containing *Metarhizium anisopliae* for the control of Cigarrinha (*Mahanarva fimbriolata*) infestations in sugarcane.

## 15.3 BioAgri (http://www.bioagri.se)

Lantmännen BioAgri AB is established since 1996 in Sweden. The main focus of the company is to control the seed-borne diseases of grains. The company developed and marketed three biopesticides and a biofertilizer. Biopesticides are based on *Pseudomonas chlororaphis* strains. Their products are available all over the Europe for 10 years. Brief information about their products is given below:

### 15.3.1 Cedomon®

Cedomon is a biopesticide that contains *P. chlororaphis* as an active ingredient. However, it also contains other ingredients, such as rapeseed oil, which supports its application. This product is available since 1997, and it is effective against several types of seed-borne diseases on barley and oats except barley loose smut caused by *Ustilago nuda*. In some countries, treated seed has also been approved for use as feed.

Cedomon®-treated seed can be stored, transported, and handled in the same manner as normal seed. The product can be used in most treatment equipment. The recommended dosage is 7.5 L/ton of seed. Shelf life of the product at 4–8 °C is up to 8 weeks after delivery and 3 weeks at room temperature. Treated seeds can be stored for up to 1 year. Cedomon® is currently approved for use in Sweden, Finland, Norway, Denmark, Poland, Lithuania, and Italy. A large portion of Swedish grain seed is treated with Cedomon®. Since its launch, a total of approximately two million hectares (in several countries) have been sown with Cedomon®-treated seed.

### 15.3.2 Cerall®

Cerall® is also a biopesticide with *P. chlororaphis* and water. It has been developed so as to be specially adapted to bare seeds such as wheat. It is effective against the seed-borne diseases of wheat such as common wheat bunt (*Tilletia caries*), wheat leaf spot (*Septoria nodorum*), and *Fusarium* (*Fusarium* spp.) in wheat. It is also effective against the seed-borne diseases caused by *Ascochyta* spp. The recommended dosage is 10 L/ton of seed. Its shelf life is same as for Cedomon. Cerall® is currently approved for use in Sweden, Finland, Switzerland, Lithuania, and Austria.

### 15.3.3 Cedress®

This product also contains the same ingredient as Cerall (*P. chlororaphis* and water), but it has been developed specially for the treatment of pea seeds. Recommended dosage and shelf life are same as for Cerall. Cedress® is currently approved for use in Sweden.

### 15.3.4 Amase®

It is a biofertilizer, based on *Pseudomonas azotoformans* and organic and inorganic nutrients. It is a dry product for growth stimulation of plants and can be mixed in peat-based substrates or in irrigation water. It is easily absorbed by the roots to provide long-term growth-promoting effects. It helps plants to quickly produce a large and strong root system, more resistant to stress and grow faster. It is recommended to be used for potted plants and nursery plants. Pine, spruce seedlings, cucumber, lettuce, tomato, peppers, eggplant, cabbage, and broccoli have shown good results with this product.

## 15.4 Fertibio (www.fertibio.com)

Fertibio is a Spanish company which makes biofertilizers and biopesticides as well as organic fertilizers and pesticides which are biodegradable. Their bio-products are described below.

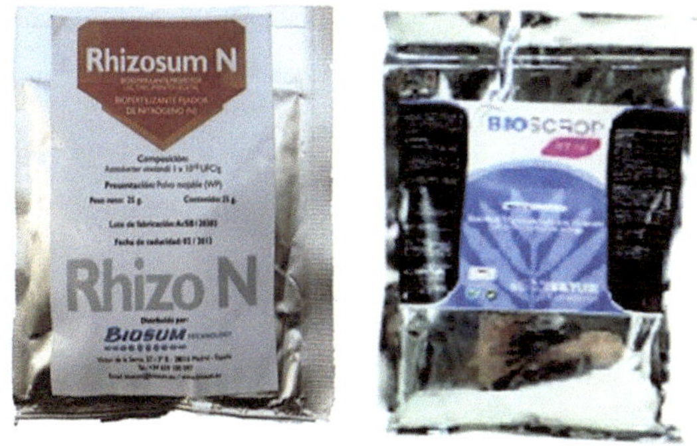


**Fig. 15.1** Bioformulations of Fertibio, Spain; biopesticide Bioscrop BT16 and biofertilizer Rhizosum N

### 15.4.1 Bioscrop BT16

It is a biopesticide based on *Bacillus thuringiensis* var. *kurstaki* (16 million IU/g) (Fig. 15.1). It is active against lepidopteran larvae and beetles. δ-Endotoxin synthesized by this bacterium is ingested by insect and causes injuries and paralysis of the digestive tract. Insect dies within 24–48 h. It is applicable to cotton, citrus, cauliflower, deciduous fruit trees, horticultural brassicas, olives, pepper, banana, and tomato. It is dissolved in water and applied as foliar spray (0.5–1.5 kg dissolved in 800–1,000 L and applied on 1 ha).

### 15.4.2 Rhizosum® N

The biofertilizer product is named as Rhizosum N (Fig. 15.1). It contains nitrogen-fixing bacteria but genera are not disclosed. The advancement of technology allows the crops to get nitrogen from the atmosphere. Recommended doze supplies 30–50 units of nitrogen per hectare, depending on soil moisture and season.

## 15.5 Symborg (www.symborg.com)

Symborg SL is a Spanish company which sells its product all over Europe and USA. The company has launched three products (Fig. 15.2) described below.

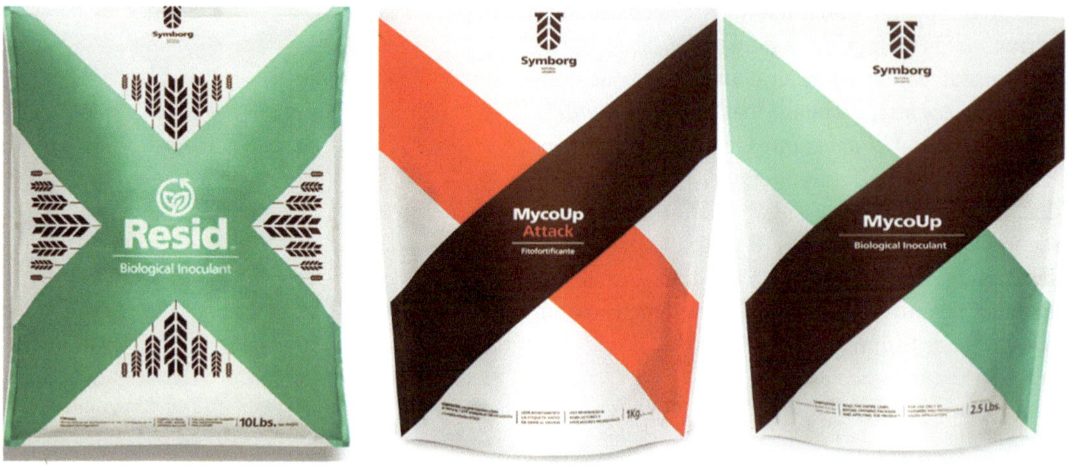


**Fig. 15.2** Bioformulations of Symborg, Spain

### 15.5.1 MycoUp

It is a biological inoculant based on *Glomus iranicum* var. *tenuihypharum*, a mycorrhizal fungus ($1.2 \times 10^4$ propagules/100 ml substrate). It is a root colonizer that boosts plant growth by more efficient water and nutrient absorption, improving plant vigor. It is recommended for vegetables, fruits, and woody crops at the rate of 2 kg/ha.

### 15.5.2 MycoUp Attack

This biological inoculant is a combination of MycoUp ($1.2 \times 10^4$ propagules/100 ml substrate) and "Attack." Attack is an organic complex, with MycoUp; it promotes the microbiotic activity of the soil, stimulates mycorrhizal growth, and activates the plant's defense system against nematodes and other phytopathogens. It is recommended for vegetables and woody crops at the rate of 2–3 kg/ha. Diluted solution of the product should be applied close to root system by injection or drip irrigation.

### 15.5.3 Resid

It is a biological product which contains *G. iranicum* var. *tenuihypharum* as its active ingredient with two mineral clay substrates, bentonite and smectite with a concentration of $0.5 \times 10^3$ propagules in 100 ml substrate. It is recommended for grains as seed coatings at the rate of 5 kg/100 kg seeds. Resid increases agricultural yields by optimizing fertilizer, water and soil inputs, tolerance to drought and salinity, and protection against fungal root diseases; contributes to soil regeneration through mycelium network; restricts the loss of $CO_2$ by recapturing it and converting it into fungal biomass; and produces no harmful residues.

## 15.6 Biagro (http://www.biagrosa.com.ar)

Biagro S.A. was established in 1984 in Argentina. Legume inoculants are their main products.

### 15.6.1 Nodulest 10

The soybean inoculant "Nodulest 10" is prepared by mixing a pure culture with sterilized *Sphagnum* peat moss. The product contains two different strains of *Bradyrhizobium japonicum* (USDA 138 for USA, 532C for Canada) depending on the country where it is used. The product has very high concentration of *B. japonicum* ($>2 \times 10^{10}$/ml). Therefore, a small quantity of carrier (peat moss) is used per bushel (60 lbs) of soybean seed. This characteristic allows the inoculant with lesser risk of blockage of seeders. Biagro S.A. is actively engaged in the search and selection of more efficient rhizobial genotypes to be used under suboptimal soil and climatic conditions.

### 15.6.2 Liquid PSA

It is a commercial formulation that contains *Pseudomonas aurantiaca* strain SR1. It is registered with Argentina's National Service for Agricultural Health (SENASA) for wheat growth promotion.

## 15.7 Labiofam (www.labiofam.cu)

It is a scientific institution of Cuba producing pharmaceutical products, biopesticides, and biofertilizers as well. Biopesticides are for killing mosquito larvae and rats. Biofertilizers under the name of Nitrofix and Bioenraiz are available. Nitrofix contains *Azospirillum brasilense*, and Bioenraiz contains phytohormones extracted from *Rhizobium*. Information about application of these biofertilizers or any detail about these products is not shared on the website.

## 15.8 Flozyme (http://www.flozyme.com/agriculture)

Flozyme Inc. was founded in 2012 in Atlanta, USA, and offers the solution for cleaning wastewater and agriculture. One of their biofertilizer products is described below.

### 15.8.1 Inogro

It is a cocktail of more than 30 microbes, selected for their abilities to rehabilitate soil and make it more productive. The product is organic, sustainable, environment friendly, and highly compatible with organic farming practices. It is suspended in 12 % humic acid carrier and stabilized at pH 7, due to which it effectively works in varied climates and soil conditions. These humates are high in organic matter and micronutrients, including Ca, Mg, Zn, Mn, etc. and also act as carbon sources for the microbes. This microbial formulation unlocks bound nitrogen in the soil and absorbs nitrogen from the air, solubilizes bound phosphate, and makes it available to plants for optimal growth and development. It is highly competent in soil and also promotes a healthy soil pH. It captures additional water and available nutrients from naturally occurring organic matter in the soil that enables plants to better sustain adverse environmental conditions. This product is claimed to inhibit plant pathogens, enhance the plant's natural defense mechanisms by *Trichoderma* strains (present in product), and increase the plant's resistance to pests.

Inogro is shown to increase crop yields by 20–400 % and decrease the use of chemical fertilizer from 50 % to 100 %. The product also reduces water requirements (~20 %) and pesticide usage. Crops have shown higher nutrient content, faster and earlier germination, earlier maturation, and ability to withstand stresses. During greenhouse trials, increased yields of 301 % for rice, 400 % for tomatoes, 127 % for soybeans, 86 % for peas, 258 % for okra, 234 % for peanuts, and over 100 % for garden beans and wonder bush beans have been shown.

## 15.9 AgriBiotics Product, Inc. (www.agribioticproducts.com)

The product is developed by Michigan State University (MSU) and manufactured by BioSoil Enhancers, Inc. Mississippi. Product brand is described below.

### 15.9.1 AgriBiotic Microbics with SumaGrow

SumaGrow is a combination of several microbes including bacteria and fungi (Fig. 15.3). It is a polymicrobial inoculant, a cocktail of more than 30 microbes with multiple functions. These microbes include nitrogen-fixing microbes isolated from leguminous and nonleguminous crops. Liquid humate is used as a carrier for the microbes. These microbes work under aerobic,

**Fig. 15.3** A bioformulation developed by Michigan State University, USA

non-aerobic, acidic, and alkaline conditions, therefore useful for multiple crops under multiple conditions. The SumaGrow microbial formulations improve nutrient and water uptake, enhance root and plant growth and crop yield, improve plant efficiency to use solar energy, reduce stresses of transplanting and drought, increase cation exchange capacity, improve soil health and water retention, help release micronutrients and trace elements, stimulate germination, increase healthy decomposition of organic matter, and reduce putrifaction.

The company claims innovation as compared to competitors due to four reasons, First, it contains multiple groups of organisms (up to eight), second, the product has six functions vs. two functions, third, it retains viability over a long period of time at ambient temperature, and fourth, it is concentrated at $10^{17}$ vs. $10^{8}$–$10^{10}$ CFU/ml. The product has demonstrated the ability to increase crop yields by 20–200 plus % while decreasing fertilizer usage from 50 to 100 % and reduces water (estimated 20 plus %) and pesticide usage. Experiments performed by MSU showed a 301 % increase in rice yield, 400 % increase in tomato yield, 127 % in soybeans, 86 % in peas, 258 % in okra, 234 % in peanuts, and over 100 % in garden beans and wonder bush beans in greenhouse.

Recommended application dose is one gallon per acre, diluted in the desired amount of water. For second application, foliar spray is recommended. It can be applied with irrigation water and with liquid nitrogen and herbicides but not with fungicides. Product should be applied without any fertilizer or with half dose of fertilizer.

## 15.10 Mapleton Agri Biotec Pty Limited (mabiotec.com)

It is an Australian company with three major products. This company is distributing its products in Australia, USA, UK, Europe, North and South Africa, Turkey, South America, and several other countries.

### 15.10.1 TwinN

It is a freeze-dried microbial inoculum packed under vacuum which provides it a longer shelf life. For application, initially, it is dissolved in small amount of water and later in large amount. It contains a consortium of microbes which includes nitrogen fixers, phosphate solubilizers, and growth hormone producers. These microbes can live in rhizosphere, root, shoot, and leaves as endophytes. It is an inoculant for crops, pastures, and trees. This product can be applied through irrigation system, sprinkler, spray, etc., depending on the crops.

### 15.10.2 CataPult

It contains the vesicular-arbuscular mycorrhizal (VAM) fungus *Glomus intraradices* and two species of *Bacillus* with its unique release and catch technology. It is available in granular form. It is easily applied at planting stage at the rate of 1.75 kg/ha. The P-solubilizing *Bacillus* species in CataPult colonize the roots and rhizosphere and solubilize P that is unavailable to the plant. In many soils, large amount of applied P ends up in bound form that can be accessed by CataPult "release and catch" technology. The *Bacillus* species stimulate a more vigorous root system and root hairs and suppress a number of root pathogens. VAM colonizes the roots and a network of hyphae extend beyond the root hair zone.

These hyphae collect P, N, Ca, Mg, and micronutrients and deliver them to the plant.

### 15.10.3 Nitroguard

Information about this product is not released.

## 15.11 EM Pro-agriculture

During 1980, EM technology was developed in Japan. EM is a naturally fermented liquid probiotic solution which contains "effective microorganisms" including lactic acid bacteria, yeast, and phototrophic bacteria. These are non-pathogenic and not genetically modified organisms (non-GMO). It is a cost-effective technology and beneficial for agriculture. EM works by improving the soil ecology to assist in plant growth and health.

EM microbes secrete vitamins, organic acids, minerals, and antioxidants, increase humus content of the soil, and provide better environment for other microorganisms and plants for better growth. Photosynthetic microbes, another constituent of EM, have powerful detoxifying, antioxidative, and antientropic properties which improve the poor soil. Plants inoculated with EM had shown higher yields. It improves soil structure and nutrient availability, lowers disease pressure, and improves the quality of produce and storage life. EM can be applied on seeds, at the time of transplantation, as foliar spray and for soil treatment.

## 15.12 Bio Power Lanka (www.biopowerlanka.com)

It is a company based in Sri Lanka. Among their agri-related bio-products, three of them contain microbes.

### 15.12.1 Bio Vaccine

It is a biofungicide that contains *Trichoderma viride* (Fig. 15.4). It protects the plant from rot and wilt diseases. It destroys the pathogenic fungi including *Pythium*, *Rhizoctonia*, and *Fusarium* spp. that cause root rot, stem rot, seed rot, fruit rot, and wilt diseases. *T. viride* grows like coils around the pathogen and degrades the cell wall of the pathogenic fungi by secreting a wide variety of enzymes including celluloses and chitinases. This process is known as mycoparasitism – one fungus killing the other fungus by limiting its growth and metabolic activity. *T. viride* also induces systemic resistance and prepares the plants to destroy the pathogens. The product helps the root system to

**Fig. 15.4** A biofungicide and a liquid biofertilizer launched by BioPower, Sri Lanka

increase uptake level of nutrients and moisture which will improve the tolerance to stressful growth conditions. The product contains $2 \times 10^{8}$ spores/ml. It is recommended to dilute and apply at nursery and transplantation stage to control soilborne diseases.

### 15.12.2 Bio Gold

Bio Gold is a liquid formulation that contains native isolates of compatible microorganisms such as *Azotobacter chroococcum* and *Pseudomonas fluorescens* (Fig. 15.4). *A. chroococcum* is a nitrogen-fixing soil bacterium which improves the soil properties by secreting polysaccharides and provides nitrogen to the crop in a balanced way. This bacterium also contributes to drought and disease resistance by improving soil's physical properties and secreting various growth-promoting substances. *P. fluorescens* is a potential bionematicide and used for cardamom, potato, and other vegetable crops, fruits, and cereal crops. Secondary metabolites produced by this bacterium are highly effective to control rot and wilt diseases of plants and nurseries. This bacterium also secretes various organic acids that help in the solubilization of insoluble phosphorus, and hence, the phosphorus availability to crops is increased. The product can be applied to all agricultural and horticultural crops by spraying around the root zone, as a foliar spray, by drip irrigation, and as a seed inoculant.

### 15.12.3 Bio Phos®

It is a liquid formulation which contains *Bacillus megaterium* – a phosphate-solubilizing bacterium. The use of Bio Phos® and Eppawala rock phosphate (ERP) mixture for plantation crops can reduce the recommended ERP doze by 50–75 % to the fact that available phosphorus content is increased. ERP+ Bio Phos® mixture could be used for annual crops to supply the recommended phosphorus requirement, as a substitute of Triple Super Phosphate (TSP).

## 15.13 AgriLife (www.agrilife.in)

The company is based in Hyderabad, India, and their products include biofertilizers, biopesticides, biostimulants, and other agri-related products. Among biopesticides, three types of products are available based on botanicals, microbes, and nanoparticles. Those which contain microbes have been discussed in this chapter.

### 15.13.1 Biofertilizers

AgriLife has launched 15 biofertilizers (Fig. 15.5), based on nitrogen fixing, phosphate solubilizing, potassium, ferrous, sulfur, silica and zinc mobilizing bacteria, manganese-solubilizing fungus, and vesicular arbuscular mycorrhizae (VAM). Each biofertilizer has a

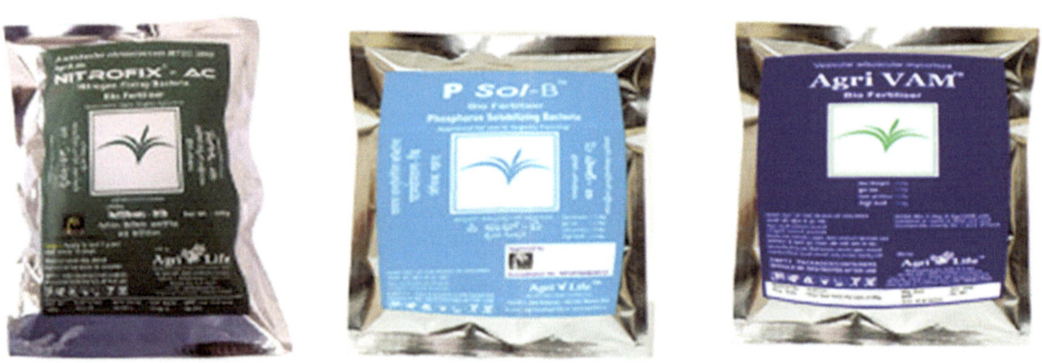


**Fig. 15.5** Biofertilizer formulations of AgriLife, India

single bacterial strain; consortium is not used in any biofertilizer. For each nutrient, a specific biofertilizer is available.

#### 15.13.1.1 Agrilife Nitrofix

These are nitrogen-fixing biofertilizers that contain *A. chroococcum* (MTCC 3853), *A. vinelandii* (NCIM 2821), *Acetobacter diazotrophicus* (MTCC 1226), *Azospirillum lipoferum* (NCIM 2908), and *Rhizobium japonicum* (NCIM 2743). Each biofertilizer is recommended for a different crop and with a different mode of application depending on crop. However, all of these are available as carrier-based powders containing $5 \times 10^7$ CFU/g or in liquid form $1 \times 10^8$ CFU/ml.

#### 15.13.1.2 P Sol B

These are phosphate-solubilizing biofertilizers and contain *Pseudomonas striata (NCIM 2847)*, *Bacillus polymyxa* (NCIM 2188), and *Bacillus megaterium* (NCIM 2087). All of these are available as carrier-based powder containing $5 \times 10^7$ CFU/g or in liquid form $1 \times 10^8$ CFU/ml. These biofertilizers can be applied on seed, seedling, and soil or through drip irrigation.

#### 15.13.1.3 Agri Life AgriVAM

It is based on spores and fragments of VAM (*Glomus* species) with vermiculite as carrier. It has 100 infective propagules/g. It helps in water absorption, phosphorus solubilization, and macro- and micronutrient availability and improves drought tolerance and resistance to soilborne fungal pathogen. It can be applied to soil, seed, and nursery bed and at planting stage.

#### 15.13.1.4 Fe Sol B

It contains an autotrophic, acidophilic *Acidithiobacillus ferrooxidans* that releases iron oxidase which metabolizes ferrous. It is available as carrier-based powder or in liquid form. It can be applied to seedlings and soil or by drip irrigation.

#### 15.13.1.5 K Sol B

It contains *Frateuria aurantia* which produces organic acids and enzymes which mobilize potassium ions. It is available as carrier-based powder containing $5 \times 10^7$ CFU/g or in liquid form $1 \times 10^8$ CFU/ml and can be applied to seedlings and soil or through drip irrigation.

#### 15.13.1.6 Mn Sol B

It contains an aerobic fungus *Penicillium citrinum* which produces citric acid and oxalic acid and solubilizes manganese. It is available as wettable powder and contains $5 \times 10^7$ CFU/g, applicable to seedlings and soil and through drip irrigation as well.

#### 15.13.1.7 Si Sol B

It contains spores of *Bacillus* species releasing organic acids which play role in silicate weathering. Silica helps the plant to tolerate biotic and abiotic stresses and pest and disease attack. This product is available as wettable powder containing $1 \times 10^8$ CFU/g, applicable to seed, seedling, and soil and by drip irrigation.

#### 15.13.1.8 S Sol B

This product contains autotrophic, acidophilic bacterium, *Thiobacillus thiooxidans*, that oxidizes sulfur and secretes organic acids, bringing down the pH of soil and helpful in reclaiming alkaline soil. It is applicable to seedling and soil and by irrigation.

#### 15.13.1.9 Zn Sol B

This biofertilizer contains another strain of *T. thiooxidans* that oxidizes zinc and makes it available to the plant. This product can be applied in the same way as S Sol B.

### 15.13.2 Biopesticides

Twenty-two biopesticides have been launched by the company (Fig. 15.6). One of them is for mosquitoes; however, the rest of them are to treat agriculture-related problems. Few of them are consortium based and their active ingredient and/or details of microbes are not shared. Some of them are based on individual bacterial or fungal strains to kill nematodes and insects or treat fungal diseases. A brief introduction is given below.

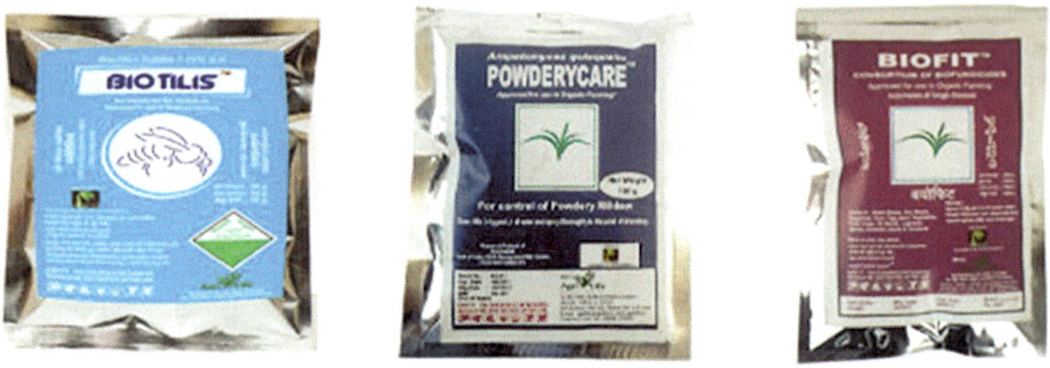


**Fig. 15.6** Biofungicides formulated and marketed by AgriLife, India

#### 15.13.2.1 BioKuprum

This product contains the fungus *Chaetomium cupreum* and protects the plants from diseases like rusts, blights, rots, and leaf spots. It is formulated as wettable powder with $2 \times 10^6$ CFU/g. It can be applied on seeds and tubers or as a foliar spray.

#### 15.13.2.2 Biotilis

This biofungicide also works as biofertilizer as it contains *Bacillus subtilis*, plant growth-promoting rhizobacteria (PGPR). As a fungicide, it targets rots, blights, wilt, leaf spot, and mildews. This product is available in powder form containing $1 \times 10^8$ CFU/g or $1 \times 10^9$ CFU/g.

#### 15.13.2.3 Downycare

This product is specifically formulated to treat downy mildew disease of plants caused by different fungal pathogens. It contains spores and mycelium fragments of fungus *Fusarium proliferatum*. Powder formulation containing $1 \times 10^8$ CFU/g is recommended to be used as foliar spray.

#### 15.13.2.4 Ecosom TV

*Trichoderma viride* containing bioformulation protects the plants from soilborne and seed-borne fungal pathogens. With $2 \times 10^6$ CFU/g, it can be applied as seed dressing, in the soil, or by drip irrigation.

#### 15.13.2.5 Ecosom TH

*Trichoderma harzianum* containing formulation is a biofungicide and a bionematicide. It protects the plants from fruit rot and pathogenic nematodes. Formulation is available in powder form containing $2 \times 10^6$ CFU/g. It can be applied as foliar spray, by drip irrigation, by soil drenching, by root dipping, and at nursery stage.

#### 15.13.2.6 Powderycare

*Ampelomyces quisqualis* is a fungus that protects plants from powdery mildew causing plant pathogens. Powdery formulation contains $2 \times 10^6$ CFU/g and it is applied as foliar spray.

#### 15.13.2.7 Sheathguard

This product is a biofungicide, a bionematicide, and a biofertilizer as it contains the PGPR *P. fluorescens*. It kills pathogenic nematodes and protects plants from sheath blights and other fungal diseases. Formulation is used as a seed dressing or applied on nursery beds before transplanting the crops.

#### 15.13.2.8 Biofit

It is a microbial consortium with an antifungal and PGP activity. It targets the fungi which cause rots, blights, and mildews.

#### 15.13.2.9 Biorub

This product is specific for pink disease of rubber plant caused by a fungus. Formulation contains consortium of antagonistic microbes with $1 \times 10^{8}$ CFU/g. It can be applied as foliar spray, by soil application, and by stem swabbing.

#### 15.13.2.10 Diebackcare

It is a consortium-based biofungicide for dieback disease of orchards and should be applied on soil.

#### 15.13.2.11 Seedguard

It comprises PGPR and microbial cultures with antifungal properties which encourage seed germination and control seed-borne fungal diseases. It is a microbial concentrate developed for seed coating.

#### 15.13.2.12 Insecticides and Nematocides

Other than biofungicides, several nematocides and insecticides have also been launched (Fig. 15.7). They include: Bionemagon (nematocide) which contains *Bacillus firmus*; Paecilo (nematocide) which contains *Paecilomyces lilacinus*; BorerGuard (insecticide) which contains microbial consortium and protects plants from borer attacks; Lipel which contains *B. thuringiensis* var. *kurstaki* and protects crops from flying insect pests; Mealikil plus (insecticide) which contains *Verticillium lecanii* that kills mealybugs and sucking insects; Pacer (insecticide) which contains *Metarhizium anisopliae* and kills soil insects; Paecilomite (insecticide) which contains *Paecilomyces fumosoroseus* and kills many types of insects and mites; and Racer (insecticide) which contains *B. bassiana* and kills insects pests.

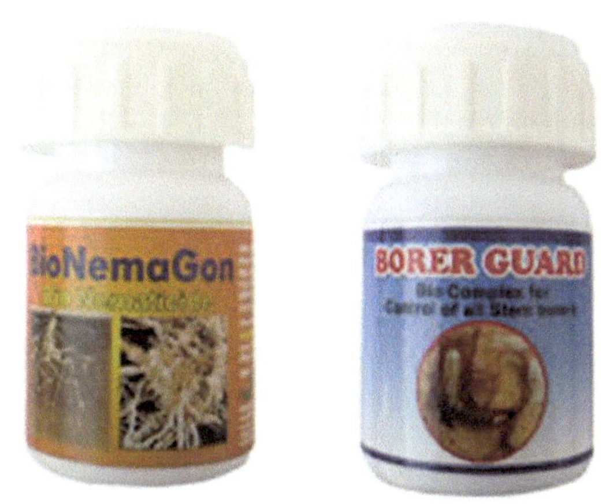


**Fig. 15.7** Bioinsecticide and bionematicides products by AgriLife, India

## 15.14 Commercially Available Biofertilizers in Pakistan

### 15.14.1 Fertibio

It is a biofertilizer launched by a private company "Microbial Biotechnologies." It contains a mixture of nitrogen-fixing, phosphate-solubilizing, and growth hormone-producing bacteria. Formulation is available in powder form, and it can be applied as a seed dressing and through irrigation as well as by sprinkling. It is recommended for rice, wheat, corn, cotton, sugarcane, and vegetables.

### 15.14.2 BioPower

A government research institute has launched a biofertilizer with the commercial name of "BioPower" in 1996. It is being produced and marketed for various leguminous and nonleguminous crops, like cotton, maize, rice, sugarcane, and wheat (Fig. 15.8). For every crop, a different consortium of PGPR has been used. It can be applied in powder form as well as in solution form, at the time of transplantation and after transplantation depending on the type and requirement of crops.

### 15.14.3 Auriga Group (www.aurigagroup.com)

Biofertilizer is one of the products manufactured by Auriga Group of industries. This product contains gram-positive phosphate-solubilizing bacteria which increase fertilizer use efficiency of added phosphate fertilizer and ensures phosphorus availability till crop matures. Plant growth hormone-producing bacteria are also included in this product that enhance the plant growth and enable the high yield. This company

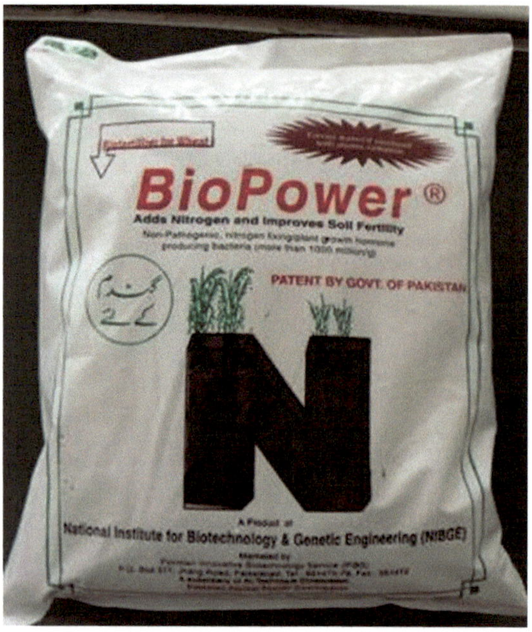


**Fig. 15.8** A biofertilizer produced by a research institute of Pakistan

is currently working on formulation of biopesticides.

## 15.15 Future of Bioformulations

Among bioformulations, biofertilizers can be divided into two major categories: nitrogen fixers and nutrient solubilizers/mobilizers. Most of the nitrogen-fixing biofertilizers contain *Azospirillum*, *Azotobacter*, and *Rhizobium*, and phosphate solubilizing have *Bacillus* and VAM. Recently, zinc, sulfur, and potash mobilizers/solubilizers have also been launched in the form of biofertilizers. Biofungicides are based on *Trichoderma*, *Bacillus*, and *Pseudomonas* strains of different species. These bacterial and fungal strains are common residents of plants rhizosphere.

For more than 100 years, biofertilizers are in use, in developed part of the world. However, it could not get the same level of popularity and usage as the chemical fertilizers still do. Among advanced countries, the major reason is inconsistency in the results of field application of biofertilizers. In the developing world, lack of awareness among farmers is the major hurdle in the growth of biofertilizer market. Looking at the current situation of fossil fuels, other alternative energy resources, high prices of chemical fertilizers, pesticides, herbicides, fungicides, and insecticides, on-time availability of these chemicals; bioformulations seem the only answer and the most effective solutions of these problems.

North America is promoting the use of bioformulations over chemical ones, due to their concern about human health- and environment-related issues. European union is promoting not only the use of bioformulations but organic farming as well through "common agriculture policy" by providing up to 30 % of payment as direct green payment. Among Asian countries, the government of China is promoting the production of bioformulations indirectly by tax holidays, exemption of VAT, and excise and agriculture tax to manufacturing industries of these products. Promoting the use of biofertilizers is part of the 5-year plan of the Indian government. The first biofertilizer, "BioPower," in Pakistan was launched by a government organization in 1996, and since then, they are selling it. Later on, several private companies started its manufacturing. Currently, at private level, Auriga is the biggest group of companies manufacturing as well as promoting the awareness and use of bioformulations among farmers.

A statistical review on the global scale of biofertilizers was published in January 2014. (http://www.grandviewresearch.com/industry-analysis/biofertilizers-industry). According to this, North America is the most dominant market of biofertilizers as it is valued at USD 420 million in 2012, followed by Europe. It reflects the fact that biofertilizers were first commercialized in North America and Europe; however, now, demand is increasing in South America and Asia Pacific as well, making it third in the row, in the biofertilizers market. Among the global revenues from biofertilizers, nitrogen fixers make the major contribution of 77 %, phosphate solubilizers are at 15 %, and the rest of the biofertilizers are responsible for 8 %. The global biofertilizers market is projected to generate USD 1650 million by 2019.

Looking at the current scenario, the future of bioformulations seems very bright. Due to awareness about hazardous effects of chemical formulations on human health and environment, acceptance of these formulations by general public has been increased. Now, it is time that governments should interfere, and with the help of regulatory authorities, the use of these formulations should be enforced in countries where farmers are not using them at large scale. Researchers should also play their role by searching for more diverse bacterial and fungal organisms rather than sticking to only two to three traditional genera.

## References

http://www.biagrosa.com.ar
http://www.bioagri.se
http://www.flozyme.com/agriculture
http://mabiotec.com
www.agribioticproducts.com
www.agrilife.in
www.aurigagroup.com
www.biopowerlanka.com
www.fertibio.com
www.labiofam.cu
www.novozymes.com
www.symborg.com

# Regulation of Biopesticides: Global Concerns and Policies

# 16

Naveen Kumar Arora, Maya Verma, Jai Prakash, and Jitendra Mishra

## Contents




**Abstract**

In the last few years, biopesticides have emerged as suitable alternatives to synthetic chemical pesticides. They are cheaper and pose no threat to agroecosystems. This is why their demand and production are also increasing at global level. The law and policies regulating their use and development vary from country to country, and in-depth analysis shows that there is no uniform regulatory model that can simplify their regulation and registration process. Although by the effort of some global agencies such as the International Organization for Biological Control (IOBC), European and Mediterranean Plant Protection Organization (EPPO) and Organization for Economic and Co-operative Development (OECD), some flexibility to biopesticide regulation have been provided, but in comparison to chemical pesticides, which have firm market and established nonoverlapping laws, biopesticides lag behind. This chapter provides comprehensive details of regulation systems adopted around the globe and to address shortcomings of existing system; besides this emphasis is also given to adopt innovative practices that could pave way for regulations which are simpler and more universal.



N.K. Arora (✉) • M. Verma • J. Prakash • J. Mishra
Department of Environmental Microbiology, School for Environmental Sciences, Babasaheb Bhimrao Ambedkar University, Lucknow 226025, Uttar Pradesh, India
e-mail: nkarora_net@rediffmail.com






N.K. Arora et al. (eds.), *Bioformulations: for Sustainable Agriculture*,
DOI 10.1007/978-81-322-2779-3_16

## 16.1 Introduction

Governments around the globe want to reduce the use of chemicals in agriculture due to their harmful effects on environment, and in this regard, they proposed some rules to minimize the use of chemical pesticides. Currently, various countries have diverse policies and laws to promote the use of biological pesticides or biopesticides. Concerns for green and safer food systems have risen globally, and countries have starkly imposed laws regarding the use of synthetic chemicals in food commodities and other allied products (Singh and Arora 2016). In this situation biopesticides as better alternatives gained much attention (Mishra et al. 2015). Biopesticides are mostly based on pathogenic microorganisms and target-specific, ecofriendly and effective solutions for pest problems in crop production (Gupta and Dikshit 2010). Although the use of biopesticides is practically simple and have advantages over chemicals, it has been observed that regulatory systems governing their use and production are somehow more complicated than chemical pesticides (Plimmer 1993). In spite of this discrepancy, the overall market of biopesticides has increased and numbers of registered products are also increasing (Sinha and Biswas 2008). At global level there are many policies and laws which not only promote biopesticide use but also regulate them. Several countries are also trying to overcome difficulties and creating a favourable business environment for commercialization of biopesticides. Clear and transparent administrative processes will provide flexibility to existing investors and at the same time also attract new investors in the biopesticide industry. However, in some countries, appropriate administrative guidelines have not been made available through regulations, and this creates problems in introduction of new biological products in the market (Ochieng 2015). Amongst the major challenges faced by biopesticide companies, registration part is the biggest one. It has been found that inappropriate evaluation methods and disproportionate data requirements often result in a long and lengthy registration process that is unnecessarily very costly and time consuming (Ehlers 2006; AGBR 2015). This discourages especially many small and medium enterprises from applying for such lengthy and costly registration processes (AGBR 2015). In this chapter we have tried to summarize the various regulation processes involved in biopesticide registration around the globe and to further address the problem related to intricate process of registration.

## 16.2 Regulation of Biopesticides: A Legal Framework

From global regulatory and registration perspective, the most accepted definition of the term biopesticide is by the US Environmental Protection Agency (USEPA) (Libman and MacIntosh 2000). The Division of Biological and Pollution Prevention (BPPD) of USEPA separates the term 'biopesticides' into three broad categories: microbial pest control agents, biochemical pesticides and genetically engineered organisms (GEOs). But due to its transgenic nature, GEOs are not recognized by the European Union (EU) as biopesticides. The current term used in the EU in place of biopesticides is biocontrol agents (BCAs). In most of the countries, these are regulated by governmental agencies for their authenticity and use in the agricultural field. As these products mostly contain live organisms or their derived products, hence, there are problems in regulatory processes and also hurdles in import and export (Mishra et al. 2015). It has also been observed that often, registration cost is very high. Microbial biopesticides have been commercially available for over 30 years, but they have less than 1 % share in the global market for agricultural crop production (Hajek 2004). There is a need to develop novel techniques to improve or modify existing biopesticides that will accelerate their higher production (Leng et al. 2014). Data of current sale levels of biopesticides suggest that generally, registration costs amount to 40 % of annual sales (ACP 2004). The testing cost for product efficacy to fulfil registration requirements is also very high.

Biopesticide regulation is a very important process but it is also very complex and dynamic field. According to Guest (2015), the development of predictive and efficient regulation processes for biopesticide is an important issue, and it must be ensured that the biopesticide product is safe and consistent without limiting commercialization, whereas market growth is also influenced by political and societal pressures (AGBR 2015). The data requirements in a dossier form are again a cumbersome process in the development of biopesticides. Ravensberg (2011) discussed some of the reasons related to unsuccessful submission of data for biopesticide development in European context but some of these are globally reported:

1. Requirements of data are not clearly defined and often adapted same as for in case of chemicals.
2. Guidance documents and test methods are also set up as for chemical substances.
3. End points of risk assessments are not clearly established.
4. Regulators and risk assessors lack expertise.

Waage (1997) also opposed to follow the chemical pesticide model for regulating biopesticide and argued that the entire pesticide regulatory process does not want to adapt the new opportunities which biopesticides provide.

Although biopesticides are used globally, the regulation processes and authorities involved vary at regional or national levels (Table 16.1). Various regulatory bodies and agencies such as International Organization for Biological Control (IOBC), European and Mediterranean Plant Protection Organization (EPPO) and Organization for Economic and Co-operative Development (OECD) have from time to time tried to resolve the registration hurdles faced by countries but the success is limited. The application of newer biopesticides may require new regulatory and economic challenges due to their highly diverse nature, which must be addressed jointly by the social and natural scientists, policymakers and the industries (Kumar 2015). The main aims of the regulations are, firstly, protection of human and environment (for safety reasons) and, secondly, the characterization of products and thereby ensure that manufacturers supply consistent and reliable quality products (Chandler et al. 2011).

## 16.3 Regulations Worldwide

The registration of biopesticides is perhaps the most intricate part around the world. There has been a rise noticed in the recent past for registered biopesticide products but this could be more if harmonization of registration process happens globally. Various types of authorities and regulations are emerging to regulate the biopesticides but very little flexibility is provided. This section summarizes the regulations being followed around the globe in different countries and continents.

### 16.3.1 Asia

Asian market presents a large opportunity for biopesticides with growing and emerging economies such as China and India and established one such as Japan (http://www.reportsnreports.com/reports/319510-global-biologicals-market-researchreport.html). Besides these other Asian countries, Pakistan, Bangladesh and South Korea are also emerging in the field of biopesticides.

#### 16.3.1.1 China

In 1997, the Regulation on Pesticide Administration law was introduced to govern pesticide production and use in China, and the law mandates the registration of biopesticides before entering the market (Kabaluk et al. 2010). The Chinese Ministry of Agriculture (MOA) along with other ministries is responsible for registration, production and commercial management of pesticides (Fang 2014). The Institute for the Control of Agrochemicals, Ministry of Agriculture (ICAMA 2008) is the national regulating authority which supervises the pesticide registration including biopesticides. Only registered and authorized companies (no research institutes, universities or other research groups) under the General Administration of Quality Supervision,

**Table 16.1** Biopesticide regulations at glance

| S. no. | Name of the country | Major regulatory actions/bodies | Contacts for further information |
|---|---|---|---|
| 1. | India | The Insecticide Act (1968) regulates biopesticide for its induced production and application. CIB and RC work under this act as highly powerful bodies to regulate biopesticides. ICAR also has role in biopesticides regulation. | www.cibrc.nic.in, http://agricoop.nic.in/imagedefault/policy/NCF3.pdf |
| 2. | China | The Regulation on Pesticide Administration (1997) was introduced to regulate pesticide/biopesticide and agrochemical market. Biopesticides are regulated by MOA in addition with other Ministries. ICAMA is the national regulating authority, which supervises the pesticide registration. | ICAMA (2008) and Fang (2014) |
| 3. | South Korea | Registration of pesticides at first level is governed by AMD and RDA, and then it is evaluated by PSED and NAAS. The AMD has a council which is concerned about safety of agrochemical. The microbial pesticide regulations, 2005, have guidelines for biopesticides, and in 2009 it worked as acts for management of agricultural chemicals (RDA 2009). | http://www.mordorintelligence.com/industry-reports/south-korea-biopesticides-market-industry, www.rda.go.kr |
| 4. | Pakistan | The PARC is responsible for monitoring the pesticides (Jabbar and Mallick 1994) and PBR (2005) regulates the marketing or trading of living microorganisms or any related products for any purpose. PBR, Rule 21, regulates the GEM, cells or products and states that these products should be approved by NBC. | Jabbar and Mallick (1994) and PBR (2005) |
| 5. | Kenya | Kenya's pesticide regulatory authority is PCPB which is supported by a range of stakeholders such as the UK DFID. The Agriculture Act (Cap 318) is responsible for the imports and exports of live organisms. The KEPHIS Act, Cap 512 2013, and Plant Protection Act regulate the imports and exports of live organisms. Pest Control Products Act also participates in registration and regulations for microbial pesticides. | Wabule et al. (2004), Kimani (2014), KEPHIS Act, cap 512/2013 |
| 6. | Tanzania | In Tanzania, the regulatory body for the pesticide regulation is named as TPRI, under the Ministry of Agriculture. | Stadlinger et al. (2013) |
| 7. | Nigeria | In Nigeria NAFDAC is an agency under the Federal Ministry of Health which is responsible for the control of the manufacture, sale and distribution of fertilizers, biofertilizer and biopesticides. | https://vmapnafdac.wordpress.com/ |
| 8. | UK | In the UK, the main regulatory body for biopesticide is CRD/PSD. The CRD is a new Directorate of the HSE which regulates pesticides, biocides, detergents, and chemicals under REACH. 'Biopesticide scheme' has also important role in regulation of biopesticides. | http://www.pesticides.gov.uk/, http://www.hse.gov.uk/pesticides/ |

(continued)

**Table 16.1** (continued)

| S. no. | Name of the country | Major regulatory actions/bodies | Contacts for further information |
|---|---|---|---|
| 9. | Ukraine | In Ukraine, the MSIPP is responsible for the registration of biopesticides and all other plant protection products. Most of these organizations fall under the research institutes of UAAN and NANU and the latter of which are governed by Ministry of Education and Science. | www.msp.ua/state-registration-of-pesticides-ukraine.htm |
| 10. | Netherlands | Like biopesticide the scheme in the UK, the Genoeg scheme in the Netherlands worked for biopesticide regulation. Genoeg is a 'bottom-up' process whereby a coalition of agencies and other interested parties are involved in creating new regulations. This scheme was developed to enhance the biopesticide production by efficient regulatory actions. | www.genoeg.net |
| 11. | Germany | At national level the regulation in Germany (EC) No. 1107/2009 is implemented through the GPPA 2012. The PPA covers plant protection products, adjuvants and plant resistance improvers. In 2003 PAN launched the OISAT programme in Germany and the aim of this service is to limit the use of hazardous pesticides by providing safer alternatives. | EC (2009), German Plant Protection Act 2012 |
| 12. | USA | USEPA is the main authority to regulate the use, sale and distribution of biological pesticides. USEPAs have mainly three parts: (1) FIFRA 1947, (2) FFDCA 1938 and (3) FQPA 1996. US-BPIA promotes industry standards for biopesticides. BPIA is working with the EPA to register biopesticides and also worked as a leading source of reliable information for biopesticides. | www.epa.gov/pesticides/biopesticides, http://www.pesticides.gov.uk/environment.asp, http://www.biopesticideindustryalliance.org/ |
| 13. | Canada | PMRA regulates biopesticides, which involves in pesticide evaluation, registration and developed joint guidelines for biopesticides with EPA, OECD. The regulation of non-indigenous macrobials is administered by CFIA under the authority of the plant protection act 1990.The Agriculture and Agri-Food Canada Pest Management Centre is committed to improving access to biopesticides as part of its Pesticide Risk Reduction Program. | www.hc-sc.gc.ca, AAFC (2009), http://www.inspection.gc.ca, CFIA (2013) |
| 14. | Cuba | In 2007, a formal process for registration was published and four institutions CNSV, CICA, CNSB EQDNCP are regulating biopesticides, and one or more involved with registration process. | Gaceta Oficial No. 016 Cuba (2007) |

(continued)

**Table 16.1** (continued)

| S. no. | Name of the country | Major regulatory actions/bodies | Contacts for further information |
|---|---|---|---|
| 15. | Colombia | In Colombia PALO are responsible for regular inspection on biopesticide trade and conduct regular direct inspection on quality control. Colombia has a specific regulation for biopesticides, which was established in 1994 and updated in 2004. Presently only acute toxicological studies are required and reviewed by The Ministry of Health. | Cotes (2010) |
| 16. | Australia | Biopesticides regulation is governed by NRS which is a partnership programme between the Australian state and territory governments. Currently COAG is most considerable regulatory body. The regulatory system is controlled by an authority named as APVMA 1992, which was previously known as NRA | http://apvma.gov.au/ King et al. (2013) |
| 17. | New Zealand | Biopesticides mainly microbial pesticides are regulated by HSNO Act, for their registration. Once approved by HSNO, biopesticides also need to be registered under ACVM Act 1997. | www.nzfsa.govt.nz/acvm, Pottinger and Morgan (2008), http://www.ermanz.govt.nz/, Agricultural Compounds and Veterinary Medicines Act, 1997 |

Inspection and Quarantine of the People's Republic of China have the right to apply for the registration of pesticides (Kabaluk et al. 2010). The Chinese Ministry of Forestry encourages the use of biopesticide by providing funds for pest control in forest. Many local government bodies encourage farmers to use biopesticides for safe agricultural practices and crop quality.

#### 16.3.1.2 India

The Indian government has various rules and regulations for promoting registration and production of biopesticides on large scale. The National Agricultural Technology Project (NATP) led the Integrated Pest Management (IPM) project (1998–2005), and the National Farmer Policy (2007) also promoted the use of biopesticides in agriculture. The Insecticide Act (1968) induced the higher production and application of biopesticides by simplifying the registration and regulation process of biopesticides. The Central Insecticides Board (CIB) and the Registration Committee (RC) worked under this act (www.cibrc.nic.in) and both are highly powerful bodies for biopesticide regulation (Kabaluk et al. 2010). CIB is the Apex Advisory Committee having experts of all disciplines/fields concerned. Based on the OECD guidelines, the CIB has streamlined the guidelines and data requirements not only for registration but also minimum infrastructural facilities for the production of biopesticides (NAAS 2013). The RC grants registrations, after scrutinizing and verifying claims with respect to their bio-efficacy and safety to human beings and animals. The National Agricultural Research System plays a leading role in promoting biopesticides. This system includes many institutes of Indian Council of Agricultural Research (ICAR) (www.icar.org.in/) and various agricultural universities in the country (Rabindra 2005).

#### 16.3.1.3 Japan

Japan was originally amongst the pioneers in terms of biopesticide use. In the last few years, Japanese research in biocontrol has resulted in the identification and characterization of several new insect pathogen delivery systems and formulation development. Each year sales of biopesticides in Japan are over $9 million (Kunimi 2007). In Japan, the Japanese Agricultural Standard (JAS) Law (2000) governs agricultural chemicals, fertilizers and soil improvement substances, and biopesticide formulation is also included in this list.

#### 16.3.1.4 Pakistan

The Pakistan Agricultural Research Council (PARC) is assigned the task of monitoring the pesticides (Jabbar and Mallick 1994). The Pakistan Biosafety Rules, 2005 (PBR 2005, rule 11), proposed that a licence is required from the Pakistan EPA for marketing or trading of living microorganisms, substances or cells or any related products for any purpose in the country. PBR, Rule 21, regulates the genetically engineered microorganism (GEM), cells or products and states that these products should be approved before being introduced in the market by the National Biosafety Committee (NBC).

#### 16.3.1.5 South Korea

In South Korea the registration of pesticides at the first level is governed by the Agromaterials Management Division (AMD) and the Rural Development Administration (RDA), and then it is evaluated by Pesticide Safety Evaluation Division (PSED) and National Academy of Agricultural Science (NAAS). The AMD has a dedicated council for agrochemical safety for the final decision and for reporting to the applicant's dossier (http://www.mordorintelligence.com/industry-reports/south-korea-biopesticides-market-industry). The microbial pesticide regulations, 2000, were replaced in 2005 having new guidelines for biopesticides, and in 2009 it was transferred to the acts for management of agricultural chemicals (RDA 2009).

#### 16.3.1.6 Other Countries

Bangladesh has some rules and regulations for biopesticide commercialization in the market. Any person in Bangladesh can import biopesticides if they have documents such as import licence, repack licence, wholesale licence, repacking factory, environment licence, etc. Besides these documents, product label must be approved by the registration authority with the trade name of the pesticide, genetic name, doses, pest, crop, first aid, toxic label, name of principle, waiting period, manufacture, expiry date and price for biopesticide commercialization (Ahsan 2012).

Thailand has made significant progress in promoting the production, application and registration of biopesticides (Grzywacz 2003). The Department of Agriculture (DOA) under the Ministry of Agriculture and Cooperatives (MOAC) has been promoting the use of biopesticides by launching various programmes such as organic farming, IPM and good agricultural practice (GAP) (Panuwet et al. 2012). In Vietnam various state research institutions such as the National Institute of Plant Protection (NIPP) and Vietnam Agricultural Science Institute (VASI), in Hanoi, have a role in biopesticide business (Jakel 2003).

### 16.3.2 Africa

Several countries in Africa use various guidelines to develop systems for registration and regulation of biopesticides in pest and disease management. Some African countries are taking a proactive role in developing capacity for regulating microbial pesticides. In 2012, six country representatives from Kenya, Uganda, Ethiopia, Tanzania, Nigeria and Ghana in the West African region undertook a regional stocktaking of the regulatory environments as part of the Commercial Products (COMPRO II) project, which is managed by the International Institute of Tropical Agriculture (IITA). The project's aim is to improve the regulation of biofertilizers and biopesticides (Simiyu et al. 2013).

#### 16.3.2.1 Kenya

Kenya has developed biopesticide-specific registration regulations, representing a proportional and reasonable system that correctly assesses the safety and risks associated with microbial pesticides (Kabaluk et al. 2010). Kenya's pesticide regulatory authority is the Pest Control Products Board (PCPB) which is supported by various stakeholders such as the UK Department for International Development (UK DFID). The Agriculture Act Chapter (Cap 318) is responsible for the imports and exports of live organisms. The Kenya Plant Health Inspectorate Service (KEPHIS) Act, Cap 512/2013, and Plant

Protection Act are working for the imports and exports of live organisms. The Pest Control Products Act is also involved in registration and regulations for microbial pesticides (Kimani 2014).

#### 16.3.2.2 Nigeria

In Nigeria the National Agency for Food and Drug Administration and Control (NAFDAC) is an agency under the Federal Ministry of Health, which is responsible for the control of the manufacture, sale and distribution of fertilizers, biofertilizer and biopesticides (https://vmapnafdac.wordpress.com/).

#### 16.3.2.3 South Africa

South Africa has laws and guidelines for regulations of registered, sold and used biological control agents. The registrations of biological remedies are regulated by the Department of Agriculture, Forestry and Fisheries (DAFF) under the Act 36 of 1947 (www.daff.gov.za) (DAFF 2010).

#### 16.3.2.4 Tanzania

In Tanzania, the regulatory body for the pesticide regulation is named as Tropical Pesticides Research Institute (TPRI), under the Ministry of Agriculture (Stadlinger et al. 2013). It has a mandate to register or deregister pesticide products under the country's Plant Protection Act. The field trial findings are submitted for the Pesticide Approval and Registration Technical Subcommittee (PARTS), and if it is approved, it is submitted to the National Plant Protection Advisory Committee for review (http://www.farmchemicalsinternational.com/crop-protection/africa-registration-guide-kenya-drives-to-harmonize-pesticide-registration).

#### 16.3.2.5 Uganda

The Uganda National Agricultural Chemicals Board (UNACB) and the Agricultural Chemical Control Technical Committee (ACCTC) have the mandate to build capacity to inspect, certify agrochemical trade in Uganda (http://www.farmchemicalsinternational.com/crop-protection/africa-registration-guide-kenya-drives-to-harmonize-pesticide-registration).

### 16.3.3 European Union

It is the second largest continent on the basis of biopesticide production and use. In EU, microorganisms, botanicals and pheromones were regulated under the Directive 91/414/EEC (EU 1991), which was originally developed for chemical pesticides (Regnault-Roger et al. 2012). The Directive 91/414 was amended by 2001/36/EC (EC 2001) and 2005/25/EC (EC 2005) to add the specific requirements for microorganisms, whereas new legislation regarding plant protection was further added in the EU in 2009, and due to this, the following four legislations: (1) Regulation (EC) No 1107/2009, (2) Directive 2009/128/EC, (3) Directive 2009/127/EC and (4) Regulation (EC) No 1185/2009, are also included. The new Regulation (EC) No 1107/2009 applies in all member states from 2011 and replaces Directive 91/414/EEC (Meeussen 2012). The registration of biopesticides in EU countries seems to be more difficult than the rest of world as the dossier is submitted along with toxicological and environmental testing; it also requires efficacy evaluation. Under Regulation (EC) No 1107/2009, registrations of products are made by three zones following geographic and climatic criteria (Hauschild 2012). These zones are Zone A (North): Denmark, Estonia, Latvia, Lithuania, Finland and Sweden; Zone B (Central): Belgium, Czech Republic, Germany, Ireland, Luxembourg, Hungary, the Netherlands, Austria, Poland, Romania, Slovenia, Slovakia and the UK; and Zone C (South): Bulgaria, Spain, Greece, France, Italy, Cyprus, Malta and Portugal.

Applicants have to submit the dossier for the registration of plant protection products (PPPs) to a 'zonal rapporteur member state' (zRMS), which evaluates the dossier. Recently Regulation (EC) No 283/2013 implements Regulation (EC) No 1107/2009 for establishing data-related issues (EC 2013).

#### 16.3.3.1 Germany

Germany is included in the largest central zone of 13 member states. At national level Regulation (EC) No 1107/2009 is implemented through the German Plant Protection Act (GPPA) 2012. This underwent a major revision in 2012. The Plant Protection Act (PPA) covers plant protection products, adjuvants and plant resistance improvers. In 2003 the Pesticide Action Network (PAN) launched the Online Information Service for Non-chemical Pest Management in the Tropics (OISAT) in Germany for limiting the use of hazardous pesticides by providing safer alternatives to poor farmers (Bissdorf 2008).

#### 16.3.3.2 Netherlands

The Netherlands is also inclined to increase the use of biopesticides, and like the biopesticide scheme in the UK, the Genoeg scheme in the Netherlands worked for biopesticide regulation. Genoeg scheme is a 'bottom-up' process whereby a coalition of agencies and other interested parties is involved in creating new regulations (www.genoeg.net). The Genoeg scheme in the Netherlands was developed to enhance the biopesticide production by efficient regulatory action (http://www.pesticides.gov.uk/environment.asp).

#### 16.3.3.3 Russia

In Russia the state registration of microbial pesticides is regulated by the Russian Agricultural Control (RAC). RAC not only administers the registration of pesticides but also regulates their use, production, sale, transportation, storage, disposal, advertising, import and export (Kabaluk et al. 2010). The All-Russian Institute for Plant Protection (VIZR) in St. Petersburg, falling under the Russian Agricultural Academy (RAN), is involved in biopesticide research and development, but it is also involved in various aspects of the registration process (Kabaluk et al. 2010).

#### 16.3.3.4 Ukraine

In Ukraine, the Main State Inspection of Plant Protection (MSIPP) is responsible for the registration of biopesticides and all other plant protection products. Most of organizations fall under the research institutes of the Ukrainian Agricultural Academy of Sciences (UAAN), the National Academy of Sciences (NANU) or the various agricultural universities, the latter of which are governed by the Ministry of Education and Science (Kabaluk et al. 2010). The two best known institutions conducting biopesticide research and development are the Institute of Plant Protection (UAAN) and National University of Life and Environmental Sciences of Ukraine (National Agrarian University of Ukraine) (www.msp.ua/state-registration-of-pesticides-ukraine.htm).

#### 16.3.3.5 United Kingdom (UK)

In the UK, the main regulatory body responsible for plant protection products, including biopesticides, is the Chemicals Regulatory Directorate (CRD)/Pesticide Safety Directorate (PSD) (http://www.hse.gov.uk/pesticides/). The CRD is a new Directorate of the Health and Safety Executive (HSE) which is responsible for the regulation of pesticides, biocides, detergents, and chemicals under the Registration, Evaluation, Authorization and Restriction of Chemicals (REACH) Regulation (http://www.pesticides.gov.uk/guidance/industries/pesticides/advisorygroups/pesticidesforum). PSD is an agency of the Department for Environment, Food and Rural Affairs (DEFRA) and deals with the registration of agricultural pesticides (DEFRA 2006). The UK regulatory system was developed according to a chemical pesticide model, and this may have acted as a barrier to biopesticide commercialization (ACP 2004). The biopesticide scheme was developed as an important project in 2003, and the main aim of this project was to enhance the biopesticide production (http://www.pesticides.gov.uk/environment.asp). This scheme has introduced the role of biopesticide champion in 2006 (Chandler et al. 2011) for biopesticide registration and regulation.

### 16.3.4 North America

This continent has the highest share in biopesticide market. The North American Free Trade Association (NAFTA) is a regulatory authority in both the USA and Canada committed to joint reviews and work sharing of biopesticide evaluation (NAFTA 2009).

#### 16.3.4.1 Canada

In Canada definitions and registration requirements are closely aligned with those used by the USEPA. Microbials, semiochemicals and nonconventional pest control products (NCPCPs) are principally regulated by the Health Canada's Pest Management Regulatory Agency (PMRA 2007) under the federal authority of the Pest Control Products Act 2002. The regulation of nonindigenous macrobials is administered by the Canadian Food Inspection Agency (CFIA) under the authority of the Plant Protection Act 1990. Invertebrate biological controls are not registered by the PMRA. PMRA regulates all pesticides including biopesticides in Canada. It is involved in pesticide evaluation and registration and developed joint guidelines for biopesticides with EPA and OECD (AGBR 2015). The PMRA website also provides comprehensive information on pest control products for registrants, regulators, researchers and the general public (www.hc-sc.gc.ca). Agriculture and Agri-Food Canada Pest Management Centre is committed to improve access to biopesticides as part of its Pesticide Risk Reduction Programme (AAFC 2009).

#### 16.3.4.2 United States (USA)

The USA has significant proportion of the global biopesticide market. In the USA the EPA shows a comprehensive and complex regulatory system for registration and regulation process of biopesticides (USEPA 2010), and this system requires unique properties for registration than other regulatory systems (Harman et al. 2010; Chandler et al. 2011). The regulation of biopesticides is a responsibility of the Office of Chemical Safety and Pollution Prevention (OCSPP) and the Office of Pesticide Programs (OPP), and there are three divisions within OPP that are involved in the registration of pesticides: (1) Antimicrobial Division, (2) Registration Division and (3) Biopesticides and Pollution Prevention Division (BPPD) (Matthews 2014). Since biopesticides cause fewer risks than chemical pesticides, EPA generally requires much less data to register a biopesticide than to register a conventional pesticide (Kumar 2012). USEPAs have mainly three parts: (1) the Federal Insecticide, Fungicide and Rodenticide Act (FIFRA) 1947; (2) the Federal Food, Drug and Cosmetic Act (FFDCA) 1938; and (3) the Food Quality Protection Act (FQPA) 1996. In the USA BPIA (Biopesticide Industry Alliance) promotes industry standards for biopesticides by communicating the value of biopesticides in agriculture and other markets (http://www.biopesticideindustryalliance.org/). It also collaborates with the biopesticide regulating authorities in order to ensure transparent and appropriate registration and regulatory requirements.

### 16.3.5 South America

#### 16.3.5.1 Argentina

The regulatory institutions in Argentina are as follows: the Servicio Nacional de Sanidad y Calidad Agroalimentaria (SENASA); Vegetal National Committee of South Cone (COSAVE); Secretary of Agriculture, Livestock, Fish and Food (SACPyA); and National Administration of Drugs, Food and Medical Technology (ANMAT). The 'Coordination of Agrochemical and Biological Products' department of SENASA issues the information relating to pesticide registration, restrictions, commercialization and use of agrochemicals and biological products (www.senasa.gov.ar). The National Agriculture Department and Environmental Policy Secretary (NADEPS) is involved in the regulatory system, and the National Agriculture Department (NAD) and Environmental Policy Secretary (EPS) are also involved in the regulation process in Argentina.

#### 16.3.5.2 Brazil

In Brazil the registration of all the pesticides is done by the Brazilian Ministry of Agriculture, Livestock and Food Supply (MAPA). National Health Surveillance Agency (ANVISA) and the Brazilian Institute of Environment and Renewable Natural Resources (IBAMA) scrutinize the process, while in urban areas and public health campaigns, products are registered with ANVISA, after scrutiny by MAPA and IBAMA. In Brazil, a decree establishing the criteria for registration of BCAs for organic agriculture was approved in 2009, and biocontrol products based on macro- and microorganisms are considered to be plant protection products with the regulations for registration and use as for chemical pesticides. In 2007, the Brazilian Biocontrol Manufacturers' Association (ABCBIO) was created to organize this sector (abcbio.org.br).

#### 16.3.5.3 Colombia

In Colombia, the Provincial Agricultural Legislation Office (PALO) is responsible for regular inspection of biopesticide trade and conducts regular direct inspection on quality control. Colombia has a specific regulation for BCAs (biopesticides) since 1994, which was updated in 2004. Presently only acute toxicological studies are required and reviewed by the Ministry of Health (Cotes 2010).

#### 16.3.5.4 Cuba

In Cuba a specific IPM Act (1982) has induced the use of biopesticides (Pérez and Vázquez 2002). The biopesticide registration process was developed according to standards and guidelines of the three authorities named as the (i) Food and Agriculture Organization (FAO 1988), (ii) OECD (1996) and (iii) USEPA (1996) (Díaz 2003). In 2007, a formal process for registration was published (Gaceta Official No. 016 Cuba 2007) and four institutions are regulating biopesticides and one or more involved with registration process:

(i) The Central Registrar of Pesticides (CNSV) – approves and registers new active ingredients and new products.
(ii) The Centre for Environmental Inspection and Control (CICA) – regulates environmental aspects for production of biopesticides.
(iii) The National Centre for Biological Safety (CNSB) – regulates research, production, trials, releases, import and export of biopesticides and production facilities.
(iv) External Quarantine Department of the National Centre of Plant Health (EQ-DNCPH) – regulates the import and export of materials under quarantine (including biopesticides).

### 16.3.6 Oceania

#### 16.3.6.1 Australia

In Australia, the regulation of biopesticides is governed by the National Registration Scheme (NRS) for the application and commercialization. Currently, the Council of Australian Governments (COAG) is the most important regulatory body amongst various regulatory bodies running in Australia (King et al. 2013). The regulatory system of biopesticides is done by the Australian Pesticides and Veterinary Medicines Authority 1992 (APVMA 1992) which has a role in registration of biopesticides having unique data (http://apvma.gov.au/), and this authority was previously named as National Registration Authority (NRA). In 2005 some guidelines were published with additional requirements for microbial pesticides, and these guidelines consider evaluation of potential hazards of microbes such as toxicity, pathogenicity, host range and impact on native flora and fauna (APVMA 2005).

#### 16.3.6.2 New Zealand

In New Zealand biopesticides mainly microbial pesticides are regulated by the Hazardous Substances and New Organisms (HSNO) Act 1996. Release and importation of all new organisms, including microbes for biocontrol, have been regulated under this act since 1998, and prior to 1998, unknown microbial species must have been approved by the Environmental Risk Management Authority (ERMA) before commercialization, even if the species were isolated from material within the country (Pottinger and Morgan 2008; ERMA 2010). Once approved by HSNO, biopesticides also need to be registered under the Agricultural Compounds and Veterinary Medicines (ACVM) Act 1997 (www.nzfsa.govt.nz/acvm), which has replaced the Pesticides Act 1979. The approval of the existing microbial products for their identity and purity for safe importation is done by the Ministry of Agriculture in New Zealand.

Table 16.1 summarizes the regulations of biopesticides in various countries around the globe.

## 16.4 Limitations of Current Regulation Systems

There are many systemic problems arising in regulation of biopesticides, and one main problem is that it is based on the conventional chemical pesticide models (Greaves 2009). According to Chandler et al. (2008), in the EU system, the regulatory failure arises from the application of an inappropriate synthetic pesticide model and lack of regulatory innovation. Besides this, registration and commercialization of biopesticides and their evaluation is also a lengthy process. EU system takes much time to complete registrations, and the industry criticizes that the present registration period, especially for microbial biological control, is costly and time consuming (Bailey et al. 2010). For example, the mean time taken by the EU was 75 months compared with 28 months in the USA (Hokkanen and Menzler-Hokkanen 2008). The US system provides flexibility, and in some cases, it invites applicant for pre-submission meeting where the applicant is advised on which studies are necessary, based on preliminary data and literature available (Mubyana-Jhon and Taylor 2015).

Biopesticide production system is underdeveloped and underutilized in Asia because there are several constraints such as technical, social and institutional that limit commercial production of new biopesticides (NAAS 2013). In developing countries like in Asia and Africa, there is a problem of quality control because of which the trust of farmers is lacking. This situation can be corrected only by proper regulatory system. Although biopesticide production and application in India is in progress but in comparison to chemical pesticides, the growth is not up to mark. In a study Rabindra (2005) estimated that the current production of microbial pesticides meets less than 10 % of the identified need. Nearly 500 biopesticides are registered by CIB and available in the Indian market, but quality control is a major issue related to most of the products (NAAS 2013).

Mensink and Scheepmaker (2007) suggest that even though data requirements are increasingly transparent and harmonized for more efficient regulatory procedures, there is lack of sufficient guidance on evaluation and use of biological products to conduct a premarket assessment of the environmental safety. Regulatory agencies realize that biopesticides are fundamentally different than chemical pesticides and should not be compared with the same standards of safety and efficacy; it is challenging to design a system for evaluation that is equally fair to both biopesticides and chemical pesticides (Bailey et al. 2010). One difficulty is that the regulatory processes assess single products, while microbial pesticides are very complex and diverse in nature (Hubbard et al. 2014). Ravensberg (2011) gave recommendations on how to build a dossier and which sources, in addition to the guidelines, can be used to better understand what exactly the authorities require. A data requirement in the USA and Canada is somewhat common except that PMRA requires a complete dossier on efficacy and phytotoxicity data, and the EPA does not.

Another complex issue is surrounding the regulation of biopesticides having multiple modes of action. For example, species of the fungus *Trichoderma*, which are used as biopesticides against soilborne plant pathogenic fungi, are able to parasitize plant pathogenic fungi in the soil (Gupta and Dikshit 2010); they also produce antibiotics (Ghisalberti and Sivasithamparam 1991; Vey et al. 2001) and fungal cell wall-degrading enzymes (Bech et al. 2015). *Trichoderma* compete with soilborne pathogens for carbon, nitrogen and other factors (Limón and Codón 2004), and they can also promote plant growth by the production of auxin-like compounds (Vinale et al. 2008; Nega 2014). Some *Trichoderma* products have been sold on the basis of their plant growth-promoting properties, rather than as plant protection products (Nega 2014), and so have escaped scrutiny from regulators in terms of their safety and efficacy (Bailey et al. 2010). Similar is the case with *Pseudomonas*. Fluorescent *Pseudomonas* can be used as plant growth-promoting agent as well as in biocontrol (Negi et al. 2005; Mehnaz 2013; Tewari and Arora 2014, 2016). However, there are no particular regulatory mechanisms to check this. There are several technological and policy gaps in effective utilization of biopesticides which need to be addressed properly in order to minimize the use of chemical pesticides and to promote the use of biopesticides (Kumar 2015).

## 16.5 Innovation for Better Regulation Schemes

Innovation in current regulation system of biopesticides is required as an important need for higher production of agrobiologicals at global level (Arora et al. 2012). The regulation situation at present varies across countries; some countries have developed system, some having progress in regulatory framework, and few have no proper regulations for biopesticides (Simiyu et al. 2013). The appropriate regulatory system is that which has concern about public safety and environmental impact, i.e. it must be ecofriendly, scientific and technologically innovative for pest control (Greaves 2009). Biopesticides have not been used more widely; the solution of this matter may be lowering registration costs and eliminating efficacy (Greaves 2009). Another appropriate plan for better regulation of biopesticides is that the governments can make regulations globally by organizing meetings, workshops, and conferences regarding uplifting the status of biopesticides (Mishra et al. 2015).

Some factors that may promote the necessary regulatory innovations ranging from government intervention are exogenous pressure and some within regulatory bodies are endogenous pressure (Greaves 2009). Various institutions have done some preliminary research about the industrialization of biopesticides and institutional changes may be significant; however, no systematic reports have appeared so far; therefore, there is requirement of guidelines to promote the collaboration of enterprises and research institutes (Leng et al. 2014).

Harmonization of biopesticide regulatory guidelines in the world is very important for the innovative approach for the production and commercialization of biopesticides, and OECD is important for harmonization at international level (Holm et al. 2005). The OECD and the World Health Organization (WHO) have an influence on pesticide regulation and their role is very important (NAAS 2013). The OECD project on biopesticides was initiated in 1999 to help its member countries to harmonize the methods and approaches used to assess biological pesticides (Sigman 2005). Currently the OECD has 34 member countries, and more than 70 developing and transition economies are engaged in working relationships with the OECD. The working group of OECD on pesticides includes the (i) Registration Steering Group (RSG), (ii) Risk Reduction Steering Group (RRSG) and (iii) Biopesticides Steering Group (BPSG) (http://www.oecd.org/chemicalsafety/pesticides-biocides/). The BPSG has achieved a lot of progress towards harmonization and work sharing through the development of guidance and working documents (Richards and Kearns 1997). The headquarters of OECD group is situated in Paris, France, and assists EU governments for assessing biopesticide risks to humans and the environment very carefully (http://www.biopesticideindustryalliance.org/).

Globally, the regulation of pesticides has engaged the attention of OECD, Food and Agriculture Organization (FAO) and EU both generally and particularly (Greaves and Grant 2011; FAO 2012). EPPO is an intergovernmental organization in Paris, funded by contributions from the member governments (www.eppo.int). In 2010, the IOBC of noxious animals and plants reviewed the rapid expansion of the application of microbial pesticides and advancement in their regulatory systems worldwide (IOBC 2010). The major steps towards the promotion of harmonization for biopesticide regulation and facility developments for work sharing between governments have been taken by various bodies including OECD, North American and European (AGBR 2015). The BPSG of OECD, the FAO, the European Commission (EC), the IOBC, the EPPO, the North American Plant Protection Organization (NAPPO) and the NAFTA have worked as most important intergovernmental organizations for biopesticide regulation and innovation (AGBR 2015).

## 16.6 Conclusion and Future Prospects

Commercialization of biopesticides is rapidly growing all over the world (Kumar 2012), but in absence of effective regulations and some other constraints (Arora 2015), the growth is not as expected. Effective regulation can also prevent spurious biopesticides in the market (NAAS 2013). Regulation of biopesticides is an obstacle in production and commercialization of these products. As is evident from aforesaid discussion, regulation criterion varies according to countries. To resolve this barrier, it is recommended to develop a common regulatory system. Moreover, it is also required to improve communication and information exchange between regulatory bodies, scientists and industries to review the possible risks associated with microbial biopesticides. Guidelines used to evaluate biopesticide efficacy, field testing and quality also need to be revised, as most of the time, it is conducted by non-expert persons especially in developing countries. Regulatory bodies are required to be established to not only ensure fast track registration of biopesticide products with justified regulations and transparent procedures but also promote the adoption of new safer technologies in the development of commercial products. The criterion adopted in policymaking for regulation should be identical for all countries and proposed according to nature of agrobiologicals and not as for chemical pesticides.

## References


ACP (2004) Advisory Committee on Pesticides. Final report of the sub-group of the advisory committee on pesticides on: alternatives to conventional pest control techniques in the UK: a scoping study of the potential for their wider use. Advisory Committee on Pesticides, York. http://www.pesticides.gov.uk/uploadedfiles/Web_Assets/ACP/ACP_alternatives_web_subgrp_report.pdf

AAFC (2009) Agriculture and Agri-Food Canada. Biopesticides: strategies for discovery, development, and adoption. www.agr.gc.ca

ACVMA (1997) Agricultural Compounds and Veterinary Medicines Act. www.nzfsa.govt.nz/acvm

AGBR (2015) Agrow Global Biopesticide Regulations. Commodity analysis. Available at www.agra-net.com

Ahsan N (2012) Status of pesticide management in Bangladesh, In: APPPC regional workshop on enhancement of regional collaboration in pesticides regulatory management, Chiang Mai. http://www.fao.org/3/a-i3323e/i3323e01.pdf

Arora NK (ed) (2015) Plant Microbes Symbiosis: applied facets. Springer, India

Arora NK, Tewari S, Singh S, Lal N, Maheshwari DK (2012) PGPR for protection of plant health under saline conditions. In: Maheshwari DK (ed) Bacteria in agrobiology: stress management. Springer, Berlin, pp 239–258

APVMA (1992) Australian Pesticide and Veterinary Medicine Authority. Registration guide. www.apvma.gov.au/registration/index.php

APVMA (2005) Australian Pesticide and Veterinary Medicine Authority. Guidelines for the registration of biological agricultural products. www.apvma.gov.au/publications/guidelines/docs/bioagprod.pdf

Bailey KL, Boyetchko SM, Langle T (2010) Social and economic drivers shaping the future of biological control: a Canadian perspective on the factors affecting the development and use of microbial biopesticides. Biol Control 52:221–229

Bech L, Busk PK, Lange L (2015) Cell wall degrading enzymes in trichoderma asperellum grown on wheat bran. Fungal Genome Biol 4:116

Bissdorf JK (2008) How to grow crops without endosulfan. In: Webber C (ed) How to live and work without endosulfan. Pesticide Action Network (PAN), Hamburg

CFIA (2013) Canadian Food Inspection Agency. Industry notice – fertilizer regulations amendments enable focus on safety. http://www.inspection.gc.ca

Chandler D, Davidson G, Grant WP, Greaves J, Tatchell GM (2008) Microbial biopesticides for integrated crop management: an assessment of environmental and regulatory sustainability. Trends Food Sci Technol 19:275–283

Chandler D, Bailey AS, Tatchell GM, Davidson G, Greaves J, Grant WP (2011) The development, regulation and use of biopesticides for integrated pest management. Philos Trans R Soc Lond B: Biol Sci 366(1573):1987–1998

Cotes AM (2010) Registry and regulation of biocontrol agents on food commodities in South America. Int Symp Biol Control Postharvest Dis: Chall Oppor 905:301–306

Cuba (2007) Gaceta Oficial de la Republica de Cuba, Ministerio do Justicia No. 016 Extraordinario de 16 de abril de 2007:77–84. www.gacetaoficial.cu

DEFRA (2006) Department for Environment, Food and Rural Affairs. The royal commission on environmental pollution report on crop spraying and the health of residents and bystanders – government response. http://www.defra.gov.uk/environment/rcep/index.htm

DAFF (2010) Department of Agriculture, Forestry and Fisheries. Act No. 36 of 1947, Guidelines on the data required for registration of biological/biopesticides remedies in South Africa, Republic of South Africa

Díaz LH (2003) Registration of biological pesticides in Cuba. In: Wabule MN, Ngaruiya PN, Kimmins FK, Silverside PJ (eds) Proceedings of the pest control products board/Kenya agricultural research institute/ dept. for international development crop protection programme workshop, Nakuru. Natural Resources International Ltd, Kenya, pp 117–133

EC (2001) European Commission. Directive 2001/36/EC of 16 May 2001 amending council directive 91/414/ EEC concerning the placing of plant protection products on the market. Off J Eur Communities 164:1–38

EC (2005) European Commission. Council directive 2005/25/EC of 14 march 2005 amending annex VI to directive 91/414/EEC as regards plant protection products containing micro-organisms. Off J Eur Union L 90:1–34

EC (2009) European Commission. Regulation (EC) No 1107/2009 of the European parliament and of the council of 21 October 2009 concerning the placing of plant protection products on the market and repealing council directive 79/117/EEC and 91/414/ EEC. Off J Eur Union L 309: 1–50

Ehlers RU (2006) Einsatz der biotechnologie im biologischen Pfl anzenschuz. Schnreihe Dtsch Phytomed Ges 8:17–31

ERMA (2010) Environmental Risk Management Authority. Investigating biological control and the HSNO Act. http://www.ermanz.govt.nz/

EU (1991) European Union. Council directive 91/414/ EEC of 15 July 1991 concerning the placing of plant protection products on the market. Off J Eur Commun L 230:1–32

EC (2013) European Commission. No. 283/2013 setting out the data requirements for active substances, in accordance with regulation (EC) No. 1107/2009 of the European parliament and of the council concerning the placing of plant protection products on the market. Off J Eur Union L 93:1–84

Fang L (2014) Development and management of biopesticides in china. https://pesticide.chemlinked.com/news/newsagrochemical-news/development-and-management-biopesticides-china

FAO (1988) Food and Agriculture Organization. Guidelines on the registration of biological pest control agents. Food and Agric Org of the Uni Nat, Rome

FAO (2012) Food and Agriculture Organization. Guidance for harmonizing pesticide regulatory management in Southeast Asia RAP publication 2012/13. Food and Agriculture Organization of the United Nations, Regional Office for Asia and the Pacific, Bangkok, p 480

Ghisalberti EL, Sivasithamparam K (1991) Antifungal antibiotics produced by *Trichoderma* spp. Soil Biol Biochem 23(11):1011–1020

Greaves J (2009) Biopesticides, regulatory innovation and the regulatory state. Public Policy Adm 24(3): 245–264

Greaves J, Grant WP (2011) The development, regulation and use of biopesticides for integrated pest management. Phil Trans R Soc B 366:1987–1998

Grzywacz D (2003) Development and registration of biopesticides in Asia, In: Registration for biocontrol agents in Kenya, Proceedings of the PCPB/KARI/ DIDCPP workshop, Natural Resources International Limited, Aylesford, pp 101–109

Guest PJ (2015) Global biopesticide regulations – challenges, opportunities and future prospects. Available at https://www.agra-net.net/agra/agra-europe/policy-and-legislation/environment/analysis-global-biopesticide-regulations-challenges-opportunities-and-future-prospects-477586.htm

Gupta S, Dikshit AK (2010) Biopesticides: an ecofriendly approach for pest control. J Biopesticides 3(1): 186–188

Hajek A (2004) Natural enemies: an introduction to biological control. Cambridge University Press, Cambridge

Harman GE, Obregon MA, Samuels GJ, Lorrito M (2010) Changing models for commercialization and implementation of biocontrol in the developing and developed world. Plant Dis: Int J Appl Plant Pathol 94(8): 928–993

Hauschild R (2012) Safety and regulation of microbial pest control agents and microbial plant growth promoters: Introduction and overview. In: Sundh I, Wilcks A, Goettel MS (eds) Beneficial microorganism in agriculture, food and the environment: safety assessment and regulation. CAB International, Wallingford, pp 67–71

Hokkanen H, Menzler-Hokkanen (2008) Deliverable 24: cost, trade-off and benefit analysis, In: Regulation of biological control agents: final activity report. Christian-Albrechts-University, Kiel

Holm RE, Baron JJ, Kunkel DL (2005) World horticulture in crisis. In: The BCPC international congress proceedings: 2005, British Crop Protection Council, Alton 1: 31–40

Hubbard M, Hynes RK, Erlandson M, Bailey KL (2014) The biochemistry behind biopesticide efficacy. Sustain Chem Proc 2(1):18

ICAMA (2008) Institute for the Control of Agrochemicals, Ministry of Agriculture. Pesticide manual, the institute for the control of agrochemicals. Ministry of Agriculture, China. www.agrolex.com.cn

IOBC (2010) International Organization for Biological Control. Global newsletter; Issue 88. http://www.iobc-wprs.org/pub/

Jabbar A, Mallick S (1994) Pesticide and environment situation in Pakistan, vol 19, Working paper series. Sustainable Development Policy Institute, Islamabad

Jakel T (2003) International symposium on biopesticides for developing countries. Turrialba, Costa Rica, p 191

Japanese Agriculture Standard (JAS) Law (2000) Notification No. 1005 of the ministry of agriculture, forestry and fisheries. www.maff.go.jp/e/jas/specific/pdf/1401qa-organic-plants.pdf

Kabaluk JT, Antonet MS, Mark SG, Stephanie GW (2010) The use and regulation of microbial pesticides in representative jurisdictions worldwide. IOBC Global. www.IOBC-Global.org

Kimani V (2014) Bio-pesticides development, use and regulation in Kenya regional experts, workshop on development, regulation and use of bio-pesticides in East Africa, Nairobi

King J, Alexander F, Brodie J (2013) Regulation of pesticides in Australia: the great barrier reef as a case study for evaluating effectiveness. Agric Ecosyst Environ 180:54–67

Kumar S (2012) Biopesticides: a need for food and environmental safety. J Biofertil Biopestici 3:e107

Kumar S (2015) Biopesticide: an environment friendly pest management strategy. J Biofertil Biopestici 6:e127

Kunimi Y (2007) Current status and prospects on microbial control in Japan. J Invertebr Pathol 95:181–186

Leng P, Zhang Z, Pan G, Zhao M (2014) Applications and development trends in biopesticides. Afr J Biotechnol 10(86):19864–19873

Libman GN, Maclntosh SC (2000) Registration of biopesticides. In: Charles JF, Delécluse A, Nielsen-le Roux C (eds) Entomopathogenic bacteria: from laboratory to field application. Kluwer Academic Publishers, Springer, Dordrecht, pp 333–353

Limón MC, Codón AC (2004) Biocontrol mechanisms of *Trichoderma* strains. Int Microbiol 7:249–260

Matthews KA (2014) Regulation of biopesticides by the environmental protection agency. In: Coats JR (ed) Biopesticides: state of the art and future opportunities, vol 18, Symposium series, 1172. Sidley Austin, Washington, DC, pp 267–279

Meeussen J (2012) OECD guidelines and harmonization for microbial control agents. In: Sundh I, Wilcks A, Goettel MS (eds) Beneficial microorganism in agriculture, food and the environment: safety assessment and regulation. CAB International, Wallingford, pp 308–321

Mehnaz S (2013) Secondary metabolites of *Pseudomonas aurantiaca* and their role in plant growth promoter. In: Arora NK (ed) Plant microbe symbiosis: fundamentals and advances. Springer, New Delhi, pp 373–392

Mensink BJWG, Scheepmaker JWA (2007) How to evaluate the environmental safety of microbial plant protection products: a proposal. Biocontrol Sci Technol 17:3–20

Mishra J, Tewari S, Singh S, Arora NK (2015) Biopesticides: where we stand? In: Arora NK (ed) Plant microbes symbiosis: applied facets. Springer, New Delhi, pp 37–75

Mubyana-John T, Taylor J (2015) Nematophagous fungi: regulations and safety. In: Askary TH, Martinelli PRP (eds) Biocontrol agents of phytonematodes. CABI, Wallingford, pp 203–213

NAFTA (2009) North American Free Trade Association. Increasing adoption of biopesticides – project RR08-04-0909. www.hc-sc.gc.ca

NAAS (2013) National Academy of Agricultural Sciences. Biopesticides- quality assurance. http://naasindia.org/Policy%20Papers/Policy%2062.pdf

National Farmers Policy (2007) Department of Agriculture and Cooperation, Ministry of Agriculture Government of India, India. https://india.gov.in/national-policy-farmers-2007-department-agriculture-and-cooperation

Nega A (2014) Review on concepts in biological control of plant pathogens. J Biol Agric Healthc 4(27):33–54

Negi YK, Garg SK, Kumar J (2005) Cold-tolerant fluorescent *Pseudomonas* isolates from Garhwal Himalayas as potential plant growth promoting and biocontrol agents in pea. Curr Sci 89:2151–2156

Ochieng JRA (2015) Towards a regulatory framework for increased and sustainable use of bio-fertilizers in Kenya, MA Thesis, Centre for Advanced Studies in Environmental Law and Policy (CASELAP), University of Nairobi

OECD (1996) Organization for Economic and Cooperative Development. Data requirements for registration of biopesticides in OECD member countries: survey results, Environmental monograph no. 106. OECD, Paris

PBR (2005) Pakistan Biosafety Rules. National biosafety guidelines notification no. F.2(7) 95-Bio pp 1–138

Panuwet P, Siriwong W, Prapamontol T, Ryan PB, Fiedler N, Robson MG, Barr DB (2012) Agricultural pesticide management in Thailand: situation and population health risk. Environ Sci Policy 17:72–81

Pérez NC, Vázquez LLM (2002) Ecological pest management. In: Funes F, Bourque M, Pérez N, Rosset P (eds) Sustainable agriculture and resistance: transforming food production in Cuba. Food First Books, New York, pp 109–143

Plimmer JR (1993) Regulatory problems associated with natural products and biopesticides. Pesticide Sci 39 (2):103–108

PMRA (2007) Pest Management Regulatory Agency. PRO2007-02: consultation document on guidelines for the registration of low-risk biochemicals and other non-conventional pesticides. www.hc-sc.gc.ca

Pottinger B, Morgan ED (2008) An overview of regulatory processes under the hazardous substances and new organisms act 1996. In: Butcher MR, Walker JTS, Zydenbos SM (eds) Future challenges in crop protection: repositioning New Zealand's primary industries for the future. The New Zealand PSS, Hastings, pp 27–36, publicationsdivision.nic.in/others/India_2010.pdf

Rabindra RJ (2005) Current status of production and use of microbial pesticides in India and the way forward. In: Rabindra RJ, SS Hussaini, Ramanujam B (eds) Microbial biopesticide formulations and applications Technical Document no. 55. Project Directorate of Biological Control, pp 1–12

Ravensberg WJ (2011) A roadmap to the successful development and commercialization of microbial pest control products for control of arthropods. Springer, Dordrecht, pp 171–233

RDA (2009) Rural Development Administration. The acts for management of agricultural chemicals. www.rda.go.kr

Regnault-Roger C, Vincent C, Arnason JT (2012) Essential oils in insect control: low-risk products in a high-stakes world. Ann Rev Entomol 57:405–424

Richards J, Kearns P (1997) Work on micro-organisms within the OECD's environmental health and safety programme. In: Evans HF (ed) Microbial insecticides: novelty or necessity? BCPC symposium proceedings no. 68. Coventry BCPC, Farnham, pp 223–226

Sigman R (2005) OECD's vision for the regulation of crop protection products. In: Proceedings of the BCPC International Congress-Crop Sci & Technol 2005, Glasgow, pp 329–336

Simiyu NSW, Tarus D, Watiti J, Nang'ayo F (2013) Effective regulation of bio-fertilizers and bio-pesticides: a potential avenue to increase agricultural productivity, Policy series, COMPRO II, IITA

Singh R, Arora NK (2016) Bacterial formulations and delivery systems against pests in sustainable agro-food production. Module in Food Sciences, Elsevier, pp 1–11

Sinha B, Biswas I (2008) S&T for rural India and inclusive growth, potential of biopesticide in Indian agriculture vis-a-vis rural development, Ind Sci Technol, www.nistads.res.in/indiasnt**2008**/t6rural/t6rur17.htm

Stadlinger N, Mmochi AJ, Kumblad L (2013) Weak governmental institutions impair the management of pesticide import and sales in Zanzibar. Ambio 42(1): 72–82

Tewari S, Arora NK (2014) Multifunctional exopolysaccharides from *Pseudomonas aeruginosa* PF23 involved in plant growth stimulation, biocontrol and stress amelioration in sunflower under saline conditions. Curr Microbiol 69(4):484–494

Tewari S, Arora NK (2016) Fluorescent Pseudomonas sp. PF17 as an efficient plant growth regulator and biocontrol agent for sunflower crop under saline conditions. Symbiosis. doi:10.1007/s13199-016-0389-8

USEPA (1996) United States Environmental Protection Agency. Prevention, pesticides and toxic substances, Series 885-microbial pesticides test guidelines. EPA, Washington, DC

USEPA (2010) United States Environmental Protection Agency. Regulating biopesticides. http://www.epa.gov/opp00001/biopesticides/index.htm

Vey A, Hogland RE, Butt TM (2001) Toxic metabolites of fungal biocontrol agents. In: Butt TM, Jackson C, Morgan N (eds) Fungi as biocontrol agent: progress, problem and potential. CAB International, Bristol, pp 311–346

Vinale F, Sivasithamparam K, Ghisalberti EL, Marra R, Woo SL, Lorito M (2008) *Trichoderma*- plant-pathogen interactions. Soil Biol Biochem 40:1–10

Waage JK (1997) Biopesticides at the crossroads: IPM products or chemical clones? In: Evans HF (ed) Microbial insecticides: novelty or necessity? BCPC symp proc, 68, pp 11–20

Wabule MN, Ngaruiya PN, Kimmins FK, Silverside PJ (2004) Registration for biocontrol agents in Kenya. In: Proceedings of the PCPB/KARI/DIDCPP workshop, Natural Resources International Limited, Aylesford, p 230